The Palgrave Handbook of Socio-ecological Resilience in the Face of Climate Change

Sunil Nautiyal · Anil Kumar Gupta ·
Mrinalini Goswami · Y. D. Imran Khan
Editors

The Palgrave Handbook of Socio-ecological Resilience in the Face of Climate Change

Contexts from a Developing Country

Editors
Sunil Nautiyal
Centre for Ecological Economics
and Natural Resources
Institute for Social and Economic
Change
Bengaluru, India

Mrinalini Goswami
Centre for Ecological Economics
and Natural Resources
Institute for Social and Economic
Change
Bengaluru, India

Anil Kumar Gupta
National Institute of Disaster
Management, Ministry of Home Affairs,
Government of India
Delhi, India

Y. D. Imran Khan
Centre for Ecological Economics
and Natural Resources
Institute for Social and Economic
Change
Bengaluru, India

ISBN 978-981-99-2205-5   ISBN 978-981-99-2206-2 (eBook)
https://doi.org/10.1007/978-981-99-2206-2

© The Editor(s) (if applicable) and The Author(s), under exclusive license to Springer Nature Singapore Pte Ltd. 2023

This work is subject to copyright. All rights are solely and exclusively licensed by the Publisher, whether the whole or part of the material is concerned, specifically the rights of translation, reprinting, reuse of illustrations, recitation, broadcasting, reproduction on microfilms or in any other physical way, and transmission or information storage and retrieval, electronic adaptation, computer software, or by similar or dissimilar methodology now known or hereafter developed. The use of general descriptive names, registered names, trademarks, service marks, etc. in this publication does not imply, even in the absence of a specific statement, that such names are exempt from the relevant protective laws and regulations and therefore free for general use.
The publisher, the authors, and the editors are safe to assume that the advice and information in this book are believed to be true and accurate at the date of publication. Neither the publisher nor the authors or the editors give a warranty, expressed or implied, with respect to the material contained herein or for any errors or omissions that may have been made. The publisher remains neutral with regard to jurisdictional claims in published maps and institutional affiliations.

This Palgrave Macmillan imprint is published by the registered company Springer Nature Singapore Pte Ltd.
The registered company address is: 152 Beach Road, #21-01/04 Gateway East, Singapore 189721, Singapore

Foreword

More than 3.3 billion people across the globe are living in conditions vulnerable to climate change. In this context, effective adaptation mechanisms and resilience building play critical roles in safeguarding the life and livelihoods of vulnerable populations. The Sixth Assessment Report of the Intergovernmental Panel on Climate Change (IPCC) highlights that Climate-resilience building is more inclusive and can integrate all levels and sectors of society and government. The IPCC also states that governments should be provided with scientific information at all levels so that they can use it to develop climate policies. The diversity of contexts in terms of climate change impacts, exposure, and adaptive capacity make adaptation planning a very knowledge-intensive exercise which demands consideration of cases from local to global levels. Thus, sharing of information and knowledge from a range of sources including traditional practices in adaptation, institutional actions on immediate disaster risk reduction, transformational adaptive strategies, etc., is expected to help in adaptation planning and resilience building.

This publication is an outcome of a truly interdisciplinary event on climate change where pathways for disaster risk reductions have been discussed. The compilation of micro to macro level case studies aims at disaster risk reduction through building socio-ecological resilience to climate change. The chapters highlight the impacts of climate change in varied ecological and socioeconomic settings, both indigenous and institutional adaptation mechanisms and important areas that need interventions, and provoke discussion in the broader policy arena. It is expected to benefit local to global actors with capacities to act at the individual, institutional, and policy-making levels as well.

It must be a challenging task for the editors to assemble diverse experiences and shreds of evidence on the complex subject of climate change impacts and adaptation to place it in a framework useful for readers of different disciplines. It is commendable work on their part where scientific articles of varied perspectives including climate-smart village, impact monitoring, traditional knowledge, sustainable urban planning, etc., have been given a reader-friendly format with a stock of useful information. I appreciate the work of the editorial team and hope their effort will be helpful for academicians, practitioners, policymakers, and readers in general who are working towards mitigation and adaptation of climate change impacts in India and abroad. I also hope that the discussions put forward in this book will encourage more research and development towards building a climate-resilient sustainable society.

Dr. Nisha Mendiratta
Advisor and Head Women
in Science and Engineering-KIRAN
(WISE-KIRAN)
Department of Science &
Technology (DST), Govt. of India
New Delhi, India

Preface

India is one of the most risk-prone countries with respect to climate change, where geo-climatic variations of the country make the population vulnerable to climate-related disasters in varying degrees, intensities, and patterns. Along with its revised and stringent targets for greenhouse gas emission reduction, India's priority has been to mitigate the impacts of climate change because a significant portion of its population depends on climate-sensitive sectors. The extent of effects on different types of systems is determined by geophysical and climatic settings, socioeconomic conditions of people, and man-made interventions. Recognising the dynamic relationships between adaptation and sustainable development, policy-making mechanisms, political will, and community aspirations is crucial for managing and mitigating adverse impacts. In view of this, the current book *Socio-ecological Resilience in the Face of Climate Change; Contexts from a Developing Country* is compiled to share experiences and knowledge on impacts, adaptation and risk reduction strategies, communities' responses to impacts of climate change, and best practices. This inter and trans-disciplinary book will provide novel insights into climate change risk reduction. The five sections of the book include research articles based on real-world situations and case studies across diverse ecological zones of India, as well as reflections on policy impacts. The book explores arrays of evidence on the impacts of climate change and adaptation in different contexts, including natural resources, livelihoods, urban development, use of traditional knowledge, and strategies and approaches prevalent in making resilient communities. The methodological approaches and techniques of the articles are also diverse, likewise.

Although the chapters are from diverse disciplines, all the chapters have the common goal of addressing the concerns of climate change. The findings and discussion are presented in a simpler way within the framework of scientific communication, which is expected to be beneficial for researchers, scholars, students, and practitioners working on any of the dimensions of climate change. The chapters are intended to be easy to understand, supported by science, and rational simultaneously.

The editors anticipate a significant contribution towards evidence-based policy-making and bottom-up problem-solving actions through this book. A small piece of empirical insight from a remote landscape, either on impact assessment or adaptation efforts, will certainly explore more avenues for scientific research in terms of validation, upscaling, policy inclusion, and many more.

We hope the readers will be benefitted from the insights from interdisciplinary knowledge on "society-ecology-economy" interface for better comprehension of challenges and strategy prospects for building resilience against changing climate.

Bengaluru, India Sunil Nautiyal
New Delhi, India Anil Kumar Gupta
Bengaluru, India Mrinalini Goswami
Bengaluru, India Y. D. Imran Khan
September 2023

Acknowledgements

This book is an outcome of the three-day International Workshop cum Training Programme on "Green Growth Strategies for Climate Resilience and DRR: Policies, Pathways and Tools" held during 26 to 28 November 2020. The workshop was jointly organised by the Institute for Social and Economic Change (ISEC), Bengaluru and the National Institute for Disaster Management (NIDM), New Delhi, In Honour of Prof. M. V. Nadkarni, an economist and philosopher well acquainted with both Western and Indian philosophy, the first head of CEENR and Former Vice Chancellor of Gulbarga University and Honorary Visiting Professor, ISEC.

We extend our sincere thanks to Prof. Sukhadeo Thorat (Chairman, BOG, ISEC), Prof. D. Rajasekhar (Director, ISEC), Prof. S. Madheswaran, (Former Director, ISEC), Dr. Akhilesh Gupta (Secretary-SERB & Sr. Advisor-DST GoI), Shri Kamal Kishore (Member Secretary, NDMA), Shri Taj Hassan (IPS, Former Executive Director, NIDM, Govt. of India), Maj. Gen. M. K. Bindal (Former Executive Director, NIDM, Govt. of India), Prof. R. S. Deshpande, Prof. K. N. Ninan, Prof. P. G. Chengappa, Prof. K. G. Saxena, Prof. K. S. Rao, Prof. R. K. Maikhuri and Prof. Parmod Kumar for their kind encouragement and support.

The profound insights shared within the pages of this book owe their existence to the scholarly authors who have contributed their expertise and perspectives. These contributions are vital in nurturing solution-oriented dialogues on climate-induced concerns in our modern world. We acknowledge the immense value brought by the CEENR team of ISEC, and team of the ECDRM division of NIDM, whose support played a pivotal role in shaping the vision, structure, and execution of this event. We are thankful to the administrative staff of ISEC and NIDM for their support and cooperation in organising this conference as well as for coming up with this fruitful outcome. A special mention of appreciation goes to Lynnie Sharon, Production Editor at Palgrave Scientific Publishing Services, whose unwavering help, support, and assistance have been invaluable throughout the publication process. We also express our gratitude to the series editors for their meticulous review of the manuscript. The production team's dedicated efforts and efficient support during the creation of this volume have not gone unnoticed and are deeply appreciated.

This book stands as a testament to the collaborative spirit, dedication, and shared vision that have guided this endeavour from its inception to its culmination. With profound gratitude, we acknowledge the contributions of all who have made this initiative a resounding success.

Best Regards,
September 2023

Sunil Nautiyal
Anil Kumar Gupta
Mrinalini Goswami
Y. D. Imran Khan

Message from Prof. M. V. Nadkarni
(as delivered at ISEC, Bengaluru on 26-11-2020)

Prof. Sukhadeo Thorat, Chairman, Prof. V. K. Malhotra, Prof Madheswaran, Prof. Anil Kumar Gupta, Prof. Sunil Nautiyal, Prof. R. S. Deshpande, Prof. K. N. Ninan, Prof. K. S. Rao, Dr. S. Manasi, Dr. Sweta Baidya, Dr. Mrinalini Goswami, Dr. Y. D. Imran Khan, Ms. Fatima Amin, Mr. Mohan Kumar, and respected participants, I extend my hearty greetings and thanks to you all for kindly attending this session. I miss the pleasure of meeting you in person, but I am glad to be connected online and in person with at least some of you.

I am really overwhelmed by your affection and regard in arranging this felicitation as a part of a very apt and relevant international Workshop for whatever little I have done. It is a tremendous honour for me. I am deeply grateful to you all, particularly to Prof. Sunil Nautiyal, who had planned this for long with great enthusiasm, loving care, and commitment. I am equally grateful to the National Institute of Disaster Management, its Executive Director, Major General Bindal and Prof. A. K. Gupta for kindly collaborating with CEENR in organising this workshop, and of course, to my esteemed good friend Prof. Madheswaran for unstinted and hearty support.

I am especially happy and grateful that Prof. Sukhadeo Thorat is chairing this session. I remember with immense pride that he was my student in Marathwada University at Aurangabad in the early 1970s, and he had impressed all his teachers with his brilliance. He rose to great heights, just as we had expected.

I would be failing in my duty if I don't recall Prof. V. K. R. V. Rao on this occasion. The starting of the Ecological Economics Unit in ISEC as early as in 1981 owes entirely to his wide vision and foresight. I was only an instrument in his hands, inspired by him.

I made an entry into ecological economics through my work in agricultural economics. I am therefore particularly happy that the workshop is on Green Growth Strategies for Climate Resilience and Disaster Risk Reduction including, particularly in agriculture, which is a significant problem faced by humanity, especially in developing countries.

Along with Prof. R. S. Deshpande, I had started working on environmental issues much earlier to starting the Ecological Economics Unit, particularly on the uncertainty in rainfall and its impact, though he was not a part of the Unit. In the Unit itself, I had a few other colleagues who also contributed to the research output of the Unit greatly. I should mention Dr. H. Ramachandran, Dr. K. N. Ninan, Dr. M. Venkata Reddy, Dr. G. Satyanarayana Shastri, Dr. Syed Ajmal Pasha, and Shri B. G. Kulkarni—a talented cartographer. Our work covered both rural and urban environmental issues. Though Karnataka received more attention, our work was not confined to it. I had unstinted cooperation and support from all the colleagues and the Directors of the Institute with whom I worked, Prof. L. S. Venkataramanan, Prof. G. Thimmaiah, Shri T. R. Satish Chandran, and Dr. P. V. Shenoi. Many books and articles in reputed journals were published by the faculty of Ecological Economics. We successfully worked on several projects sponsored by the Central Water Commission, the Ford Foundation, ICSSR, and the Government of Karnataka. Many Ph.D. students were guided by the faculty during this period. The Unit became known both in the country as a whole and internationally as a centre for research and research guidance in ecological or environmental economics.

We had a good support of the ISEC administrative staff, including stenographers like Shri Arun Kumar who was a valuable asset. Along with other Units at ISEC, this was also renamed as a Centre subsequently. I should happily add that thanks to seminars and workshops conducted by our Unit, interest in ecological or environmental economics spread to other Units as well, and this became a major area of work in ISEC as a whole. This happy situation continues to this day.

I am proud to mention that my student, Prof. K. N. Ninan who succeeded me as the Professor and Head of the Unit, distinguished himself by becoming the President of the Indian Society of Ecological Economics, and served in the United Nations committees. Prof. R. S. Deshpande, another distinguished student, also did me proud by becoming Fellow of the Indian Society of Agricultural Economics and Director of ISEC. There is no greater happiness for an academician than to see his students becoming distinguished Professors making valuable contributions to research and turning out good students themselves. I am happy that other students of mine have also done very good work, like Prof. M. Ravichandran, Prof. G. V. Joshi, Prof. Sunita Raju, Dr. Syed Ajmal Pasha, Prof. Ramakrushna Panigrahi, and Dr. S. Manasi.

I must also note how happy I am with the exemplary work being done today by the Centre for Ecological Economics and Natural Resources (the present name of the old Ecological Economics Unit), under Prof. Sunil's leadership. Thanks to him and his colleagues, the Centre has acquired a good reputation both nationally and internationally. Prof. Sunil is very enterprising and has raised the needed resources to have an adequate research team. I wish them continued success in achieving greater heights and recognition.

I am, however, very unhappy with the situation of environment both in India and the world at large. I am nearing 82 years in age now, and as I am growing older, I am becoming more and more disappointed with the gravity of emerging environmental problems. Our tragedy is that moral growth is not taking place with economic growth. When we become more powerful, we should be more responsible and understanding. That is not happening. What pains me is that environmental problems are known, there has been tremendous amount of research work on them, and we also know the solutions to these problems, but there is no will both on the part of the states and the people at large to implement these solutions. It has become urgent to sharpen this will. This urgency prompted me to make a tiny effort by composing a long poem on environmental ethics, consisting of 113 verses in Sanskrit along with their translation also in English verse, entitled *Parisara-neeti-shatakam*. I have no illusions that this work would contribute towards creating the will to solve environmental problems, but I wrote it out of a sense of exasperation and anguish. I would like to summarise its message briefly.

I feel that we should resolve to adopt five principles or *Pancha-sheel* towards solving the environmental problems. The principles are:

1. Aim at economic growth only within the sustainable capacity of the environment. Specifically, it means three things: (a) exploit natural resources within the capacity to recycle, not beyond; (b) achieve carbon neutrality as soon as possible, setting a target year, by which carbon emissions do not exceed the earth's capacity to sink or absorb them; (c) don't reduce forest or green cover any further, since forests cannot be replaced; if any forest or tree cover rich in biodiversity is expected to suffer as a result of a development project, then give up that project for God's sake; it is ignoring this principle which caused immense flooding and damage in Kodagu and Kerala.
2. Search for green technology is alright, but technology is not all. Restraint on wants and wasteful consumption is also essential. Where one light is enough, switching on two is a waste. Washing cars and front yards of houses with a hosepipe is a criminal waste of water. Prudence in the use of natural resources, especially water and air as a sink, is urgently needed. Let me give an example to show the limits of technology. A heart surgeon does not rely only on surgery; he or she advises the patient to discipline his life with more exercise and eat less. This policy holds for the environment too.
3. Every individual is important, no one is dispensable. Even if a minority of people become victims of a development project, and beneficiaries are large in number, please see that all the victims are compensated, rehabilitated, and resettled satisfactorily.

4. Be honest and thorough in making environmental assessments of impacts of projects, and don't take up projects which damage the environment and forests.
5. Decentralise the economy, and to the best possible extent, prefer small-scale labour-intensive projects, and prioritise producing for local markets using local materials.

I had concluded my long poem on environmental ethics with a prayer. I will conclude my talk also by reciting the same prayer.

नाशोद्यतमनुष्येभ्यः सुबुद्धिं देहि धीश्वर ।
यैः सर्गं प्रति तादात्म्यं सामरस्यमुपेक्षितम् ॥११३॥

Its translation in English is:

Oh Lord! Give good sense
to people bent upon destruction,
Who have ignored oneness
and harmony with Creation! (113)

Thank you very much.[1]

---

[1] Prof. M. V. Nadkarni is a renowned economist by training and profession, and has a deep interest in other social sciences as well as in religion and philosophy. His contribution to Economics has been mainly in the areas of Agricultural Price Policy, Agrarian Change, Problems of drought Prone Areas, Farmers' Movements, Political Economy of Forest Use and Management, Economics of Pollution Control, and Integrating Ethics into Economics. He was a Professor at the Institute for Social and Economic Change (ISEC), Bangalore, from 1976 to 1999, and is presently an Honorary Visiting Professor at the same Institute since 2002. The three-day International Workshop cum Training Programme on "Green Growth Strategies for Climate Resilience and DRR: Policies, Pathways and Tools" organised at ISEC, took the opportunity to felicitate Prof. Nadkarni and acknowledge his scholarly contribution.

Contents

1   Climate Change and Resilient Society in Contemporary World: Ecology, Economy and Society Interface in Indian Perspective   1
Sunil Nautiyal, Anil Kumar Gupta, Mrinalini Goswami, and Y. D. Imran Khan

2   Community-Based Disaster Reduction for Climate Resilience in Developing World   11
Fatima Amin, Sweta Baidya, and Anil Kumar Gupta

Part I   Natural Resources and Livelihoods

3   Climate Sensitivity, Forest Ecosystems, and Regional Challenges: Exploring Opportunities to Reduce Disaster Risks in Central Himalayas   23
Shalini Dhyani, Deepak Dhyani, Rakesh Kadaverugu, Paras Pujari, and Rakesh Kumar Maikhuri

4   Measuring Climate Change Impact on Crop Yields in Southern India: A Panel Regression Approach   39
Rajesh Kalli and Pradyot Ranjan Jena

5   Impact of Climatic Variations on Crop Yield in Chamoli District, Uttarakhand, India   53
Sunil Nautiyal, Satya Prakash, Mrinalini Goswami, and A. Premkumar

6   A Comparative Assessment of Farmers' Perception on Drought and Related Impacts in Western Part of Odisha   71
Devi Prasanna Swain, Arunasis Goswami, Bhabesh Chandra Das, Debasis Ganguli, and Madan Mohan Mahapatra

7 Indian Pastoralism Amidst Changing Climate and Land Use: Evidence from Dhangar Community of Semi-arid Region of Maharashtra 85
Dada R. Dadas

8 Forest Fire Characterization with Relation to Meteorology and Topography Parameters in Madhya Pradesh, India 99
Tapas Ray, Satyam Verma, and M. L. Khan

9 Coexistence and Conflict—Case Study on Colonial Waterbirds in Southern India 111
Vaithianathan Kannan

10 Estimation of Rural Drinking Water Supply in Southern India: A Contingent Valuation Method 125
A. Xavier Susairaj and A. Premkumar

Part II Adaptation—Approaches and Mechanisms

11 Assessment of Climatic Risk and Adaptation Measures in Rural India: A Case Study of Vanvasi Village, Maharashtra 141
Sujit Chauhan and Madhuri Pal

12 Climate-Smart Agricultural Practices and Technologies in India and South Africa: Implications for Climate Change Adaption and Sustainable Livelihoods 161
Juliet Angom and P. K. Viswanathan

13 Flood Resilience of Pokkali Rice: A Study of the 2018 Flood in Kerala 197
Sweta Baidya, Parvathi Radhakrishnan, and Anil Kumar Gupta

14 The Role of Gharbari Model to Support Climate-Smart Agriculture 213
Kumudini Mishra

15 Understanding the Mental Models that Promote Water Sharing for Agriculture Through Group Micro-Irrigation Models in Maharashtra, India 229
Upasana Koli, Arun Bhagat, and Marcella D'Souza

16 Water Budgeting for Sustainable Development: A Micro Level Study on Climate-Resilient Agriculture 257
Gireesan K

17 Making of a Climate Smart Village: A Study on Meenangadi Gram Panchayat in Kerala (India) 277
Jos Chathukulam and Manasi Joseph

18	Nanominerals: A Climate Smart Tool in Livestock Feeding—A Review Partha Sarathi Swain, S. B. N. Rao, D. Rajendran, Sonali Prusty, and George Dominic	299
19	Climate Change Impacts on the Higher Altitude Forests of Indian Himalayan Regions: Nature-Based Solutions for Climate Resilience and Disaster Risk Reduction Anwesha Chakraborty and Purabi Saikia	313

Part III   Traditional Knowledge in Climate Resilience

20	Traditional Ecological Knowledge *Versus* Climate Change Adaptation: A Case Study from the Indian Sundarbans Sneha Biswas and Sunil Nautiyal	331
21	From Resilience to Vulnerability: Indigenous Agri-Food Systems of Wayanad District V. Swaran, C. S. Chandrika, and Chubamenla Jamir	351
22	Ethnobotanical Knowledge and Floristic Diversity of South East Indian Coastal Region Nallakaruppan Nagaraj, Veluchamy Chandra, Sekaran Manoj, Nachiappan Kanagam, Sunil Nautiyal, Thiagarajan Kalaivani, and Chandrasekaran Rajasekaran	367

Part IV   Urban Sustainability and Resilience

23	Urban Green Spaces for Environmental Sustainability and Climate Resilience Amit Kumar, Pawan Ekka, Manjari Upreti, Shilky, and Purabi Saikia	389
24	Urban Civic Services Delivery and Climate Change Challenges: A Study of Two Indian Cities Ramakrishna Nallathiga and Kala Seetharam Sridhar	411
25	Climate Change and Water Insecurity: Who Bears the Brunt? (A Case of Yelenahalli Village, Bengaluru) Aakanksha Srisha and S. Yogeshwari	429
26	People's Awareness, Perceptions and Attitudes on Green Buildings: *A Study in Bengaluru* S. Manasi, Channamma Kambara, and N. Latha	447
27	Decentralisation in the Urban Sphere: Successes and Failures—(A Note from Experience) Kathyayini Chamaraj	471

Part V Policy Issues and Future Strategies

28 **Integrating Climate Resilience in Sectoral Planning: *Analysis of India's Agriculture Disaster Management Plan*** — 481
Sanayanbi Hodam, Richa Srivastava, Anil Kumar Gupta, and Kirtiman Awasthi

29 **Scalable Adaptation Model for Sustainable Agriculture Livelihoods Under Changing Climate: A Case Study from Bihar and Madhya Pradesh** — 499
Ravindra S. Gavali, V. Suresh Babu, Krishna Reddy Kakumanu, Shrikant V. Mukate, Y. D. Imran Khan, Basavaraj Patil, Utkarsh Ghate, and V. Srinivasa Rao

30 **Delineating Health Sector Resilience in Post COVID-19 Pandemic in the Backdrop of Changing Climate and Disasters** — 527
Atisha Sood, Anjali Barwal, and Anil Kumar Gupta

31 **Climate Change and Environment: Holistic Approaches Towards Climate Resilience** — 537
Usha Swaminathan, G. B. Reddy, Inocencio E. Buot, and Siva Ramamoorthy

32 **Green Finance for a Greener Economy** — 553
Meenakshi Rajeev and Oisikha Chakraborty

33 **Climate Change: A Major Challenge to Biodiversity Conservation, Ecological Services, and Sustainable Development** — 577
Shilky, Subhashree Patra, Pawan Ekka, Amit Kumar, Purabi Saikia, and M. L. Khan

34 **Green Social Work; A Call for Climate Action** — 593
Jose John Mavely, Rosemariya Devassy, and Kiran Thampi

Part VI Conclusion

35 **Epilogue** — 605
Sunil Nautiyal, Anil Kumar Gupta, Mrinalini Goswami, and Y. D. Imran Khan

Index — 621

Notes on Contributors

Fatima Amin has completed her master's from University of Kashmir and has dedicated her life towards exploration and experimentation. Currently she is working as young professional at NIDM under ECDRM Division. She is also pursuing her Ph.D. in Disaster Management from CEDM, GGS IP University. Her area of interest includes Integrated Risk Management strategies, floods risk management, sustainable development goals, environmental studies, global climate change, and natural resources management.

Juliet Angom currently serves as a Ph.D. scholar under Amrita Vishwa Vidyapeetham. She is exploring the integration of climate smart-resilient agriculture (CS/RA) as a strategic approach to addressing climate change. Her research and teaching interests relate to; Sustainable Development, Climate Change, Food Security, Agriculture, Ecosystems, Agrarian Transformation, Rural livelihoods, Aspects of Agri-Technology, Institutional Policies and Governance, Globalisation and its impacts on Environment.

Kirtiman Awasthi is a climate change and sustainability professional working as a Senior Advisor at GIZ India. He has spearheaded project management, strategic planning, policy advice, collaborative partnership & resource mobilisation covering issues of climate adaptation & resilience, climate risk management, climate finance, natural resource management and sustainable development. He has been advising national and state-level partners to strengthen and operationalise the State Action Plan on Climate Change (SAPCCs), including capacity building. Kirtiman Awasthi has been part of the UNFCCC processes, where he covered issues of adaptation, loss & damage and capacity building and supported the Durban Platform process. He is an alumnus of Wageningen University, the Netherlands.

V. Suresh Babu has a doctoral degree in Agronomy and is an Associate Professor at the North Eastern Regional Centre of the National Institute of

Rural Development and Panchayati Raj, Guwahati. He specialised in climate-resilient agriculture, management of rainfed agriculture, integrated farming system, agroforestry systems, and convergence with flagship programmes. He has published several books and articles in national and international publications. He has also been associated with several research and evaluation studies.

Sweta Baidya has done her Ph.D. in Oceanography from CSIR-National Institute of Oceanography. She has worked extensively on the Paleo-climate, especially the paleo-monsoon shift from Last Glacial Maximum to Holocene. Later she has worked on Desertification and Land degradation in Jawaharlal Nehru University. She also has an experience of working on Glaciology in Ministry of Earth Sciences. Formerly she was a faculty of Disaster Management in Jamia Millia Islamia University. Later she worked in National Institute of Disaster Management as consultant. At present she is a senior consultant in National Disaster Management Authority.

Anjali Barwal is working as a Consultant at National Institute of Disaster Management, Ministry of Home Affairs, New Delhi. She has done her M.Phil. and Ph.D. in Energy and Environment from Devi Ahilya Vishwavidyalaya, Indore. She has published more than 35 research articles, book chapters, policy papers, training manuals, etc.

Arun Bhagat is a Senior Researcher (Irrigation Water Management) at Watershed Organisation Trust (WOTR), Pune India. He has an M.Tech. degree in Irrigation water management and a Ph.D. in Irrigation and Drainage Engineering. He has published more than 15 publications in International and National journals and is also a recipient of multiple National level Fellowships for scientific research.

Sneha Biswas has earned her Ph.D. degree in Development Studies from the Institute for Social and Economic Change, Bengaluru. Her areas of interest include socioeconomic vulnerability, coastal vulnerability, disaster studies, climate change, community and local adaptations. She is currently working as a Research Fellow at Karnataka Evaluation Authority, Government of Karnataka.

Inocencio E. Buot Jr. is a Professor of Botany and curator in the Plant Biology Division Herbarium (PBDH) and current head of the Plant Systematics Laboratory at the Institute of Biological Sciences (IBS), a Center of Excellence for Biology since 1984, and University of the Philippines Los Banos (UPLB). His laboratory is now focusing on biodiversity conservation and plant systematic studies. He has published more than 150 articles in journals. He has received many prestigious awards for his research contribution.

Anwesha Chakraborty completed her M.Sc. from the Department of Life Sciences, Presidency University, Kolkata in the year 2021.

Oisikha Chakraborty has completed her M.A. in Financial Economics from the University of Hyderabad. She has worked on various projects at the Ministry of External Affairs, Institute for Social and Economic Change, and Ernst & Young. She has previously contributed to the book *Belt and Road Initiative China's Global Business Footprint*, published by the Centre for Business and Economic Research (CBER), UK. She is currently working as a research associate at the School of Economics, University of Hyderabad.

Kathyayini Chamaraj has been a freelance journalist for 35 years, writing mainly on development issues, urban governance, unorganised labour, primary education, and child labour. She has been the Executive Trustee of CIVIC-Bangalore since 2005. CIVIC works on implementing the 74th Constitutional Amendment or Nagarapalika Act in letter and spirit to make urban governance more transparent, accountable, and participatory. CIVIC-Bangalore also works with the urban poor on their Rights to Food, Health, Education, Housing, and Social Security and advocates for policy on these issues.

Veluchamy Chandra has completed M.Sc. in Biotechnology from Dr. N.G.P. Arts & Science College, Coimbatore. She is currently pursuing Ph.D. in marine microbiology at Department of Biotechnology, SBST, VIT, Vellore. Her research interests are coastal microbiome, Anti-Microbial Resistance, and antibiotic degradation especially by marine bacteria.

C. S. Chandrika is a renowned author of Malayalam literature and a Social Scientist, Development practitioner, and Public intellectual based in Kerala, India. She has a multidisciplinary academic background with a doctoral degree in Fine Arts. She is currently a member of the Anti-Sexual Harassment committee of the Kerala Biodiversity Board and has been a member of the committee constituted by the Wayanad district administration for the social and environmental impact assessment for all major development programmes in the district since 2018.

Jos Chathukulam is Sri. Ramakrishna Hegde Chair Professor on Decentralization & Development, Institute for Social and Economic Change (ISEC) Bengaluru, India. He is also the Director of Centre for Rural Management (CRM), Kottayam, Kerala, India. He has extensively researched on decentralisation and local governance across Indian states, Latin America, Nuba Mountains in South Kordofan, Sudan, Africa, and Cuba. His areas of interest are democratic decentralisation, multi-level planning, policy studies, rural development, administrative innovations, and Kerala Studies.

Sujit Chauhan is a Ph.D. candidate in the Department of Humanities and Social Sciences (HSS), Indian Institute of Technology, Bombay. His research focuses on agricultural technology adoption, disadoption, and climate change adaptation. Sujit holds an M.Phil. degree in Planning and Development from IIT Bombay with a specialisation in economics.

Dada R. Dadas has completed his M.Phil. in Social Work from Tata Institute of Social Sciences, Mumbai. He is a Research Associate at the Ecosystem-based Adaptation project of Watershed Organisation Trust (WOTR), Pune. His research focuses on the livelihood of pastoral nomads and role of nature-based solutions to combat climate change. He has more than eight years of work experience across academia and research. He was previously associated with SARATHI (a Government of Maharashtra research institute) and Karve Institute of Social Service, affiliated with Savitribai Phule Pune University.

Bhabesh Chandra Das is currently working as Associate Professor at the College of Veterinary Science and Animal Husbandry, REWA, MP and is well-recognised in the field of Veterinary Extension. He has served in OUAT, and during his tenure, he has published many Books, Research, and Review papers. His contribution in extension education is instrumental in teaching and research.

Rosemariya Devassy is a professional social worker and Counsellor at Rajagiri Family Counselling Centre (A Government of India-funded project for Women and Children). Besides, she has a year of experience in counselling after completing her Post Graduation in Social Work. Her expertise lies in Child Development, Child Rights (UNCRC), and Family functioning. Highly interested in analysing situations and formulating, implementing, and reviewing appropriate intervention programmes.

Deepak Dhyani is presently working as a Senior Researcher associated with NTFP-EP, Indonesia-funded project with COPAL, an NGO based in Uttarakhand India. He has 20 years of experience of working on conserving Lesser Known underutilised wild edible plants and Strengthening Community Conservation Initiatives for integrating livelihood and conservation in Uttarakhand, India.

Shalini Dhyani is a Senior Scientist with the Critical Zone Group of the Water Technology and Management Division in CSIR-NEERI, India. She is Asia Vice Chair for IUCN CEM (Commission on Ecosystems Management) and Steering Committee Member; IPBES Lead Author of the Global thematic assessment on Sustainable Use of Wild Species (2018–2021) and IPBES Asia Pacific regional assessment of biodiversity and ecosystem services in Asia Pacific (2015–2018).

George Dominic has a Master's (M.V.Sc) and doctoral degree (Ph.D.) in the Discipline of Animal Nutrition. He is well-experienced in Ruminant Nutrition portfolio with special interest in research work pertaining to plant secondary metabolites, rumen microbial profiling, and climate change. Currently, he is working as a faculty member in the Department of Animal Nutrition, Rajiv Gandhi Institute of Veterinary Education & Research, Puducherry.

Marcella D'Souza is the Director of the WOTR Centre for Resilience Studies (W-CReS) and former Executive Director of Watershed Organization Trust

(WOTR), India from 2006 till March 2019. She was the Coordinator, Women's Promotion in the bilateral Indo-German Watershed Development Programme. Gaps in understanding the local needs in different agroecological regions led Marcella to establish the trans-disciplinary applied research unit: the WOTR Centre for Resilience Studies (W-CReS), to provide evidence-based lessons from practice to policy. Marcella has co-authored research papers and methodologies (tools) to help upscale good practices in natural resources management.

Pawan Ekka completed his M.Sc. from the Institute of Environment & Sustainable Development, Banaras Hindu University, Varanasi in the year 2020. Presently doing his Ph.D. research in the Department of Environmental Sciences at the Central University of Jharkhand, India, with research interest in forest ecology, urban green spaces, and sustainable development.

Debasis Ganguli presently working as an Associate Professor in Department of Veterinary and Animal Husbandry Extension Education, WBUAFS, Kolkata. Academically, Dr. Ganguli has B.V.Sc, M.V.Sc, Ph.D. He has many research and review articles to his name and has guided many M.V.Sc students as well.

Ravindra S. Gavali is Professor and Head of the Centre for Natural Resource Management, Climate Change and Disaster Mitigation at National Institute of Rural Development and Panchayati Raj (NIRDPR), Hyderabad. He received his Ph.D. in Environmental Science from Jawaharlal Nehru University, New Delhi. Currently, he works on national and international training, research, and consultancy programmes related to Natural Resource Management, Sustainable Development, Applications of Remote Sensing & GIS, Watershed Management, Biodiversity Conservation and Governance, Climate Change and Disaster Management.

Utkarsh Ghate did Ph.D. in botany with 30 years of experience in biodiversity & agri-marketing, agri-technology & IPR (patents, etc.), especially of herbs. He has 33 research publications in peer-reviewed journals, including international ones and an encyclopedia. He visited 12 countries as conference speaker and was member of Govt. of India's Planning Commission task force (2011) on forestry & Green India Mission (2014–2019). He promoted 6 farmers-owned producer organizations (FPOs) that won United Nations environment awards & he directed national biodiversity award 2016 winner the Gram Mooligai (village herbs). company. Presently he is ecologist at Bharatiya Agro industries Foundation (BAIF), Pune.

K. Gireesan is the Director, School of Government, MIT World Peace University, Pune. Previously, he worked with the Rajiv Gandhi National Institute of Youth Development, Ministry of Youth Affairs and Sports, Govt. of India, Sriperumbudur, and Centre for Rural Management, Kottayam. As a National

Evaluator, he has vast experience in monitoring, evaluation, and impact assessment studies for the Ministry of Rural Development and Ministry of Panchayat Raj, Govt. of India, in different parts of the country. He is the author/co-author of several books and research reports. In addition, he has written several research papers in national/ international journals and edited books.

Arunasis Goswami is a Professor at the Department of Vety. & Anim. Hus. Ext. Education, Faculty of Veterinary & Animal Sciences, and Director of Research, Extension and Farms (Actg.) West Bengal University of Animal and Fishery Sciences, Dr. Goswami has served in many administrative capabilities.

Mrinalini Goswami is a researcher in ecological science and is associated with Centre for Ecological Economics and Natural Resources, Institute for Social and Economic Change, Bengaluru. She received her Ph.D. in Environmental Science for her research on sustainability of ecosystem-based livelihoods in peri-urban landscapes. She has been engaged in research on varied topics, including environmental sustainability, climate change, natural resource management, water management, and urban development. Dr. Goswami also has substantial work experience in the development sector, which has motivated her towards interdisciplinary approach in environmental research.

Anil Kumar Gupta is a disaster mitigation & crisis management professional, Ph.D. (1995), Post-doctorate (CSIR, NEERI 1996), he possesses interdisciplinary expertise, with institutional development-administration & management, coordinated several international/national projects, over 180 publications, including 10 books and 60 papers, guided Ph.D. research. His special Contribution were in preparation of National DM Plan (including mitigation & response plan), National HR Plan for DRM, Perspective Plan for NIDM's strengthening, National strategy on climate change, NAP on Chemical Disasters, international cooperation with GIZ, UNEP, UNDP, ISET-US, UNU-EHS, UNESCO, CKDN, ICIMOD, NORAD, World Bank, etc., and coordination with States/UTs for DM capacity building/training matters. Areas include risk/Vulnerability analysis, DMP, PDNA, CCA-DRR, housing safety & local emergency preparedness, DM planning, governance, etc.

Sanayanbi Hodam is a doctorate in agricultural engineering and has worked on Quantitative Assessment of Vulnerability to Floods for changing climate in Northeast India. She has worked with National Institute of Hydrology, Roorkee and ICAR-Barapani, and Meghalaya. She has worked with NIDM for preparation of NEFC-DMP for the MoEFCC and for preparation of NADMP for the MoAFW. She is now presently working as Assistant Professor in North Eastern Regional Institute of Water and Land Management, Ministry of Jal Shakti.

Y. D. Imran Khan is currently working as a Senior Researcher and Thematic lead (Ecology) in the WOTR-Centre for Resilient Ecosystem (W-CReS),

Watershed Organization Trust (WOTR), Pune. He received his Ph.D. in Zoology (Ecology) from Bangalore University. He worked and associated with various research organisations such as Centre for Ecological Economics and Natural Resources (CEENR), Institute for Social and Economic Change (ISEC), Bengaluru; Tata Institute of Social Sciences, Mumbai; and Centre for Natural Resource Management (CNRM), National Institute of Rural Development & Panchayati Raj (NIRDPR), Hyderabad. Dr. Khan has published scientific articles in national and international peer-reviewed journals, as well as contributed to book chapters and authored books. His research focuses on biodiversity conservation, natural resource management, socio-ecology, nature-based solutions, and green livelihoods.

Chubamenla Jamir is currently Co-Lead for Mountain Agriculture in the Himalayan Universities consortium. She has served in multiple institutes, including TERI School of Advanced Studies, Stockholm Environment Institute (York, UK), Environment Department of University of York (UK), Global Centre for Food Systems Innovation (USAID), and UN Sustainable Development Solutions Network, South Asia Regional Centre, in the past. She has an M.Phil. in Environmental Sciences from Jawaharlal Nehru University, New Delhi and a doctoral degree from the University of York, UK.

Pradyot Ranjan Jena is currently working as an Associate Professor at National Institute of Technology Karnataka. Dr. Jena's core research areas of expertise are impact evaluation, environmental impact valuation, and climate change. He has worked across continents and has experience in rural livelihoods in many developing countries. He is currently the Associate Editor of Frontiers in Climate: Climate and Economics. He published several books and peer-reviewed journal articles.

Manasi Joseph is a Researcher at the Centre for Rural Management (CRM), Kottayam, Kerala, India. She holds a postgraduate certificate in New Media—Journalism, Oakville, Ontario, Canada. Her research interests include Public and Policy Administration, Health and Nutrition, Gender Studies, Development Communications, and Paradigms and Counter Culture. Mainstream Weekly.

Rakesh Kadaverugu is a senior scientist at the Cleaner Technology and Modeling Division of CSIR National Environmental Engineering Research Institute, India. He has more than ten years of experience in environmental systems modelling on different scales ranging from lab-scale prototypes to socio-environmental systems. He has been working on species habitat modelling to project the future potential habitats of sensitive tree species and the uncertainty and sensitivity analysis in such projections.

Krishna Reddy Kakumanu is an Associate Professor at the Centre for Natural Resource Management, Climate Change and Disaster Mitigation at the National Institute of Rural Development and Panchayati Raj, Hyderabad. He has vast experience in the economics of conjunctive use of water, energy pricing, and climate change impact assessment and adaptation in India's agriculture and water sectors. He has 15 years of research experience and published several research papers in international journals and books.

Thiagarajan Kalaivani works as Associate Professor Grade-2, School of Bio-Sciences and Technology, VIT, India. She has expertise in Biochemistry and plant secondary metabolites. Her research interests are Secondary metabolites from Medicinal and Aromatic Plants, Toxicology, and Bioremediation. She is fellow in FSEDI and FLS (Lon). She has published over 50 research articles in reputed journals, 20 book chapters, and edited books.

Rajesh Kalli is currently working as an Assistant Professor in School of Commerce and Management Studies, Dayananda Sagar University, Bangalore. He is awarded Ph.D. from National Institute of Technology Karnataka, Surathkal. He has published articles in several national and international Journals. He also attended conferences in various international events—Kumamoto University, Japan and 30th International Conference on Agriculture Economists. His area of research is on management and economics.

Channamma Kambara is Assistant Professor, Centre for Research in Urban Affairs (CRUA). Her research interests are on factors affecting the marginalised and vulnerable sections, mainly women, informal sector viz., street vendors, migration, public spaces in urban areas, sustainable/resilient urbanisation, and related issues.

Nachiappan Kanagam has completed his M.Tech. in Biotechnology from SRM University and serving as Assistant Professor at Sri Venkateswara College of Engineering, Sriperumbudur. Pursuing her Ph.D. in the Department of Biotechnology, SBST, VIT, Vellore. Her research interests are Phycoremediation, Algology, Plant tissue culture, and Environmental biotechnology.

Vaithianathan Kannan is a Zoologist and Wildlife Biologist working as a Scientist with the Gujarat Institute of Desert Ecology, Bhuj-Kachchh, Gujarat. He was associated with the Bombay Natural History Society, Mumbai & AVC College (Autonomous), Tamil Nadu and various other NGOs. He is a member of the IUCN/SSC Pelican Specialist Group (Old World). He has also taken up a voluntary position within the Old World Pelican Specialist Group titled "Taxon and Regional Coordinator". To his credit, he has published various research articles on wildlife in national and international journals.

M. L. Khan is a Professor, and he is a plant ecologist and conservation biologist. He works in the fields of biodiversity conservation, eco-restoration, ecosystem services, carbon sequestration, forest fire ecology. He has contributed 205 peer-reviewed papers in prestigious journals such as Nature, PNAS (USA), Nature Ecology and Evolution, Nature Scientific Report, Critical Review in Biotechnology, Science of the Total Environment.

He has also contributed 60 book chapters. Three books have been authored/edited by him.

Upasana Koli is currently working as a Research Associate at WOTR-Centre for Resilience Studies. She has done M.A. In Economics, with over four years of experience undertaking research, particularly in the semi-arid region of Maharashtra, India. Her research contributes to studying the behavioural aspects of the farming community towards water and agriculture, subjective concepts of well-being and resilience building, and economic valuation of intervention on agricultural output.

Amit Kumar is working as an Assistant Professor at the Central University of Jharkhand, India and published numerous research articles in various journals and books of international and national repute. He is a member of multiple academic societies, including IUCN-Commission on Ecosystem Management (South Asia), National Association of Geographer-India (NAGI), Global Forest Biodiversity Initiatives (GFBI), USA, *etc.*, and the recipient of the Young Scientist Award-2020 of The Society for SCCSE.

N. Latha is working as a Research Associate at the Centre for Research in Urban Affairs, Institute for Social and Economic Change, Bangalore. She has more than ten years of research experience.

Madan Mohan Mahapatra now works as Assistant Director in Animal Resources Development Department with Govt. of Odisha. Dr. Mahapatra possesses a Master's degree in Veterinary and Animal Husbandry Extension along with Post Graduate Diploma in Management. He has done research on curbing distress migration of bonded labourers through goat rearing, which has been well-appreciated in different circles. He has authored and co-authored many publications.

Rakesh Kumar Maikhuri is a Professor, Department of Environmental Sciences, HNB Garhwal University, Srinagar Garhwal. He was Scientist "G" is an ecologist with over 40 years of experience in the interphase area of ecology-natural resource management and sustainable development in the central Himalayan region. For his outstanding scientific contribution, he has received many national and international awards.

S. Manasi is currently working as an Associate Professor at the Centre for Research in Urban Affairs, Institute for Social and Economic Change, Bangalore. She has long-time research experience in urban ecology, Natural Resource Management, land resources, livelihoods, and pollution abatement with a specific focus on water, sanitation, and waste management. She has completed many collaborative research studies and authored several articles in journals and edited books.

Sekaran Manoj has completed M.Sc. in Botany from St Joseph college Trichy. He joined VIT as a Project Assistant in the DST-funded project. Currently pursuing a Ph.D. in plant biology in the Department of Biotechnology, SBST, VIT, Vellore. His research interests are phenology of macroalgae and in vitro propagation and elicitation of seaweed secondary metabolites.

Jose John Mavely, a Pre-Doctoral Fellow, Rajagiri College of Social Sciences (Autonomous), is a social practitioner who aims to extend his talents and skills in social welfare and social policies. His area of expertise lies in Corporate Social Responsibility, Stakeholder Engagement, and Project Management. Highly interested in providing versatile ideas for enhancing stakeholders' well-being and designing social projects to help people meet their basic needs.

Kumudini Mishra is a Sociologist with over 15 years of experience in the teaching, counselling, training, and development sector. Currently working in the Planning & Convergence Department of Odisha as a Micro Planning Livelihood Expert. She looks after micro planning and ensures decentralised plan procedures and systems are established through full participation of local communities in the district. Spearheaded the effective implementation of different Livelihood Projects/Schemes in the district.

Shrikant V. Mukate is an independent researcher with a background in Hydrology, water quality, ecology, river management, and climate change. He is a DST-INSPIRE fellowship (Govt. of India) awardee and completed his Ph.D. from the School of Earth Sciences, SRTM University Nanded in Maharashtra, India. He also completed Post-Doctoral Research at the National Institute of Technology Karnataka, Surathkal. He worked at the National Institute of Rural Development and Panchayati Raj, Hyderabad and Centre for Resilience Studies of Watershed Organisation Trust, Pune. He has published more than 30 research papers in national and international peer-reviewed journals.

Nallakaruppan Nagaraj has served as a Project Assistant in the DST sponsored project. Currently, he is pursuing a Ph.D. in Plant Biology at Department of Biotechnology, School of Bio Sciences and Technology, Vellore Institute of Technology. His research interests are conservational plant biology, Natural products research, *in vitro* propagation, and production of plant secondary metabolites.

Ramakrishna Nallathiga is a Research Scholar pursuing Ph.D. at the Centre for Research in Urban Affairs, Institute of Social and Economic Change, Bengaluru. His career spanning over more than two decades involved teaching, research, and consulting at different academic and/or policy research institutes. He currently serves as a Faculty Member at National Institute of Construction Management and Research, Pune.

Sunil Nautiyal (AvHF, JSPSF, ZF, FNIE) is a Professor at the Centre for Ecological Economics and Natural Resources (CEENR), Institute for Social and Economic Change (ISEC), Bengaluru and Honorary Professor at the University of Ladakh. Presently Dr. Nautiyal serving as Director,

GB Pant National Institute of Himalayan Environment (NIHE), Kosi—Katarmal, Almora, Uttarakhand, India. His specialisation includes areas of Natural Resource Management and Conservation, Socioeconomic and Ecological Approaches for Sustainable Development, Production System Analysis, Protected Area Management and HWC&C, Land Use Land Cover Change Analysis, Sustainable Livelihood Development, Ecological Modelling, GIS and Remote Sensing for Landscape Research, Urban-rural interface and Sustainability, Climate Change, Socio-ecological Development (Socio-Ecologist). Dr. Nautiyal has published over 170 scientific papers/articles and 14 books. Dr. Nautiyal is a recipient of several national/international prestigious fellowships/awards. He is on the Editorial Board of 06 international journals and board of reviewers for more than 70 international/national journals. Dr. Nautiyal has been working in close coordination with scientists in India, Germany, UK, Australia, Japan, Bangladesh, and the USA. His research contribution has large policy implications in the area of socio-ecological development and biodiversity conservation.

Madhuri Pal is a research student at the Graduate School of Energy Science at Kyoto University. Her research focuses on environment and energy economics. She was a Research Officer at the NITI Aayog in Delhi. She was previously associated as a faculty member with the University of Delhi and the National Institute of Technology, Calicut. She completed her M.Phil. in Planning and Development from IIT Bombay.

Basavaraj Patil is an Agronomist, and he has more than five years of research experience in the area of Agronomy and Rural Development. Dr. Patil has completed his Ph.D. from the University of Agricultural Science, Bangalore.

Subhashree Patra completed her M.Sc. from the Department of Biodiversity and Conservation of Natural Resources, Central University of Odisha, India, in 2020. Presently, she is working on her Ph.D. thesis on Disturbance impacts on vegetation dynamics at the Central University of Jharkhand, India.

Satya Prakash is a Researcher at the Centre for Ecological Economics and Natural Resources in the Institute for Social and Economic Change, Bengaluru. He earned his M.Tech. in Geoinformatics from the Central University of Jharkhand. His research interests include remote sensing and Geographic Information Systems (GIS), management of natural resources and climate change.

A. Premkumar is currently working as an Assistant Professor at Department of Economics, Sacred Heart College (Autonomous), Tirupattur, Tamil Nadu. He received his Ph.D. in Economics from Thiruvalluvar Universty, Tamil Nadu. He has written and published several articles in UGC Care listed journals and chapters in edited books. He has presented many research papers in international and national conferences.

Sonali Prusty is currently working as an Assistant professor, Animal Nutrition, College of Veterinary Science and Animal Husbandry, Anjora, Durg (C.G.),

India, Dr. Prusty has been Awarded with many awards like Merit certificate for academic excellence in 2013–14 academic session for Ph.D. (Animal Nutrition), Pashudhan Samriddhi India Award-2022, Dr. S. K. Ranjhan award for best doctoral thesis, etc. She has published many articles in highly reputed national and international journals.

Paras Pujari is a Principal Scientist at National Environmental Engineering Research Institute. He is an expert environmental geophysicist with more than 30 years of experience. He has been instrumental in TECO Indo-Italian collaboration and mobility of researchers. He has significantly contributed to the emerging area of biogeophysics.

Parvathi Radhakrishnan is an Environmentalist and Disaster Management professional working with the Kerala Institute of Local Administration (KILA) and Local Self Government for Disaster Risk Reduction and Climate Action. She has a strong academic working background with M. S. Swaminathan Research Foundation in the study of wetlands and river systems. Based on the 2018 floods, she studied vulnerability and adaptive capacity during flood and landslide disasters in Kerala along with the University of Kerala. She has built a strong academic background in disaster management and climate science during her career with collaborations with the National Institute of Disaster Management (NIDM) and ESSO-Indian National Center for Ocean Information Services (INCOIS).

Chandrasekaran Rajasekaran is presently serving as a Professor (Grade-2), School of Bio Sciences and Technology, VIT, India. He is a fellow in various scientific bodies FSEDI, FLS (Lon), FRSB (UK), FISPP, FASCh., FNABS. He is authors more than 80 research publications in reputed journals, 30 book chapters, many manographs, and edited books.

Meenakshi Rajeev is currently the Reserve Bank of India Chair Professor and Head, Centre for Economic Studies and Policy, at the Institute for Social and Economic Change, Bangalore. She has graduated from IIT Kanpur in Statistics and did her Ph.D. in Economics from the Indian Statistical Institute, Kolkata. She has worked as a faculty and taught in the University of California at San Diego, Central Michigan University, Centre for Studies in Social Sciences, Kolkata, Presidency College, Kolkata, and in a large number of universities in USA, UK, Germany, France, and Norway, including Cornell University, University of Essex, Central Bank of Norway to name a few. She has published more than 100 articles in reputed journals and as working papers from India and abroad and has over 10 books and monographs to her credit. She is in several committees of the Government.

D. Rajendran is presently working as Principal Scientist, Animal Nutrition Division, ICAR- NIANP, has published many research and review articles in reputed journals. He has a vast experience in the field of Animal Nutrition pertaining to nanominerals supplementation, Feed resources, Mineral Nutrition. His contribution to the feed formulation through excel based app for ruminants is a noteworthy achievement.

Siva Ramamoorthy is a Professor and the Dean of the School of Biotechnology at Vellore Institute of Technology, India. His Ph.D. thesis was on plant genetic diversity, completed at Bharathidasan University, India. He completed postdoctoral research at Ben Gurion University (Israel) and Gyeongsang National University (South Korea). Dr. Siva is an IUCN species commission member for medicinal plants. He was nominated as a Fellow of the Linnean Society (FLS), Fellow of the Royal Society of Biology (FRSB), and Fellow of the Royal Society of Chemistry (FRSC), and he has published around 150 research articles in reputed journals with high impact factor. He has received many awards for his research contribution. Dr. Siva has been recognised as one of the top 2% scientists in the world as per a survey of Stanford University and Mendeley.

S. B. N. Rao presently working as Principal Scientist and Head, Animal Nutrition Division, ICAR- NIANP, has published more than 75 peer-reviewed publications. He has a vast experience of 32 years in the field of animal nutrition. He has worked in the area of Goat Nutrition, Feed resources, Mineral Nutrition, Environmental Pollutants.

V. Srinivasa Rao has 20 years of experience in climate change research, climate-resilient agriculture interventions, biodiversity conservation, and livelihood-based programmes. He is passionate about working with rural communities on agriculture interventions.

Tapas Ray is a research scholar with interests in forest fire ecology. He is currently working as a Senior Research Fellow at the IIFM Bhopal, M.P. India. He received his Ph.D. in 2021 from Dr. Hari Singh Gour University Sagar, M.P. His doctoral research focused on the ecological impacts of forest fires on the dry deciduous forests of the Panna Tiger Reserve, Madhya Pradesh.

G. B. Reddy is a Lawyer turned academician, columnist, and administrator. Teaching law since 1991. Specialised in Constitutional Law and IPRs. He has authored 23 books and published 74 articles on Law. On the academic and governing bodies of many institutions, including Central Universities and National Law Schools. Possesses rich experience as Principal, Chairperson of BoS, Head of Department and Dean, Faculty of Law, and the Director of the University Legal Cell and University Foreign Relations office.

Purabi Saikia is working in the broad areas of plant ecology and presently working as an Assistant Professor at the Central University of Jharkhand, India. She is an active member of the IUCN Commission on Ecosystem Management, South Asia, and Global Forests Biodiversity Initiatives (GFBI), USA. She is a recipient of several national awards and recognitions, and published several articles in various journals and books of national/international repute, including one in Nature.

Shilky completed her M.Sc. from the Department of Environmental Sciences, Forest Research Institute, Dehradun, in the year 2019. She has also worked as a young professional at the National Institute of Disaster Management for 16 months. Presently, she is working on her Ph.D. thesis on biomass estimation and carbon sequestration in the urban vegetation of Delhi-NCR at the Central University of Jharkhand, India.

Atisha Sood is a public health professional with a Master's Degree in Public Health. She has been working as a Research Associate at National Institute of Disaster Management, New Delhi for over three years. She is also pursuing her Ph.D. in Public Health from SRMIST Chennai. To her credit, she has published various research articles, thematic papers, policy briefs, training manuals and book chapters, etc.

Kala Seetharam Sridhar is a Professor at the Institute of Social and Economic Change, Bengaluru. She has authored or edited several books on urbanisation, with her most recent one on the Global South: Perspectives and challenges, Oxon: Routledge, 2021. She has published papers in international journals of high repute. She is consistently in the top 10% of authors globally on Social Science Research Network (SSRN) by total and all-time downloads.

Aakanksha Srisha is a first-year student of M.Sc. Economics and Social Sciences at Bocconi University in Milan, Italy. She has completed a few research internships in the fields of macroeconomics and development studies. Her research interests lie at the intersection of environmental, financial, and public economics.

Richa Srivastava has done her Masters' in Environmental Sciences from Wageningen University, The Netherlands and has more than 18 years of experience of working on Project Management, Research, Knowledge Management, and Community Outreach in the field of Environment, Agriculture, and Development. She is an experienced Coordinator with a demonstrated history of working in the Program Development Industry. She is presently working as Research Consultant with NIDM for developing National Environment Climate and Forest Disaster Management Plan (NEFCDMP) for the MoEFCC.

A. Xavier Susairaj is an Associate Professor and Head of the Post Graduate and Research Department of Economics, Sacred Heart College (Autonomous), Tirupattur. He is a visiting fellow at Chonbuk national university, South Korea, Fulbright scholar in 2010 at Florida International University in Miami, USA. He has published four books and authored numerous research articles in international reputed and UGC Care listed journals. He is also a chief editor of *Journal of Social Sciences and Management Research*.

Devi Prasanna Swain Swain has a Bachelor's Degree in Veterinary and Animal Sciences (B.V.Sc. and A.H) and a Master's (M.V.Sc) and doctorate degree (Ph.D.) in the Discipline of Veterinary and Animal Husbandry Extension Education. Dr. Swain has conducted researches on filed oriented problems and his research on impact of Livestock in drought affected western Odisha is a well-appreciated. He has authored and co-authored many publications.

Partha Sarathi Swain is a Recipient of Mrs. Saroj Jakhmola Award for Best Ph.D. Thesis in Animal Nutrition by Animal Nutrition Association. Dr. Swain is a Veterinary Doctor and done his specialisation in the field of Animal Nutrition. Presently working under Govt. of Odisha. He has worked in the field of nanominerals, and non-conventional feed resources. He has authored more than 20 publications in reputed journals.

Usha Swaminathan is working at Vellore Institute of Technology (VIT) in the School of Social Sciences and Languages as an Associate Professor of Law. She obtained her UG and PG in the field of Law from Osmania University, Hyderabad. She practised as a lawyer initially and later carried out her Ph.D. work on IPR and Traditional Knowledge from VIT. She also holds an M.B.A. (HR) and M.Phil. in English. She carries out multidisciplinary research work on Legal studies with Biodiversity, Marketing, and allied management areas and has published articles in journals of repute and book chapters.

V. Swaran is a generalist currently focusing at the intersection of Agri-Food Systems and Science, Technology and Society (STS). He is currently a Research Associate with the Hyderbad-based resource support organisation and think-tank Watershed Support Services and Activities Network (WASSAN). Previously he has worked with different communities, including the Thar Desert region of Western Rajasthan and the Western and Eastern Ghats Mountain ranges. He is currently pursuing his Ph.D. from the Centre for Technology Alternatives for Rural Areas (CTARA) at IIT Bombay, on the evolution of millets in India since nineteenth century from a Technology and Development perspective.

Kiran Thampi is an Assistant Professor in the Department of Social Work at Rajagiri College of Social Sciences and Assistant Director of the Office of International Relations. He is currently the project lead, partnering with Israel and Australia in the International project funded by IASSW. His research and publication interests include Social Audit, Participatory project planning & management, Mental Health of professionals and youth, and International Social Work.

Manjari Upreti completed her M.Sc. in Remote Sensing and GIS from Amity University, Noida in 2020. Presently, she is working on her Ph.D. thesis on urban ecological hazard assessment in Delhi- NCR at the Central University of Jharkhand, India.

Satyam Verma is an ecologist with a research interest in fire ecology, climate change, and landscape ecology. He is currently working as an Assistant Professor at the Department of Environmental Sciences, SRM University-AP, Amaravati, Andhra Pradesh, India. He was a Fulbright fellow at the University of Washington Seattle and worked as a Project Scientist at the Wildlife Institute of India, Dehradun. He received his Ph.D. in 2017 from Pondicherry University. His doctoral research focused on the ecological impacts of fires on the tropical forests of the Western Ghats.

P. K. Viswanathan serves as a Professor (Economics and Sustainability) at Amrita School of Business, Amritapuri, Kollam, Kerala. He obtained Ph.D. in Economics in 2001 from the Institute for Social and Economic Change, Bengaluru, affiliated to the Indian Council of Social Science Research. He was a postdoctoral fellow at the Asian Institute of Technology, Bangkok, Mahidol University Thailand, and Chinese University of Hong Kong (CUHK). He teaches at the M.B.A. and Ph.D. programmes, including courses on Public Policy, International Business, Business Ethics, Sustainability, and Rural Livelihoods. Dr. Viswanathan has authored 6 books and published about 50 research articles in peer-reviewed international and national journals and books published by Sage, Routledge, Springer, etc.

S. Yogeshwari holds a position as an Assistant professor in the department of Economics, Yeshwanthpur Campus, Christ (Deemed to be University), Bengaluru, since 2018. She received her Ph.D. from the Institute for Social and Economic Change (ISEC), Bengaluru. She is particularly interested in researching themes related to agrarian change and its dynamics, environment and cities. She has published articles and book chapters with reputable publishers.

List of Figures

Fig. 3.1	(i) Prospective (threshold < p < 1), plausible (threshold < p < 0.5), and highly plausible (0.5 < p < 1) zones for *Q. leucotrichophora's* existence in the Uttarakhand state of IHR in the current settings. (ii) Prospective (threshold < p < 1), plausible (threshold < p < 0.5), and highly plausible (0.5 < p < 1) zones of *Q. leucotrichophora's* existence in the Uttarakhand state of IHR as per future IPCC RCPs for 2050 and 2070 years (*Source* Dhyani et al. 2020a, b, c, d)	29
Fig. 3.2	Existence possibility of *Hippophae salicifolia* in the present scenario over Uttarakhand along the riparian buffers of river Ganga and its tributaries (*Source* Dhyani et al. 2018)	30
Fig. 5.1	Study area map of the Chamoli district, Karnataka, India	55
Fig. 5.2	Rainfall variation from 1980 to 2019	59
Fig. 5.3	Maximum temperature variation from 1980 to 2019	59
Fig. 5.4	Minimum temperature variation from 1980 to 2019	60
Fig. 5.5	Anomalies change in *Kkharif* crops for Paddy, Soyabean, Finger millet (ragi) and small millets crops	63
Fig. 5.6	Anomalies change in *Kharif* crops for Wheat crops	63
Fig. 6.1	Map of Odisha showing different districts and study area	73
Fig. 6.2	Interaction with respondents	73
Fig. 7.1	Presence of pastoralist communities in Maharashtra and India	88
Fig. 8.1	Location of study area, Madhya Pradesh	102
Fig. 8.2	Monthly fires (MODIS) recorded from 2012 to 2021. [GL = Grasslands, DBF = Deciduous Broadleaf Forests, SA = Savannas, MF = Mixed Forests, WS = Woody Savannas, EBF = Evergreen Broadleaf Forests, ENF = Evergreen Needleleaf Forests]	105
Fig. 8.3	Monthly fires (VIIRS S-NPP) recorded from 2012 to 2021. [GL = Grasslands, DBF = Deciduous Broadleaf Forests, SA = Savannas, MF = Mixed Forests, WS = Woody Savannas, EBF = Evergreen Broadleaf Forests, ENF = Evergreen Needleleaf Forests]	106

Fig. 8.4	Forest fire hotspot map of Madhya Pradesh prepared by integrating all forest fire points (2012–2021): (a) represented MODIS fire hotspot and (b) represented VIIRS S-NPP fire hotspot	107
Fig. 8.5	All forest fire counts (MODIS and VIIRS S-NPP) during forest fire season (FMAMJ) and non-fire season (JASONDJ) from 2012 to 2021	108
Fig. 9.1	Map showing the study sites of colonial waterbirds breeding in southern India	115
Fig. 9.2	Number of literature published on coexistence, conflicts and interaction (*Source* König et al. 2020)	119
Fig. 10.1	Location of the study area	128
Fig. 11.1	Conceptual framework of the study (*Source* Authors' own construct)	143
Fig. 11.2	Location of study area	148
Fig. 11.3	The timeline of agricultural preparation and harvest in Vanvasi village (*Source* Authors' illustration [based on field survey, 2018])	150
Fig. 11.4	The rainfall trend during 2012–2018 (*Source* Taluka Krishi Adhikari Office, Jawhar)	151
Fig. 11.5	Tools used in Agriculture	156
Fig. 11.6	Storage system of the village (*Source* Authors' field survey, 2018)	157
Fig. 11.7	The sources of irrigation (*Source* Authors' field survey, 2018)	157
Fig. 12.1	Map of Asia showing the location of India	166
Fig. 12.2	Map of Africa showing the location of South Africa	167
Fig. 12.3	Synopsis of climate-smart and resilient agricultural strategies (*Source* Viswanathan et al. 2020; Angom et al. 2021; Andrieu et al. 2019; Ali et al. 2014; Abegunde et al. 2020; Adger et al. 2013; Aggarwal et al. 2018)	169
Fig. 13.1	Showing the annual rainfall data of Ernakulam (2014–2018) (*Source Indian Meteorological Department*)	207
Fig. 13.2	Showing crop production and average rainfall in Ernakulam district (*Source* Department of Economics and Statistics, Government of Kerala and Indian Meteriological Department)	208
Fig. 13.3	Showing total survived Pokkali after a flood from the survey in Kumbalangi	208
Fig. 13.4	Comparison of the overall rice production before and after the Kerala Flood 2018	210
Fig. 13.5	Showing the comparison of farmland changes according to the number of respondents after flood in the two study regions	210
Fig. 14.1	Stages in Gharbari model (*Source* Field visit)	218
Fig. 14.2	Demonstration of climate-resilient farming gharbari models (*Source* Field visit)	221
Fig. 14.3	Lessons from the field, the Case study of Jayanti Bhoi (*Source* Field visit)	223
Fig. 14.4	Innovative practices in Gharbari model	224
Fig. 15.1	Location map of the study area	233

Fig. 15.2	Theoretical framework by Kolkman et al. (2007)	241
Fig. 15.3	Thematic segregation of narrative data	244
Fig. 15.4	Tigalkheda GMI group mental model	247
Fig. 15.5	Ranmala GMI group mental model	248
Fig. 15.6	Bhangadewadi GMI group mental model	250
Fig. 20.1	Geographical location of the study area	335
Fig. 20.2	Social-ecological system, adopted from Ostrom (2009)	336
Fig. 20.3	Livestock composition of the study area (*Source* 20th Livestock Census 2019, Ministry of Animal Husbandry and Dairying, Government of India)	342
Fig. 21.1	Location of Wayanad district in Kerala	353
Fig. 21.2	Elevation profile of Wayanad district	353
Fig. 21.3	Time series of monthly average rainfall in Wayanad district between 2004 and 2013	361
Fig. 21.4	Time series of rainfall trend in May between 2004 and 2013 in Wayanad district	361
Fig. 22.1	Visualization of seasonal indices	376
Fig. 22.2	Distribution of plants in coastal sand dunes **a** Palm trees and sand dune pattern, **b** *Ipomea pescarprae*, **c** *Spinifex littoreus*	377
Fig. 22.3	Different seaweed collection **a** *Dictyota cervicornis*, **b** *Sargassum cineserum*, **c** *Padina boergerenii*, **d** *Gracillaria corticata*	378
Fig. 23.1	Sustainable adaptation strategies for the development of UGSs	397
Fig. 23.2	Avenue plantations in Ranchi city, Eastern India, with pedestrians and stormwater management	398
Fig. 23.3	Different types of green spaces in Bangalore city based on EVI **A** Avenue plantation, **B** Botanical Garden, **C** Forests, **D** Roadside greenery, **E** Green strips, and **F** Parks	400
Fig. 23.4	Dimensions of urban green space (*Source* Milvoy and Roué-Le Gall 2015)	402
Fig. 26.1	Motivation (*Source* Primary survey 2021)	453
Fig. 26.2	Green features adopted—Office spaces (*Source* Primary survey 2021)	454
Fig. 26.3	Optimum use of energy	454
Fig. 26.4	Water conservation methods (*Source* Primary survey 2021)	455
Fig. 26.5	Measures for wastewater management (*Source* Primary survey 2021)	455
Fig. 26.6	Measures for effective indoor air quality (*Source* Primary survey 2021)	457
Fig. 26.7	Temperature inside the office	457
Fig. 26.8	Awareness on benefits (*Source* Primary survey 2021)	458
Fig. 26.9	Green features addressing environmental issues (*Source* Primary survey 2021)	459
Fig. 26.10	Health benefits (*Source* Primary survey 2021)	460
Fig. 26.11	Perceptions of promoting green buildings (*Source* Primary survey 2021)	461
Fig. 26.12	Motivation to purchase or construct a green building (*Source* Primary survey 2021)	462

Fig. 26.13	Electricity saving techniques (*Source* Primary survey 2021)	463
Fig. 28.1	Framework for CCA-DRR integration into sectoral plan based on NADMP experience	485
Fig. 28.2	Benefits of NADMP for achieving specific SDGs	492
Fig. 29.1	The inter-connectedness of community, climate change, and adaptations and their linkage with the SLACC project	501
Fig. 29.2	Stakeholder of the SLACC project	503
Fig. 29.3	Location map of the SLACC implementation areas	504
Fig. 29.4	Interventions classification system	511
Fig. 29.5	Illustrates all selected SDG goals and their targets that call for action in the SLACC project. As the interventions are classified into four (Production, Ecology, Technology, and Knowledge and Finance) types, SDGs can also be distributed per the respective categories	523
Fig. 30.1	Direct and indirect effects of climate change on health and wellbeing	530
Fig. 30.2	Determinants of health systems resilience framework	531
Fig. 30.3	Four resilience elements of highly effective country responses	533
Fig. 32.1	Green Bond Issuance in a selected emerging market, 2020 (USD million) (*Source* Compiled from International Finance Corporation 2020)	560
Fig. 32.2	Total global climate finance flows from 2011 to 2020 (USD billion) (*Source* Compiled from Buchner et al. (2021) and Buchner et al. (2019))	561
Fig. 32.3	Total green bond and green loan issuance in selected Asian countries (2018–2020) in USD billion (*Source* Compiled from Climate Bonds Initiative (2019, 2020 and 2021) ASEAN Sustainable Finance State of the Market. Available via https://www.climatebonds.net/files/reports/cbi_asean_sotm_2019_final.pdfhttps://www.climatebonds.net/files/reports/asean-sotm-2020.pdfhttps://www.climatebonds.net/files/reports/asean_sotm_18_final_03_web.pdf Accessed on 12 Nov 2022	565
Fig. 32.4	Year wise green finance investment and the gap between actual and estimated green finance investment required to meet the current Nationally Determined Contributions (NDCs) in India *Note* The required green finance from 2016 to 2030 was estimated to be INR 11 lakh crores per year (*Source* Khanna et al. (2022))	566
Fig. 32.5	India's annual green bond issuance (USD billion) and its percentage change from 2014 to 2020 (*Note* 2020 figure is calculated up to August; *Source* India's transition to a high performing, low emission, energy-secure economy, USAID 2021)	567
Fig. 33.1	Ecosystem services provided biodiversity for climate change resilience (*Source* Locatelli 2016)	581
Fig. 33.2	Climate change and its effect on different levels of biodiversity	583
Fig. 33.3	Policies on climate-compatible development and sustainable development (*Source* Alemaw and Simatele 2020)	585

Image 13.1	**A** Picture of natural growth of mangroves in Pokkali wetland at Kumbalangi study region. **B** Pokkali field at the time of prawn culture in the North Paravur study area. **C** At Kumbalangi study area during seasonal shrimp farming in Pokkali field. **D** Picture of Pokkali wetland at the time of shrimp culture at North Paravur	199

List of Tables

Table 2.1	Global Initiatives Towards Community-Based Disaster Risk Management (CBDRM)	16
Table 4.1	Descriptive statistics for rice, maize and finger millet datasets	44
Table 4.2	Fixed effect regression result of rice crop	45
Table 4.3	Fixed effect regression result of maize crop	45
Table 4.4	Fixed effect regression result of finger millet crop	45
Table 4.5	Root Mean Square Error (RMSE) for three crops (Rice, Maize and Finger Millet)	48
Table 4.6	Fixed effect regression result of rice and maize yields in dry-land cultivation	49
Table 5.1	Data used in the present study	56
Table 5.2	Sen's slope values for seasonal and annual precipitation	60
Table 5.3	Sen's slope values for seasonal and maximum temperature	60
Table 5.4	Sen's slope values for seasonal and minimum temperature	61
Table 5.5	Sen's slope values for *kharif* crops	61
Table 5.6	Sen's slope values for *rabi* crops	61
Table 5.7	Correlation of crop yield (Paddy, Finger millet (ragi), Small millets, Soyabean and Wheat) with climatic variables in different crop seasons	62
Table 5.8	Multivariate regression analysis of detrended crop yields	64
Table 5.9	Unit root test	65
Table 5.10	Bound test	66
Table 5.11	Granger causality test	66
Table 6.1	Responses on effects of drought on economy of agriculture (n = 194)	75
Table 6.2	Responses on effect of drought on economy of animal respondents	78
Table 6.3	Responses on effect of drought on economy of agricultural labour	80
Table 8.1	Forest fire datasets and various meteorological variables used in this study	103
Table 8.2	MODIS fire counts in different vegetation from 2012 to 2021	104

LIST OF TABLES

Table 8.3	VIIRS S-NPP fire counts in different vegetation from 2012 to 2021	105
Table 8.4	Comparative evaluation of monthly forest fire recorded for MODIS and VIIRS S-NPP (2012–2021)	106
Table 8.5	Pearson Correlation (r) among forest fire counts and meteorological variables	108
Table 9.1	Characteristic features of breeding colonial waterbirds at the villages in southern India	116
Table 10.1	Socio-economic characteristics	129
Table 10.2	Water supply	130
Table 10.3	Cost for water system	131
Table 10.4	Explanatory variables included in the WTP function	133
Table 10.5	Tobit analyses of WTP for improved quality of drinking water	135
Table 10.6	Willingness to pay in rupees by the households	135
Table 11.1	Cropping pattern in Vanvasi village	149
Table 11.2	The pattern of crop rotation in Vanvasi village	152
Table 11.3	The practice of fertilizers and pesticides in Vanvasi village	153
Table 11.4	Crop Calendar of Vanvasi village	156
Table 13.1	Dataset used for the study, represents the set of data on Pokkali cultivation including production, income, expenditure, and the overall net profit from Pokkali farming in Kumbalangi (KL) and Ezhikara (EZ)	205
Table 14.1	Impact of the Gharbari model	219
Table 15.1	GMI Tigalkheda, Ranmala and Bhangadewadi group details	236
Table 15.2	Interviewed group details	243
Table 16.1	Water budgeting analysis in Sembedu VP for the year 2018	267
Table 16.2	Water budgeting analysis in Sembedu VP for the year 2011	269
Table 16.3	Comparative analysis of water budgeting in Sembedu VP	270
Table 17.1	Details of activities undertaken under MGNREGS in Meenangadi (2015–2021)	292
Table 18.1	Instances of Nano minerals supplementation in animal and poultry	306
Table 19.1	Details of the area statistics and forest types in the forest stretch in the states of IHRs	316
Table 19.2	Total geographic area (sq. km) and area under very dense forest (VDF), moderately dense forest (MDF), and open forest (OF) of Himalaya states of India as per ISFR 2019	319
Table 20.1	Biodiversity of Indian Sundarbans	337
Table 20.2	Crop calendar of the study area as described by the farmer for the year 1990	339
Table 20.3	Crop calendar of the study area as described by the farmer for the year 2019	339
Table 20.4	Major areas of usage of TEK in	346
Table 20.5	Climate induced changes and adaptation strategies among different livelihood options	347
Table 21.1	Some of the traditional food reported to be consumed in the present study	357
Table 21.2	Economic status and application of traditional knowledge	358
Table 21.3	Perceptions of changes in local climate	360

Table 22.1	Geographical and floristic details of South East Indian Coastal line	370
Table 22.2	Physicochemical parameters of water and soil samples from South East Indian Coastal line	373
Table 22.3	Seasonal family-wise distribution of the angiosperms	376
Table 22.4	Seasonal family-wise distribution of algal species	377
Table 22.5	Season wise distribution of bioresources with their varying physicochemical parameters	380
Table 22.6	Utilisation of bioresources by the fisherman community in treating major and minor diseases	383
Table 23.1	Water retention for traditional standard roof vs. green roof	397
Table 24.1	Summary of comparative civic service status and vulnerability of study cities	425
Table 25.1	KSPCB classification of lakes into five classes	436
Table 25.2	Sample structure	437
Table 25.3	Tests of normality	439
Table 25.4	Wilcoxon signed ranks test on price per Kl and quantity in Kl	439
Table 25.5	Generalised estimating equation results	441
Table 25.6	Kruskal–Wallis test results	442
Table 28.1	Stakeholder involvement at various levels	488
Table 28.2	Contents of NADMP, 2020	491
Table 28.3	Strengthening climate resilience through Gupta et al. (2020)	493
Table 29.1	Area-wise components and activities to identify the inter-relationship for the effective implementation of program	502
Table 29.2	Components and activities of SLACC project	506
Table 29.3	Crop-wise area, cost, yield, income profit details of SLACC and Non-SLACC farmers in Bihar State	520
Table 29.4	Crop-wise area, cost, yield, and income profit details of SLACC and Non-SLACC farmers in Madhya Pradesh State	521
Table 31.1	Indian judiciary response over the decade towards climate resilience	545
Table 33.1	Effect of climate change on ecosystem and ecosystem services	582

CHAPTER 1

Climate Change and Resilient Society in Contemporary World: Ecology, Economy and Society Interface in Indian Perspective

Sunil Nautiyal, Anil Kumar Gupta, Mrinalini Goswami, and Y. D. Imran Khan

1.1 BACKGROUND

Climate change has shown visible impacts on different spheres of our planet, affecting the structures and functions of all the sub-systems. It has the potential to alter or eliminate certain ecosystem services drastically. It can exhibit severe consequences in the viewpoint of delivery of ecosystem services, for which substitutes usually are costly or unavailable to vulnerable societies. Climate-related disasters affect the economy, ecology and society. Insufficient knowledge and uncertainty of the impacts of extreme climate events and long-term changes in climatic parameters on people and ecosystems are perilous to society. The varying extent of impacts on different types of systems is determined by geophysical and climatic settings, socioeconomic conditions of people and man-made interventions. India and other developing countries, where a significant portion of their population depends on climate-sensitive sectors, should be highly cautious about the impacts of climate change.

Vulnerability is defined as the degree to which a system is susceptible to and unable to cope with the adverse effects of climate change and weather

---

S. Nautiyal (✉) · M. Goswami · Y. D. Imran Khan
Centre for Ecological Economics and Natural Resources,
Institute for Social and Economic Change, Bengaluru, India
e-mail: sunil@isec.ac.in

A. K. Gupta
National Institute of Disaster Management, Ministry of Home Affairs, Government of India, Delhi, India

© The Author(s), under exclusive license to Springer Nature Singapore Pte Ltd. 2023
S. Nautiyal et al. (eds.), *Palgrave Handbook of Socio-ecological Resilience in the Face of Climate Change*,
https://doi.org/10.1007/978-981-99-2206-2_1

extremes (IPCC). The vulnerability is a function of three components: exposure, susceptibility and adaptive capacity. First, it is essential to understand the nature, duration and extent to which a system or any component of a system is exposed to the variations of climatic parameters and extreme climatic events. The second element determining the vulnerability is the sensitivity of the system exposed to climate change, i.e. how the system responds to changing climate. Finally, coping or adaptation strategies are essential for reducing communities' vulnerability to the impacts of climate change. As a response to current climate change for reducing climate-induced vulnerability, adaptation measures and enabling conditions mostly provide adjustments of existing systems (IPCC AR6 2022).

There are several adaptation options and strategies available, whether traditionally evolved or designed with modern knowledge of science and technology. However, the effectiveness of those adaptation measures depends on a range of social, economic and ecological factors. Socioeconomic factors such as income, market access, types and size of farms, age, gender and education of decision-making individuals, access to credits, land tenure, skill development training for livelihood diversification, livelihood assets, access to information and extension services, etc., have a significant influence on the adoption of adaptation measures mostly among farming communities (Piya et al. 2013; Kumar and Sidana 2018; Marie et al. 2020; Datta and Behera 2022a, b). Small and marginal farmers face difficulty in selecting effective adaptation measures. There are also variations in adaptation mechanisms and their level of adoption across geographical locations and socio-ecological, political, institutional and cultural settings (Piya et al. 2013; Eriksen et al. 2011; Reidsma et al. 2010). Other than those directly influencing factors, there are other dimensions to the mechanisms of adaptation to climate change which determine households' choice of adaptation options. Facilitation of adaptations through policies and programmes often influences farmers to deviate from the traditional strategies to cope with climate change and reduce the adverse impacts (Piya et al. 2013). With the changing environmental conditions, many communities have organically learned to adjust to the changing system; such strategies include crop diversification, mixed cropping, crop rotation and agroforestry (Loria and Bhardwaj 2016). Therefore, any policy initiative or formulation of an adaptation scheme should assess the existing adaptation mechanisms and their effectiveness. In this context, information and data from different biophysical and socioeconomic settings, in any form but scientifically validated, have tremendous relevance and should be advocated for their use in impactful policy development. Some of the influencing factors act as barriers or constraints to adopting coping strategies; those include lack of knowledge about technology, lack of finance and credit availability and inadequate training and demonstrations about climate-resilient technologies (Kumar and Sidana 2018; Sam et al. 2020). Farmers' perceptions of climate change also influence adaptation significantly (Gbetibouo 2009); however, in many instances, low access to information, household income, etc., act as decisive factors for households who well perceive the tangible impacts of climate change (Udmale et al. 2014; Datta and Behera 2022a, b). Availability of farming assets and access to credits are two critical criteria that help the farmers' decision-making on adopting climate-smart alternatives. In Indian context, improvement in institutional support, access to subsidies and credits, extension support, field demonstration of technologies, awareness building, upscaling of locally effective measures and validation of traditional practices and prospecting for their upscaling, etc., are essential for building a climate-resilient society (Singh 2020; Tanti et al. 2022; Datta and Behera 2022a, b).

Developing countries, including India, need to understand the underlined factors influencing the resilience of society under the changing climate. These factors are diverse and contextual to local environmental, socio-cultural and governance aspects. It is pertinent to consider the dynamic interactions between adaptation and sustainable development (Castells-Quintana et al. 2018), policy-making structures, political willingness and communities' aspirations. Therefore, the development of an understanding on impacts and coping strategies at the local level, as well as policy initiatives at various tiers of policy formulation is important. It has been realised that the sharing of knowledge is of paramount significance, which includes risk assessment, status of access to resources and adaptive capacity, level and quality of institutional support, and above all, the delineation of needs which are relevant for building a resilient community and a resilient nation as well. In view of this, the current book on "Socio-ecological Resilience in the face of Climate Change Contexts from a Developing Country" attempts to corroborate case studies across different agro-ecological zones, urban settings and socio-ecological systems to highlight the concerns to be addressed and learning from best practices and way forwards in resilience building. The chapters of the book are grouped into five parts: Part-I: Natural Resources and Livelihoods; Part-II: Adaptation—Approach and Mechanisms; Part-III: Traditional Knowledge in Climate Resilience; Part-IV: Urban Sustainability and Resilience; Part-V: Policy Issues and Future Strategies.

1.2 Natural Resources and Livelihoods

During the last decade, loss due to natural disasters in monetary terms is more than USD 187 billion per year, which is associated with a displacement of 24 million people per year worldwide (UNISDR 2014a). Natural hazards are inevitable; we can only work towards eliminating those we cause, minimising those we exacerbate and reducing vulnerability to damage (Abramovitz et al. 2011). Understanding the system in terms of health, structure and function is essential to minimise vulnerability. Formulating effective strategies for disaster risk reduction demands impact assessment of ecosystem services in terms of quality and quantity along with changing characteristics of social and economic sub-systems. Current status and past trends are crucial to make suitable decisions on mitigation and adaptation of climate change. The assessment and monitoring of an environmental system may be conducted to understand the quality and quantity with respect to driving force, pressure, state of environmental components and possible impacts. Prediction and development of early warning systems based on those assessments are foundations of disaster risk reduction. Informed decision-making and forecasting are primarily based on environmental monitoring. In addition to hazard forecasting, impact forecasting is very critical and is an emerging dimension in risk reduction. Along with the assessment of hazard-related parameters, impact forecasting needs to consider social systems and the structures that support them (Merz et al. 2020). Exposure and vulnerability to hazards are dynamic in space and time, requiring comprehensive appraisal using advanced technology to boost traditional knowledge. Recognising those anomalies across space and time help in developing localised and contextualised approach for minimising vulnerability. In Spatio-temporal assessment based on earth's observation has been established as very effective in environmental decision-making. On the other hand, biophysical findings supported by socioeconomic research provide a holistic validation of decisions for effective implementation. Investigations of the stressor and their extent of influence help in vulnerability profiling of a community or ecosystem. In this section of the book, scientific articles cover a wide range of topics starting from the use of remote sensing and GIS for biophysical change to people's perception and response analysis. This book section addresses a wide range of topics involving climate change impact assessment, natural resource-dependent livelihoods and resource status monitoring. The topics covered are landscape change assessment, livelihood vulnerability, water scarcity, crop productivity, food security, etc.

1.3 Adaptation—Approaches and Mechanisms

Climate change adaptation is the adjustment in any human and natural systems in response to climate change impacts through regulation of damage and exploitation of beneficial opportunities (IPCC 2013). Reduction in exposure, reduction in damage to property, proper management of resources, improved preparedness and enhancement of communities' resilience are the core components of disaster risk reduction (UNISDR 2014b). Coherence between policies on adaptation to climate change and disaster risk reduction is the cornerstone for a resilient society and ecosystem. However, both fields work in their own dimensions with common goals through different sets of actors, institutions, policy frameworks and patterns (Schipper 2009). The scientific community has emphasised integrating both fields for effective policy development (Forino et al. 2017). Climate change was recognised as a driver of disaster risk (The Sendai Framework for Disaster Risk Reduction 2015–2030). Both Sustainable Development Goals and Paris Agreement (Nations Framework Convention on Climate Change) pronounced the importance of strengthening resilience and adaptive capacity for disaster risk reduction, thereby highlighting the linkages between climate change adaptation and disaster risk reduction. The disjuncture of knowledge, skills and planning at the national, state and local levels is also evident in policy formulation and implementation (Nautiyal et al. 2021). India being an economy dominated by agricultural activities, historically a large number of policies focussed on yield improvement and food security.

Nevertheless, in the last decade, predominantly facilitated by climate change adaptation and mitigation initiatives, India is evolving to develop and implement policies on sustainable agriculture and green growth. With technological advancements and institutional interventions for climate change adaptation, it is essential to comprehend the adoption pathways, behaviour, social and cultural norms of communities, knowledge and awareness. Therefore, the shreds of evidence from the field are crucial in this regard which can contribute towards bottom-up formulation of policies for building a resilient society. Moreover, there is a strong need to develop synergies among policies across multiple sectors—rural development, urban planning, agriculture and climate change. This chapter on policy, adaptation and disaster risk reduction includes a range of scientific articles on climate-resilient agriculture, climate-smart village, ecosystem-based adaptation and community-based adaptation for better management of resources to cope with the impacts of climate change.

1.4 Traditional Knowledge in Climate Resilience

In India, almost 70% of the population is directly or indirectly dependent on natural resources to acquire their livelihoods. In rural areas, two scenarios can be observed, i.e. (1) Rural communities are highly vulnerable to extreme events. (2) The Traditional knowledge and coping skills learned from the generations are considered to be making them more resilient (Maru et al. 2014; Imperiale and Vanclay 2016). The vulnerability assessment of a community is critical to understand the resilience of the society under changing climatic conditions. In Indian socio-ecological systems, local and traditional governance in natural resource management, the role of women in environmental conservation and traditional knowledge in resource use, agricultural practices as well as climate-related decisions play a significant role in building a resilient society. On the contrary, low skills, lack of alternative livelihoods and poverty push the rural communities to greater vulnerable status. Changing socioeconomic anchorage and growing aspirations do exhibit impacts on the resilience of communities. The articles in this section provide evidence from Indian societies on how societal norms, knowledge and their willingness contribute towards disaster risk reduction under changing climatic conditions. The findings and recommendations have the potential to provide a new dimension to the policy domain and an opportunity for the development of a climate-resilient, low-carbon, sustainable and inclusive society.

1.5 URBAN SUSTAINABILITY

Urbanisation is a growing phenomenon in developing countries like India, and it profoundly affects cities' spatial distribution, resource consumption, environmental impact; and sustainability also plays a crucial role in a circular economy which flows from rural to urban and vice-versa. The services and livelihood opportunities pull the rural population to the urban areas. This migration is exacerbated due to push factors like poverty, economic stagnation and extreme weather events (Rana and Parves 2013). From the perspective of developing countries like India, this push–pull paradigm is a complex phenomenon due to its dualistic (opportunities and challenges) nature (Girard 2003). Besides this, in recent periods and past, climate change and its direct consequences, such as floods, droughts and heat waves, have had adverse repercussions on physical infrastructures, transport systems, communication, energy and water supply, livelihoods across the social groups and the provision of ecosystems goods and services. The impact within the urban agglomeration may vary from one urban area to another urban area due to the environmental, social, geographical and political dynamics.

Under this, building resilience to extreme events in urban areas is challenging, particularly in the rapidly growing cities in developing countries. However, building resilient urban areas can be achieved by bringing all stakeholders under one roof at multiple scales and integrating resilience into the urban master plans by creating innovative paths. The articles in this section provide an overview and address the challenges of services, management and governance-related issues under the changing climate.

The main objectives of this volume are: (i) to bring together scientific knowledge from various disciplinary studies on climate impacts, adaptation and building resilient society; (ii) to share experiences and know-how; (iii) to document the pathways for climate change adaptation strategies, barriers and possible solutions to adopt different adaptation options; (iv) to facilitate discussion and develop understanding among a wider audience from science, policy and practice on the vulnerability of different socioeconomic sections in varied ecological settings to climate-related risks with special reference to natural resource-based livelihoods, water and human-wildlife interaction; (iv) to collate and place forward initiatives for disaster risk reduction and their effectiveness to identify strategic requirements for policy recommendation. The book covers the diversity in ecology and socio-economy with their corresponding concerns driven by climate change to delineate scopes for socio-ecological sustainability. It has a micro–macro perspective to socio-ecology under changing climate depending on the demand arising out of the concerns and their extent. The field-based reflections in the book which our book is one of the essential aspects for the benefit of societies and supporting major stakeholders in taking insights related to the real-world situation for implementing the doable policy recommendations for achieving the desired goals.

We hope that the collection of research papers in this book will help develop better strategies to build resilient societies for a sustainable future.

References

Abramovitz J, Banuri T, Girot PO, Orlando B, Schneider N, Spanger-Siegfried E, Switzer J, Hammill A (2011) Adapting to climate change: Natural resource management and vulnerability reduction. The International Institute for Sustainable Development (IISD).

Datta P, Behera B (2022a) Factors Influencing the Feasibility, Effectiveness, and Sustainability of Farmers' Adaptation Strategies to Climate Change in The Indian Eastern Himalayan Foothills. Environmental Management 70(6):911–25.

Datta P, Behera B (2022b) Climate Change and Indian Agriculture: A Systematic Review of Farmers' Perception, Adaptation, and Transformation. Environmental Challenges 4:100543.

David Castells-Quintana, MdPL-U, McDermott TKJ (2018) Adaptation to climate change: A review through a development economics lens, World Development, 104:183-96, ISSN 0305-750X. https://doi.org/10.1016/j.worlddev.2017.11.016

Eriksen S, Aldunce P, Bahinipati CS, Martins RD, Molefe JI, Nhemachena C, O'brien K, Olorunfemi F, Park J, Sygna L, Ulsrud K (2011) When not every response to climate change is a good one: Identifying principles for sustainable adaptation. Climate and development.

Forino G, von Meding J, Brewer G, Van Niekerk D (2017) Climate change adaptation and disaster risk reduction integration: strategies, policies, and plans in three Australian local governments. International Journal of Disaster Risk Reduction 100–8.

Gbetibouo GA (2009) Understanding farmers' perceptions and adaptations to climate change and variability: The case of the Limpopo Basin, South Africa. The International Food Policy Research Institute.

Girard, LF (2003) Introduction. In: R. Girard et al. (Eds.), The human sustainable city: Challenges and perspectives from the Habitat Agenda. ASHGATE. https://doi.org/10.1016/j.envc.2022.100498

Imperiale AJ, Vanclay F (2016) Experiencing local community resilience in action: learning from post-disaster communities. *Journal of Rural Studies* 47:204–219. https://doi.org/10.1016/j.jrurstud.2016.08.002

IPCC (Intergovernmental Panel on Climate Change) (2013): The physical science basis. In: TF Stocker, D Qin, G-K Plattner, M Tignor, SK Allen, J Boschung, A Nauels, Y Xia, et al. (Eds.), Contribution of Working Group I to the Fifth Assessment Report of the Intergovernmental Panel on Climate Change. Cambridge: Cambridge University Press

IPCC (Climate Change (2022): Impacts, adaptation, and vulnerability. In: H-O Pörtner, et al. (Eds.), Contribution of Working Group II to the Sixth Assessment Report of the Intergovernmental Panel on Climate Change (Vol. In press). Cambridge University Press.

Kumar S, Sidana BK (2018) Farmers' perceptions and adaptation strategies to climate change in Punjab agriculture. Indian Journal of Agricultural Sciences, 88(10):1573–81.

Loria N, Bhardwaj SK (2016) Farmers' response and adaptation strategies to climate change in low-hills of Himachal Pradesh in India. Nature Environment & Pollution Technology 15(3).

Marie M, Yirga, F, Haile, M, Tquabo F (2020) Farmers' choices and factors affecting adoption of climate change adaptation strategies: evidence from north-western Ethiopia. Heliyon, 6(4):e03867.

Maru YT, Smith MS, Sparrow A, Pinho PF, Dube OP (2014) A linked vulnerability and resilience framework for adaptation pathways in remote disadvantaged communities. Global Environmental Change 28:337–350. https://doi.org/10.1016/j.gloenvcha.2013.12.007

Merz B, Kuhlicke C, Kunz M, Pittore M, Babeyko A, Bresch DN, Domeisen DI, Feser F, Koszalka I, Kreibich H, Pantillon F(2020) Impact forecasting to support emergency management of natural hazards. Reviews of Geophysics 58(4):e2020RG000704.

Nautiyal S, Goswami M, Khan, Y.D I, Prakash S, Kishan R, Gupta A.K, Bindal M, and Baidya S (2021) Climatic Variations and Agricultural Landscape: A Study on Policies and Practices for Resilience. National Institute of Disaster Management, New Delhi. ISBN No. 978-93-82571-57-5.

Piya L, Maharjan KL, Joshi NP (2013) Determinants of adaptation practices to climate change by Chepang households in the rural Mid-Hills of Nepal. Regional environmental change: 437–47. Purna Chandra Tanti, Pradyot Ranjan Jena, Jeetendra Prakash Aryal, Dil Bahadur Rahut.

Rana M, Parves M (2013) Urbanization and sustainability: Challenges and strategies for sustainable urban development in Bangladesh. Environment, Development and Sustainability 237–56.

Reidsma P, Ewert F, Lansink AO, Leemans R (2010) Adaptation to climate change and climate variability in European agriculture: the importance of farm level responses. European Journal of Agronomy 32(1):91–102.

Sam AS, Padmaja SS, Kächele H, Kumar R, Müller K (2020) Climate change, drought and rural communities: Understanding people's perceptions and adaptations in rural eastern India. International Journal of Disaster Risk Reduction 101436.

Schipper EL (2009) Meeting at the crossroads?: Exploring the linkages between climate change adaptation and disaster risk reduction. Climate and Development 1(1):16–30.

Singh S (2020) Farmers' perception of climate change and adaptation decisions: A micro-level evidence from Bundelkhand Region, India. Ecological Indicators, 116:106475.

Udmale P, Ichikawa Y, Manandhar S, Ishidaira H, Kiem AS (2014) Farmers' perception of drought impacts, local adaptation and administrative mitigation measures in Maharashtra State, India. International Journal of Disaster Risk Reduction 10:250–69.

Tanti PC, Jena PR, Aryal JP (2022) Role of institutional factors in climate-smart technology adoption in agriculture: Evidence from an Eastern Indian state. Environmental Challenges 100498.

UNISDR (United Nations International Strategy for Disaster Reduction) (2014a) Integration of disaster risk reduction and climate change adaptation in SAARC region: Implementation of the Thimphu statement of climate change—A comprehensive study of the policy, institutional landscape and resource allocation for disaster risk reduction and climate change adaptation in South Asia (Disaster prevention, preparedness & management, linkages with CCA). New Delhi 8:100543, ISSN 2667-0100. https://doi.org/10.1016/j.envc.2022.100543

UNISDR (United Nations International Strategy for Disaster Reduction) (2014b, October 20) Development of the post-2015 framework for disaster risk reduction. Zero draft submitted by the co-Chairs of the Preparatory Committee. http://www.wcdrr.org/preparatory/post2015. Accessed 31 Jan 2015.

CHAPTER 2

Community-Based Disaster Reduction for Climate Resilience in Developing World

Fatima Amin, Sweta Baidya, and Anil Kumar Gupta

2.1 INTRODUCTION

As long as there is life on the globe, disasters will undoubtedly continue to impact human existence. No nation is resistant to these global disasters, regardless of their geo-climatic location or level of preparedness. As a result, every human culture on this planet has experienced disasters at some point of time. Additionally, in the latter quarter of the twentieth century, the frequency and number of natural disasters and their effects have significantly grown. Despite of the development of advanced technologies, the modern world is extremely vulnerable to disasters of hydro-meteorological and geophysical origins. Already there is serious concern about the rise in catastrophic events and an increase in global warming may result in more climate-related disasters in the coming decades.

Developing nations are far more negatively impacted by natural disasters than the developed nations. In 2001, developing nations accounted for 95% of all disaster-related deaths worldwide of which approximately 90% were natural disasters (Ariyabandu and Wickramasinghe 2003; Ginige & Amaratunga, 2009). It is estimated that by the year 2025, 80% of the world's population will live in developing countries and up to 60% of them will be highly vulnerable to floods, cyclones, and earthquakes (Moin, 2001). The effects of disastrous

---

F. Amin (✉) · S. Baidya · A. K. Gupta
National Institute of Disaster Management, Ministry of Home Affairs, Government of India, Delhi, India
e-mail: aminfatima011@gmail.com

© The Author(s), under exclusive license to Springer Nature Singapore Pte Ltd. 2023
S. Nautiyal et al. (eds.), *Palgrave Handbook of Socio-ecological Resilience in the Face of Climate Change*,
https://doi.org/10.1007/978-981-99-2206-2_2

events are complex due to the additional weight from ecological debasement, destitution, landlessness, political and social imbalances. It is proven that the impact of any disaster is not the same for all the strata of the society. Poor or marginalized groups have less ability to survive the disasters, which leaves them abandoned and powerless against disastrous events (McEntire, 2005).

The tsunami of December 26, 2004, devastated nine nations on a large scale, leaving the common people and governments in ruins. South Asia, mostly comprising of developing countries, is the world's most vulnerable region to both natural and human-induced disasters. This region faces different kinds of natural hazards like flood, earthquake, cyclone, etc. In the decade 1992 to 2001, this subcontinent lost 96,285 lives due to various types of disasters (International Federation of Red Cross and Red Crescent Societies [IFRCRCS], 2002).

Residents of the local community are the first to respond to a calamity. Because the local administration reportedly lacks the technical and human resources for community-level disaster monitoring. Therefore, is unable to fully identify or map potential local hazards or develop the necessary disaster management plans, and as a result, many disasters, particularly in remote areas, go unreported. The basis of a sustainable disaster administration approach and practice is the people group, especially vulnerable community members. A small failure either in the strategy or methods of disaster management may leave terrible impacts on the lives of enormous number of people of a particular zone. There is a lack of coordination between the Government and the public, especially the affected community, during the time of disaster. The rescue operation system and the relief program take a long time to reach the affected people and to the place where the disaster has taken place (Public Accounts Committee, 2015).

2.2 AN OVERVIEW

The idea of Disaster Risk Reduction (DRR) has gone through a thorough evolution. During the decades from 1960 to 1970, disasters were assumed as extreme events, but the 1980s saw a strong emphasis on pre-disaster preparedness, which led to the United Nations (UN) International Decade of Natural Disaster Reduction (1990–1999). In 1995, the major earthquake of Hanshin, (popularly known as the Kobe earthquake), devasted one of the most disaster-prone and also disaster-prepared countries of the world. The resultant activities included long-term recovery processes, which drastically changed the role of Japan's civil society and NGOs (Shaw and Goda, 2004). The need for pre-disaster mitigation measures was emphasized at this time. Some of the biggest earthquakes, such as Taiwan and Turkey earthquake in 1999, and in 2001 Gujarat, India, earthquake brought out the requirement for mainstreaming DRR in local and national level development policies. The 2004 Indian Ocean tsunami was another decisive moment for the risk reduction school of thought. In 2005, at the 2nd World Conference on Disaster Reduction (WCDR) in Hyogo, Japan, the Hyogo Framework for Action (HFA 2005–2015) was adopted. This conference outlined a measurable framework to identify the advancement of communities and countries in DRR approaches Later, in 2015, Government of Japan organized the Third World Conference on Disaster Risk Reduction in the city of Sendai. The ideas of this Sendai Framework for Disaster Risk Reduction (SFDRR, 2015) conference were adopted by the member states for the period of 2015–2030.

2.3 Discussion

CBDRR approach is not new, from the starting of civilization people have taken community approaches to save their assets and people (Shaw, 2014). Although the terms Community Based Disaster Risk Management (CBDRM) and Community Based Disaster Risk Reduction (CBDRR) are used in a substitutable manner but there is a thin line of difference between them. CBDRM focuses on a comprehensive perspective on risk reduction-related tasks by communities, before, during and after a disaster, whereas CBDRR targets more on pre-disaster community risk reduction initiatives (Shaw, 2012) community-based approaches are competent in fixing the shortcoming in top-down approaches to development planning and disaster management, which often avoid community needs and dimensions. These top-down approaches are normally unable to reach out the most vulnerable groups of any community due to absence of community involvement. In the present digital age, with the advancement of technology, social media-based virtual communities have a greater impact on DRR. There are two different kinds of virtual communities that are relevant to DRR: (1) the learning forum, where scholars and practitioners discuss and debate new discoveries in DRR; and (2) community voices, where individuals and groups voice their concerns about DRR-related issues. In the present generation "Z", both are considered as equally important looking into the digital networking and cloud sharing. Several stakeholders have a particular responsibility for community planning, actions, and management. At the national and municipal levels, governments and civil society organisations (CSOs) both play significant roles in DRR.

Role of Stakeholders

Various Stakeholders normally the educational institutes, civil societies, and other NGOs come up with various best practices, educational and public awareness materials, etc. Recently, the National Governments of many of Asian countries have started taking initiatives and commitments. But policy differences always exist between the national and local governments. The Hyogo Framework for Action (HFA) has attracted attention at the international, regional, and national levels but has yet to reach the local level with additional details. As a result, the Asian Disaster Reduction and Response Network (ADRRN), a civil society organisation, launched the "Road to Sendai" a drive to develop communities, and community-based organizations about the HFA and the World Conference on Disaster Risk Reduction, in March 2015.

National and International Best Practices for Resilience Toward Tsunami 2004

Case Study from Tamil Nadu, India

Private sectors and academic/research academies are considered non-traditional stakeholders in the local communities and they have crucial roles to play. Following the Indian Ocean tsunami, the University of Madras, Tamil Nadu, India, created a higher education program and established a community college for poor students. The majority of the students came from fishing communities in nearby tsunami-affected coastal villages. In collaboration with the village council, the district office, and non-governmental organizations (NGOs), the college assigned students to specific community-based activities such as village and coastal alerts, map making, delineating vulnerable areas of villages, and identifying rescue routes (Krishnamurthy and Kamala, 2015). The sustainability of these kinds of efforts are to be maintained and in this particular case, it was included in the disaster management curriculum for all college students to make it compulsory for all the students to participate every year. As a result, periodic monitoring is also ensured.

Case Study from Malaysia

Toward Tsunami resilience, Malaysia implemented a Community Based Disaster Risk Management, Program to improvise disaster preparedness. The program was initiated in 2008 for two years. As per the plan, the local communities were asked to list out the hazards and vulnerabilities bugging their livelihoods and to identify the possible strategies for DRR. Disaster Management committees for respective areas were formed along with the task force to conduct disaster-specific response activities. The program also included School Safety Initiatives along with ensuring the building standards, sensitization workshops, and mock drills (Mercy Malaysia, 2022).

National and International Best Practices for Resilience Toward Cyclone/Hurricane

Case Study from Odisha, India

The loss of 10,000 lives in the Super Cyclone of 1999 propelled the Disaster Risk Reduction activities of the State of Odisha, India. The Community Risk Reduction measures included building of multipurpose cyclone shelters equipped with community kitchens and life saving equipment and announcement vehicles. These shelters provide all necessary emergency services. They have also prepared evacuation maps for cyclone and tsunami and have reliable tsunami alert office, along with two tsunami ready model villages who perform mock drills on the regular basis. As per World Bank, Odisha has a commendable community outreach system through which emergency communication can be established during the disasters. There are 450 cyclone shelters along with trained maintenance committee for rescue and relief activities. During Cyclone Amphan Odisha was able to successfully evacuate 200,000 people to a safe location which is one of the biggest evacuations in human history, resulting in the "zero casualty" (Bose, 2020).

Case Study from Pacific
Pacific countries have given various examples of best practices and opportunities for integration which includes but not limited to the ideas like a. holistic approach, b. using multi-sectoral and multi-stakeholder teams and expertise, c. Genuine community participation, d. avoiding fragmentary policy approaches, etc. Table 2.1 shows some of the initiatives toward CBDRM.

2.4 Way Forward

The community's diversity is one of CBDRR's key traits. Diversity exists on various levels, depending on geographic location, customs, culture, habits, and so on. CBDRR's first step is to understand and appreciate this diversity. As an entry point into the community, a change agent is required. Sustainability is the major hurdle faced in CBDRR. Normally after any disaster some government and Non-Government agencies work toward CBDRR for as long as resources are available. Disasters can be development opportunities, and it is critical to consider how the disaster recovery process can be central to the CBDRR's long-term viability.

The institutionalization of CBDRR efforts at the local governance level is critical. DRR is a responsibility shared by all citizens, both individually and collectively. Citizens of all the communities have to take up their respective responsibilities for safeguarding their resources. This has to be done for all types of major and minor disasters. With its changing characteristics the CBDRR approach is acknowledged both in local and national policies and activities, but the CBDRR approaches must be disaster specific and custom made as per the requirements of the local community. Social media, crowd sourcing, AI, and other technologies can have great role to play to successfully implement CBDRR.

Constant monitoring, effective early warning, and last mile connectivity is the key to CBDRR and CBDRM. Climate Change Adaptation at the community level can bridge the gap to resist, absorb, accommodate, and recover at the soonest.

Table 2.1 Global Initiatives Towards Community-Based Disaster Risk Management (CBDRM)

Project/initiative	Donor	Location	Implementing agency/organisations	Activities	Aims and objectives
Samoa Red Cross Community Based Health and First Aid (CBHFA) Program	International and National Red Congress Societies	Samoa (nation-wide)	Samoa National Red Cross Society and government partner ministries	Education and community awareness relating to the specific needs of the community, using Red Cross's Vulnerability and Capacity Assessment (VCA) tool. Specific attention paid to disaster and climate change related issues and needs. Inclusion of government ministries to allow for follow up of additional activities	To assess the specific vulnerability of the village and develop a targeted response to educate people in ways to overcome and become more aware of the risks in their lives
Navua Local Level Risk Management	UNDP Pacific Centre	Navua, Fiji	UNDP, SOPAC, Red Cross, National Disaster Management Office (NDMO)	Education and community awareness for pre-existing early warning flood system in addition to multi-stakeholder involvement in long term community awareness activities	Using the Local Level Risk Management (LLRM) approach, capacity building with the community, NGOs and local authorities in terms of risk sensitisation and disaster risk sensitive development project
GEF-SGP Community Based Adaption (CBA)	GEF/AusAID	Global: 10 pilot countries including Samoa	Small Grant Programms (SGP) and United Nations Development Programme (UNDP)	Enhancing Community resilience to climate change via community education and awareness, coupled with "hard solutions" such as shoreline protection	Enhancing Community resilience and ecosystems upon which they depend via a "results based approach" including community adaptation priorities (United Nations Development (UNDP), 2008)

Project/initiative	Donor	Location	Implementing agency/organisations	Activities	Aims and objectives
Building Disaster Response and Preparedness in the Pacific	AusAID	Fiji, Samoa, Kiribati, Vanuatu	Caritas Samoa and Australia, Caritas Oceania and Pacific	Education and community awareness with the aim being to change behaviour to incorporate better preparedness for disasters in everyday living	To raise awareness and educate key Catholic people in disaster risk reduction in order to pass this information on to the wider community (Caritas Australia, 2008)
WWF Coastal Resilience	GEF	Fiji, India, East and West Africa	WWF, USP, SOPAC, Fiji Met Service	Community consultation coupled with scientific evidence to devise strategy to manage coastal mangrove ecosystems	To develop a "generalisable" approach to addressing coastal resilience across similar habitats (i.e. mangroves), and maintaining intact support the connectivity between mangroves and coral reefs
Samoa Disaster Risk Reduction and Awareness Workshop	UNESCO, SOPAC, World Bank	Samoa	NDMO, multitude of other government agencies, NGOs, Red Cross	Education and community awareness relating to disasters. Follow up activities with assistance of government ministries including potential "hard solution" depending on the needs of the community	To strengthen village understanding of current vulnerable and capacity, risk reduction measures and consequently formulating a village Response Plan Booklet for all households. To also have a village simulation to test the response of village to a disaster
Pacific Community-Focused Integrated Disaster Risk Reduction (PCIDRR)	National Council of Churches (NCCA), AusAID	Fiji, Solomon Islands, Tonga, Vanuatu	PCIDRR Team, NCCA, NDMO, Adventist Development and Relief Agency (ADRA)	Disaster management training, development of Community Disaster Plan and disaster response practice via simulation exercise	To create better awareness and understanding of disaster risks at the community level and to identify means to enhance resilience to these risks. Creation of Community Disaster Plan, training of people in village in disaster response (National Council of Churches Australia [NCCA], 2007)

(continued)

Table 2.1 (continued)

Project/initiative	Donor	Location	Implementing agency/organisations	Activities	Aims and objectives
Climate Change and Food Security	FAO	Samoa	Women in Business for Development Inc (WIBDI)	Education community awareness relating to food security, nutrition and sustainable livelihoods. Provision of seeds and piggeries as start-up resources for identified family in need of assistance	To target the most vulnerable people in communities and assist them in developing their own sustainable livelihoods. The approach includes assisting families reduce their dependence on remittances from family members overseas by becoming self-sufficient and growing their own food, and possibly growing enough to provide and additional source of income

Source Gero et al. (2011)

REFERENCES

Ariyabandu, M. M., & Wickramasinghe, M. (2003). *Gender dimensions in disaster management—A guide for South Asia.* ITDG South Asia. https://doi.org/10.3362/9781780445465

Bose, S. (2020). Following the Odisha Example for Developing Community Based Disaster Management in India. Observer Reaserch Foundation, Prevention Web, UNDRR. https://www.preventionweb.net/news/following-odisha-example-developing-community-based-disaster-management-india

Gero, A., Meheux, K. and Dominey-Howes, D. (2011). Integrating community based disaster risk reduction and climate change adaptation: examples from the Pacific. *Natural Hazards and Earth System Sciences, 11,* 101–113, 2011 www.nat-hazards-earth-syst-sci.net/11/101/2011/; https://doi.org/10.5194/nhess-11-101-2011

Ginige, K., Amaratunga, D., & Haigh, R. (2009). Mainstreaming gender in disaster reduction: why and how? *Disaster Prevention and Management: An International Journal, 18*(1), 23–34.

IFRCRCS. (2002). World Disasters Report focus on Reducing Risk, Chemin Des Crets, P.O. Box 372, CH-1211 Geneva 19, Switzerland

Krishnamurthy, R., & Kamala, K. (2015). Impacts of higher education in enhancing the resilience of disaster prone coastal communities: A case study of Nemmeli Panchayat, Tamil Nadu, India. In Shaw, R. (Ed.), *Recovery from the Indian Ocean Tsunami: A ten year journey* (pp. 361–380). Tokyo, Japan: Springer.

McEntire. (2005). Why vulnerability matters: Exploring the merit of an inclusive disaster Reduction Concept. *Disaster Prevention and Management, 14*(2), 206–222.

Moin. (2001). *Disasters and development. Disaster mitigation: Experiences and reflections* (pp. 22–30). New Delhi: Prentice- Hall of India.

National Council of Churches Australia [NCCA]. (2007). *Executive summary—PCIDRR program.* NCCA.

Public Accounts Committee. (2015). Disaster preparedness in India. *Twenty-Fifth Report, Ministry of Home Affairs, Lok Sabha Secretariat.* New Delhi, agrahayana, December 2015.

SFDRR. (2015). Sendai framework for disaster risk reduction, 2015–2030. Retrieved from http://www.preventionweb.net/files/43291_sendaiframeworkfordrren.pdf

Shaw, R. (2012). *Community based disaster risk reduction.* Bingley, UK: Emerald Publisher.

Shaw, R. (2014). Kobe earthquake: Turning point of community based risk reduction in Japan. In Shaw, R. (Ed.), *Community practices for disaster risk reduction in Japan* (pp. 21–31). Tokyo, Japan: Springer. https://www.mercy.org.my/programme/post-tsunami-project-community-based-disaster-management-cbdrm/

Shaw, R., & Goda, K. (2004). From disaster to sustainable community planning and development: The Kobe Experiences. *Disaster, 28*(1), 16–40.

PART I

Natural Resources and Livelihoods

CHAPTER 3

Climate Sensitivity, Forest Ecosystems, and Regional Challenges: Exploring Opportunities to Reduce Disaster Risks in Central Himalayas

Shalini Dhyani, Deepak Dhyani, Rakesh Kadaverugu, Paras Pujari, and Rakesh Kumar Maikhuri

3.1 INTRODUCTION

In last few decades, there has been a remarkable rise in global and recognition of forest ecosystem services because of their role in safeguarding nutritional and livelihood security of local communities and reducing disaster risks (Dhyani and Dhyani 2016a, 2016b; Ammer 2019). Conserving biodiversity and protecting ecosystem health is fundamental to ensure unhindered flow of ecosystem services for developing resilience and human well-being (Jetz et al. 2019). Ecosystem services or nature's contribution to people are important to link ecosystem functioning with human well-being (Diaz et al. 2015). Land degradation, biodiversity loss, and climate change have emerged

S. Dhyani (✉) · R. Kadaverugu · P. Pujari
CSIR-National Environmental Engineering Research Institute (NEERI), Nagpur, Maharashtra, India
e-mail: shalinidhyanineeri@gmail.com; s_dhyani@neeri.res.in

D. Dhyani
Society for Conserving Planet and Life (COPAL), Dehradun, Uttarakhand, India

R. K. Maikhuri
Department of Environmental Sciences, HNB Garhwal University, Srinagar Garhwal, Uttarakhand, India

© The Author(s), under exclusive license to Springer Nature Singapore Pte Ltd. 2023
S. Nautiyal et al. (eds.), *Palgrave Handbook of Socio-ecological Resilience in the Face of Climate Change*,
https://doi.org/10.1007/978-981-99-2206-2_3

as the most threatening global environmental concerns (IPBES 2018a; Dhyani et al. 2022a, b). More than 29% of land worldwide is labeled as "land degradation hotspots, and extensive loss of biodiversity and ecosystems may lead to ecosystem collapse (Sato and Lindenmayer 2018; Dhyani et al. 2022a, b). Mountains as high value but threatened ecosystems, and key biodiversity areas (KBA) support more than 20% of the global and millions living in downstream areas (Sharma et al. 2019). The varied ecosystem benefits of the Indian Himalayan Region (IHR) support millions living in upstream areas for various subsistence requirements (Misra et al. 2008b; Dhyani et al. 2011; Dhyani and Dhyani 2016a; Dhyani 2018). In last few decades the ecosystems in IHR have suffered rapid degradation ensuing in tremendous biodiversity loss affecting the flow of ecosystem services and human well-being (Wang and Stone, 2019). Loss of ecosystem services due to rapid degradation in Himalayan highlands because of changes in species niches, leading to loss of species habitats, and upward or outward shifting of species are additional probable changes anticipated in the region (Chaturvedi et al. 2011; Chakraborty et al. 2018; Krishnan et al. 2019; Kumar et al. 2019; Dhyani et al. 2020c). In their inaccessible mountain terrains, Central Himalayas are a habitat to rich biodiversity and ecosystems, with many being threatened because of several drivers of biodiversity and ecosystem loss and their cumulative impacts. Climate change followed by changes in land use land cover, deforestation, degradation, forest fragmentation, encroachment by invasive plants, demographic changes, local socioeconomic conditions, and migration have emerged as critical drivers of loss of biodiversity in the IHR (Locatelli et al. 2017; Payne et al. 2017; Dhyani and Dhyani 2020). There is clear evidence that climate change will alter the habitat suitability for various keystone and indicator species that will further erode the living conditions for locals and increase disaster risks (Dhyani et al. 2018, 2019, 2020c). IHR is a key indicator region for global climate change; hence, early warnings need to be observed, recorded, analyzed, and well understood and integrated with the region's conservation and climate adaptation planning programs (Chaturvedi et al. 2011; Krishnan et al. 2019). Disaster preparedness capacity is necessary to be developed for Central Himalayas where the frequency and intensity of climate extreme events and disasters are increasing with relief, reclamation, and restoration opportunities being limited or insufficient because of the remoteness of the areas (Dhyani and Thummarukuddy 2016). Himalayan region is observed to be substantially warmer than consistent global warming impacts reported from other regions of the country that have also been endorsed by regional climate models with projections that temperature and rainfall will upsurge in near future (Kulkarni et al. 2013). The amount and frequency of extreme weather events have severely affected the forest and other natural ecosystems across the region (Chaturvedi et al. 2011). Understanding the influences of climate change on Himalayan ecosystems and their ecosystem services is critical to reducing disaster risks and ensuring long-term human well-being. Disturbing projections about the plausible alternative scenarios for mountain ecosystems and species habitat suitability will be relevant to stimulate global and national climate and conservation policy dialogues (Dawson et al. 2011). IPCC reports followed by IPBES Asia Pacific regional assessment, global assessment, and Hindukush Himalaya Assessment (HKH assessment) have endorsed that changing climate will severely affect Himalayan forests and biodiversity elements (IPBES 2018b; IPCC 2019; Krishnan et al. 2019; Sharma et al. 2019). The 4 × 4 assessment report by MoEF&CC, Govt. of India has also stressed and endorsed the views of IPCC, IPBES, and HKH assessment on forests of IHR. The affected natural ecosystems of IHR will greatly impact the co-occurrence and co-management of forests by and for the local communities (Mondal and Zhang 2018). Forests in IHR are priority ecosystems for conservation under the National Action Plan for Climate Change of India, 2008 under the sub-mission of National Mission on Sustaining Himalayan Ecosystem (NMSHE) (Pandve 2009). However, crucial task for ecologists at current time is to establish probable approaches and advantages to moderate, reduce, and reverse this massive loss (Dawson et al. 2011). Managing the sustainability of ecosystem services requires an understanding of the condition and extent of ecosystems as well as the ability to predict the impacts of alternative policy or management decisions on them. More species and biodiverse forests are required to deal with spatial and temporal heterogeneity, ecosystem processes, and climatic uncertainties. Involving indigenous and local communities (IPLC) in ecosystem conservation by site-specific restoration efforts that include climate uncertainty in its core has emerged as an important enabling concern for socioeconomics, governance, and other mega drivers of change but also to support the long-term conservation and restoration efforts (Dhyani et al. 2020a). The study attempts to provide a broader overview of climate impacts and adaptation potential of natural forest ecosystems and its larger impact on reducing socioeconomic vulnerabilities in Central Himalayas. The chapter will explore evidence of climate change on natural forests, understand the future implications, and explore the opportunities of climate-sensitive restoration planning for increasing climate adaptations for resilience planning under the following objectives:

i. Key evidences to establish the impact of climate change on natural forest ecosystems and key stone and indicator species
ii. Crucial challenges and future implications of loss of key stone and indicator species, natural ecosystems affecting socioeconomics
iii. Relevance of unified climate-sensitive restoration arrangement for transformative shifts to restoration of high value threatened ecosystems for climate resilience

To understand the impacts of climate variability through evidence and to have a broader understanding about keystone and indicator species, we used case study approach to reflect on our findings. *Quercus leucotrichophora* (A. Camus) (*Banj* Oak) as a keystone species of natural moist temperate mixed broadleaved forests which grows luxuriantly across all 11 hill districts of Uttarakhand state in Central Himalayas from 1500 to 2300 m amsl were taken as a prominent case; whereas, *Hippophae salicifolia* (D. Don) (Sea buckthorn) as an indicator species of riparian buffers in high elevation zones of the region was considered to assess the impacts of climate change on slope stability where the species luxuriantly grows in dense patches from 2000 to 3600 m amsl. Based on the climate projections for these species, the study attempts to understand the larger challenges and future implications of loss of keystone and indicator species on natural ecosystems and socioeconomics of the region including the loss of springsheds, encroachment of *Pinus roxburgii* (Chir Pine) in *Banj* Oak growing forests, invasion of alien invasive species and increasing disaster risks especially slope destabilization and landslides. With an understanding on the impacts of climate change on future scenarios of natural forests, which might lead to socioeconomic vulnerabilities in the Central Himalayas, this study explores the opportunities through an integrated climate-sensitive restoration planning using the indigenous and traditional knowledge base to develop climate resilience. For exploring fundamental issues of loss of biodiversity, a methodical attitude for cause source exploration was employed to comprehend the crucial elements that act together through socioeconomic drivers in a multifaceted mode. Population rise, marginalization of local communities with insufficient options and opportunities resulting in disparities, out-migration, insufficient market access, power dynamics, and rampant loss of ecosystems for development in the region leading to significant societal changes to understand socioeconomic drivers were attempted to be analyzed. Case studies as an imperative methodical tool for conceptual modeling were explored to find answers to the proposed objectives. The important drive to use relevant case studies was to provide an overview of natural resources to understand the scenario for designing specific solutions to protect forest ecosystems in the region.

3.2 Study Area and Climate

Central Himalayas in Uttarakhand State of India with an area of 53,484 km^2 covers around 1.63% of the geographical area but provides home to more than 2.86% of the forest area of India. Central Himalayas with diverse forest types distributed on altitudinal gradients are biodiversity hotspots (Champion and Seth 1968; Marchese 2015; Roy et al. 2015). State has 45.43% of its geographical area under forests which can be categorized as 9.3% under very dense, 24.1% moderately dense, 12.1% open, 0.7% scrub, and 53.9% under the non-forest category. State has around 16,919 villages with many close to natural forests (Dhyani et al. 2020c). Local people dwelling in villages of Central Himalayas are reliant on natural forests for numerous sustenance burdens (Misra et al. 2008b; Dhyani et al. 2011; Dhyani and Dhyani 2016a; Dhyani 2018). Forest dependent agriculture along with cattle raising is fundamental livelihood option for locals in the region. Shift of traditional crops and cropping practices to cash or water and high external input agriculture has enhanced pressure on natural forests.

3.3 CASE STUDY

Climate Impact on Quercus Leucotrichophora Growing Forests

The first case study aims to comprehend how climate change is affecting the *Quercus leucotrichophora* A. Camus, or banj oak natural forests. In moist temperate mixed broad-leaved forests of the state, *banj* oak is a keystone species (Singh et al. 2014; Chakraborty et al. 2018; Verma and Garkoti 2019). *Quercus leucotrichophora* dominated moist temperate forests are renowned Central Himalayan forest biomes (Birdlife International 2019). Champion and Seth and others have categorized and kept these *banj* oak forests under the Himalayan moist temperate forests category (Champion and Seth 1968; Roy et al. 2015; Bahuguna et al. 2016). *Banj* oak grows widely across Hindukush Himalayas and has been reported from Central Himalayas in Uttarakhand, Kashmir, Himachal Pradesh, Nepal, Myanmar, Pakistan, and also from Sri Lanka (Manandhar 2002). *Banj* oak forms climatic climax in mixed temperate broad-leaved forests distributed on an altitudinal gradient from 1000 to 3500 m amsl with high tree, shrub, herb, and epiphyte diversity. Important associate trees in *banj* oak forests are *Rhododendron arboreum, Alnus nepalensis, Pyrus pashia, Neolitsea pallens, Prunus cerasoides, Lyonia ovalifolia, Cinnamomum tamala*, etc. *Banj* oak covers 5.24% *i.e.* 1284 km^2 geographical area of the state that makes it the most dominant tree species in Central Himalayan highlands. Although, *banj* oak is found throughout the state, it is more prevalent in the northwest and southeast regions known as Garhwal and Kumaon Himalaya, respectively. *Banj* oak forests along with high biodiversity values also provide rich soil organic matter that supports adaptive agriculture practices of the region, high water holding capacity by enhancing water recharge in catchment areas of watersheds and spring sheds and supporting local subsistence demands (Misra et al. 2008a). In last few decades, it was observed that climate change followed by continuous human interference has significantly affected regeneration and natural course of succession in *banj* oak forests (Dhyani et al. 2019). Rapid changes in regeneration, the natural course of succession is affecting stand structures, distribution pattern of trees inside forests and will largely affect the regional forest cover (Chaturvedi et al. 2011; Kumar et al. 2019). Regeneration of *banj* oak has historically reduced across Central Himalayan forests and there are clear observations of its replacement by its associates or other tree species (Thadani and Ashton 1995). Increasing frequency and intensity of forests fires in the region, along with increasing seasonal temperature, regular biomass removal by locals has significantly affected *banj* oak dominated moist temperate forests of the region (Kumar and Ram 2005; Verma and Garkoti 2019; Dhyani and Dhyani 2020). Rising forest fires have become a major factor in the slow growth of banj oak forests and the quick encroachment of *Pinus roxburghii* (chir pine) (Negi 2019). Habitat suitability modeling approach used by (Dhyani et al. 2020c) reported present potential distribution of *banj* oak in 10,387 km^2 (19.3%) followed by 7806 km^2 (14.5%) as plausible habitat having required suitable niche conditions and 2581 km^2 (4.8%) of highly suitable areas/niches (Fig. 3.1). It was found that the prospective habitat suitability for the *banj* oak by 2050 is expected to fluctuate between 1049 and 2252 km^2 [following Representative Concentration Pathway (RCP) results from 2.6, 4.5, 6.0, and 8.5] and is likely to reduce by 84% of the present habitats. Likewise, plausible habitat suitability is also expected to drop by 79% and reduction of highly plausible habitats by 99%. By 2070 potential *banj* oak habitats are expected to decline by 86%, plausible habitats by 84%, and highly plausible *banj* oak forest habitats by 99%. The observations and conclusions of the study are alarming as they project almost disappearance of significant habitats from 2581 km^2 (in present) to a meager area of 13 km^2. While the trends of loss of *banj* oak potential habitats for 2050 varied inconsistently, for 2070 drastic decline was observed in loss of these habitats resulting in remaining only 951 km^2 of habitat from present scenario. Temperature seasonality emerged as a mega driver influencing the distribution of *banj* oak by 42.1% followed by elevation that influenced distribution by 14.5%. Nevertheless, there inconsistent drift shift observed in elevation of the potential *banj* oak growing forest habitats from 2050–2070 which was between 1716 ± 230 to 1780 ± 350 m amsl (Fig. 3.1).

Climate Impact on Hippophae Salicifolia Growing Forests

Hippophae salicifolia D. Don also recognized as Sea buckthorn, *ames, amil, chuk* has multiple benefits to riparian ecosystems and marginalized local communities across Himalayas. Species grows luxuriantly from 2000 to 3600 m amsl in riparian buffer/slopes of river Ganga and its tributaries in high-altitude valleys of Central Himalayas (Dhyani et al. 2010, 2018). *H. salicifolia* as a pioneer prefers to grow on stable and enriched soil with alluvial gravel, humid, wet landslips in riparian buffers across the moist and dry temperate zones. Species has extraordinary property to grow in wide range of climatic conditions varying from –40 to 40°C. *H. salicifolia* grows through root turins and hence forms large single patches of male and female trees with extensive root system that strongly binds soil and rocks and helps in slope stabilization for reducing land, mud, and rockslides. Seabuckthorn has strong nitrogen fixing property and it is considered an ecosystem health indicator species for riparian buffers and slopes in higher Himalayan highlands (Dhyani et al. 2018). *H. salicifolia* also is known for its socioeconomic benefits,

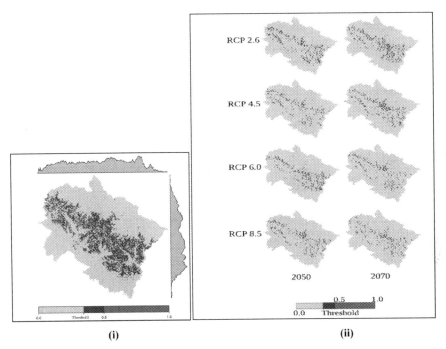

Fig. 3.1 (i) Prospective (threshold < p < 1), plausible (threshold < p < 0.5), and highly plausible (0.5 < p < 1) zones for *Q. leucotrichophora's* existence in the Uttarakhand state of IHR in the current settings. (ii) Prospective (threshold < p < 1), plausible (threshold < p < 0.5), and highly plausible (0.5 < p < 1) zones of *Q. leucotrichophora's* existence in the Uttarakhand state of IHR as per future IPCC RCPs for 2050 and 2070 years (*Source* Dhyani et al. 2020a, b, c, d)

providing nutritional and livelihood security to indigenous and local communities. In long-term research carried out by Dhyani et al. on multiple aspects of the species in 24 prominent locations of five high-altitude valley (Yamunotri, Gangotri, Niti, Bhyundhar, and Mana) of Uttarakhand it was clearly observed that human induced pressure and increasing climate sensitivity in Himalayas has significantly affected *H. salicifolia* growing pockets (Dhyani et al. 2010, 2013, 2018) (Fig. 3.2). Substantial deterioration of probable *H. salicifolia* growing pockets from remote and fragile valleys of the state was anticipated from present situation (Dhyani et al. 2018). By 2050 *H. salicifolia* habitats are projected to decline by 87.2% of the present habitats while, by 2070 marginal improvement of 30 km^2 from 2050 can be expected [following minimum to maximum Representative Concentration Pathway results from (RCP) 2.6, 4.5, 6.0, and 8.5]. Study reported 80.6% of the high plausible habitat loss of *H. salicifolia* by 2050 following 75.6% decline by 2070. Projections for upward migration of species for seeking climatically suitable habitats cannot be denied in this situation and this shift was expected from a lowest of 2813 ± 761 m

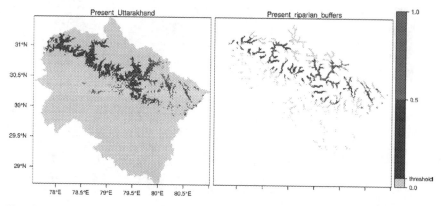

Fig. 3.2 Existence possibility of *Hippophae salicifolia* in the present scenario over Uttarakhand along the riparian buffers of river Ganga and its tributaries (*Source* Dhyani et al. 2018)

amsl to a extreme of 4278 ± 202 m amsl by 2050 followed by 4485 ± 230 m amsl by 2070. High plausible niches of the species were also anticipated to further migrate from a lowest of 2542 ± 564 m amsl to a extreme of 4280 ± 195 m amsl by 2050 and 4510 ± 250 m amsl by 2070.

3.4 Future Implications of Loss of Species, Natural Ecosystems

Geological sensitivity coupled with its socio-ecological characteristics makes the region and its inhabitants susceptible to increasing frequency of climate alteration impacts and geo risks. Remoteness of the region, its fragile terrains further enhance the risks of climate variation and reduce the resilience of native ecosystems and communities (Dhyani and Dhyani 2016b). Changes in the forest community structures due to changing climate may bring in large scale changes to the ecosystem arrangement and working (Kala et al. 2018). Decline and deforestation in *banj* oak growing moist temperate mixed broad-leaved forests and other important forest types distributed on altitudinal gradients might affect and change the biogeochemical cycles or supporting ecosystem services in the region which may further change the bioclimatic, meteorological, and also hydrological cycles (Singh et al. 1984; Negi and Joshi 2004). Depleting springsheds across Himalayas is a clear indicator of deforestation, changing rainfall patterns in the region and to restore them restoring natural forests will be crucial (Negi and Joshi 2004; Joshi et al. 2018; Shrestha et al. 2018). In the last few decade forest fires have increased in the region resulting in loss of thousands of hectares of forests along with flora and fauna every year. This increase in annual forest fires has accelerated drier conditions and with slow regeneration rates of *banj* oak encroachment of *P. roxburghii* in *banj* oak growing moist temperate forests has enhanced (Negi 2019). Changes and shift in timing of flowering and fruiting from spring to early and mid-winter season due to increasing winter temperatures reflects to the larger impact of climate variation on seed setting of forest trees and agriculture crops (Shrestha et al. 2012; Hart et al. 2014; Ramachandran and Roy 2018). *Banj* oak dominated moist temperate broad-leaved forests are the lifeline and greatest support to subsistence life style of millions dwelling across Himalayas hence, ecological loss of these forests will significantly affect the socioeconomics of the local communities. The slow growth and regeneration of species will require human interventions and appropriate silviculture efforts to make sure by planning more *banj* oak plantations and restoring degraded oak forest patches in the region using the predictions for potential ecological niches of the species. Assisted natural regeneration and reducing pressure on forest soil seed bank by regulating leaf litter from forest floor and rotational harvesting of resources from the forests (Dhyani et al. 2011; Dhyani 2018). Similarly, loss of *H. salicifolia* growing habitats across Central Himalayas is a critical challenge for the health of slopes and riparian buffers in the high-altitude areas. Traditionally *Hippophae salicifolia* provides multiple benefits in slope stabilization for reducing disaster risks, soil quality amelioration by its rapid nitrogen fixing ability, traditional usage by local communities, potential for developing value added products, and greening the high-altitude degraded areas. Species provides nutritional security and livelihood opportunities to Bhotiya tribal sub-communities during lean winter months. In last few years many pharmaceutical, nutraceutical companies have become interested in developing diverse medicinal, nutritional, and sanitary products from the fruits of the plant. The species has the potential to assist downstream communities nutritionally and financially if it is preserved and used well (Dhyani et al. 2010). *Hippophae salicifolia* as a pioneer species facilitates transition from pioneer to late successional stages of succession hence, loss of species will reduce habitat or niche suitability for several seral and climatic climax species in the zone. As the species was viewed as a promising species for slope stabilization, greening, and catastrophe risk reduction in Himalayan highlands across IHR, the loss of the species has stressed conservationists, state forest department, and protected area managers (Dhyani et al. 2013, 2018). The Central Pollution Control Board (CPCB) and the Ministry of Environment, Forests and Climate Change, Government of India, under the green India mission of the National Action Plan on Climate Change (NAPCC), considered species as an excellent support for EcoDRR (ecosystem-based disaster risk reduction) that can be effective Nature based Solutions for slope stabilization and landslide reduction (Dhyani et al. 2020d). Growing number of Hydropower projects in the fragile landscapes and ecosystems followed by continuous construction and broadening for security and strategic purposes and all weather road for enhancing tourism opportunities with debris being mostly thrown on river banks have been key constraining conditions. Remoteness of the IHR and underprivileged settings make these populations reliant on natural forests for life-sustaining requirements hence loss of natural ecosystems will enhance drudgery of upstream and downstream communities.

3.5 Integrated Climate-Sensitive Restoration Planning

Transformative revolution to improve sustainable land restoration and mainstreaming integrated climate-sensitive restoration planning requires novelties in preparation, application, and monitoring that address climate uncertainty and related emerging novel challenges for mountain ecosystem and communities. Achieving Sustainable Development Goal (SDG) 15 "Life on Land" by 2030 especially target 15.3 "Land Degradation Neutrality"; Forest land restoration under Bonn Challenge and targets of UN Decade of restoration (2021–2030) cannot be successful by scientific tools and technologies alone. As per both the case studies (*Q. leucotricohophora* growing forests across the region and *Hippophae salicifolia* growing patches) warming in Himalayas will affect the natural forest ecosystems and their vegetation; especially, the distribution of keystone and indicator species and hence, there is a growing need to plan and implement climate-sensitive restoration to protect these species, their associate species and natural forest ecosystems (Dhyani et al. 2020b). Nature based Solutions (NbS) is considered an effective umbrella approach with diverse options and opportunities for disaster risk reduction and sustaining Himalayan ecosystems and safeguarding human well-being (Dhyani et al. 2020d). Restoration to improve native biodiversity and reinstate declining ecosystem may not be very successful due to climate change and loss of habitat suitability for many native species. Hence, to ensure positive restoration outcomes understanding of tree and vegetation assemblages especially understanding of their survival and mortality due to changing climate will be a prerequisite. Restoration programs, forest department, and practitioners working across Himalayas require broader understanding of the natural ecosystems, their successional patterns to implement climate-sensitive restoration that acknowledges environmental uncertainty and accordingly frames the future restoration targets. Integrated climate-sensitive restoration approach ensures participation of all local stakeholders especially indigenous and local communities in mapping degraded lands, exploration, and identification of potential tree species that are well supported by the scientific understanding of their future habitat suitability (2050 and 2070 scenarios). The global efforts to acclimatize and alleviate adverse climate change effects can only be successful if reinforced by active local support and efforts. Forest landscape restoration and sustainable land restoration are key approaches to initiate participatory restoration of degraded forests, halt and reverse biodiversity loss to reestablish ecosystem services flow. The approach is expected to potentially enhance the outcomes of sustainable land restoration across Himalayas where community involvement is available as a key enabling condition for ensuring restoration effectiveness that ensures climate sensitivity and uncertainty is included in the planning process. Community Based Adaptation (CBA) that integrates local priorities, their knowledge, and capacities to adjust to the growing climate vulnerabilities and disaster risks especially active involvement of local communities in land managing choices, long-term goal setting, and land-use trade-offs are pertinent. Traditional and indigenous knowledge of local and indigenous communities in the region are capable to ensure sustainable use, effective local management, and community conservation of natural forests to reduce disaster risks and improve climate change adaptations. Forest landscape restoration (FLR) as per Bonn Challenge finds its relevance in the context of restoring moist temperate broad-leaved forests in the region with the ground support of 12,000 *Van Panchayats* (Community Forests) in the state (Misra et al. 2009). Indigenously supported site-specific and cost effective Ecosystem-based approaches for Disaster Risk Reduction (Eco-DRR) and climate change adaptations (EbA) have already proven their effectiveness to reduce disaster risks (Dhyani and Thummarukuddy 2016; Dhyani et al. 2020d). Innovative and traditional precautionary measures that have their own scientific credibility are real strategies to improve alternatives, develop local resilience, and help in reducing the frequency and severity of disasters. Protecting *H. salicifolia* growing areas and domestication of species in agroforestry, agri-horti, and agri-silvi-pastoral models in local cropland bunds and degrade eroded slopes can ensure *in-situ* and *ex-situ* conservation of the species and reducing pressure from natural pockets. For *Q. leucotrichophora* slow regeneration of species requires assisted natural regeneration in natural forests along with appropriate forestry and silviculture intervention that ensure more *banj* oak plantations instead of using Manipuri banj for restoring degraded oak forest patches in the region using habitat suitability maps. This approach can benefit 40 odd associate tree species growing in moist temperate mixed broad-leaved forests that face threat of existence due to climate change impacts.

3.6 Conclusion

Long-term efforts to reduce disaster risks in the region have just begun. Case studies of climate impact on *Q. leucotrichophora* growing forests across the region and *Hippophae salicifolia* growing patches clearly indicate that habitat suitability for both the species at large in Central Himalayas is in its critical phase and is expected to substantially degrade further in future following different IPCC scenarios. Although, case studies used simulations and predictions for habitat suitability in Central Himalayas for projecting climate change as a mega driver yet, it is necessary to understand and include anthropogenic interferences in the habitat suitability modeling along with additional crucial drivers (for ex. land use land cover change, forest fire, competition, invasion, demographic changes, socioeconomics). A better understanding of local knowledge practices will be necessary to further refine the approach and improve findings at local level and this will further also help in recognizing native species that are relevant and effective for local and regional environment for necessary ecosystem services. Practical approaches to protect the existing natural forest ecosystems is only possible by supporting assisted natural regeneration, reducing pressure from soil seed bank, and ensuring livelihood diversification of forest dependent marginalized communities. Himalayan forests support subsistence provisions of millions living upstream and many more in downstream areas. Given the sensitivity of the context and issue, the approaches of protection need to be fundamentally wide-ranging and more customized, area specific in nature and that is where there is need to localize the global and regional climate models to understand their more local impact. Integrated climate-sensitive restoration planning is expected to streamline efforts that pursue to disclose the associations between local socio-ecological and economic interlinkages and their interdependencies. Distinguishing the intricacy and the diversity of factors that govern the dynamics of forest landscape restoration efforts can only be transformed and revolutionized by enabling and instigating more dialogue, involvement of communities in improving health of nature forest ecosystems for developing societal resilience to climate change.

Acknowledgements The authors thank the Knowledge Resource Centre (KRC) of CSIR-NEERI, Nagpur for the plagiarism and similarity check.

Conflict of Interest Authors declare no conflict of interest.

References

Ammer C (2019) Diversity and forest productivity in a changing climate. New Phytologist 221:50–66. https://doi.org/10.1111/nph.15263

Bahuguna VK, Swaminath MH, Tripathi S, Singh TP, Rawat VRS, Rawat RS (2016) Revisiting forest types of India. IFRE 18:135–145. https://doi.org/10.1505/146554816818966345

Birdlife International (2019) Endemic Bird Areas factsheet: Western Himalayas. http://datazone.birdlife.org/eba/factsheet/124. Accessed 18 Sep 2019

Chakraborty A, Joshi PK, Sachdeva K (2018) Capturing forest dependency in the central Himalayan region: Variations between Oak (*Quercus* spp.) and Pine (*Pinus* spp.) dominated forest landscapes. Ambio 47:504–522. https://doi.org/10.1007/s13280-017-0947-1

Champion HG, Seth SK (1968) A revised survey of the forest types of India. Manager of Publications, Delhi

Chaturvedi RK, Gopalakrishnan R, Jayaraman M, Bala G, Joshi NV, Sukumar R, Ravindranath NH (2011) Impact of climate change on Indian forests: A dynamic vegetation modeling approach. Mitig Adapt Strateg Glob Change 16:119–142. https://doi.org/10.1007/s11027-010-9257-7

Dawson TP, Jackson ST, House JI, Prentice IC, Mace GM (2011) Beyond predictions: Biodiversity conservation in a changing climate. Science 332:53–58. https://doi.org/10.1126/science.1200303

Dhyani D, Dhyani S, Maikhuri RK (2013) Assessing anthropogenic pressure and its impact on Hippophae salicifolia pockets in Central Himalaya, Uttarakhand. J Mt Sci 10:464–471. https://doi.org/10.1007/s11629-013-2424-z

Dhyani D, Maikhuri RK, Misra S, Rao KS (2010) Endorsing the declining indigenous ethnobotanical knowledge system of Seabuckthorn in Central Himalaya, India. J Ethnopharmacol 127:329–334. https://doi.org/10.1016/j.jep.2009.10.037

Dhyani S (2018) Impact of forest leaf litter harvesting to support traditional agriculture in Western Himalayas. 59:473–488

Dhyani S, Dhyani D (2016a) Significance of provisioning ecosystem services from moist temperate forest ecosystems: Lessons from upper Kedarnath valley, Garhwal, India | SpringerLink. https://doi.org/10.1007/s40974-016-0008-9. Accessed 12 Sep 2019

Dhyani S, Dhyani D (2016b) Strategies for reducing deforestation and disaster risk: Lessons from Garhwal Himalaya, India. In: Renaud FG, Sudmeier-Rieux K, Estrella M, Nehren U (eds) Ecosystem-based disaster risk reduction and adaptation in practice. Springer International Publishing, Cham, pp 507–528

Dhyani S, Dhyani D (2020) Local socio-economic dynamics shaping forest ecosystems in Central Himalayas. In: Roy N, Roychoudhury S, Nautiyal S, Agarwal SK, Baksi S (eds) Socio-economic and eco-biological dimensions in resource use and conservation: Strategies for sustainability. Springer International Publishing, Cham, pp 31–60

Dhyani S, Kadaverugu R, Dhyani D, Verma P, Pujari P (2018) Predicting impacts of climate variability on habitats of Hippophae salicifolia (D. Don) (Seabuckthorn) in Central Himalayas: Future challenges. Ecol Inf 48:135–146. https://doi.org/10.1016/j.ecoinf.2018.09.003

Dhyani S, Bartlett D, Kadaverugu R, Dasgupta R, Pujari P, Verma P (2020a) Integrated climate sensitive restoration framework for transformative changes to sustainable land restoration. Restoration Ecol 28:1026–1031. https://doi.org/10.1111/rec.13230

Dhyani S, Bartlett D, Kadaverugu R, Dasgupta R, Pujari P, Verma P (2020b) Integrated climate sensitive restoration framework for transformative changes to sustainable land restoration. Restoration Ecol https://doi.org/10.1111/rec.13230

Dhyani S, Kadaverugu R, Pujari P (2020c) Predicting impacts of climate variability on Banj oak (*Quercus leucotrichophora* A. Camus) forests: Understanding future implications for Central Himalayas. Reg Environ Change 20:113. https://doi.org/10.1007/s10113-020-01696-5

Dhyani S, Karki M, Gupta AK (2020d) Opportunities and advances to mainstream nature-based solutions in disaster risk management and climate strategy. In: Dhyani S, Gupta AK, Karki M (eds) Nature-based solutions for resilient ecosystems and societies. Springer, Singapore, pp 1–26

Dhyani S, Maikhuri RK, Dhyani D (2011) Energy budget of fodder harvesting pattern along the altitudinal gradient in Garhwal Himalaya, India. Biomass Bioenergy 35:1823–1832. https://doi.org/10.1016/j.biombioe.2011.01.022

Dhyani S, Maikhuri RK, Dhyani D (2019) Impact of anthropogenic interferences on species composition, regeneration and stand quality in moist temperate forests of Central Himalaya. Trop Ecol 60:539–551. https://doi.org/10.1007/s42965-020-00054-0

Dhyani S, Santhanam H, Dasgupta R, Bhaskar D, Murthy IK, Singh K (2022) Exploring synergies between India's climate change and land degradation targets:

Lessons from the Glasgow Climate COP. Land Degradation and Development. https://doi.org/10.1002/ldr.4452

Dhyani S, Sivadas D, Basu O, Karki M (2022) Ecosystem health and risk assessments for high conservation value mountain ecosystems of South Asia: A necessity to guide conservation policies. Anthr Sci. https://doi.org/10.1007/s44177-022-00010-8

Dhyani S, Thummarukuddy M (2016) Ecological engineering for disaster risk reduction and climate change adaptation. Environ Sci Pollut Res 23:20049–20052. https://doi.org/10.1007/s11356-016-7517-0

Diaz S, Demissew S, Joly C, Lonsdale WM, Larigauderie A (2015) A Rosetta Stone for nature's benefits to people. PLoS Biol 13:e1002040. https://doi.org/10.1371/journal.pbio.1002040

Hart R, Salick J, Ranjitkar S, Xu J (2014) Herbarium specimens show contrasting phenological responses to Himalayan climate. PNAS 111:10615–10619. https://doi.org/10.1073/pnas.1403376111

IPBES (2018a) The IPBES assessment report on Land Degradation and Restoration. Secretariat of the Intergovernmental Science-Policy Platform on Biodiversity and Ecosystem Services, Bonn, Germany

IPBES (2018b) Summary for policymakers of the regional assessment report on biodiversity and ecosystem services for the Americas of the Intergovernmental Science-Policy Platform on Biodiversity and Ecosystem Services. IPBES Secretariat, Bonn, Germany

IPCC (2019) Summary for policymakers. In: Pörtner H-O, Roberts DC, Masson-Delmotte V, Zhai P, Tignor M, Poloczanska E, Mintenbeck K, Alegría A, Nicolai M, Okem A, Petzold J, Rama B, Weyer NM (eds) IPCC special report on the ocean and cryosphere in a changing climate. In press, p 35

Jetz W, McGeoch MA, Guralnick R, Ferrier S, Beck J, Costello MJ, Fernandez M, Geller GN, Keil P, Merow C, Meyer C, Muller-Karger FE, Pereira HM, Regan EC, Schmeller DS, Turak E (2019) Essential biodiversity variables for mapping and monitoring species populations. Nat Ecol Evol 3:539–551. https://doi.org/10.1038/s41559-019-0826-1

Joshi B, Negi GCS, Rawal R, Joshi R, Sharma S, Rawat DS, Bhattacharjee A (2018) Opportunities for forest landscape restoration in Uttarakhand, India using ROAM. https://www.researchgate.net/publication/328212174_Opportunities_for_forest_landscape_restoration_in_Uttarakhand_India_using_ROAM. Accessed 30 Sep 2019

Kala R, Bhavsar D, Kumar A, Roy A, Rawat L (2018) Quantification of potential area of incursion of pine in oak forest in western Himalaya using fuzzy classification technique. J Appl Remote Sens 12:1. https://doi.org/10.1117/1.JRS.12.026032

Krishnan R, Shrestha AB, Ren G, Rajbhandari R, Saeed S, Sanjay J, Syed MdA, Vellore R, Xu Y, You Q, Ren Y (2019) Unravelling climate change in the Hindu Kush Himalaya: rapid warming in the mountains and increasing extremes. In: Wester P, Mishra A, Mukherji A, Shrestha AB (eds) The Hindu Kush Himalaya assessment: Mountains, climate change, sustainability and people. Springer International Publishing, Cham, pp 57–97

Kulkarni A, Patwardhan S, Kumar KK, Ashok K, Krishnan R (2013) Projected climate change in the Hindu Kush–Himalayan region by using the high-resolution regional climate model PRECIS. Mountain Research and Development 33:142–151

Kumar A, Ram J (2005) Anthropogenic disturbances and plant biodiversity in forests of Uttaranchal, central Himalaya. Biodiversity and Conservation 14:309–331. https://doi.org/10.1007/s10531-004-5047-4

Kumar M, Kalra N, Khaiter P, Ravindranath NH, Singh V, Singh H, Sharma S, Rahnamayan S (2019) PhenoPine: A simulation model to trace the phenological changes in *Pinus roxhburghii* in response to ambient temperature rise. Ecological Modelling 404:12–20. https://doi.org/10.1016/j.ecolmodel.2019.05.003

Locatelli B, Lavorel S, Sloan S, Tappeiner U, Geneletti D (2017) Characteristic trajectories of ecosystem services in mountains. Frontiers in Ecology and the Environment 15:150–159. https://doi.org/10.1002/fee.1470

Manandhar NP (2002) Plants and people of Nepal. Plants and people of Nepal

Marchese C (2015) Biodiversity hotspots: A shortcut for a more complicated concept. Global Ecol Conserv 3:297–309. https://doi.org/10.1016/j.gecco.2014.12.008

Misra S, Dhyani D, Maikhuri R (2008a) Sequestering carbon through indigenous agriculture practices. In: Undefined. http://indiaenvironmentportal.org.in/files/Sequestering%20carbon.pdf. Accessed 25 Dec 2020

Misra S, Maikhuri R, Kala C, Rao K, Saxena K (2008b) Wild leafy vegetables: A study of their subsistence dietetic support to the inhabitants of Nanda Devi Biosphere Reserve, India. J Ethnobiology Ethnomedicine 4:15. https://doi.org/10.1186/1746-4269-4-15

Misra S, Maikhuri RK, Dhyani D, Rao K (2009) Assessment of traditional rights, local interference and natural resource management in Kedarnath Wildlife Sanctuary. Int J Sustain Develop World Ecol 16:6. https://doi.org/10.1080/13504500903332008. Accessed 27 Sep 2019

Mondal PP, Zhang Y (2018). Research progress on changes in land use and land cover in the western Himalayas (India) and effects on ecosystem services. Sustain 10(12):4504. https://doi.org/10.3390/su10124504

Negi GCS (2019) Forest fire in Uttarakhand: Causes, consequences and remedial measures. Int J Ecol Environ Sci 45:31–37

Negi GCS, Joshi V (2004) Rainfall and spring discharge patterns in two small drainage catchments in the Western Himalayan Mountains, India. The Environmentalist 24:19–28. https://doi.org/10.1023/B:ENVR.0000046343.45118.78

Pandve HT (2009) India's national action plan on climate change. Indian J Occup Environ Med 13:17–19. https://doi.org/10.4103/0019-5278.50718

Payne D, Spehn EM, Snethlage M, Fischer M (2017) Opportunities for research on mountain biodiversity under global change. Curr Opinion Environ Sustain 29:40–47. https://doi.org/10.1016/j.cosust.2017.11.001

Ramachandran RM, Roy PS (2018) (PDF) Vegetation response to climate change in Himalayan hill ranges: a remote sensing perspective. In: ResearchGate. https://www.researchgate.net/publication/323722532_Vegetation_response_to_climate_change_in_Himalayan_hill_ranges_a_remote_sensing_perspective. Accessed 1 Oct 2019

Roy PS, Roy A, Joshi PK, Kale MP, Srivastava VK, Srivastava SK, Dwevidi RS, Joshi C, Behera MD, Meiyappan P, Sharma Y, Jain AK, Singh JS, Palchowdhuri Y, Ramachandran RM, Pinjarla B, Chakravarthi V, Babu N, Gowsalya MS, Thiruvengadam P, Kotteeswaran M, Priya V, Yelishetty KMVN, Maithani S, Talukdar G, Mondal I, Rajan KS, Narendra PS, Biswal S, Chakraborty A, Padalia H, Chavan M, Pardeshi SN, Chaudhari SA, Anand A, Vyas A, Reddy MK, Ramalingam M, Manonmani R, Behera P, Das P, Tripathi P, Matin S, Khan ML, Tripathi OP, Deka J, Kumar P, Kushwaha D (2015) Development of Decadal (1985–1995–2005) Land use and land cover database for India. Remote Sens 7:2401–2430. https://doi.org/10.3390/rs70302401

Sato CF, Lindenmayer DB (2018) Meeting the global ecosystem collapse challenge. Conserv Lett 11:e12348. https://doi.org/10.1111/conl.12348

Sharma E, Molden D, Rahman A, Khatiwada YR, Zhang L, Singh SP, Yao T, Wester P (2019) Introduction to the Hindu Kush Himalaya Assessment. In: Wester P, Mishra A, Mukherji A, Shrestha AB (eds) The Hindu Kush Himalaya Assessment: Mountains, climate change, sustainability and people. Springer International Publishing, Cham, pp 1–16

Shrestha R, Desai J, Mukherji A, Dhakal M, Kulkarni H, Mahamuni K, Bhuchar S, Bajracharya S (2018) Protocol for reviving springs in the Hindu Kush Himalayas: A practitioner's manual

Shrestha UB, Gautam S, Bawa KS (2012) Widespread climate change in the Himalayas and associated changes in local ecosystems. PLoS ONE 7:e36741. https://doi.org/10.1371/journal.pone.0036741

Singh JS, Rawat YS, Chaturvedi OP (1984) Replacement of oak forest with pine in the Himalaya affects the nitrogen cycle. Nature 311:54–56

Singh V, Thadani R, Tewari A, Ram J (2014) Human Influence on Banj Oak (*Quercus leucotrichophora*, A. Camus) Forests of central Himalaya. J Sustain Forest 33:373–386. https://doi.org/10.1080/10549811.2014.899500

Thadani R, Ashton PMS (1995) Regeneration of banj oak (*Quercus leucotrichophora* A. Camus) in the central Himalaya. Forest Ecol Manag 78:217–224. https://doi.org/10.1016/0378-1127(95)03561-4

Verma AK, Garkoti SC (2019) Population structure, soil characteristics and carbon stock of the regenerating banj oak forests in Almora, Central Himalaya. Forest Science and Technology

CHAPTER 4

Measuring Climate Change Impact on Crop Yields in Southern India: A Panel Regression Approach

Rajesh Kalli and Pradyot Ranjan Jena

4.1 INTRODUCTION

Anthropogenic greenhouse gas emissions have caused variation in climate patterns and occurrence of adverse climate events (floods *and* droughts). This shift in climate patterns has direct impact on biodiversity. Past studies in the recent literature, estimating the economic impact of climate change, have turned the spotlight on evaluating climate change impact on several indicators (agriculture yields, Gross Domestic Product, financial markets, farmer's behaviour, migration, water, mortality) (Bandara and Cai 2014; Carleton 2017; Kalli and Jena 2022; Jena et al. 2022). Agriculture is the primary sector among most of the developing economies which acts as the major source of food security. With the limited availability of resources to adapt, developing nations are highly susceptible to climate change. The situation is grimmer as most of the developing economies are under tropical regions, where future projections of climate change indicate devastating distress on agricultural sector. The climate in India can be classified as tropical wet and

---

R. Kalli
School of Commerce and Management Studies, Dayananda Sagar University, Bangalore, Karnataka, India
e-mail: rajukalli.nitk@gmail.com

P. R. Jena (✉)
School of Management, National Institute of Technology Karnataka, Surathkal, Mangalore, Karnataka, India
e-mail: jpradyot@gmail.com

© The Author(s), under exclusive license to Springer Nature Singapore Pte Ltd. 2023
S. Nautiyal et al. (eds.), *Palgrave Handbook of Socio-ecological Resilience in the Face of Climate Change*,
https://doi.org/10.1007/978-981-99-2206-2_4

dry climate. The annual mean temperature showed a significant warming trend of 0.51 °C for the period 1901–2007 (Revadekar et al. 2012). The long-time trend indicated decreased dispersion of south-western and northeast rainfall in the peninsular India (Varadan et al. 2015). This significant climate trends with uneven patterns of rainfall and compounded with rising temperature will have an adverse effect on the crop yields.

Climate change has significant impact on arid and semi rid region (Kalli and Jena 2020). In such countries, the survival of the population will depend on the effective adaptation to climate change. The information on the climate change impact on the crop specific will help to identify suitable adaptation to mitigate the climate change effect. Karnataka belongs to tropical zone in India which follows semi-arid climate. The state lies in the south-western part of India with an area of 1.92 lakh km^2 and population of sixty-one-million. In the present study, we fix the growing season to months from June to September as kharif is the predominantly significant cultivation period. Several episodes of high intensity drought and significant variation in monsoon rainfall have been observed in Karnataka (Guhathakurta et al. 2015). Agriculture in Karnataka is highly dependent on the monsoon rainfall and highly susceptible to climate change (Kalli and Jena 2021); 26.5% area is irrigated while the remaining depends on rainfed cultivation (Bhende 2013). The state has a varied topographical character ranging from coastal plains to gentle slopes and the heights of the Western Ghats. There is need to perform consistent assessment of climate change impact on crop yields. This analysis will exhibit the impact of climate change in arid and semi-arid region. The dearth of studies in assessing climate change impact at regional scale can be noted. In this context, the present assessed the region-specific climate sensitivities on crop yields using statistical technique. The analysis from the present study contributes to the literature pertaining to climate change impact assessment on agriculture and also to formulate policy measures in the region.

4.2 Theoretical Background

Process-based and statistical models are two different approaches used to quantify climate change impact on agriculture. Process-based models refer to experimental based method to understand the physiological process of plant growth under different scenarios (Mall et al. 2006). Process-based models have rich theoretical background, which are constructed in the specific field scale analysis and further calibrated to future projections. There are chances of high uncertainty in the results of the process-based models due to greater number of model parameters and inefficiency to represent the natural process. Production function approach or statistical approach employs historical dataset to estimate climate change impact on agriculture production, and to predict the impact at global, national and regional scales. This approach relies on production function, where the output is maximized with a given set of inputs. This input–output process is established with an empirical framework between yield and climate factors. Statistical models account for variation in the climate and the region-specific characteristics. The theoretical background behind this approach is the plant growth or final agriculture yield that depends on several input parameters such as water, soil, economic inputs and climate variables that are employed as explanatory variables.

Statistical models have been widely used in assessing the climate change impact on crop yields (Arshad et al. 2017; Lobell and Burke 2010; Poudel and Kotani 2013; Wang et al. 2014). Lobell and Burke (2010) studied the climate change effect on maize yields using 20,000 maize trials dataset in Africa. The result indicated that each degree day above 30 °C reduces the final yield by 1% and 1.7% in optimal rainfed and drought conditions. Poudel and Kotani (2013) measured climate change impacts on the rice and wheat yields in Nepal. The result indicated that a 1 °C rise in summer temperature caused the rice yield to decline by 0.10% in low altitude regions, and by 0.001% in the mid-latitude region. Contrastingly, wheat yields in mid and high-altitude regions declined by 0.003% and 0.051% respectively. Wang et al. (2014) estimated climate change impact on crop yield using a fixed effect panel regression model. The result indicates that variations in temperature had a positive impact on wheat (1.3%) and rice (0.4%) yield, but negative impact on maize (12%) yield. Contradictory results were found in China, where climate change has a significant positive impact on rice yield due to the cold climate in the study region higher latitude region (Zhou et al. 2013). However, there have been only a few studies that have explained the yield sensitiveness to climate change in Indian context (Auffhammer et al. 2006; 2012; Gupta et al. 2017; Pattanayak and Kumar 2014). Krishna Kumar et al. (2004) studied the impact of climate change using correlation methodology. The study used rainfall as a factor, and also the sea surface temperature anomalies to determine the impact on a large scale. The result concluded that there is a significant relationship between agriculture yields and the summer monsoon rainfall. Auffhammer et al. (2012) estimated the impact of climate change on rice production, predominantly in nine major rice-growing states in India. The estimated result showed that the decline of 1% rainfall in Jun–Sep would result in a decline of 0.20% of the yield of rice. No other climatic variables were significant. Pattanayak and Kumar (2014) estimated weather sensitivity of rice yield in India, using district level data for the period 1967–2007. The study indicated that maximum temperature declined the rice yield by 3.2% with every 1 °C rise in temperature. Gupta et al. (2017) estimated the climate change impact on wheat yields during rabi season. The result from the study indicated that a 1 °C rise in average daily maximum and minimum temperatures tends to lower yields by 2–4% each.

4.3 Materials and Method

Climate Variables

The key climate variables used in study include temperature and rainfall. The information on temperature and rainfall was obtained from Indian Meteorological Department (IMD). IMD collects rainfall data from the rain gauge stations, and temperature data from surface observatories across India. We use a fine scale daily grid dataset of 0.25° × 0.25° for rainfall and 1° × 1° for temperature that was interpolated from unevenly distributed station-wise data using Shepard interpolation method (Pai et al. 2014; Srivastava et al. 2009). For rainfall, the grid points within the district boundaries were averaged daily and then summed up for the season. For temperature, the present study explores possibilities of adopting growing degree days as an indicator to discern the effect of climate change on crop yields. Growing degree days is employed as the temperature indicator, which captures the nonlinear effects of temperature over the growing season (D'Agostino and Schlenker 2016). The effect of heat on relative plant growth is cumulative over time, and that yield is proportional to total growth. The cumulative effect of temperature over time shows that temperature effect is additive over the growing season.

Degree days are constructed with the cumulative sum of temperature observations between two bounds. The upper and lower bounds of temperature are rooted in agronomy. The upper and lower bounds of temperature in the present study are fixed based on the crop and the climate characteristics of the region. The present study area comprises distinct climate features: the observed daily maximum temperature is above 42 °C among the northern regions whereas in southern region, the highest recorded temperature is 38 °C. Due to this varied climate within the state, degree days with a lower and higher threshold of 8 °C and 34 °C is developed as the plant growth diminishes below or above these threshold levels (Luo 2011). To formulate degree days, the upper bound threshold will be differenced from the actual maximum temperature of a particular day, and then summed over the growing season (Gupta et al. 2017). If the daily maximum temperature is found to be higher than the upper bound threshold in a day, then the growing degree days for that day is 26 °C. Crops may have different thresholds. However, most of the crops below 8 °C cannot absorb heat, and above 34 °C, the plant growth diminishes. Thus, growing degree days are the special case of our model where the above bounds are represented as an example below.

$$g(h) = \begin{cases} 0 \text{ if } h \leq 8 \\ h - 8 \text{ if } 8 < h < 34 \\ 26 \text{ if } 34 \leq h \end{cases}$$

Agriculture Variables

The study uses district level data from Directorate of Economics and Statistics, Karnataka (DES, Karnataka). DES, Karnataka publishes annual reports on the estimates of principal crops at district level. The data on agriculture yield is collected through cross-cutting experiments conducted each year during the harvest season. Stratified multistage random sampling has been adopted to accumulate data with village as the primary level, and later aggregated to district level. These reports describe the in-depth district level data on area, production and the yield for all major crops. The dependent variable is the average yield generated per hectare of land. Rice, maize and finger millet are the three major kharif crops considered for the study. The districts under maximum cultivation of each crop have been considered in the study. For rice, 14 districts that constitute more than 90% of area have been considered. Similarly, 10 major districts that largely cultivate finger millet and maize cultivation are used in the study.

Empirical Framework

The study focused on the climate change effect on agriculture yields of three important crops, i.e., rice, maize and finger millet. Fixed effect panel regression model was adopted to estimate the climate change impact on the crop yields. Panel models provide a large number of observations that identify nonlinear response function, which is crucial in identifying the causal relationship between climate and agriculture productivity (Blanc and Schlenker 2017). The temperature has been modelled as additively substitute over the time. This can be referred as the nonlinear modelling of temperature, used to assess the relationship between the heat stress and the crop yields. The current study validates the assumption by establishing statistical relationship between the cumulative distribution of the temperature and the yields. The empirical model is given by:

$$\log Y_{it} = c_i + \beta_0 + \beta_1 \text{Rain}_{it} + \beta_2 \text{GDD}_{it} + \beta_3 \text{Irrigation}_{it} + \beta_4 \text{Fertiliser}_{it} + \gamma_t + \varepsilon_{it}$$

where Y_{it} is yield in district i and year t. c_i is the district fixed effect which controls for time invariant factors. γ_i is the linear trend which captures the growth in the input use and capital expenditure. Rain_{it} is total rainfall in kharif season, GDD_{it} is the cumulative distribution of the temperature referred as degree days for the kharif season, Irrigation_{it} refers to the proportion of irrigated land in kharif season, Fertiliser_{it} refers to the average fertilizer used per hectare of land. In this specification, the logged crop yield was regressed on degree days and rainfall. Additional input variables with climate variables were introduced in the regression to assess the relative effect of climate change impact on the crop yields. The study further probes to estimate the impact of climate change on dry land cultivation. The districts which constitute less irrigation under each crop were estimated separately to understand the climate change impact on crop yields in absence of irrigation.

4.4 Results

Impact of Climate Change on Crop Yields

Table 4.1 presents the summary statistics of the three crops and variables aggregated at state level. The estimated results of fixed effect panel regression for Rice, Maize and Finger Millet are presented in Tables 4.2, 4.3 and 4.4. Two specifications of estimation have been modelled: Model 1 includes only climate variables and Model 2 includes climate variables and other input variables. The overall estimates for all the three crops indicate significant negative impact of temperature on crop yield in monsoon season. Table 4.2 presents the coefficients of rice estimation and corresponding standard errors. The result from the Model 1 indicates that each additional degree day above 34 °C decreases the rice yield. The coefficient of growing degree days resulted with − 0.065 (6.5%) signifying that each additional degree day above 34 °C decreases the rice yield by 6.5%. In case of rainfall, the coefficient was positive and had significant effect on the rice yields. This indicates that increase in summer monsoon rainfall will benefit rice cultivation in kharif season. However, with the addition of input variables to Model 2, the effect of temperature decreased from 6.5% to 5.1%. The other non-climate input variable reduced temperature effect, indicating the importance of adaptation. As expected, the input variables (irrigated land and fertilizer) had a significant positive impact on rice yield, indicating that adaptation plays a significant role in reducing the climate change impact on rice yields.

Table 4.1 Descriptive statistics for rice, maize and finger millet datasets

Variable	Mean	Std. Dev.	Min	Max
Rice crop				
Yield (kg/hectare)	2524.95	737.72	369.00	4249.00
GDD (thousands)	25.19	2.27	21.62	30.06
Rain (cm)	128.55	115.42	11.42	466.70
Irrigated land (%)	59.45	40.53	0.00	100.00
Fertilizer (kg/hectare)	123.09	62.32	15.12	334.31
Maize crop				
Yield (kg/hectare)	3039.28	664.13	1124.00	4945.00
GDD (thousands)	259.49	18.9	216.20	300.60
Rain (cm)	65.3	51.07	11.27	277.62
Irrigated land (%)	43.04	35.50	0.00	100.00
Fertilizer (kgs/hectare)	115.42	49.85	27.66	284.81
Finger millet crop				
Yield (kg/hectare)	1618.29	446.07	616.00	3069.00
GDD (thousands)	248.77	18.58	216.20	283.80
Rain (cm)	73.92	61.23	15.01	321.44
Fertilizer (kg/hectare)	123.12	63.18	24.11	388.74

Table 4.2 Fixed effect regression result of rice crop

Variable	Model 1	Model 2
GDD	−0.065* (0.036)	−0.051* (0.027)
Rainfall	0.0011** (0.0005)	0.0009** (0.0004)
Irrigation	–	0.007** (0.003)
Fertilizer	–	0.0009* (0.0005)
Linear trend	–	0.002 (0.004)
Constant	9.258*** (0.885)	8.355*** (0.588)
No. of observation	294	294
R^2	0.72	0.76

Note Dependent variable (Yield) is expressed in natural logarithm and all exogenous variables are expressed in their physical form. Values in parentheses include robust standard errors
*Significant at 10%; **Significant at 5%; ***Significant at 1%

Table 4.3 Fixed effect regression result of maize crop

Variable	Model 1	Model 2
GDD	−0.10*** (0.031)	−0.05** (0.026)
Rainfall	0.0007* (0.0004)	0.0013*** (0.0003)
Irrigation	–	0.005*** (0.001)
Fertilizer	–	0.002*** (0.0006)
Linear trend	–	−0.01*** (0.003)
Constant	10.67*** (0.84)	9.1*** (0.73)
No. of observation	210	210
R^2	0.37	0.22

Note Dependent variable (Yield) is expressed in natural logarithm and all exogenous variables are expressed in their physical form. Values in parentheses include robust standard errors
*Significant at 10%; **Significant at 5%; ***Significant at 1%

Table 4.4 Fixed effect regression result of finger millet crop

Variable	Model 1	Model 2
GDD	−0.18*** (0.029)	−0.15*** (0.031)
Rainfall	0.0008 (0.0008)	0.0006 (0.0008)
Fertilizer		0.0018*** (0.0003)
Linear trend	–	−0.006 (0.003)
Constant	11.87*** (0.78)	10.98*** (0.80)
No. of observation	210	210
R^2	0.43	0.46

Note Dependent variable (Yield) is expressed in natural logarithm and all exogenous variables are expressed in their physical form. Values in parentheses include robust standard errors
*Significant at 10%; **Significant at 5%; ***Significant at 1%

Table 4.3 includes the estimated coefficients and corresponding standard errors for maize yields. As mentioned earlier, a similar method of temperature construct and two different specifications have been modelled to assess the climate change effect on maize yields. The estimates of Model 1 showed a negative response of maize yield, indicating that each additional degree day above 34 °C declines the maize yield by 10%. This loss in maize yield is associated with heat stress that causes reduction in soil moisture, which in turn affects the cellular process of plants. Further, the effect of rainfall was modest (0.007), and was found to have significant positive effect on maize yields. The other non-climate input variables were added to the estimation to ensure the effect of climate variables in presence of suitable adaptation. The result varied with the addition of control variables in Model 2. Both irrigation and fertilizer were found to be statistically significant, indicating that increase in the application of fertilizer and the additional water supply enhances the maize yields significantly. However, with the addition of control variables in Model 2, the coefficient of temperature decreased from 10% to 5%, indicating that input variables minimized the effect of temperature on plant growth. The control variables had a significant impact on maize yields, as irrigation and fertilizer reduced the effect of temperature on maize yields almost by 50%. With suitable adaptation strategies, the effect of temperature rise on maize cultivation can be controlled to certain extent.

Table 4.4 presents the estimates of finger millet yield. Similar to Rice and Maize yields, two specifications were modelled to estimate the climate change impact on millet yields. Model 1 with climate variables and Model 2 with climate variables and input variables. A significant decline in the crop yield with response to temperature has been observed in millet yields. In Model 1, the coefficient of maximum temperature resulted in a negative coefficient of −0.18, indicating a decline in millet yields by 18% with 1 °C increase in day temperature. This huge negative impact on millet yield is associated with loss in the water moisture due to heat stress during the monsoon period. The estimated coefficient of rainfall was positive but insignificant in nature, which failed to represent a strong causal association with the crop yield. In Model 2, we add fertilizer and linear trend to find the significant impact on crop yield. The result indicates that input variables had a significant role in reducing the climate effect on millet yields. The effect of maximum temperature declined to 15% from 18%, with the addition of the input variables. Fertilizer had a significant positive impact with modest effect on millet yields. Linear trend resulted in negative response to millet yields representing a negative trend in terms of other inputs.

The result of the regression shows that temperature effect on finger millet cultivation is highly vulnerable. Finger millet is a drought tolerant crop, and is cultivated as rainfed cultivation. This could be one possible reason for the high negative effect of temperature stress on millet yields when compared to rice and maize yields. However, in case of drought and warmer climate, irrigation acts as a potential advantage to millets. In such cases, irrigation may further diminish the effect of heat stress on millet yields. Being a drought-prone crop, finger millet can withstand higher optimum temperature. The extent of yield loss in millet yields is higher when compared to previous crop estimates in Indian scenario. There may be several factors associated with this high extent of loss in millet yields. In modern agriculture, maximizing grain yield leads to increase in the cultivation of rice, maize and wheat; and that replaces more nutrient-rich crops such as millets, barley and sorghum. The transition of modern agriculture has led to wide adaptation practices to maximize yield. Some adaptation practices that focus on maximizing yield are—application of higher input of fertilizer and pesticides, irrigation and use of heat resistant seeds, etc. Though all these adaptation strategies are uniform for all crops, they are more focused on crops that earn higher yield but demand lower labour than on crops with nutritional and functional characteristics. Besides this progress towards adaptation, focus on maximizing the agriculture output during green revolution caused a significant decline in the paradigm of millet cultivation. The notion of the farmer's cultivation choice shifted based on the modified consumption pattern.

Out of Sample—Yield Forecasts

The above estimated results have confirmed the negative effect of temperature on the crop yields. Hence, we evaluate the prediction power of the models using out-of-sample forecast method. To validate the model, out of 21-year data, we train 18-year data for estimation, and validate with 3 years of yield prediction. Two models which have been estimated previously to gauge climate change impact on agriculture yield for all the three crops (see Tables 4.2, 4.3 and 4.4) have been considered for out-of-sample forecasts. Model 1 shows the predicted error from the estimation with only climate variables (temperature, and rainfall); Model 2 shows the same with both climate and control variables. Table 4.5 represents Root Mean Square Error (RMSE) was estimated for two models for each of the three crops. District-wise log yields have been predicted for the three years, and prediction errors—which are the difference between actual and predicted yield. For rice and Maize, the lowest RMSE out of two models was from Model 2 with 0.282 and 0.272 respectively. However, in case of finger millet Model 1 had lowest RMSE of 0.630. The adaptation in case of rice and maize is high when compared to other crops. Thus it becomes significant that we include all possible covariates while estimating and forecasting the climate change impact on yields. The importance of adaptation for rice and maize yield was evidenced in the model estimation and the better fit model was the Model 1, which included both input variables and climate variables. However, crops like millets are less focused in terms of adaptation, which are highly cultivated in marginal lands for self-consumption. Millets are drought crops which are cultivated in rainfed regions and farmers do not strongly follow adaptation procedure similar to

Table 4.5 Root Mean Square Error (RMSE) for three crops (Rice, Maize and Finger Millet)

Crop	Model 1	Model 2
Rice	0.421	0.282
Maize	0.301	0.272
Finger millet	0.630	0.664

Source Own calculation

other cash crops. This was evidenced from our estimation, where the model with only climate variables was better fit for future projections.

Impact of Climate Change on Rice and Maize Yields in Dry-Land Cultivation

Past studies have assessed the impact of climate change on crop yields. However, only a few focused on disaggregated assessment of climate impact on rainfed cultivation (Schlenker et al. 2007). The major constraint in the evaluation of climate risk on irrigated and rainfed cultivation is the availability of data. Irrigation acts as a supplement source for the plant growth during deficit rainfall, and acts as one of the major adaptations in modern era. The well-known fact is that irrigation offsets the effect of climate change to certain extent. However, the trade-off assessment between the irrigated and non-irrigated land would benefit to distinguish the potential effect of climate change. Results from the studies undertaken in Africa show that irrigated lands have higher revenues when compared to dry land (Kurukulasuriya et al. 2006). Though the aggregate results on rice and maize yields show a dominant effect of temperature and the potential impact of heat stress is reduced under optimal adaptation. To have a better insight into this, we consider the districts which are highly dependent on the rainfall (dryland cultivation) to estimate the climate change effect. Out of 14 districts under rice cultivation, 8 districts account for 90% of irrigated land whereas in other 6 districts, the average irrigated land is 15%. Similarly, out of 10 districts under maize cultivation, 6 districts were totally dependent on rainfall for cultivation. Table 4.6 presents the results of climate impact on rice and maize yield under dryland cultivation. Rice yield exhibited higher negative effect of temperature on rice yield, indicating that each additional degree day above 34 °C reduces the final yield by 18% under dry land cultivation. The negative effect of temperature on dry land cultivation is highly devastating when compared to cultivation of rice under irrigation. There is a limited less evidence in understanding the mechanism of temperature effect on dry land cultivation using regression framework. However, the estimates from the crop simulation models show the temperature effect to be modest. Contrastingly, historical dataset from the present study shows large effect of temperature on dry land cultivation. Similarly, for maize yields the negative effect of temperature resulted in 13% (−0.13) decline in maize yield with each additional degree day above 34 °C.

Table 4.6 Fixed effect regression result of rice and maize yields in dry-land cultivation

Variable	Rice yield	Maize yield
GDD	−0.18** (0.070)	−0.13** (0.046)
Rainfall	0.0008 (0.0005)	0.001* (0.0007)
Linear trend	0.009 (0.003)	−0.010 (0.005)
Constant	11.81*** (1.69)	11.38*** (1.19)
Observation	126	126
R^2	0.62	0.32

Note Dependent variable (yield) is expressed in natural logarithm and all exogenous variables are expressed in their physical form. Values in parentheses include robust standard errors
*Significant at 10%; **Significant at 5%; ***Significant at 1%

The impact of climate change varies with the region and crops yield. The present study which focused particularly on rice, maize and millet yields in dryland cultivation resulted with 18%, 15% and 18% yield loss, which is much higher than previous estimations in Indian scenario when considered all the crops and sub divisions. In Indian context nearly 5 to 10% loss can be noted from the previous estimates, while majority of these studies focus primarily on large geography. The loss among crop yield not only varies with respect to climate factors but also with the crop phenology, effectiveness of adaptation and soil conditions. This extent of loss could be associated with several other factors. In modern agriculture, adaptation practices play a key role. Several adaptation practices are followed to maximize yield by higher inputs of fertilizer, water harvesting and use of heat resistant seeds. Though all these adaptation strategies are uniform for all crops but are limited to major crops. Irrigation is the major adaptation followed to mitigate climate change, however, in the absence of irrigation the negative effect is higher. The surprising result is finger millet being the drought crop, which is especially cultivated in dryland was resulted with higher negative effect. One can note that the cultivation of finger millet has been gradually decreased over a period of time due to manifest in altered consumption and demand level of consumers. Significant growth in rice and maize cultivation has resulted in a shift in land use pattern. This shift in the land use pattern is also accompanied by more fertile land being allocated to other cash crop cultivation. The cultivation of millet yields in the marginal land could be possible reason for large damage in the millet yield with response to rise in temperature.

4.5 Conclusion

We have assessed empirically the impact of climate variability on crop yield using panel data. A time series of 21 years data on three crop yields (rice, maize and finger millet) at district level has been used to estimate the impact of climate change on agriculture. The study extensively focused on the Kharif season, which is the important cropping season in the state of Karnataka. The negative impact of temperature can be highly noted from the estimation, while the rainfall has significant positive impact on the crop yields. The result indicated that the temperature affected the crop yield significantly: 9% loss in case of rice, 5% loss in case of maize and nearly 15% loss in case of finger millet was noted from the above estimation. Though the loss associated with millet yield is high when compared to that with other crops. The disaggregated result from the dryland cultivation of rice and maize also shown a significantly higher impact. The loss in rice and maize cultivation in dryland was nearly 18% and 13%. These results reveal that rice cultivation has a significant negative impact in the absence of irrigation. Though the damage resulted is due to increase in the frequency of extreme temperature, while there could be several other factors associated with this extent of loss in crop yields.

Irrigation plays a major role in the rice and maize cultivation, decrease in the kharif rainfall will have additional impact on the crop yields. Further, higher utilization of the irrigation in kharif season may significantly impact on the followed up seasons, which would cause water scarcity during summer season. Additionally, one can note the negligible cultivation of finger millet under irrigated land. This could be the possible reason for the higher negative effect of temperature stress on millet yields when compared to rice and maize yields. Further, the government incentives also play a major role in farmer's choice of crop, the revenue and institutional support is high for rice and other cash crop cultivation. The other possible reasons could be shift in consumption patterns, which has led to significant shift in most of fertile land to other cash crops and decline in the area under millet. This research acts as a stepping stone for further in-depth study on finger millet crop. Finger millets being a sustainable crop in dry conditions should be promoted as a possible alternative adaptation strategy in the long run to reduce the dependency on irrigation. Irrigation capacity is high among rice-growing districts in south Karnataka. However, the shift towards economically profitable crops like rice has led to unsustainable use of groundwater and thereby worsen the water scarcity situation in future. In present situation, government and non-governmental organizations encourage rice and wheat cropping system. This paradigm is resulted with utilization of irrigation to few main crops, where the water requirement for crops like rice is much higher than millets. The adoption of irrigation management to millet production may intensify higher productivity and reduce the impact of climate change on millet yields. The government support in cereal cultivation to mitigate the risk involved in climate change will also benefit equity towards environment.

REFERENCES

Arshad M, Amjath-Babu TS, Krupnik TJ, Aravindakshan S, Abbas A, Kachele H, Müller K (2017) Climate variability and yield risk in South Asia's rice–wheat systems: Emerging evidence from Pakistan. Paddy and Water Environment 15(2): 249–261

Auffhammer M, Ramanathan V, Vincent JR (2006) Integrated model shows that atmospheric brown clouds and greenhouse gases have reduced rice harvests in India. Proceedings of the National Academy of Sciences of the United States of America 103(52): 19668–19672

Auffhammer M, Ramanathan V, Vincent JR (2012) Climate change the monsoon and rice yield in India. Climatic Change 111(2): 411–424

Bandara JS, Cai Y (2014) The impact of climate change on food crop productivity, food prices and food security in South Asia. Economic Analysis and Policy 44(4): 451–465

Bhende MJ (2013) Agricultural profile of Karnataka state. Agricultural Development and Rural Transformation. Centre Institute for Social and Economic Change, Bangalore.

Blanc E, Schlenker W (2017) The use of panel models in assessments of climate impacts on agriculture. Review of Environmental Economics and Policy 11: 258–279

Carleton TA (2017) Crop-damaging temperatures increase suicide rates in India. Proceedings of The National Academy of Sciences of The United States of America 114(33): 201701354.

D'Agostino AL, Schlenker W (2016) Recent weather fluctuations and agricultural yields: implications for climate change, Agricultural Economics (United Kingdom) 47: 159–171

Guhathakurta P, Rajeevan M, Sikka DR, Tyagi A (2015) Observed changes in south-west monsoon rainfall over India during 1901–2011. International Journal of Climatology 35: 1881–1898

Gupta R, Somanathan E, Dey S (2017) Global warming and local air pollution have reduced wheat yields in India. Climatic Change 140(3): 593–604

Jena PR, Majhi B, Kalli R, Majhi R (2022) Prediction of crop yield using climate variables in the south-western province of India: a functional artificial neural network modeling (FLANN) approach. Environment Development and Sustainability 1–24

Kalli R, Jena PR (2020) Impact of climate change on crop yields: Evidence from irrigated and dry land cultivation in semi-arid region of India. Journal of Environmental Accounting and Management 8(1): 19–30

Kalli R, Jena PR (2021) Combining agriculture, social and climate indicators to classify vulnerable regions in the Indian semi-arid regions. Journal of Water and Climate Change 13(2): 542–556

Kalli R, Jena PR (2022) How large is the farm income loss due to climate change? Evidence from India. China Agricultural Economic Review 14(2): 331–348

Krishna Kumar K, Rupa Kumar K, Ashrit RG, Deshpande N R, Hansen JW (2004) Climate impacts on Indian agriculture. International Journal of Climatology 24(11): 1375–1393

Kurukulasuriya P, Mendelsohn R, Hassan R, Benhin J, Deressa T, Diop M, Eid HM, Fosu KY, Gbetibouo G, Jain S, Mahamadou A, Mano R, Kabubo-Mariara J, El-Marsafawy S, Molua E, Ouda S, Ouedraogo M, S ene I, Maddison D, Seo SN,

Dinar A. (2006) Will African agriculture survive climate change? The World Bank Economic Review 20(3): 367–388

Lobell DB, Burke MB (2010) On the use of statistical models to predict crop yield responses to climate change. Agricultural and Forest Meteorology 150(11): 1443–1452

Luo Q (2011) Temperature thresholds and crop production: A review. Climatic Change 109(3–4), 583–598

Mall RK, Singh R, Gupta A, Srinivasan G, Rathore LS (2006) Impact of climate change on Indian agriculture: A review. Clim. Change 78(2–4): 445–478.

Pai DS, Sridhar L, Badwaik MR, Rajeevan M (2014) Analysis of the daily rainfall events over India using a new long period (1901-2010) high resolution (025° × 025°) gridded rainfall data set. Climate Dynamics 45(3–4): 755–776

Pattanayak A, Kumar KSK (2014) Weather sensitivity of rice yield: Evidence from India. Climate Change Economics 05(04): 1450011

Poudel S, Kotani K. (2013) Climatic impacts on crop yield and its variability in Nepal: Do they vary across seasons and altitudes? Climatic Change 116(2): 327–355

Revadekar JV, Kothawale DR, Patwardhan SK, Pant GB, Rupa Kumar K (2012) about the observed and future changes in temperature extremes over India. Natural Hazards 60(3): 1133–1155

Schlenker, W., Hanemann, WM., and Fisher, AC (2007) Water availability degree days and the potential impact of climate change on irrigated agriculture in California. Climatic Change 81(1): 19–38

Srivastava A, Rajeevan M, Kshirsagar S (2009) Development of a high resolution daily gridded temperature data set (1969–2005) for the Indian region. Atmospheric Science Letters 10: 249–254

Varadan RJ, Kumar P, Jha G K, Pal S, Singh R (2015) An exploratory study on occurrence and impact of climate change on agriculture in Tamil Nadu, India. Theoretical and Applied Climatology 1–18.

Wang P, Zhang Z, Song X, Chen Y, Wei X, Shi, P, Tao F (2014) Temperature variations and rice yields in China: Historical contributions and future trends. Climatic Change 124(4): 777–789

Zhou Y, Li N, Dong G, Wu W (2013) Impact assessment of recent climate change on rice yields in the Heilongjiang reclamation area of north-east Chin. Journal of The Science of Food and Agriculture 93(11) 2698–2706

CHAPTER 5

Impact of Climatic Variations on Crop Yield in Chamoli District, Uttarakhand, India

Sunil Nautiyal, Satya Prakash, Mrinalini Goswami, and A. Premkumar

5.1 INTRODUCTION

Climate change is a long-term temporal change phenomenon that has negative impact on socio and ecological systems. Fast-developing countries and growth in extensive use of technology are highly vulnerable to climate change. The climate change impacts are realized in terms of changes in social and ecological sub-systems, affecting food availability, life and livelihoods of the population. Climate change occurs when there is a significant change in maximum temperature or minimum temperature change, annual precipitation changes, irregular pattern of rainfall and intense heavy rainfall events over a period of time. The rise in temperature, variations in rainfall, decreasing groundwater, flooding due to heavy rainfall, drought, soil erosion, sea level rise, melting of glacier, cyclone and fog are strong indications of climate change phenomenon. The adverse impacts of climate change on agriculture are global concerns. In developing economies, poor communities are mostly reliant on climate-sensitive activities including agriculture for their livelihoods (Kyei-Mensah et al. 2019). It has been widely observed that countries with weaker economies have negative

---

S. Nautiyal (✉) · S. Prakash · M. Goswami
Centre for Ecological Economics and Natural Resources, Institute for Social and Economic Change, Nagarabhavi, Bengaluru, India
e-mail: sunil@isec.ac.in

A. Premkumar
Department of Economics, Sacred Heart College (Autonomous), Tirupattur 635 601, Tamil Nadu, India

© The Author(s), under exclusive license to Springer Nature Singapore Pte Ltd. 2023
S. Nautiyal et al. (eds.), *Palgrave Handbook of Socio-ecological Resilience in the Face of Climate Change*,
https://doi.org/10.1007/978-981-99-2206-2_5

impacts of climate change on agricultural production (Praveen and Sharma 2020). The reason behind these consequences are lack of advancement of technology and resources to mitigate the adverse effects of climate change on agriculture and also due to the high share of population dependent on agriculture for their livelihood (Nath and Behera 2011). Hence, these consequences increase the severity of disparities in cereal yields between developed and developing countries (Fischer et al. 2005).

Climate change can adversely affect the food security through agriculture production. Food security has four dimensions viz.—food availability, food accessibility, food utilization and food stability. This impact is also visualized on human health, livelihood assets, food production and distribution channels (FAO 2008). A range of studies found that climate change had caused adverse effect on declining rate of agricultural productivity or crop yield (Lal and Bruce 1999). Studies on growth and yield responses of soybean in Madhya Pradesh, India to climate variability and change, used the CROPGRO model to analyse the impact of thermal and moisture stress with assessing the observed climatic elements on the extent of losses in growth and yield of soyabean crop. Nageswararao et al. (2018) have studied the impact of climatic variations on *rabi* crops over Northwest India. They found that the productivity of different *rabi* crops is mostly influenced by variations in local temperatures. Kumar et al. (2011) have analysed the decline in irrigated land for maize, wheat and mustard in North Eastern and coastal regions, and rice, sorghum and maize in Western Ghats, which might have impacts of climate change in terms of productivity loss. In a study in Tamil Nadu, findings show the decreasing rate of rice productivity up to 41% due to increase in temperature up to 40 °C (Geethalakshmi et al. 2011). Research on yield change due to climate change for two main cereals crops i.e., rice and wheat using crop simulation model, indicates the significant reduction in crop yields (Kumar and Parikh 2001). The literature reviewed provides an overview of negative impacts of climate change on crop production in India. The key highlights of the review show that most of the crops in India are affected by heavy rainfall and increase in temperature. There are some additional factors which affect food security, such as soil quality, technology, loss of farmland, pests and diseases, water stress, poverty. Climate change, agriculture productivity, food security and poverty are directly linked with each other (Hollaender 2010).

The present study focused to understand the effects of variations in climatic parameters on crop yields using various statistical models. The climatic parameters were analysed from 1980 to 2019. Further, the relationships between crop yields and climatic parameters were analysed using regression model for climate and crop yield anomalies.

5.2 Study Area

Chamoli district lies in the central Himalaya and bounded by Uttarkashi in northwest, Rudraprayag in west, Pithorgarh and Bageshwar in east and Almora in south. It is the second largest district of Uttarakhand state. Geographically the district is bounded by latitude N 29°15′00″ and 31°00′00″ and longitude E 79°15′00″ and 80°00′00″ (Fig. 5.1). Alaknanda and Pinder are the major rivers of Chamoli district. The geographical area of the district is 7604 sqkm. The district is divided into 10 tehsils namely Joshimath, Chamoli, Karnaprayag, pokhari, Garsain, Tharali, Narayanbagar, Jilasu, Adibadri and Ghat. According to the Census 2011 data, the total population of Chamoli district is 3,91,605.

The district has diverse and distinct ecological and geographical features. The economy of the district is mainly dependent on horticulture, agriculture and animal husbandry. The major agriculture crop produced in this district are Paddy (Rice), Wheat, Finger millet (Ragi), Barley, Mustard, Pigeon pea (Arhar), Black gram split (Urd) and soyabean, etc., and major fruits cultivated such as Apple, Citrus, Walnut, Peach and Pear. Cattle, buffaloes, goat and sheep are major livestock of Chamoli district.

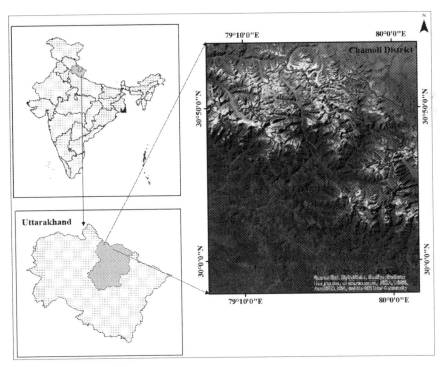

Fig. 5.1 Study area map of the Chamoli district, Karnataka, India

Data Used in Present Study

The climate data, specifically precipitation, maximum temperature and minimum temperature data were collected from Indian Meteorological Department (IMD) from 1981 to 2019. The rainfall data has been collected at 0.25° × 0.25° of spatial resolution whereas maximum and minimum temperature are available at 1.0° × 1.0° of resolution. Moreover, the crop yield data were collected from Area and Production statistics, Ministry of Agriculture and Farmers Welfare (https://aps.dac.gov.in/APY/Public_Report1.aspx) for paddy, wheat, small millets and finger millet (ragi) crop (Table 5.1).

5.3 METHODOLOGY

The time series data for selected variables were taken into consideration for analysis. Primarily, the trend analysis was conducted using MAKESENS tool for all the considered variables and examined using nonparametric trend test methods. MAKESENS tool is used for estimating trends in the time series data with annual values. The Mann–Kendall trend test for trend and Sen's slope estimate were used to analyse the trend test, Approach is nonparametric rank-based approach and robust against the influence of extremes. Mann–Kendall test with a 99% confidence limit was used as a monotonic trend test. In the testing process, the null hypothesis (H_0) is no trend; then, the alternate hypothesis (H_1) is that there is a trend from which dataset is drawn. The H_0 will be rejected if $p \leq 0.1$. This approach was used for non-normally distributed time series data. Trend analysis is evaluated for all climatic variable's rainfall, maximum and minimum temperature and crop yield on crop seasonal basis.

Hence, the trend analysis was examined by using Sen's slope method. To analyse the change per year the Sen's nonparametric method is used. The slope can be estimated using the equation:

$$Q_i = \frac{x_j - x_k}{j - k} \qquad (5.1)$$

where $j > k$

Since x_j and x_k are data values, Sen's estimator of slope is the median of these N values of Q_i. If N is odd then Sen's estimator is calculated as $Q_{med} = Q(N+1)/2$, if N is even then Sen's estimator is calculated as $Q_{med} =$

Table 5.1 Data used in the present study

Data used	Resolution	Year	Source
Temperature	1°	1981–2019	http://www.imdpune.gov.in/
Rainfall	0.25°	1981–2019	http://www.imdpune.gov.in/
Crop yield	–	1997–2019	https://aps.dac.gov.in/APY/Public_Report1.aspx

[QN/2 + Q(N + 2)/2]/2. Hence Q_{med} is tested at $100(1-\alpha)$ % by two-sided confidence interval and the slope estimate is obtained by nonparametric approach based on normal distribution.

Further, the time series econometric analysis is used to identify the effects of climate change on crops yield in Chamoli district. In the first stage, the unit root test is used to check the level of stationarity in the time series. Testing of stationarity is pioneered by Dickey and Fuller (Dickey and Fuller 1979; Fuller 1985). Therefore, Augmented Dickey-Fuller (ADF) and Phillips Perron (PP) tests were used to check the stationarity. The null hypothesis states that no long-run relationship exists among the variables. Hence, the bound test was used to determine a long-run relationship in the second stage. According to the value of F-statistics, in the first case, we accept the null hypothesis when there is no long-run relationship (if the value is lower than I (0)). In the second case, we reject the null hypothesis when there is a long-run relationship (if the value is greater than I(1)). The Granger Causality test is a technique used to check the usefulness of time series in forecasting i.e., if the variable "X" is found to be helpful for predicting variable "Y" then it can be inferred as "X" Granger-cause "Y" (Granger 1969). In the third stage, the Granger Causality test is used to examine the direction of causality among the variables.

5.3.1 Relationship Between Climate and Crop Yield

Analysing the relationship between annual crop yield and climatic parameters, correlation coefficient and multivariate regression have been also used. The analysis was done as per seasonal basis of crop months. Pearson's correlation coefficient measures the strength of a relationship; here, between crop yield and climatic factors. By plugging their respective values into Eq. 5.2, we were able to calculate the correlation coefficient between the two variables (x and y); where, x is independent variable and y is dependent variable.

$$x = \frac{\Sigma(x-x)(y-y)}{\sqrt{\Sigma(x-x)^2(y-y)^2}} \tag{5.2}$$

In addition, multivariate regression model was used to perform the impact of climate change on crop yield. The anomalies of crop yield and climatic components are used to estimate the quantitative relationships between climate change and crop yield. The anomalies of both the parameters were developed to determine the crop yield change due to the impact of climatic influences. Hence the relationships can be expressed as follows:

$$\Delta Y = \text{constant} + (\alpha \times \Delta \text{Ppt}) + (\beta \times \Delta \text{Temp}_{min}) + (\Upsilon \times \Delta \text{Temp}_{max}) \tag{5.3}$$

Thus, ΔY represents the yield change owing to maximum *and* minimum temperature, and rainfall. α, β and Υ represents the coefficients of maximum *and* minimum temperature, and rainfall. ΔPpt represents the observed changes in precipitation, $\Delta Temp_{min}$ and $\Delta Temp_{max}$ represent the observed changes in maximum and minimum temperature.

5.4 RESULT AND DISCUSSIONS

Variations in Climatic Parameters

The graphs were plotted in MATLAB software for climatic parameters, viz.— precipitation, maximum temperature and minimum temperature (Figs. 5.2, 5.3 and 5.4). The temporal change of climatic pattern was observed from 1980 to 2019. Based on the climatic pattern of Chamoli district, year has been divided into four seasons viz. monsoon (June, July, August and September), post-monsoon (October), winters (November, December, January and February) and pre-monsoon (March, April and May). The seasons have been classified for better understanding of climate change over last 39 years. The contour plots were prepared for presenting changes in climatic pattern. Average rainfall ranges from 0 to 1800 mm. Mostly the changes are observed during the monsoon period with fluctuations of rainfall in June, July, August, September (JJAS). The change shows an increase in rainfall during monsoon season after 1998. Further, increase in rainfall during monsoon period was observed from 2010 to 2019. Observation reveals a positive indication of increase in rainfall from 1980 to 2019; however, a decrease was seen in rainfall after 2004 during winter season. High rainfall was observed in the month of June, July and August in 2013 whereas low rainfall was seen in October, November and December. Maximum temperature ranges from 0 to 25 °C; whereas, highest minimum temperature is 12 °C. The maximum temperature shows a meagre decrease in the month of June over the years. The maximum temperature has been slightly decreasing during June month over the years. The results show that June month is hottest month in Chamoli district.

Trend Analysis

The trend analysis was done using Man-Kendall and Sen's Slope to understand the seasonal and annual trend for rainfall, maximum and minimum temperature. The seasonal data were taken according to the crop months period viz., summer, *kharif* and *rabi*. Summer months are May and June, *kharif* months are July to October and *rabi* months are November to April.

Precipitation Trend

The trend was analysed from 1980 to 2019 using Mann–Kendall and Sen's slope tests. The results expressed that the precipitation is increasing over the years. In *kharif* season, the rainfall is increasing by 8.16 mm/year over the years; in *rabi* season, it is increased by 1.3 mm/year; and in summer season, it is increased by 3.5 mm/year. Whereas, annual increase in rainfall is by

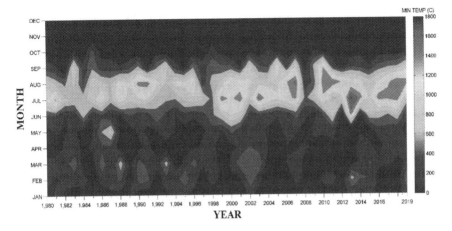

Fig. 5.2 Rainfall variation from 1980 to 2019

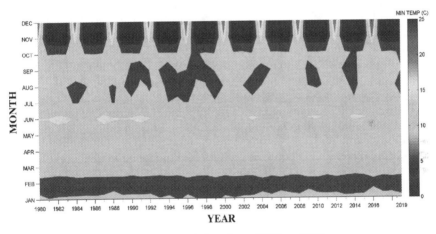

Fig. 5.3 Maximum temperature variation from 1980 to 2019

2.8 mm/year over the studied years (Table 5.2). The analysis shows a positive result of increase in rainfall over the time period.

Temperature Trend
Similarly, the temperature trend was also analysed using Mann–Kendall and Sen's slope tests; where, low fluctuations were observed in maximum and minimum temperature. In maximum temperature, the results clearly show that the only 0.04 °C was increased in summer season, 0.02 °C increased in *kharif* season month, 0.01 °C increased in *rabi* season over the studied years; whereas, annual increase is 0.02 °C. With respect to minimum temperature, the results show an increase by 0.03 °C in summer season, an increase by

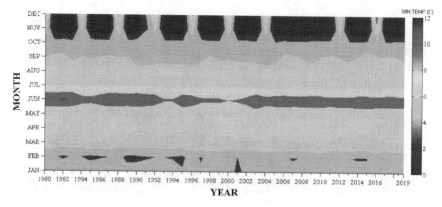

Fig. 5.4 Minimum temperature variation from 1980 to 2019

Table 5.2 Sen's slope values for seasonal and annual precipitation

Time series	Test Z	Signific.	Q	Qmin99	Qmax99
Summer	0.394771	NS	3.5	−22.8548	22.821591
Kharif	1.63548	NS	8.166667	−10.55954	31.753881
Rabi	0.507563	NS	1.333333	−6.644112	9.7255021
Annual	1.240709	NS	2.875	−5.512549	12.153996

Note NS—No Significant

Table 5.3 Sen's slope values for seasonal and maximum temperature

Time series	Test Z	Signific.	Q	Qmin99	Qmax99
Summer	2.71	**	0.044	0.003	0.082
Kharif	2.57	*	0.025	0.000	0.044
Rabi	0.93	NS	0.011	−0.033	0.064
Annual	2.03	*	0.025	−0.003	0.062

Note **, * represents 5% and 10% level of significance. NS—No Significant

0.006 °C in *kharif* season, an increase by 0.001 °C *rabi* season; whereas the annual increase is 0.01 °C per year over the studied years. Hence, it is clearly evident that the low changes were observed during 1980–2019 (Tables 5.3 and 5.4).

Table 5.4 Sen's slope values for seasonal and minimum temperature

Time series	Test Z	Signific.	Q	Qmin99	Qmax99
Summer	3.244052	**	0.03	0.01	0.047659
Kharif	1.551503	NS	0.006964	−0.00465	0.018979
Rabi	0.084627	NS	0.001667	−0.01698	0.0309
Annual	2.002849	*	0.011458	−0.00339	0.031543

Note **, * represents 5% and 10% level of significance. NS—No Significant

Table 5.5 Sen's slope values for *kharif* crops

Time series	Test Z	Signific.	Q	Qmin99	Qmax99
Paddy	1.466292	NS	4.516478	−4.912764	14.437273
Finger millet (ragi)	0.332166	NS	1.955821	−10.70542	13.090381
Small millet	1.215608	NS	19.34665	−17.52762	37.425369
Soyabean	2.59421	**	21.96742	0.0508407	37.69861

Note ** represents 5% level of significance. NS—No Significant

Table 5.6 Sen's slope values for *rabi* crops

Time series	Test Z	Signific.	Q	Qmin99	Qmax99
Wheat	1.579084	NS	8.351222	−8.823149	33.230465

Note NS—No Significant

Crop Yield Trend Analysis

The trend was analysed as per the seasonal crop months period i.e., *kharif* and *rabi* crop months period. The trend analysis shows increase in productivities; where, soyabean with as value 21.96 kg/ha, paddy by 4.51 kg/ha, small millets by 19.34 kg/ha and finger millet (ragi) productivity by 1.95 kg/ha. Significant increase in production of Soyabean and Small millets has been observed over the years. In *rabi* season, wheat is increased by 8.35 kg/ha over the years. In both the season viz., *kharif* and *rabi*, the crop yields are gradually increasing over the time period (Tables 5.5 and 5.6).

Crop Yield Relationship

The relationship was established using correlation coefficient between variables—crop yield and climate variability. The correlation coefficient was analysed with climatic parameters i.e., precipitation, maximum and minimum temperature. The results indicate that a moderate positive association was

Table 5.7 Correlation of crop yield (Paddy, Finger millet (ragi), Small millets, Soyabean and Wheat) with climatic variables in different crop seasons

Season	Crop	Correlation coefficient (r)		
		Ppt	Tmax	Tmin
Kharif	Paddy	0.30	0.15	0.16
	Finger millet (ragi)	0.03	0.03	0.008
	Small millets	0.12	0.22	0.20
	Soyabean	0.2	0.02	0.05
Rabi	Wheat	0.47	0.14	0.06

Note Tmax—Maximum Temperature; Tmin—Minimum Temperature

observed for rainfall with wheat in *rabi* season ($r = 0.47$); whereas in *kharif* season, low positive ($r = 0.3$) correlation was observed for paddy crops with precipitation. Finger millet (ragi), Small millets and soyabean are not found to have a good correlation with precipitation and temperature (Table 5.7).

Change in Yield Due to Climate Change

To examine whether the parameters (climatic variables and crop yields) are in direct relationship or not between, a multi-linear regression analysis between crop yield anomalies and climatic parameters anomalies was performed. The anomalies of each climatic parameter and crop yield were computed by analysing the difference in values from one year to the next. The crop yield anomalies for *kharif* season and *rabi* season as seen in Figs. 5.5 and 5.6, Table 5.8.

The results indicate that the model is able to describe the variations in crop yield through the regression analysis; crops ranging from 37% (0.37) in case of wheat during *rabi* season crops, 22% (0.22) in case of soyabean. Moreover, the sign of coefficients indicates the direction of change in yield versus climate variations. In case of wheat, climatic variables account for only 37% of the yield change, whereas 63% of the variations in wheat yield is explained by other influencing factors which perhaps include better crop management practices and introduction of new agro-technology. The R^2 value is 0.22 for soyabean, which represents that the impact of climatic influences accounts only 22% of the soyabean yield; whereas, 78% of the variations are explained by other influencing factors.

Time Series Analysis

The present section used time series analysis for the selected variables to determine the factors influencing the yield of crops in the study area. The

5 IMPACT OF CLIMATIC VARIATIONS ON CROP YIELD ... 63

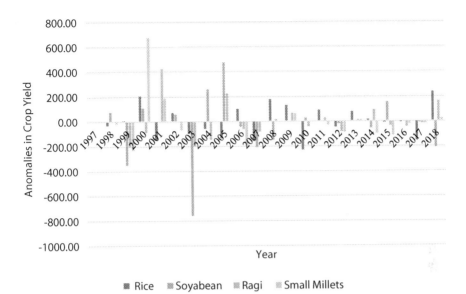

Fig. 5.5 Anomalies change in *Kkharif* crops for Paddy, Soyabean, Finger millet (ragi) and small millets crops

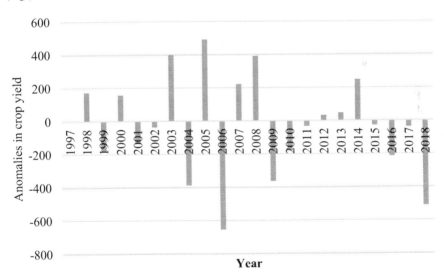

Fig. 5.6 Anomalies change in *Kharif* crops for Wheat crops

Table 5.8 Multivariate regression analysis of detrended crop yields

Season	Crop		Ppt_K	Tmax_K	Tmin_K	R^2	R
Kharif	Paddy	Coeff	0.166	−27.07	11.61	0.16	0.4
		p-value	0.128	0.90	0.26		
	Soyabean	Coeff	−0.152	−298.81	−110.55	0.22	0.47
		p-value	0.42	0.08*	0.80		
	Finger millet (ragi)	Coeff	0.06	56.74	−145.42	0.17	0.02
		p-value	0.65	0.62	0.63		
	Small Millets	Coeff	−0.34	−183.15	−39.92	0.11	0.34
		p-value	0.37	0.33	0.93		
Rabi	Wheat	Coeff	1.38	−41.89	124.15	0.37	0.60
		p-value	0.06*	0.10	0.09*		

Note * represents 10% level of significance
Tmax—Maximum Temperature; Tmin—Minimum Temperature

dependent variables are paddy, *kharif*-finger millet (ragi), *kharif*-small millet, *rabi*-wheat and the independent variables are rainfall and average temperature. Further, the variables are transformed into logarithm (ln) form to reduce the high values (Table 5.9).

In initial stage, the unit root test was conducted for the selected variables i.e., ln (paddy), ln (*kharif*-finger millet (ragi)), ln (*kharif*-small millet), ln (*rabi*-wheat), ln (rainfall) and ln (avg. temp.) to check the stationarity level by using Augmented Dickey-Fuller (ADF) test and Phillip-Perron (PP) test. The stationarity exists for all the variables in PP test on level and difference at 1% level of significance (with and without trend). As far as ADF test is concerned, the variables are stationary at 1% level in level (without trend) except the other variables: ln (paddy), ln (*kharif*-finger millet (ragi)) and ln (avg. temp.). Hence, the results of unit root test suggest to conduct Bound test and Granger Causality test for the study.

In time series analysis, bound test is used to determine the cointegration among the variables in order to avoid the spurious results in the model are presented in Table 5.10. The computed F-statistics was found to be more than upper bound critical value at 1, 2.5, 5 and 10% level of significance which rejects the null hypothesis. This confirms that there is existence of strong long run relationship between the variables in the model except *kharif*-small millet where F-statistics was found to be lesser than upper bound critical value. After establishment of long run relationship there should be causal relation among variables are expected to be at least in one direction.

The Granger Causality test is the statistical hypothesis used to identify the causal relationship between the variables. The result for Granger Causality Wald test is established in Table 5.11. The result shows that there are some unidirectional relationships among the variables in our model. Δ ln (paddy),

Table 5.9 Unit root test

Variables	ADF test				PP test			
	Level		1st Difference		Level		1st Difference	
	Without trend	With trend	Without trend	With trend	Without trend	With trend	Without trend	With trend
ln (paddy)	−2.47*	−0.95	−2.60	−3.62**	−4.43***	−5.02***	−11.12***	−12.82***
ln (kharif-finger millet (ragi))	−1.68	−2.37	−3.21**	−3.08	−5.28***	−4.99***	−6.56***	−6.06***
ln (kharif-small millet)	−3.97***	−3.89**	−3.46***	−3.86**	−4.24***	−4.86***	−10.14***	−9.95***
ln (rabi-wheat)	−4.24***	−4.86***	−3.97***	−3.86**	−4.24***	−4.86***	−10.14***	−9.95***
ln (rainfall)	−4.60***	−4.50***	−2.97**	−3.07	−4.60***	−4.50***	−9.22***	−8.86***
ln (avg. temp)	−1.36	−2.54	−3.20**	−3.06	−7.95***	−8.14***	−17.35***	−18.19***

***, **, * represents 1%, 5% and 10% level of significance

Table 5.10 Bound test

Variables	Critical Bound	Significance Level (%)				F-statistics
		10	5	2.5	1	
ln (paddy)	I (0)	3.17	3.79	4.41	5.15	17.870
	I (1)	4.14	4.85	5.52	6.36	
ln (kharif-small millet)	I (0)	3.17	3.79	4.41	5.15	1.945
	I (1)	4.41	4.85	5.52	6.36	
ln (kharif-finger millet (ragi))	I (0)	3.17	3.79	4.41	5.15	17.716
	I (1)	4.41	4.85	5.52	6.36	
ln (rabi-wheat)	I (0)	3.17	3.79	4.41	5.15	13.758
	I (1)	4.14	4.85	5.52	6.36	

Null hypothesis: No long run relationship
Note I (0) and I (1) represent lower and upper bound values

Table 5.11 Granger causality test

Dependent variable	Null hypothesis	Chi-Square	Probability	Status
Δ ln (paddy)	Δ ln (rainfall) does not cause Δ ln (paddy)	7.8391**	0.049	Reject
	Δ ln (avg. temp) does not cause Δ ln (paddy)	5.8305	0.120	Accept
Δ ln (kharif-small millet)	Δ ln (rainfall) does not cause Δ ln (kharif-small millet)	9.5128***	0.009	Reject
	Δ ln (average temp) does not cause Δ ln (kharif-small millet)	0.49118	0.782	Accept
Δ ln (kharif-finger millet (ragi))	Δ ln (rainfall) does not cause Δ ln (kharif-finger millet (ragi))	18.102***	0.000	Reject
	Δ ln (avg. temp) does not cause Δ ln (kharif-finger millet (ragi))	8.4328**	0.038	Reject
Δ ln (rabi-wheat)	Δ ln (rainfall) does not cause Δ ln (rabi-wheat)	3.3061*	0.069	Accept
	Δ ln (avg. temp) does not cause Δ ln (rabi-wheat)	1.7838	0.182	Accept

Note Δ refers to differentiation
***, **, * represents 1%, 5% and 10% level of significance

Δ ln (*kharif*-small millet), Δ ln (*kharif*-finger millet (ragi)) has a unidirectional causality to Δ ln (rainfall) that means when rainfall increases yield of paddy (5% level of significance), *kharif*-small millet (1% level of significance) and *kharif*-finger millet (ragi) (1% level of significance) also increases, but not vice versa. In case of *rabi*-wheat, rainfall and average temperature do not cause Granger Causality.

5.5 Discussion and Conclusion

Climate change has a direct impact on food systems and food security, and the need to counteract it may raise competition for agricultural resources. (FAO 2018; Mbow et al. 2019). It has been envisaged that global crop production will be reduced by more than 10% by 2050 with the potential to significantly affect global food production (Tai et al. 2014). Moreover, the increase in temperature also deteriorates grain quality and nutritional value (Kumar et al. 2016). The present research is focused on understanding the impact of climate change on crop productivity in the Chamoli district over the past 39 years. Precipitation has increased annually as well as seasonally over the past years. The maximum and minimum temperatures have also shown an increasing trend over the study period (0.025 °C & 0.011 °C annually) which is in line with other studies. Nautiyal et al. (2022) found that the minimum temperature has been rising at a rate of 0.68 °C over the study area. The crop yield variations in the study region were found to have increased over the time period. In the *kharif* season, yields of all the crops have been increasing while soybean crop yield has tremendously increased over the study area (21.96 Kg/ha). Agriculture of Uttarakhand is mostly rainfed; therefore, cultivation is concentrated in the *kharif* season (Sati and Bandooni 2018). Hence, the rate of crop productivity is higher in the *kharif* season. The analysis of crop yields with respect to climate parameters shows that crops are not significantly affected by changing precipitation, maximum and minimum temperature. Only the wheat crop has shown a positive and moderate correlation with precipitation in the *rabi* season ($r = 0.47$). Multivariate regression results are also not showing the high influence of climatic variations on various crops. Moreover, the Granger Causality test between crop yield and climatic parameters reveals a unidirectional relationship among the variables. A unidirectional relationship exists between rainfall and yield of paddy in *kharif* season. While small millet and finger millet (ragi) crops also show that the increase in yield is due to the increase in rainfall. Therefore, the development of irrigation system along with improved agronomic management practices, appropriate use of fertilizers, pest and weed control may help to improve food crop yield. However, promotion of widespread use of paddy cultivars, *kharif*-small millet cultivars and *kharif*-finger millet (ragi) cultivars will be also useful to reduce the dependence on climate variables (rainfall) in Chamoli district. The study region belongs to mountainous land where people are facing food insecurity and disappearance of traditional crop varieties. In Uttarakhand, majority of the household grows ~ 30 crop races/cultivars annually (Sati and Bandooni 2018). Whereas some crops need specialized environment to grow in mountainous region. The Uttarakhand Himalaya is ecologically fragile zone and has very low resource productivity (Rahut et al. 2010). Chamoli district shares the 6.2% of arable land in Uttarakhand state (Khanduri 2018); where cash crop like sugarcane has dominated the agricultural production in Uttarakhand, which is followed by the fruits and vegetables, wheat, paddy, finger millet (ragi), small millets and so on (Sati and Wei 2018). Crops like paddy and wheat are highly suitable for plain regions nevertheless farmers are also growing these crops in hilly region. The crop suitability in Uttarakhand state varies with altitudinal gradients. Agro-climatic conditions also differ with the altitude; hence, it provides diversity in the agricultural field. It is essential to consider the changes in climatic conditions and other vulnerability aspects such as high rate of soil erosion, low fertility, etc., for self-reliance and food security.

It is important to understand the impacts of climate change on crop productivity for food security and sustainable livelihoods. Temperate regions are highly sensitive to climate change; a minor shift in climate conditions of the mountainous regions would affect crop yield by altering soil degradation, nutrient availability, etc. (Gitz et al. 2016). The impact of climate change can be minimized through appropriate crop management and adaptation strategies. In the present study, all crops are not highly affected by climate change but for future perspective, it is essential to understand the impact of climate change on crop productivity.

REFERENCES

Dickey DA, *and* Fuller WA (1979) Distribution of the estimators for autoregressive time series with a unit root. Journal of the American Statistical Association 74(366a): 427–431

FAO F (2018) The future of food and agriculture: alternative pathways to 2050. Food and Agriculture Organization of the United Nations Rome

Fischer G, Shah M, Tubiello F, Van Velhuizen H (2005) Socio-economic and climate change impacts on agriculture: an integrated assessment, 1990–2080. Philosophical Transactions of the Royal Society B: Biological Sciences 360(1463): 2067–2083

Food and Agricultural Organization (2008) Climate change and food security: a framework document. FAO Rome

Fuller WA (1985) Nonstationary autoregressive time series. Handbook of statistics 5: 1–23

Geethalakshmi V, Lakshmanan A, Rajalakshmi D, Jagannathan R, Sridhar G, Ramaraj AP, Anbhazhagan, R (2011) Climate change impact assessment and adaptation strategies to sustain rice production in Cauvery basin of Tamil Nadu. Current Science, 342–347

Granger CW (1969) Investigating causal relations by econometric models and cross-spectral methods. Econometrica: Journal of the Econometric Society, 424–438

Gitz V, Meybeck A, Lipper L, Young CD, Braatz S (2016) Climate change and food security: risks and responses. Food and Agriculture Organization of the United Nations (FAO) Report. 110:2–4

Hollaender M (2010) Human right to adequate food: NGOs have to make the difference. CATALYST Newsletter of Cyriac Elias Voluntary Association (CEVA) 8(1): 5–6

Khanduri S (2018) Landslide distribution and damages during 2013 Deluge: a case study of Chamoli district, Uttarakhand. Journal of Geography and Natural Disasters 8(2): 1–10

Kumar KK, Parikh J (2001) Indian agriculture and climate sensitivity. Global Environmental Change 11(2): 147–154

Kumar SN, Aggarwal PK, Rani S, Jain S, Saxena R, Chauhan N (2011) Impact of climate change on crop productivity in Western Ghats, coastal and northeastern regions of India. Current Science, 332–341

Kumar A, Patel JS, Bahadur I, *and* Meena VS (2016) The molecular mechanisms of KSMs for enhancement of crop production under organic farming. In Potassium solubilizing microorganisms for sustainable agriculture (pp. 61–75). Springer, New Delhi

Kyei-Mensah C, Kyerematen R, Adu-Acheampong S (2019) Impact of rainfall variability on crop production within the Worobong Ecological Area of Fanteakwa District, Ghana. Advances in Agriculture

Lal R, Bruce JP (1999) The potential of world cropland soils to sequester C and mitigate the greenhouse effect. Environmental Science *and* Policy 2(2): 177–185

Mbow C, Rosenzweig C, Barioni LG, Benton TG, Herrero M, Krishnapillai M, Liwenga E, Pradhan P, Rivera-Ferre MG, Sapkota T, Tubiello FN (2019) Food security. In Climate Change and Land: An IPCC Special Report on Climate Change, Desertification, Land Degradation, Sustainable Land Management, Food Security, and Greenhouse Gas Fluxes in Terrestrial Ecosystems. PR Shukla, J Skea, E Calvo Buendia, V Masson-Delmotte, H-O Pörtner, DC Roberts, P. Zhai, R. Slade, S. Connors, R. van Diemen, M. Ferrat, E. Haughey, S. Luz, S. Neogi, M. Pathak, J. Petzold, J. Portugal Pereira, P. Vyas, E. Huntley, K. Kissick, M. Belkacemi, and J. Malley, Eds., Intergovernmental Panel on Climate Change.

Nageswararao MM, Dhekale BS, Mohanty UC (2018) Impact of climate variability on various Rabi crops over Northwest India. Theoretical and Applied Climatology 131(1): 503–521

Nath PK, Behera B (2011) A critical review of impact of and adaptation to climate change in developed and developing economies. Environment, Development and Sustainability 13(1): 141–162

Nautiyal S, Goswami M, Prakash S, Rao KS, Maikhuri RK, Saxena KG, *and* Banerjee S (2022) Spatio-temporal variations of geo-climatic environment in a high-altitude landscape of central Himalaya: an assessment from the perspective of vulnerability of glacial lakes. Natural Hazards Research

Praveen B, Sharma P (2020) Climate change and its impacts on Indian agriculture: an econometric analysis. Journal of Public Affairs 20(1): e1972

Rahut DB, Castellanos IV, Sahoo P (2010) Commercialization of agriculture in the Himalayas. Institute of Developing Economies (IDE), Discussion Paper 265, IDE-JETRO, Chiba, Japan Retrieved from http://ir.ide.go.jp/dspace/bitstream/2344/931/1/ARRIDE_Discussion_No.265_rahut.pdf

Sati VP, *and* Bandooni SK (2018) Forests of Uttarakhand: diversity, distribution, use pattern and conservation. ENVIS Bulletin Himalayan Ecology 26: 21–27

Sati VP, *and* Wei D (2018) Crop productivity and suitability analysis for land-use planning in Himalayan ecosystem of Uttarakhand, India. Current Science 115(4): 767–772

Tai AP, Martin MV, *and* Heald CL (2014) Threat to future global food security from climate change and ozone air pollution. Nature Climate Change 4(9): 817–821

CHAPTER 6

A Comparative Assessment of Farmers' Perception on Drought and Related Impacts in Western Part of Odisha

Devi Prasanna Swain, Arunasis Goswami, Bhabesh Chandra Das, Debasis Ganguli, and Madan Mohan Mahapatra

6.1 Introduction

Drought, a natural calamity, results from below-average rainfall in a given region, causing prolonged shortages in the water supply, whether atmospheric, surface water, or groundwater (Swain et al. 2019). Droughts can have severe concerns for water use in agriculture, as well as adverse impacts on ecosystems (Van Loon et al. 2016). The impact of drought is associated with social, economic, political, and climatological aspects (Campbell 1984). In India, an increase in the frequency of drought occurrence in states, namely Bihar, Uttar Pradesh, Karnataka, Kerala, and Maharashtra. In nine out of the past 15 years, about 100 districts of the country have witnessed a drought-like situation produced due to poor southwest monsoon. The western parts of Odisha rely more on the monsoon rain for agriculture, and almost once in two years, this

---

D. P. Swain (✉) · A. Goswami · D. Ganguli
Department of Veterinary and Animal Husbandry Extension Education, WBUAFS, Kolkata, West Bengal, India
e-mail: dr.swain07@gmail.com

B. C. Das
Department of Veterinary and Animal Husbandry Extension Education, OUAT, Bhubaneswar, Odisha, India

M. M. Mahapatra
Fisheries and Animal Resources Development Dept., Govt. of Odisha, Bhubaneswar, Odisha, India

© The Author(s), under exclusive license to Springer Nature Singapore Pte Ltd. 2023
S. Nautiyal et al. (eds.), *Palgrave Handbook of Socio-ecological Resilience in the Face of Climate Change*,
https://doi.org/10.1007/978-981-99-2206-2_6

area faces the problem of either drought or a drought-like situation. Farmers in this region are mostly marginal or sharecroppers, making them more vulnerable to these climatic calamities. Frequent droughts result in mass migrations in this region. It can have a substantial impact on the ecosystem and agriculture of the affected region and harm to the local economy. The worst affected district is Mayurbhanj followed by Bargarh, Balangir, Keonjhar, and Nayagarh districts. Western Odisha districts lack irrigation facilities. Erratic monsoon destroys the crop leaving the farmer with nothing for sustenance and a mountain of debt that he cannot repay (Special Relief commissioner's office, 01 Nov 2018.) Farmers suffered crop loss of 33% and above due to moisture stress in the studied districts (Special Relief commissioner's office, 01 Nov 2018.)

Livestock is generally more adaptable to environmental shocks as compared to crops. They are mobile, which increases survivability and may also be relatively omnivorous, thereby surviving dramatic effects on specific feed resources. Native animal varieties, particularly, are adapted to local environmental risks and use natural resources efficiently. At the same time, they are essential providers of nutrients and traction for growing crops in smallholder systems (Thornton and Herrero 2001).

Farmers are confronting problems associated with unreliable rainfall and soils of low fertility, while animal keepers can adjust grazing patterns to their modified access to pasture and water and also can take nomadic methods of living as well (Campbell 1984). Reports suggest animal husbandry is less affected than agriculture by natural calamities like drought (Swain 2019; Swain et al. 2019, 2020). As drought intensifies, pasture deteriorates, water becomes increasingly scarce, and movement is necessary to access these resources (Campbell 1981). Campbell (1984) advocated the combination of livestock and crop production to be a more successful strategy in overcoming drought-related shortages than either activity practised alone.

Hence, the study aimed to evaluate animal husbandry practices as an alternative to agriculture and agricultural labours for generating livelihood in drought-prone areas of western Odisha. Furthermore, this study aimed to assess the economy of the three categories of respondents, namely, animal husbandry farmers, agriculture farmers, and agricultural labours during drought conditions, and to suggest the best-suited farming activity among these three to the drought-prone area farmers for livelihood generation.

6.2 Materials and Methods

The western part of Odisha covers a number of districts, viz. Sambalpur, Bargarh, Kalahandi, Nuapada, Bolangir, Sonepur, Deogarh, Jharsuguda, and Sundargarh, out of which, Balangir, Kalahandi, and Nuapada face frequent droughts and simultaneously do not have any way to earn livelihood to its populace at the time of drought. Most people (nearly 70%) in these districts gain their livelihood from agriculture, animal husbandry, and agriculture labour. So, these three districts were selected for the study. However, the sample district people are more forced to distress migration to support their livelihood, and drought makes their life more vulnerable. Hence, animal husbandry is hypothesized as a drought-proof enterprise for poor and landless farmers.

The research site (site for data collection) covered three blocks from these three districts, one block from each district viz. Bangomunda block of Bolangir, Golamunda block of Kalahandi, and Boden block of Nuapada were selected for data collection. These three blocks were selected as they are drought-prone, and there is no other source of livelihood for the farmers (Fig. 6.1). Moreover, these blocks do not have significant irrigation projects or industries to provide farmers with alternative livelihoods during drought.

Based on the pilot study and discussions with the stakeholders and experts, the statements were framed. Repeated verifications and proper measures were taken to avoid vague and ambiguous responses that may distort the information flow. Close ended questions were put in the schedule to get appropriate responses. 300 respondents were interacted (Fig. 6.2) randomly from these three blocks namely, Bangomunda, Golamunda, and Boden based on their primary occupation of Agriculture, animal husbandry, or agriculture labour following stratified random sampling.

Fig. 6.1 Map of Odisha showing different districts and study area

Fig. 6.2 Interaction with respondents

Data Analysis

Statistical measures provide the investigator with an opportunity of expressing the fact in an empirical way. The data expressed in terms of percentage and frequency (n) which is no of respondents answered in favour of the statement.

$$\text{Percentage (\%)} = \frac{\text{No of respondents}}{\text{Total No. of respondents}} \times 100$$

The mean value of the collected information was also calculated. It is the arithmetic average and the result obtained when the sum of values of the individuals in the data is divided by the number of individuals in the data. Each response was given a score viz. strongly agree, Agree, Undecided, Disagree, and Strongly Disagree, respectively, 5, 4, 3, 2, and 1. The mean score was calculated after taking the total no of respondents under each response, a mean score more towards 5 score signifies that most of the respondents agreed to the statement. Mean is the simplest and relatively stable measure of a series and in enabling data to be composed. Mean is better than other averages, especially in social and economic studies where direct quantitative measurements are possible.

$$\text{M.S} = \frac{\sum fx}{N}$$

where
M.S = Mean score,
$\sum fx$ = Sum of total score obtained in individual category
N = Total number of respondents.

6.3 Results and Discussion

Effect of Drought on Economy of Agriculture Farmers

Out of 300 respondents, 194 were involved in agriculture to earn their livelihood. Hence, a total of 194 respondents were questioned on the impact of drought on their agricultural practices and depicted in Table 6.1. Most respondents strongly agreed that drought brought a substantial loss of their crops; whether there was no or negligible return from agriculture during drought, they could not get the capital invested in cropping. 74.74% of respondents strongly agreed, and 18.04% agreed that they were unable to deposit the school fees of their wards during the drought from agricultural activities, and the remaining 7.22% of respondents either disagreed with this statement or found undecided. The majority of agricultural respondents strongly agreed that they had no savings in the bank during the drought.

Almost all respondents either strongly agreed (79.38%) or agreed (18.55%) that they were forced to take loans from private money lenders at exorbitant interest rates due to the loss of crops and not a single respondent strongly

Table 6.1 Responses on effects of drought on economy of agriculture (n = 194)

Sl. No	Statement	Strongly agree	Agree	Undecided	Disagree	Strongly disagree	Mean score
1	Drought brought in substantial loss of my crops	144 (74.22)	40 (20.62)	3 (1.55)	4 (2.06)	3 (1.55)	4.64
2	There was no or negligible return from agriculture during drought	162 (83.50)	28 (14.43)	2 (1.03)	1 (0.52)	1 (0.52)	4.80
3	I could not be able to get the capital which I had invested in cropping	124 (63.92)	50 (25.78)	10 (5.15)	6 (3.09)	4 (2.06)	4.46
4	I was unable to deposit the school fees of my wards during drought from agricultural activities	145 (74.74)	35 (18.04)	7 (3.61)	4 (2.06)	3 (1.55)	4.62
5	During drought, I had no savings in the bank	162 (83.50)	28 (14.43)	2 (1.03)	1 (0.52)	1 (0.52)	4.80
6	I was forced to take loan from private money lenders at exorbitant rates of interest due to loss of crops	154 (79.38)	36 (18.55)	3 (1.55)	1 (0.52)	0 (0.00)	4.77
7	I could not repair my agricultural equipment as I earned no money from the agricultural activities	152 (78.35)	38 (19.58)	2 (1.03)	1 (0.52)	1 (0.52)	4.75
8	The village shopkeeper denied to give me goods on credit as I lost the crops during drought	145 (74.74)	35 (18.04)	7 (3.61)	4 (2.06)	3 (1.55)	4.62

(continued)

Table 6.1 (continued)

Sl. No	Statement	Response					Mean score
		Strongly agree	Agree	Undecided	Disagree	Strongly disagree	
9	I was fully dependent on quacks for the healthcare of my family members as I didn't have money from agriculture to visit a doctor	145 (74.74)	35 (18.04)	8 (4.12)	4 (2.07)	2 (1.03)	4.63
10	I couldn't be able to purchase a new bicycle for my son as I didn't get any income from agriculture	99 (51.03)	75 (38.66)	10 (5.15)	6 (3.09)	4 (2.07)	4.34
	Total mean score	46.43					

The figures in upper row of in 3rd, 4th, 5th, 6th and 7th column are frequency and lower is percentage

disagreed to this statement. Most of the agricultural respondents could not repair their agricultural equipment during drought as they had no income from agriculture. Almost all agricultural respondents agreed when they were asked if the village shopkeeper denied giving him goods on credit as they lost the crops during drought; however, 7.22% of respondents were found either undecided, disagreed, or strongly disagreed with the statement. The study revealed that majority of respondents were entirely dependent on quacks for the healthcare of their family members as they didn't have money from agriculture to visit a doctor. They couldn't purchase a new bicycle for their son because they didn't have any agriculture income. The overall mean score of the statement "Effect of drought on economy in Agriculture" was 46.43. Swain et al. (2020) reported that drought threatened household food security, erratic rainfall caused drought and subsequent loss of crops, drought caused distress sale of capital resources were the major problems by the agriculture farmers during drought, which may affect the economy of the agricultural farmers. The study signifies that the agriculture severely affects the economy of agricultural farmers.

Effect of Drought on Economy of Animal Husbandry Farmers

Out of 300 respondents, 72 had Animal Husbandry as their primary occupation. Hence total of 72 respondents were questioned on the impact of drought on their economy through animal husbandry. During the interaction, majority of respondents disagreed to the statements of (a) whether drought brought substantial loss of their Animal Resources and (b) whether there was no or negligible return from animal resources during drought (Table 6.2). The responses infer that almost all respondents disagreed that drought significantly affected the income from animal resources. It was observed that 68.05% of respondents strongly disagreed with the statement when asked "I could not get the capital which I had invested in the animal resources", and 29.17% disagreed. In contrast, only 2.78% of respondents agreed with this statement. However, not a single respondent was either undecided or strongly agreed with the above statement, assuring that return from Animal resources is least affected in drought conditions. On enquiring whether the respondent could not deposit the school fees of his wards during drought, most animal husbandry respondents went against the statement. On probing whether, during the drought, he had no savings in the bank, 55.56% disagreed, 19.44% strongly disagreed, 1.39% found undecided, 20.83% agreed, and 2.78% respondents strongly agreed to this statement. Almost all respondents either strongly disagreed (80.56%) or disagreed (13.89%) that they were forced to take loan from private money lenders at exorbitant rates of interest due to no return from animal resources, which was having a very low mean score of 1.29 indicating the level of disagreement of the respondents to the above assumption. Almost all the respondents either strongly disagreed or disagreed when they were asked (a) whether they could not purchase animal feeds and medicines as they earned no money from the Animal Husbandry activities and (b) whether the village shopkeeper denied to give them goods on credit during drought, and (c) they were fully dependent on quacks for the healthcare of his family members as he didn't have money from animal husbandry to visit a doctor. Out of 72 respondents, 62.50% strongly disagreed that they could not purchase a new bicycle for their son as they didn't get any income from animal rearing, whereas 30.55% disagreed with this statement. However, only 5.56% of respondents agreed and the remaining 1.39% were undecided about this statement. The mean score of the statement was 1.50. The overall mean score of Effect of drought on the Economy in Animal Husbandry was 16.31, which is a positive sign towards Animal husbandry practices. Swain et al. (2019) revealed that during drought animal husbandry farmers suffer from decreased productivity of animals, disturbed marketing facility for livestock and livestock products, and threatened household food security as there was low animal production. However, animal husbandry is less affected as compared to agriculture practices.

Table 6.2 Responses on effect of drought on economy of animal respondents

Sl. No	Statement	Strongly agree	Agree	Un-decided	Disagree	Strongly disagree	Mean score
1	Drought brought in substantial loss of my Animal Resources	0 (0.00)	5 (6.94)	1 (1.39)	20 (27.78)	46 (63.89)	1.51
2	There was no or negligible return from my Animals during drought	0 (0.00)	5 (6.94)	0 (0.00)	21 (29.17)	46 (63.89)	1.50
3	I could not be able to get the capital which I had invested in keeping the animal resources	0 (0.00)	2 (2.78)	0 (0.00)	21 (29.17)	49 (68.05)	1.38
4	I was unable to deposit the school fees of my wards during drought	2 (2.78)	5 (6.94)	0 (0.00)	40 (55.560)	25 (34.72)	1.88
5	During drought, I had no savings in the bank	2 (2.78)	15 (20.83)	1 (1.39)	40 (55.56)	14 (19.44)	2.32
6	I was forced to take loan from private money lenders at exorbitant rates of interest due to no return from animal resources	1 (1.39)	1 (1.39)	2 (2.78)	10 (13.89)	58 (80.56)	1.29
7	I could not purchase animal feeds and medicines as I earned no money from the Animal Husbandry activities	0 (0.00)	6 (8.33)	2 (2.78)	21 (29.17)	43 (59.72)	1.60

(continued)

Table 6.2 (continued)

Sl. No	Statement	Response					Mean score
		Strongly agree	Agree	Un-decided	Disagree	Strongly disagree	
8	The village shopkeeper denied to give me goods on credit during drought as I have income source from Animal Husbandry	0 (0.00)	1 (1.39)	1 (1.39)	35 (48.61)	35 (48.61)	1.56
9	I was fully dependent on quacks for the healthcare of my family members as I didn't have money from Animal Husbandry to visit a doctor	4 (5.56)	5 (6.94)	4 (5.56)	17 (23.61)	42 (58.33)	1.78
10	I couldn't be able to purchase a new bicycle for my son as I didn't get any income from animal rearing	0 (0.00)	4 (5.56)	1 (1.39)	22 (30.55)	45 (62.50)	1.50
	Total mean score	16.31					

The figures in upper row of in 3rd, 4th, 5th, 6th, and 7th column are frequency and lower is percentage

Effect of drought on Economy of Agricultural Labour

Out of 300 respondents, 34 farmers' primary occupation was Agricultural Labourer for earning their livelihood. Hence total of 34 respondents were questioned to assess the effect of drought on their livelihood and their responses are presented in Table 6.3. It was observed that drought brought in substantial loss as they didn't get agricultural work during drought. The majority of respondents agreed that agricultural labourers had no or negligible return from wage-earning works during drought; this signifies the dependence of agricultural labourers on climate.

The bulk of respondents (73.53%) working as agricultural labour strongly agreed that they could not deposit the school fees of their wards during drought. On the contrary, only 5.88%, either strongly disagreed or disagreed with the above statement. When asked—whether they had no savings in the bank, almost all farmers responded in favour of the statement. Majority of

Table 6.3 Responses on effect of drought on economy of agricultural labour

Sl. No	Statement	Strongly agree	Agree	Undecided	Disagree	Strongly disagree	Mean score
1	Drought brought in substantial loss for me as I didn't get agricultural work during drought	27 (79.41)	4 (11.76)	0 (0.00)	2 (5.88)	1 (2.94)	4.59
2	There was no or negligible return from wage-earning works during drought	22 (64.71)	5 (14.71)	2 (5.88)	3 (8.82)	2 (5.88)	4.24
3	I could not be able to get the money which I had expected to get from agricultural labour	18 (52.94)	12 (35.29)	2 (5.88)	1 (2.94)	1 (2.94)	4.32
4	I was unable to deposit the school fees of my wards during drought	25 (73.53)	4 (11.76)	3 (8.82)	1 (2.94)	1 (2.94)	4.50
5	During drought, I had no savings in the bank	28 (82.35)	3 (8.82)	1 (2.94)	1 (2.94)	1 (2.94)	4.65
6	I was forced to take loan from private money lenders at exorbitant rates of interest due to non-availability of labour intensive work in agriculture during drought	28 (82.35)	2 (5.88)	1 (2.94)	2 (5.88)	1 (2.94)	4.59
7	I could not repair my bicycle which is the only mode of conveyance for me, as I earned no money from agricultural labour	18 (52.94)	8 (23.53)	2 (5.88)	4 (11.76)	2 (5.88)	4.06

(continued)

Table 6.3 (continued)

Sl. No	Statement	Response					
		Strongly agree	Agree	Undecided	Disagree	Strongly disagree	Mean score
8	The village shopkeeper denied to give me goods on credit as I lost the opportunity to get labour intensive work in agriculture	20 (58.82)	8 (23.53)	1 (2.94)	2 (5.88)	3 (8.82)	4.18
9	I was fully dependent on quacks for the healthcare of my family members as I didn't have money from agricultural labour to visit a doctor	18 (52.94)	9 (26.47)	2 (5.88)	3 (8.82)	2 (5.88)	4.12
10	I couldn't be able to repair the thatched roof of my house as I didn't get any income from agricultural labour	22 (64.71)	8 (23.53)	0 (0.00)	2 (5.88)	2 (5.88)	4.35
	Total Mean Score	43.59					

The figures in upper row of in 3rd, 4th, 5th, 6th, and 7th column are frequency and lower is percentage

agricultural labourers were forced to take loans from private money lenders at exorbitant rates of interest due to non-availability of labour intensive work in agriculture during drought, as they had no work to earn their livelihood, and similar pattern of response was also observed to the statement that they could not repair their bicycle which is the only mode of conveyance, as they earned no money from agricultural labour activities. The same was the case with the statement—whether the village shopkeeper denied giving the respondent goods on credit as he lost the opportunity to get labour intensive work in agriculture. 52.94% strongly agreed and 26.47% agreed that respondents were fully dependent on quacks for the healthcare of their family members as they didn't have money from agricultural labour to visit a doctor. On enquiring whether they could not repair the thatched roof of their house as they didn't get any income from agricultural labour, 64.71% strongly agreed and 23.53% agreed to this statement. In contrast, only 11.76% of respondents strongly disagreed with the above statement. The study indicated that, as the agriculture sector is affected severely due to drought, thus the associated labourers are also affected adversely, resulting in poor economic conditions.

From the above findings, it was observed that during drought, most of the farmers working as agricultural labourers could not take care of the livelihood requirements of their family. The overall mean score of the statement, effect on drought on economy in Agricultural labour was 43.59.

The physical aspects of droughts and their effects are well known, but the often delicate and complex dynamics emanating from drought outbreaks are less understood among communities (Mogotsi et al. 2013). The nature and intensity of drought impacts vary from location to location, depending on the relative influence of various agro-climatic, geophysical, and socio-economic factors (Swain and Swain 2011). Durga et al. (2016) stated that drought brought manifold impacts on the community in Odisha; for instance, the crop loss due to drought has led to shortage and low food intake and has further put the community in a vulnerable situation. In the absence of alternative livelihood options, people have been compelled to migrate in search of jobs and exploited further (Durga et al. 2016); this is also observed in the current study. Drought affects the communities negatively, mainly through increased livestock mortalities and crop failure—especially among poorer households (Mogotsi et al. 2013). This was also observed during the study as agriculture farmers and agriculture labours were severely affected economically due to massive loss in agriculture. However, animal husbandry-associated respondents were in a better position than agriculture farmers and agriculture labours as the livestock system is less vulnerable to climatic fluctuations than agriculture. Khan et al. (2017) found that diversification of livelihood sources can increase the income of farmers, and they considered three sources of income, viz. crops, animal husbandry, and non-farm sector, which holds for the drought-prone areas as it was observed during the study that animal husbandry can curb the adverse effects of drought on agriculture and farmers can earn a livelihood by selling their livestock as a backup policy to sustain during drought. SANDRP (2016) observed that fearing inability to repay private moneylenders after deficit rainfall, farmers of rainfed and drought areas of western Odisha were reportedly falling into the clutches of labour sardars and many of the people migrated from Kalahandi, Nuapada, and Bolangir district to neighbouring states as labourers to work in brick kilns.

6.4 Conclusion

From the above study, it is evident that drought affects the agriculture economy, animal husbandry farmers and associated agricultural labourers. However, animal husbandry farmers are less affected by climatic variations like drought than agriculture and agricultural labourers. Farming practices like agriculture depend entirely on rainfall, and any minor change in rainfall intensity and pattern drastically affects their economy. As drought affects the agriculture labourers indirectly, they are also found jobless, forcing them to move to nearby districts under distress situation. This study found that animal husbandry is less affected by rainfall and that animal husbandry farmers are more economically resilient even in drought-hit years. Thus, in a drought-prone area, animal husbandry alone or in combination with other occupations can provide financial assurance to the farmers.

Acknowledgements The authors acknowledge The Dean, Faculty of Veterinary and Animal Sciences, Kolkata and The VC, WBUAFS, Kolkata for providing all necessary facilities for the study.

References

Campbell, D.J., 1981. Land-use competition at the margins in the rangelands: An issue in development strategies for semi-arid areas. *Planning African development/ edited by Glen Norcliffe and Tom Pinfold*, Croom Helm, London.

Campbell, D.J., 1984. Response to drought among farmers and herders in southern Kajiado District, Kenya. *Human Ecology, 12*(1), pp. 35–64.

Durga, N., Verma, S., Gupta, N., Kiran, R. and Pathak, A., 2016. Can solar pumps energize Bihar's agriculture? *IWMI-Tata Water Policy Research Highlight*.

Khan, W., Tabassum, S. and Ansari, S.A., 2017. Can diversification of livelihood sources increase income of farm households?—A case study in Uttar Pradesh. *Agricultural Economics Research Review, 30*(347-2017-2741), pp. 27–34.

Mogotsi, K., Nyangito, M.M. and Nyariki, D.M., 2013. The role of drought among agro-pastoral communities in a semi-arid environment: The case of Botswana. *Journal of Arid Environments, 91*, pp. 38–44.

SANDRP, 2016. *South Asia Network on dams, rivers and people, Odisha drought profile-2016*. https://sandrp.in/2016/05/20/odisha-drought-profile-2016/

Swain, D.P., 2019. Assessment of animal husbandry as an alternative source of livelihood during drought in western part of Odisha. *PhD. Thesis, submitted to the West Bengal University of Animal and Fishery Sciences*, Kolkata, West Bengal, India.

Swain, M. and Swain, M., 2011. Vulnerability to agricultural drought in Western Orissa: A case study of representative blocks. *Agricultural Economics Research Review, 24*, pp. 47–56.

Swain, D.P., Goswami, A., Das, B.C., Ganguli, D. and Mahapatra, M.M., 2019. Study on the constraints of animal husbandry farmers during drought in western parts of Odisha. *International Journal of Current Microbiology and Applied Sciences, 8*(11), pp. 1022–1029. doi: https://doi.org/10.20546/ijcmas.2019.811.120

Swain, D.P., Goswami, A., Das, B.C., Ganguli, D. and Santra, B., 2020. Constraints faced by agriculture farmers during drought in drought prone Western Odisha. *International Journal of Current Microbiology and Applied Sciences, 9*(10), pp. 2119–2125. doi: https://doi.org/10.20546/ijcmas.2020.910.258

Thornton, P.K. and Herrero, M., 2001. Integrated crop–livestock simulation models for scenario analysis and impact assessment. *Agricultural Systems, 70*(2–3), pp. 581–602.

Van Loon, A.F., Stahl, K., Di Baldassarre, G., Clark, J., Rangecroft, S., Wanders, N., Gleeson, T., Van Dijk, A.I., Tallaksen, L.M., Hannaford, J. and Uijlenhoet, R., 2016. Drought in a human-modified world: Reframing drought definitions, understanding, and analysis approaches. *Hydrology and Earth System Sciences, 20*, pp. 3631–3650. www.hydrol-earth-syst-sci.net/20/3631/2016/. doi: https://doi.org/10.5194/hess-20-3631-2016

CHAPTER 7

Indian Pastoralism Amidst Changing Climate and Land Use: Evidence from Dhangar Community of Semi-arid Region of Maharashtra

Dada R. Dadas

7.1 Introduction

Pastoralism is an age-old livelihood mode that still survives in different parts of the world. Pastoralism is defined as the herding and caring system for livestock, and it is acknowledged as a traditional production system that embodies resilience, strength, and perseverance. The practice evolved more than 6000 years ago in response to climatic and environmental variability (World Food Program, 2016). According to FAO (2021) pastoralists are people who adopt their herding system according to seasonal or spatial weather variability and the availability of rangelands for grazing. Herders are mainly involved in herding sheep, goats, cattle, buffalo, etc. Pastoralists move from one place to another, searching for pasture and water resources with their herd. They follow a nomadic, semi-nomadic, or transhumant system. Pastoralists are known for using dryland or semi-arid landscapes for grazing (World Food Program, 2016). However, a rational understanding of pastoralism of Dhangar is still neglected and is perceived as an unproductive and backward livelihood system. This colonial bias persists even to this day.

Pastoralism contributes to livelihoods, nutrition, and food security; however, this contribution is hardly recognised in any policies and

D. R. Dadas (✉)
WOTR Centre for Resilience Studies, Watershed Organisation Trust, Pune, India
e-mail: dadatiss@gmail.com

© The Author(s), under exclusive license to Springer Nature Singapore Pte Ltd. 2023
S. Nautiyal et al. (eds.), *Palgrave Handbook of Socio-ecological Resilience in the Face of Climate Change*,
https://doi.org/10.1007/978-981-99-2206-2_7

programmes. Moreover, as the pastoral livelihood system has yet to be recognised officially, finding exact data on the number of ruminants and the number of people involved in pastoralism is difficult. Nevertheless, it is assumed that pastoralists form 7% of India's population and inhabit mainly the arid and semi-arid areas of the country (Fig 7.1).

Next to agriculture, Pastoralism is the backbone of the rural system, contributing 3% to the national GDP and providing livelihoods to a significant share of the rural population. The number of practising pastoralists is estimated to be close to 13 million in India, out of which at least 1–2 million are in Maharashtra. The latter includes communities like Hatkar Dhangar, Shegar Dhangar, Ahir Dhangars, Gavli Dhangar along with Nandgavli, Bharwad, Raika, Kurumar, and others (Kishore and Köhler-Rollefson, 2020). 'The pastoral nomads to whom we are studying are the shepherds known as Hatkar Dhangar in local dialectics (Dadas, 2012). 'The word Dhangar means 'Dhang, the Sanskrit word; the literal meaning of this word is mountain/hill, and it means the people who are residing in the mountain areas. Hatkar is the clan of Shepherd Dhangars (Sonthiemer, 1997) and Dhangars-Shepherds are the pastoralists who rear sheep and goats as a livelihood source.

Though community presence is all over the state, they are mainly concentrated in western Maharashtra Dhangars, the pastoralist of Maharashtra, migrate towards the traditional routes of Western Ghat, Deccan, Mulshi, and Konkan, some of the groups also move towards the Marathwada region, and few of them migrate to local areas.

The area where Dhangars reside has a drought history and climate change has aggravated the impacts more severely. Since centuries, pastoralists like Dhangars have developed strategies to cope with temporal and spatial variability; accordingly, they have also developed livelihood strategies and resilience. Semi-arid grounds of Maharashtra offer conducive environment for the pastoralists.

Climatic and land use changes have added many vulnerabilities in the life and livelihood of the Dhangars following various migratory routes. Pastoralism is executed in semi-arid regions, so climatic and land use changes effects on sustenance of the pastoralism includes limited access to pasture and water resources. Environmental and climate changes have severely impacted the ecosystem services like grasslands, forests, availability of water resources, and the emergence of new diseases. It has badly affected pastoralists like Dhangars and sustenance of age-old pastoral livelihood system. The impact of climate and land use change on agriculture and allied sector is documented and there is an amount of literature available but there lacks the body of knowledge related to the impact of climate and land use change. Development of understanding on pastoralism amidst the climate and land use change is crucial. Therefore, an attempt has been done to highlight key concerns associated with the same.

Objective

- To study the impact of climate and land use change on the life and livelihood of Dhangar (Pastoral Nomads) community

7.2 Methodology

A study was carried out in the state of Maharashtra during COVID-19 to understand the impact of COVID-19 on life and livelihood of pastoralists. Climate change and land use change were important factors explored throughout the discussion. Since it was a COVID-19 period, carrying survey was difficult, so it was decided to use the qualitative research approach. For the study, the non-probability sampling method (purposive technique) was used for the in-depth interviews. The pastoralists who were ready to give answers were selected for in-depth interviews. This technique explored pastoralists' perceptions of climate change, land use, and land cover. A total of 20 participants, including practitioners, veterinary doctors, and community representatives, were interviewed. Open-ended questions were prepared to understand the change in land use, agricultural practices, agrarian relations, and change in climatic factors and their impact on pastoralists' life and livelihood, etc.

The duration of the each interview was about 1 hour. The interviews were conducted in the Marathi language and recorded in an audio recorder. Oral consent were taken while recording the interviews. The participants were asked open-ended questions including, but not limited to, the following questions:

- In the face of the climate change (CC), what are the challenges faced regarding access to common resources like grassland, water, and forest?
- What is the impact of CC, i.e. (i) changing rainfall patterns; (ii) increasing temperature on livestock, poultry, local breeds
- What are the coping mechanisms to CC adopted by the small ruminant/large ruminant holders in villages?
- What has changed in pastoral livelihood system? What were the triggers?

7.3 Results and Discussion

Factors Affecting Sustenance of Pastoralism/Herding

In the past two–three decades, many changes have occurred in the life of herders, which will have negative implications on the sustainability of herding. In addition, many anthropogenic and climatic factors are affecting the sustenance of herding. Various issues like recurrent climatic shocks, land-related conflicts, and infrastructure development have led to land degradation and

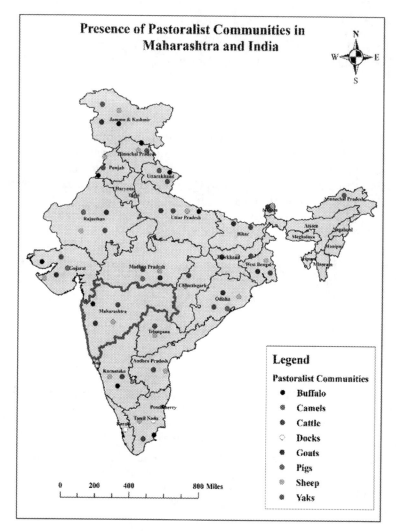

Fig. 7.1 Presence of pastoralist communities in Maharashtra and India

further access to pasture and grazing resources. Due to the constant challenges faced, Dhangars have adapted to the changing ecosystem; they are better tuned to variability, risks, and uncertainty than other groups. As a result, they have better resilience to climatic and other factors.

Climatic factors: The livestock system contributes to and is affected by climate change. Climatic changes affect the quality and quantity of pastures, water resources, and access to the same. The disturbance of the monsoon cycle, subsequent drought situation, and rising temperature are some notable climatic factors that further affect the drying up of pasture and water resources.

In addition, the health and reproductive health of the livestock will be badly affected, which will have long-term repercussions on pastoralists' food, nutrition, and livelihood security. Such distress conditions will further push pastoralists to face many socio-economic and ecological vulnerabilities. The following are the vital climatic factors that need to be discussed for better understanding:

- *Issues in access to grazing resources*:

Low rainfall and drought-related situation affect on quality and quantity of grazing resources. In the absence of grazing resources for the livestock, herders are forced to migrate in search of grazing. The study participant (practitioner), while discussing grazing issue, shared below information:

> *Since Dhangars have small ruminants, they are considered supplementary stock and remain neglected. Access to grazing land for small ruminants is not considered. The fodder issue is felt when we talk about the cow and buffalo. Grazing on commons is not encouraged.*

- *Issues in access to water for drinking and water for livestock*:

As discussed above, herders face a mounting challenge in accessing water resources for drinking and livestock due to climatic and other factors. A study participant (veterinary doctor), while discussing issues related to access to water, shared the information below:

> *Earlier pastoralists were digging water sources in the jungles etc., so it was helping for the long-term storage of the water, which was utilised to meet the water needs of the small ruminants. That system no longer exists as the entry of the pastoralists in the jungle is restricted. A real water scarcity is faced during the summer when Dhangars are on migratory routes.*

The above narrative shows that Dhangars face issues with access to water. While designing water-related policies and programmes, the programmes need to adequately consider herders' water needs in their own and migratory villages. Initiatives like WOTR's Water Stewardship Initiative are needed to address livestock needs while preparing village-level water budgeting. The upscaling of such initiatives in national and state programmes will help consider herders' water-related needs.

- *The emergence of new diseases*:

The climatic changes result in new diseases affecting livestock health and productivity. Climate change is also resulting in morbidity and mortality of ruminants. While local breeds thrive in the local climatic conditions and are more resilient to the effects of climate change, the shift towards hybrid/non-local breeds makes livestock susceptible to various diseases. Furthermore, during the outbreak of any livestock-related disease, it becomes difficult for pastoralists to access veterinary services. So this may further result in morbidity and mortality of sheep, goats, and other livestock. A study participant (veterinary doctor), while discussing issues related to disease conditions, shared the information below:

> *Regarding coping with climate change, shepherd shares that Deccanies are better at migrating and dealing with climatic conditions. For better production, a few Dhangar groups have started preferring to grow Madgyal.*[1]

- COVID-19 and pastoralism:

COVID-19 is believed to be a part of zoonosis, which relates to climatic and environmental factors. This pandemic has had far-reaching impacts on the agrarian relations of pastoralists. The pandemic restricted herders' movement, which affected access to water for drinking and moving around villages. They were stuck on migratory routes and destination areas and could not access groceries and other basic amenities. Access to the market for livestock sale was difficult as the livestock market was closed during the pandemic, resulting in substantial economic loss for herders. Even when on the move, access to government advisories around COVID precautions and self-care was severely restricted.

Factors Related to Land Use Change and Its Impact on Dhangar Community

Changes in land use badly affect access to herders' traditional grazing grounds. With infrastructure development being carried out through the construction of canals for irrigation, construction of roads and buildings, and conversion of rangeland into agriculture, the area available for grazing is getting reduced. This has exposed herders to numerous socio-economic and ecological vulnerabilities.

In addition, the land use change is leading herders to change symbiotic relations with the agriculturalists and villages on traditional migratory routes, e.g., earlier pastoralists visiting agriculture fields for manuring and farmers were welcoming pastoralists in the exchange of fodder, so this system is declining

---

[1] Madgyal is supposed to be less tolerant to disease or drought than the Deccani.

gradually and resulting in changing symbiotic relations and affecting long-term sustenance of pastoralists.

- *Change in Customary Grazing Management*:

Dhangars had a customary grazing management system in which the grazing areas were allocated mutually. Village/area-wise pastures were given to Wadas.[2] This system was helping to share resources mutually. Due to changes in land use and conversion of grazing land for other purposes, changes in customary grazing systems are observed. A study participant (pastoralist), while discussing issues related to traditional grazing management, shared the information below:

> *The equal rule was applied to every wada. Dhangars migrating to specific grazing areas had grazing rights in their respective regions. There was a rule that no one would infringe on or enter into the grazing area of others. We were not allowed to enter the villages of others, and if so, it was only allowed for the day. Breaking rules was considered an offense. Now we have left with limited grazing ground, so this system is disappearing.*

Land Conversion for Other Economic Activities
Land conversion is a significant issue today. Infrastructure development and land conversion were the key factors reported during the discussion. Those issues are affecting the livelihood of the Dhangars.

> *Land is being converted for various purposes. Getting land for grazing and tenting is difficult now, and we are still on the same land where we used to reside for years. Earlier, nothing was here except agriculture, and now there are buildings, roads, and whatnot. We are here because of the past-relations with the agriculturalist; if the grazing grounds gets over we will not have a place to graze.*

- *Degradation of rangeland and its impact on livelihood*:

Rangeland, in semi-arid areas provides various ecosystem services and has supported livelihoods for millennia. 'According to a report India presented to UNCCD, from 2005 to 2015, the country lost around 31% or 5.65 million hectares of grassland area'. Under grasslands, the total area reduced to 12.3 mha from 18 mha between 2005 and 2015 (Pandey, 2019). 'Land degradation is directly linked to food insecurity, vulnerability to climate change, and poverty.' 'Maharashtra, Karnataka, Gujarat and Uttar Pradesh have undergone severe degradation and loss of grassland ecosystems' (Misher et al., 2022).

---

[2] Wada is the temporary place where Dhangars reside.

This deprives Dhangar community access to land resources that were accessible earlier.

- *Changing cropping patterns*:

In changing times, farmers today prefer cash-crops; hence mono-cropping is taking place. This is affecting Dhangars' access to grazing ground. A study participant (pastoralist), while discussing issues about changing cropping patterns, shared the information below:

> *There is a problem as there is a shift of crops most of the farmers have shifted to sugar cane. I think these relations have been enduring for many years, and Shepherds, on their part, are continuing those relations and trying for it.*

Farmers are shifting towards horticulture, mono-cropping, etc.; hence they have limited grazing resources. Earlier farmers were more inclined towards food and fodder-related crops, so crop residue was left accessible to pastoralists on the move. However, with changing cropping patterns, the farmers are not in a position to accommodate herders for grazing and give land for manuring. With effect, herders cannot access crop residue and post-harvesting land for herders' livestock.

- *Increasing conflict between pastoralists and farmers*:

The competition over land resources has become a significant cause of conflict between herders and farmers. As a result, the symbiotic relationship between farmers and herders has significantly weakened. Recent cases associated with conflict show that grazing scarcity leads to conflict. A study participant (pastoralist), while discussing issue of farmers–shepherd conflict, shared the information below:

> *Now days grazing related problems are common; thus, conflicts are observed because grazing lands are getting shrunk, so the animals sometimes enters the crops. So it creates problems among shepherds and farmers. Since grazing issue is there, the, conflicts are common.*

- *Change in herd composition*:

Earlier, pastoralists had enough grazing area to take care of their herds. However, due to anthropogenic and climatic changes, pastoralists are reducing their herd sizes, e.g., where pastoralists earlier had a herd of more than 500 sheep, today the size has come down to 100 (area to area number may vary, but this is a common trend observed). If the situation persists, the future of herding will be in the dark. Furthermore, if the herd size goes down, the traditional manuring agriculture method will vanish, and farmers will shift to other means of fertilising their farms. In effect, they will not be in a position to welcome herders and provide grazing ground.

- *Issues in access to Tribes and Forest Dwellers Right Act-2006*:

Though the Forest Right Act has a provision of ensuring grazing rights under community forest rights for pastoralists and other dwellers, surprisingly, there is rare evidence of accessing the benefits of pastoralists under the act. Therefore, the primary purpose of this provision is not yet served. Furthermore, the criteria to access provisions seem too complicated, and proving generational access to forest areas looks too difficult for pastoralists as they do not have documentary evidence related to traditional access to grazing. Even grazing in the forest is perceived as a threat to the forest, so there is an inadequate will to support such provisions under the act.

- *Impact of wasteland development program*:

Pasturelands are defined as wastelands by the colonial government, and later the same notion is percolated in the present context. Also, wastelands are continuously being used to set the industries, so earlier wastelands were used as pastureland by pastoralists. Hence it has also restricted the movement of pastoralists.

Under the wasteland development programme, the pastureland was converted for watershed and afforestation activities. This has restricted herders' movement on grazing grounds that were historically accessible to herders. This also resulted in the shrinking of pastureland. Such land development initiatives have not adequately considered the issues of herders in accessing grazing resources; thus, it has limited their access to traditional grazing grounds. This needs more exploration but is one of the key factors restricting grazing access.

Changes in Migration Pattern

Land use change, infrastructure development, and conversion of grazing ground in agriculture is affecting access to grazing grounds for Dhangars. This is resulting in changing migration pattern. A study participant (pastoralist), while discussing the issue of changing migratory pattern, shared the information below:

> *Earlier we used to migrate towards Panvel-Kalyan. We were staying there as we had availability of post-harvest of rice. But as the city has grown, it's becoming difficult to sustain, so we stopped migrating. We were taken care, but today's*

generation is not in a position to migrate and face hardship. We had a joint family and me, and my brother was used to migrate towards Panvel. As we started staying separately, we stopped migrating toward Panvel. Still, some of the herds migrate towards Lonavala and Panvel but we don't now.

Other Factors Affecting Overall Resilience of Dhangars to Climate Change

Many development pathways are planned without understanding the rationality behind the pastoral livelihood system, so it often leads to maladaptation. The development pathways should be planned to understand pastoralists' real needs, which may work as a safety net in times of crisis like climatic and land use change. The following are the policy and other related factors affecting the sustenance of the pastoral livelihood system.

- *Lack of Proper Census Data and exclusion from government schemes*:

There is no proper census and documentation of pastoralists involved in herding, even though they own a large number of livestock. While a livestock census is conducted for the village-level ruminants, the representation of pastoralists and their herds is missed as they keep moving.

Since pastoralists keep moving, getting access to government schemes is difficult. Most of the schemes are formulated keeping in view the sedentary population. As such, pastoralists are often excluded from getting access to government schemes. There are hardly any budgetary provisions made for the welfare of herders. Livestock insurance schemes are also not herder-centric and herder-friendly, restricting their access to insurance-related schemes.

- *Changing aspirations of the young generation*:

Due to hardships associated with herding, the younger generation has started abandoning this activity, choosing other livelihoods perceived to be more rewarding. Moreover, the younger generation perceives this livelihood activity as less dignified as there is no respect for this occupation. However, due to access to education opportunities, the aspirations of the young generation are changing, which is a threat to the livelihood systems in the future.

- *Market Dynamics*:

Market dynamics have also impacted the pastoralist's way of living. With sheep herders, for example, the sheared wool was in great demand to prepare the

Ghongadi.[3] However, with the development of the textile sector, the demand for hand-woven *Ghongadis* has come down. Consequently, the *Khutekar*.[4] that used to make *Ghongadi* disappeared from these livelihood activities.

The sale of livestock (small ruminants) in the market is not carried out considering the weight of the ruminants; the rather traditional system is commonly adopted when exchanging the ruminants. This informal way of selling and buying results in a loss for the pastoralists.

- *Inadequate livestock care*:

When on the move in remote areas, vaccinating herds and getting access to veterinary services becomes difficult. During the outbreak of any disease, it becomes difficult for pastoralists to access veterinary services. So, this may further result in morbidity and mortality of sheep, goats, and other livestock. This may further cause loss to the herd, thus pastoralists. A study participant (practitioner), while discussing the issue of changing migratory pattern, shared the information below:

> *In terms of livestock care government system chooses to be a little blind toward small ruminants because for too many years, they believe that only cattle and cattle matter. There are diseases, and I think the problem is that the government is not dealing adequately. Last year there was blue tongue among sheep. It took a long time for the government, and they did not answer the shepherd.*

7.4 Suggestions

- Developing an early warning system about weather and disease condition, especially on various migratory routes of Dhangars
- Disseminating Climate advisories through the FarmPrecise Application developed by WOTR[5] to provide access to knowledge/advisories, translation of simplistic advisories into timely action(Utility), and finally building capacity/resilience against the impacts of climate change
- Promoting one health approach among Dhangars
- Promoting climate-resilient local breeds of livestock to cope with climatic shocks
- Designing special insurance to overcome climatic and livelihood shocks
- Improving water and pasture accessibility on migratory routes during drought years

---

[3] Local blanket made up with sheep wool.

[4] Khutekar is the sub-group of Dhangar in Maharashtra, traditionally involved in weaving of Ghongadi.

[5] A Farm Precise App is developed by WOTR provides advisories to farming community and livestock rearers, so such advisories help pastoralists to cope with climatic changes.

- Declaring pasture and grazing zones and protecting them from developmental projects
- Since pastoralists are dependent on grassland, prompting them to work for protecting grassland. Grassland works as a carbon sink, so it will help in mitigating climate change
- Sustaining market links in the source and destination areas
- The role of pastoral livelihood system in climate change adaptation and in biodiversity conservation must also be recognised.

7.5 Conclusion and the Way Forward

The overall trend shows that socio-political, geographical, economic, and climatic factors affect herders' age-old livelihood. Moreover, these factors have also negatively impacted natural resources such as grasslands, forests, and water; this is an alarm for the livelihood of Dhangars and their way of life.

The Dhangars of the Deccan plateau mainly depends on Savannah type of grasslands. Grasslands are essential as they are carbon sequesters, so protecting grasslands will help in going towards climate action. Therefore, the involvement of Dhangars and other rangeland users in the planning and policy processes will allow embracing their perspectives and make more inclusive policies. In this context, the state's response in addressing issues of pastoralists is needed.

Indigenous knowledge plays a crucial role in coping with climate change and variability. Therefore, while addressing adaptation needs, considering the traditional knowledge of Dhangar pastoralists becomes a vital strategy. This may help adaptive capacities used by pastoralists to cope with drought and climatic changes.

> Ecosystem-based adaptation is the use of biodiversity and ecosystem services as part of an overall adaptation strategy to help people to adapt to the adverse effects of climate change (Convention on Biological Diversity, 2009).

Herders have developed resilience by coping with temporal and spatial variability despite inadequate policy support. This is also known as one of the nature-based solutions. Therefore, a reasonable and sustainable way forward lies in adopting the Ecosystem-based Adaptation (EbA) approach to build more resilience among herders. The EbA approach will help improve the stakeholders' adaptive capacity through training, constant monitoring, livelihood diversification, and institution building (both customary and formal). Also, national-level programmes like National Livestock Mission, and Rashtriya Gokul Mission-2014, need to include the EbA approach for the sustainable development of pastoralists and livestock holders. In the present situation, the Department of Animal husbandry and dairying, Govt. of India intends to initiate a particular cell for pastoralists, so such special programmes are required to address the present climatic and land use-related issue associated with mobile pastoralism.

The EbA approach with sustainable land management (SLM) can reduce, and halt rangeland degradation, maintain land productivity, and reverse the impacts of climate change on rangeland. This will contribute to national and international commitments like LDN, SDGs, etc. Global Agenda on Sustainable Livestock (GASL) is a partnership of stakeholders committed to the sector's sustainable development, which aims to address food security and health, equity and growth, and resources and climate. This offers an opportunity for the overall development of the livestock sector. The UN's announcement of the International Year on Rangelands and Pastoralists 2026 is a significant opportunity to promote the value of the pastoral livelihood system. To encourage the EbA approach, there is a strong need for political will and policy-level commitments.

References

Convention on Biological Diversity. (2009). *Connecting biodiversity and climate change mitigation and adaptation: Key messages from the report of the second Ad Hoc technical expert group on biodiversity and climate change.* https://www.cbd.int/doc/publications/ahteg-brochure-en.pd

Dadas, D. (2012). *Livelihood among pastoral nomads: Continuity and change.* A Study with reference to Dhangar community of Maharashtra.

FAO. (2021). *Pastoralism-making variability work, FAO animal production and health paper no. 185.* https://doi.org/10.4060/cb5055en

Kishore, K. and I. Köhler-Rollefson. (2020). *Accounting for pastoralists in India.* League for Pastoral Peoples and Endogenous Livestock Development, Ober-Ramstadt, Germany.

Misher, C., Majgaonkar, I., Malhotra A., Samrat, A. Nair, S., Sethiya, P., Godbole, M., Jagadeesh, N., Vanak, A.T. (2022). Savannah grassland conservation in Maharashtra for people, climate and biodiversity: A policy brief. Centre for Policy Design, Ashoka Trust for Research in Ecology and the Environment, *Savannah Grassland Conservation in Maharashtra _ Policy Brief* (arest.in).

Pandey, K. (2019). *India lost 31% of grasslands in a decade, down to earth.* Accessed on 11th Nov., 2022. https://www.downtoearth.org.in/news/agriculture/india-lost-31-of-grasslands-in-a-decade-66643.

Sontheimer, G et al. (1997). *King of hunters, warriors and shepherds: Essays on Khandoba, New Delhi.* Indira Gandhi National Centre for Arts.

World Food Program. (2016). Pastoralism in the age of climate change, *Pastoralism in the Age of Climate Change | by World Food Program USA | Age of Awareness | Medium.*

CHAPTER 8

Forest Fire Characterization with Relation to Meteorology and Topography Parameters in Madhya Pradesh, India

Tapas Ray, Satyam Verma, and M. L. Khan

8.1 INTRODUCTION

Forest fire has become a major hazard and more frequent in the last few decades worldwide and is a critical issue between the biosphere and atmosphere interface (Cochrane, 2003). Every year, millions of hectares of forest areas are destroyed by fire, leading to the loss of human and wild animal life, massive economic damage by wood and non-woody forest products, and the release of greenhouse gases (Ray et al., 2019). The fifth assessment report of the Inter-governmental Panel on Climate Change (IPCC) has also reported that annual carbon emissions from forest fire range between 2.5 billion to

---

T. Ray (✉) · M. L. Khan
Forest Ecology and Eco-Genomics Lab, Department of Botany, Dr. Hari Singh Gour Vishwavidyalaya (A Central University), Sagar, Madhya Pradesh 470003, India
e-mail: tapasray1892@gmail.com

M. L. Khan
e-mail: khanml61@gmail.com

T. Ray
Centre for Sustainable Forest Management and Forest Certification, Indian Institute of Forest Management (IIFM), Bhopal, Madhya Pradesh 462003, India

S. Verma
Department of Environmental Science, School of Engineering and Sciences, SRM University-AP, Amaravati, Andhra Pradesh 522240, India
e-mail: satyamverma69@gmail.com

© The Author(s), under exclusive license to Springer Nature Singapore Pte Ltd. 2023
S. Nautiyal et al. (eds.), *Palgrave Handbook of Socio-ecological Resilience in the Face of Climate Change*,
https://doi.org/10.1007/978-981-99-2206-2_8

4.0 billion tons of CO_2. The ecological impact of fires on forest ecosystems, especially tropical, temperate, and boreal forest, received global attention (Nasi et al., 2002). Globally approximately 98 million hectares of forest were affected by the fire in 2015, which comprises 3% of the global forest area (FAO, 2020). The occurrence was mainly in the tropical domain, where about 30% of tropical forests were degraded by fire or logging between 2000 and 2012 (Coomes et al., 2017).

In India, severe forest fires occur in many forest types, particularly dry deciduous forests. More than 36% forest cover of the country has been estimated to be prone to frequent forest fires as per the long-term analysis performed by the FSI, about 10.66%, 11.61%, 13.19%, and 64.54% of areas of forest cover are under extremely to very highly, highly, moderately, and less fire-prone respectively (ISFR, 2021). States under North-Eastern Region showed the highest tendency of forest fire and fall under extremely to very high forest fire zone. In Madhya Pradesh, nearly 0.43% and 6.10% of areas of forest cover are under extreme forest fire zone and very high forest fire zone, respectively (ISFR, 2021). It can be caused by natural or anthropogenic activities; however anthropogenic activities are dominant in India. Approximately 90% of forest fires are caused by human activities (Roy, 2003). Apart from anthropogenic activities the meteorological parameters, topography, and landscape fuel condition also play an essential role in their uncontrolled expansion (Prentice et al., 2011). There is also some problematic aspect to the forest fire associated with the local microclimate condition, e.g., relative humidity, wind velocity, and soil moisture. The summer season plays a major role in the increased number of forest fire events. The rapid spread of fire is mainly caused by the availability of burnable materials, strong summer temperature, and limited moisture content (Sannigrahi et al., 2020). Fire behavior is also influenced by slope, aspect, and elevation (Jaiswal et al., 2002). Over the last few years, forest fires in India have received greater attention because of their ecological and economic impacts (Vadrevu et al., 2010).

Remote sensing data and GIS techniques are powerful tools that assess forest fire trend analysis and fire risk zone analysis and also help in better visualization and understanding of the causes of forest fires (Thompson et al., 2015). Several studies have been carried out in developed countries on forest fire events and their relationship with different environmental parameters. Tian et al. (2014) studied forest fire events in China and suggested that the forest fire would increase in the future due to the variation in rainfall and temperature. Wotton et al. (2010) studied in Canada and suggested an increase in forest fire events of approximately 25% by 2030. Pinol et al. (1998) also studied in Europe and suggested that a warming climate would lead to an increase in forest fire counts.

In India, the need for more study on the forest fire regime and its relationship with meteorology and topography parameters creates a conspicuous research gap for implementing policy. In this regard, it needs to identify vulnerable areas that come under forest fire hotspots. Therefore, parameters that play an important role in the forest fire need to be quantified timely with the proper advanced technology (Lamat et al., 2021). The present study area is the Madhya Pradesh state of India, it is known as one of the forest fire-prone states due to the climatic variation as well as dry deciduous forest types, which together make fire-prone areas. Therefore, the study has utilized forest fire data from Madhya Pradesh and analyzed it in the GIS domain for better visualization and spatial and temporal pattern of the forest fire.

8.2 Materials and Methods

Study Area

The study area is spread over the entire Madhya Pradesh (MP) state with a geographical area of 3,08,252 km^2, it is the second largest state of the country, comprising 9.38% of the geographical area of the country. Madhya Pradesh lies between 21° 17′–26° 52′ N and 78° 08′–82° 49′ E and has a subtropical type climate with fire season from February to June and non-fire season from July to January (Fig. 8.1). The annual minimum and maximum temperature vary between 17–19 °C and 31–33 °C in the winter and summer respectively, with the annual average temperature of Madhya Pradesh ranging from 22.5 to 25 °C. The average annual rainfall varies from 800 mm to ~ 1800 mm and the state receives the highest precipitation from July to September. The elevation varies from 72 to 1317 m.

Datasets Used

The data earth observation satellites are now widely used to monitor and management of forest fire activities. Data used, attributes, and their resolution is shown in Table 8.1.

1. MODIS active fire location from 2012 to 2021 (https://earthdata.nasa.gov/learn/find-data/near-real-time/firms/active-fire-data).
2. VIIRS S-NPP active fire location from 2012 to 2021 (https://earthdata.nasa.gov/learn/find-data/near-real-time/firms/active-fire-data).
3. MCD12Q1 data was collected from NASA Earthexplorer for the year 2019 at 500-meter spatial resolution (https://earthexplorer.usgs.gov).
4. Meteorological data was collected from NASA Prediction of Worldwide Energy Resources (POWER) (https://power.larc.nasa.gov/data-access-viewer).

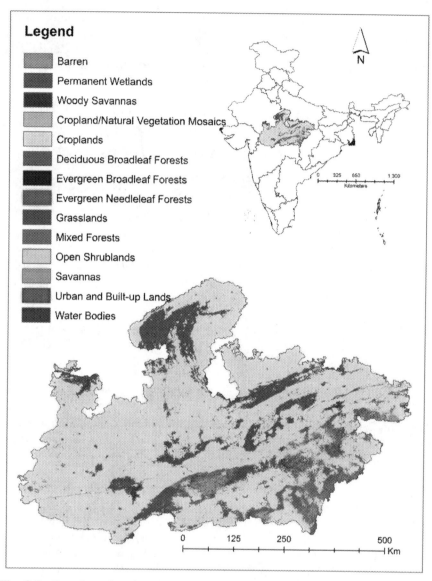

Fig. 8.1 Location of study area, Madhya Pradesh

Methods

The administrative boundary of the state was downloaded from DIVA-GIS. Analyze spatial and temporal trends of forest fires between 2012 and 2021. The spatial pattern of the fire density map was evaluated using the point density module of the Spatial Analyst tool in ArcGIS software. The created fire density map was classified from the dense forest fire to the least forest

Table 8.1 Forest fire datasets and various meteorological variables used in this study

Datasets used	Attributes	Resolution
MODIS	Forest fire detection	1000 m
VIIRS S-NPP	Forest fire detection	375 m
MCD12Q1	Land use Land cover	500 m
NASA Prediction of Worldwide Energy Resources (POWER)	Meteorology	Datasets such as relative humidity, wind velocity, soil moisture, temperature, and precipitation are available on a daily basis

fire. Through this process, the map was generated showing forest fire hotspots area for both MODIS fire points and VIIRS S-NPP fire points separately. Hotspot means areas with a high forest fire magnitude. Monthly forest fire counts from 2012 to 2021 were estimated. The active fire counts were divided into two periods viz; forest fire season (February, March, April, May, and June (FMAMJ)) and non-fire season (July, August, September, October, November, December, and January (JASONDJ)) for each year during the 10-year study period. Forest fires are very closely associated to change in the different types of variables such as relative humidity, wind velocity, soil moisture, temperature, and precipitation. Therefore, in the present study, means relative humidity, wind velocity, soil moisture, minimum, and maximum temperature, and precipitation from 2012 to 2021 on a monthly basis were downloaded and exported into the ArcGIS for further analysis. The above five meteorological variables were evaluated month and year-wise. The relationship between meteorological variables and forest fire was evaluated based on Pearson correlation to find statistical associations with a forest fire. Karl Pearson correlation (KPC) value was calculated to find the statistical associations between different variables and forest fire.

8.3 Results and Discussion

The spatial distribution of forest fire and its association with meteorological variables are important as it helps in the control, prevention, and mitigation of forest fire. Last ten years data on forest fire was analyzed in the GIS domain to understand the forest fire spatial pattern in Madhya Pradesh.

Forest Fire Recorded Based on MODIS and VIIRS S-NPP from 2012 to 2021

A significant number of fires were found in all vegetation classes between 2012 and 2021 (Tables 8.2 and 8.3). The year 2021 is showing the highest number

Table 8.2 MODIS fire counts in different vegetation from 2012 to 2021

Vegetation type	Number of fire incidence									
	2012	2013	2014	2015	2016	2017	2018	2019	2020	2021
GL	2241	606	433	312	1454	1947	2241	1298	633	3514
DBF	418	162	135	45	556	729	723	407	73	995
SA	207	43	35	37	214	428	384	114	27	339
MF	143	62	46	57	419	470	431	113	19	315
WS	33	2	7	3	34	62	73	19	5	59
EBF	0	0	0	1	1	5	1	0	0	0
ENF	0	0	0	0	0	0	0	0	0	0
Total	3042	875	656	455	2678	3641	3853	1951	757	5222

GL = Grasslands, DBF = Deciduous Broadleaf Forests, SA = Savannas, MF = Mixed Forests, WS = Woody Savannas, EBF = Evergreen Broadleaf Forests

of forest fire occurrences followed by 2017, 2018, and 2016 while the least forest fire recorded in 2015 followed by 2020, 2014, and 2013. It was also observed that precipitation was very less in the years 2017, 2018, and 2021 which may be an important factor for the high forest fire. Kumari and Pandey (2020) also reported that the forest fire increased with decreased precipitation in Jharkhand. Of the total forest fire between 2012 and 2021, grasslands accounted for the highest percentage (56.72%) of fire occurrences followed by deciduous broadleaf forests (20.01%), savannas (12.07%), mixed forests (9.48%), woody savannas (1.66%), evergreen broadleaf forests (0.02%), and evergreen needleleaf forests (0.001%). About 56.72% of the fire was recorded in grasslands in 2021 followed by deciduous broadleaf forests (20%). Month-wise analysis shows that the highest fire points were recorded in March, April, and May collectively contributing to more than 92% of fire based on the observation from 2012 to 2021 (Figs. 8.2 and 8.3). Overall, 10 years of analysis show an increasing trend of forest fires in Madhya Pradesh.

Comparison Between MODIS and VIIRS S-NPP Fire

The comparison of forest fire incidences recorded by MODIS and VIIRS S-NPP shows that VIIRS recorded the highest number of fire counts (Table 8.4). The monthly forest fire incidences recorded between 2012 and 2021 had shown a significant change in the number of fires by MODIS and VIIRS. This change may be attributed to the difference in the spatial resolution of their thermal bands. The thermal band of VIIRS has a 375 m resolution per pixel, while MODIS has a 1000 m resolution per pixel. This higher resolution of VIIRS enables it to detect fire that MODIS overlooks. Ten years of data on forest fire by both VIIRS and MODIS also show an increasing trend of forest fire in Madhya Pradesh.

Table 8.3 VIIRS S-NPP fire counts in different vegetation from 2012 to 2021

Vegetation type	Number of fire incidence									
	2012	2013	2014	2015	2016	2017	2018	2019	2020	2021
GL	15,074	5575	4495	3375	14,209	15,429	16,651	10,715	5721	22,633
DBF	3641	2308	1912	1009	6284	6612	6295	3887	1038	8134
MF	1848	1024	770	987	5092	5444	4530	1414	375	3804
SA	1808	670	529	531	2876	4053	3799	1363	354	3680
WS	298	93	78	163	595	667	697	236	83	567
EBF	0	0	0	5	1	37	6	7	0	1
ENF	0	0	0	0	0	0	0	2	0	1
Total	22,669	9670	7784	6070	29,057	32,242	31,978	17,626	7571	38,821

GL = Grasslands, DBF = Deciduous Broadleaf Forests, SA = Savannas, MF = Mixed Forests, WS = Woody Savannas, EBF = Evergreen Broadleaf Forests

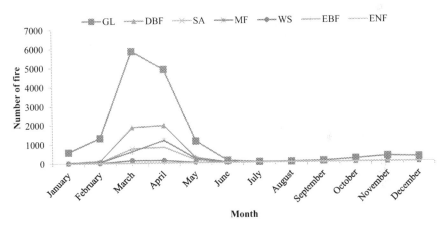

Fig. 8.2 Monthly fires (MODIS) recorded from 2012 to 2021. [GL = Grasslands, DBF = Deciduous Broadleaf Forests, SA = Savannas, MF = Mixed Forests, WS = Woody Savannas, EBF = Evergreen Broadleaf Forests, ENF = Evergreen Needleleaf Forests]

Forest Fire Hotspot Analysis

Forest fire hotspots analyze for both MODIS and VIIRS fire points separately (Fig. 8.4). Raster density values of forest fire were generated with the help of the point density of the Spatial Analyst tool in ArcGIS based on utilizing all fire points from 2012 to 2021 in Madhya Pradesh. In Fig. 8.4, high-density values are represented by the color dark red and low values by light pink. Most of the southeast area of Madhya Pradesh was affected by the forest fire. District-wise analysis was done and found that the highest fire counts were recorded in the Balaghat district followed by Raisen, Sidhi, and Betul districts.

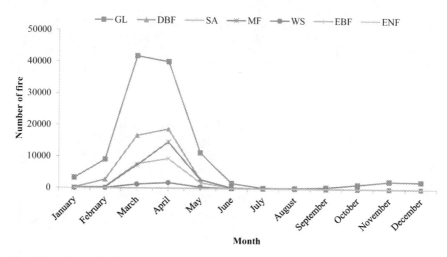

Fig. 8.3 Monthly fires (VIIRS S-NPP) recorded from 2012 to 2021. [GL = Grasslands, DBF = Deciduous Broadleaf Forests, SA = Savannas, MF = Mixed Forests, WS = Woody Savannas, EBF = Evergreen Broadleaf Forests, ENF = Evergreen Needleleaf Forests]

Table 8.4 Comparative evaluation of monthly forest fire recorded for MODIS and VIIRS S-NPP (2012–2021)

Month	MODIS	VIIRS S-NPP
January	661	3596
February	1522	12,423
March	9205	74,640
April	8962	84,515
May	1771	19,070
June	151	2190
July	25	214
August	28	168
September	62	484
October	167	1359
November	304	2424
December	272	2402
Total	23,130	203,485

Seasonal Analysis of Forest Fire

Spatio-temporal patterns of forest fire counts were calculated during fire season (FMAMJ) and non-fire season (JASONDJ) in Madhya Pradesh (Fig. 8.5). Forest fire count density sharply increased in 2016, 2017, 2018, and 2021 during the fire season, whereas in 2018 during the non-fire season. The highest fire counts were recorded during the fire season (FMAMJ). Chances of rainfall are very low during fire season (FMAMJ) and it also helps decrease soil

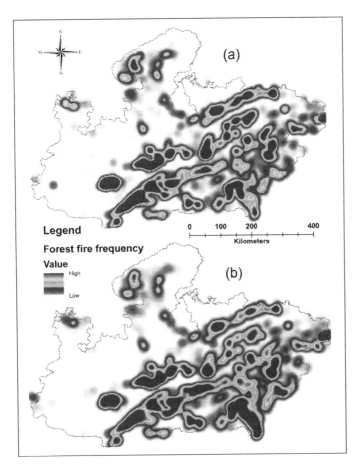

Fig. 8.4 Forest fire hotspot map of Madhya Pradesh prepared by integrating all forest fire points (2012–2021): (**a**) represented MODIS fire hotspot and (**b**) represented VIIRS S-NPP fire hotspot

moisture. Thus, low precipitation and low soil moisture create a drier environment conducive to fire ignition during the fire season (FMAMJ). In addition, in central India, most deciduous trees shed their leaves by the end of January and therefore the availability of dry leaf litter and dry grasses is significantly high during fire season (FMAMJ) (Reddy et al., 2017).

Forest Fire with Meteorological Analysis

Statistical associations based on Pearson correlation were analyzed among forest fires and selected meteorological variables (Table 8.5). In Table 8.5, MODIS forest fire counts show a significant positive correlation with relative humidity ($r = 0.755$, $p = 0.005$), wind velocity ($r = 0.786$, $p = 0.002$),

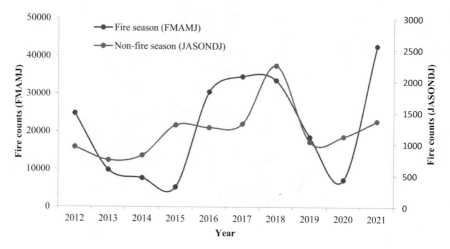

Fig. 8.5 All forest fire counts (MODIS and VIIRS S-NPP) during forest fire season (FMAMJ) and non-fire season (JASONDJ) from 2012 to 2021

Table 8.5 Pearson Correlation (r) among forest fire counts and meteorological variables

	RH	WV	T_{max}	T_{min}	SM	P	MODIS	VIIRS
RH	1	0.982**	0.964**	0.972**	0.929**	0.841**	0.755**	0.762**
WV	0.982**	1	0.989**	0.934**	0.962**	0.884**	0.786**	0.793**
T_{max}	0.964**	0.989**	1	0.915**	0.976**	0.914**	0.819**	0.828**
T_{min}	0.972**	0.934**	0.915**	1	0.857**	0.735**	0.650*	0.650*
SM	0.929**	0.962**	0.976**	0.857**	1	0.972**	0.906**	0.910**
P	0.841**	0.884**	0.914**	0.735**	0.972**	1	0.969**	0.974**
MODIS	0.755**	0.786**	0.819**	0.650*	0.906**	0.969**	1	0.998**
VIIRS	0.762**	0.793**	0.828**	0.650*	0.910**	0.974**	0.998**	1

Note RH, relative humidity (fraction); WV, wind velocity (m/s); T_{max}, maximum temperature (°C); T_{min}, minimum temperature (°C); SM, soil moisture (kg/m^2); P, precipitation (mm/day); MODIS, MODIS fire counts; VIIRS, VIIRS fire counts
**Correlation is significant at the 0.01 level (2-tailed)
*Correlation is significant at the 0.05 level (2-tailed)

temperature ($r = 0.819$, $p = 0.001$), soil moisture ($r = 0.906$, $p = 0.000$), and precipitation ($r = 0.969$, $p = 0.000$). The VIIRS fire counts also show a positive correlation with relative humidity ($r = 0.762$, $p = 0.004$), wind velocity ($r = 0.793$, $p = 0.002$), temperature ($r = 0.828$, $p = 0.001$), soil moisture ($r = 0.910$, $p = 0.000$), and precipitation ($r = 0.974$, $p = 0.000$).

8.4 Conclusion

Analysis of ten years of forest fire occurrence from 2012 to 2021 indicates an increasing trend in Madhya Pradesh and also present study evaluated the frequency of forest fires and their correlation with various meteorological parameters was observed. The study revealed that meteorological parameters are strongly correlated with forest fires. It was found that precipitation has the highest impact on forest fire events. The analysis of forest fire hotspots and their relationship with meteorological parameters give a better comprehension of future forest fire events that will help in the control, prevention, and mitigation of forests. A special focus should be made on the fire hotspot areas to alert about fire trends to the nearest respective administrative headquarters that will help to take action to control the extent of fire damage. The application of remote sensing and GIS techniques can be scientifically used to study forest fire. In addition, there is an urgent need to implement a forest fire policy by the government. Thus, an integrated approach would help with its prevention and mitigation.

Acknowledgements The author is thankful to USGS and DIVA-GIS for providing free access to forest fire data.

Conflict of Interest The author declares no conflict of interest.

References

Cochrane, M. A. (2003). Fire science for rainforests. *Nature, 421*(6926), 913–919. https://doi.org/10.1038/nature01437

Coomes, D. A., Dalponte, M., Jucker, T., Asner, G. P., Banin, L. F., Burslem, D. F. R. P., Lewis, S. L., Nilus, R., Phillips, O. L., Phua, M. H., *and* Qie, L. (2017). Area-based vs tree-centric approaches to mapping forest carbon in Southeast Asian forests from airborne laser scanning data. *Remote Sensing of Environment, 194*, 77–88. https://doi.org/10.1016/j.rse.2017.03.017

FAO. (2020). Global Forest Resources Assessment 2020. In *Global Forest Resources Assessment, key findings*. FAO. https://doi.org/10.4060/ca8753en

ISFR. (2021). *India State of Forest Report (Chapter 5 Forest Fire Monitoring)* (p. 140).

Jaiswal, R. K., Mukherjee, S., Raju, K. D., *and* Saxena, R. (2002). Forest fire risk zone mapping from satellite imagery and GIS. *International Journal of Applied Earth Observation and Geoinformation, 4*(1), 1–10. https://doi.org/10.1016/S0303-2434(02)00006-5

Kumari, B., *and* Pandey, A. C. (2020). MODIS based forest fire hotspot analysis and its relationship with climatic variables. *Spatial Information Research, 28*(1), 87–99. https://doi.org/10.1007/s41324-019-00275-z

Lamat, R., Kumar, M., Kundu, A., *and* Lal, D. (2021). Forest fire risk mapping using analytical hierarchy process (AHP) and earth observation datasets: a case study in the mountainous terrain of Northeast India. *SN Applied Sciences, 3*(4), 1–15. https://doi.org/10.1007/s42452-021-04391-0

Nasi, R., Dennis, R., Meijaard, E., Applegate, G., *and* Moore, P. (2002). Forest fire and biological diversity. *Unasylva*, *53*, 36–40.

Pinol, J., Terradas, J., *and* Lloret, F. (1998). Climate warming, wildfire hazard, and wildfire occurrence in coastal eastern spain. Kluwer Academic publischers. Netherlands. *Climatic Change*, *38*, 345–357. http://citeseerx.ist.psu.edu/viewdoc/download?doi=10.1.1.457.4445andrep=rep1andtype=pdf

Prentice, I. C., Kelley, D. I., Foster, P. N., Friedlingstein, P., Harrison, S. P., *and* Bartlein, P. J. (2011). Modeling fire and the terrestrial carbon balance. *Global Biogeochemical Cycles*, *25*(3), 1–13. https://doi.org/10.1029/2010GB003906

Ray, T., Malasiya, D., Dar, J. A., Khare, P. K., Khan, M. L., Verma, S., *and* Dayanandan, A. (2019). Estimation of greenhouse gas emissions from vegetation fires in Central India. *Climate Change and Environmental Sustainability*, *7*(1), 32. https://doi.org/10.5958/2320-642x.2019.00005.x

Reddy, C. S., Jha, C. S., Manaswini, G., Alekhya, V. V. L. P., Vazeed Pasha, S., Satish, K. V., Diwakar, P. G., *and* Dadhwal, V. K. (2017). Nationwide assessment of forest burnt area in India using Resourcesat-2 AWiFS data. *Current Science*, *112*(7), 1521–1532. https://doi.org/10.18520/cs/v112/i07/1521-1532

Roy, P. S. (2003). Forest fire and degradation assessment using satellite remote sensing and geographic information system. *Satellite Remote Sensing and GIS Applications in Agricultural Meteorology*, 361–400.

Sannigrahi, S., Pilla, F., Basu, B., Basu, A. S., Sarkar, K., Chakraborti, S., Joshi, P. K., Zhang, Q., Wang, Y., Bhatt, S., Bhatt, A., Jha, S., Keesstra, S., *and* Roy, P. S. (2020). Examining the effects of forest fire on terrestrial carbon emission and ecosystem production in India using remote sensing approaches. *Science of The Total Environment*, *725*, 138331. https://doi.org/10.1016/j.scitotenv.2020.138331

Thompson, M. P., Haas, J. R., Gilbertson-Day, J. W., Scott, J. H., Langowski, P., Bowne, E., *and* Calkin, D. E. (2015). Development and application of a geospatial wildfire exposure and risk calculation tool. *Environmental Modelling and Software*, *63*, 61–72. https://doi.org/10.1016/j.envsoft.2014.09.018

Tian, X. R., Zhao, F. J., Shu, L. F., *and* Wang, M. Y. (2014). Changes in forest fire danger for south-western China in the 21st century. *International Journal of Wildland Fire*, *23*(2), 185–195. https://doi.org/10.1071/WF13014

Vadrevu, K. P., Eaturu, A., *and* Badarinath, K. V. S. (2010). Fire risk evaluation using multicriteria analysis-a case study. *Environmental Monitoring and Assessment*, *166*(1–4), 223–239. https://doi.org/10.1007/s10661-009-0997-3

Wotton, B. M., Nock, C. A., *and* Flannigan, M. D. (2010). Forest fire occurrence and climate change in Canada. *International Journal of Wildland Fire*, *19*(3), 253–271. https://doi.org/10.1071/WF09002

CHAPTER 9

Coexistence and Conflict—Case Study on Colonial Waterbirds in Southern India

Vaithianathan Kannan

9.1 Introduction

Human–Wildlife Conflict (HWC) emerged as a research area in the late 1990s and has developed rapidly since around 2005. The term encompasses a range of negative interactions between humans and wildlife species of various taxa, giant carnivores and elephants. Since 2000, there have been numerous scientific publications on this subject. Most research is applied to solve problems, including conflict trends and patterns and their determinants, spatial risk modelling and prediction, and assessment of the effectiveness of designed solutions by conservation biologists seeking to develop knowledge on the research front. In addition, numerous studies have examined local attitudes and perceptions of conflict and the animal species involved. Many researchers have focused on the entanglement between human–wildlife conflicts. Research has begun to treat HWC management as a governance issue. Thus there is an increasing amount of literature that applies social science research methods to understand the socioeconomic, political and cultural contexts in which human–wildlife conflicts are embedded. Recently, there has been growing

V. Kannan (✉)
Gujarat Institute of Desert Ecology, Bhuj, Gujarat, India
e-mail: kannan.vaithianathan@gmail.com

AVC College (Autonomous), Mayiladuthurai, Tamil Nadu, India

Bombay Natural History Society, Mumbai, India

© The Author(s), under exclusive license to Springer Nature Singapore Pte Ltd. 2023
S. Nautiyal et al. (eds.), *Palgrave Handbook of Socio-ecological Resilience in the Face of Climate Change*,
https://doi.org/10.1007/978-981-99-2206-2_9

interest in academic and public debates on replacing the narrative of 'human–wildlife conflict' with the history of 'human–wildlife coexistence'. But the conceptualisation of 'conflict' and 'coexistence' leaves essential questions.

In India, the breeding sites of colonial waterbirds of the order Pelecaniformes and Ciconiiformes (i.e., pelicans, darters, cormorants, egrets, herons, ibises, spoonbills and storks) are often associated with villages. The occurrence of these nesting colonies (collectively and commonly termed as heronries) near human habitation is due to tolerance or/and protection extended by villagers due to religious or sentimental reasons. Compilation of data on heronries in the early 1990s (Subramanya 1996, followed by subsequent publications, some with updated data: Subramanya 2001, 2005a, 2005b) based primarily on literature review and questionnaires circulated to ornithologists and birdwatchers, reported a total of 533 heronries (existing or defunct) in India with about 53% of these located in or around human habitation. This statistic shows village communities' role in conserving breeding sites of heronry species in India. However, due to the 'erosion of traditional beliefs,' changing lifestyles, and their related reasons, the support or tolerance by local communities towards village heronries is waning Manakadan and Kannan (2003) in the case of the Spot-billed Pelican *Pelecanus philippensis*). The data collated by Subramanya (1996) revealed that as much as 32.5% of the 533 known heronries had already been lost by the early 1990s, many of these village heronries.

After this inventory, and concerning southern India, only the breeding sites of the Spot-billed Pelican have been subject to systematic surveys (Manakadan and Kannan 2003; Kannan and Manakadan 2005). Otherwise, no systematic study of other heronry species has been undertaken in southern India. The information available on these species to date (and earlier) is primarily based on accounts of birdwatchers. Dependence on secondary data and untrained birdwatchers, especially in large heronries, tend to be prone to errors due to non-adherence to systematic counting procedures, and wrong timing of counts, besides constraints of time and workforce required for conducting intensive surveys and counts (Urfi et al. 2005).

Furthermore, given India's significant human population growth and the scenario of solid development in India since the beginning of the twenty-first century, it is necessary to carry out a systematic census to assess the current situation and identify conservation issues they face. In addition, the prospects of heronries need to be evaluated so that sites with the potential for long-term conservation of herons can be pre-selected for conservation initiatives for specific management and protection of each location. Indeed, it is very likely that many of the extant herons will not survive in the future, based on what has been lost in the recent past and considering the increasing pressure on natural resources due to a growing population. Indeed, it is very likely that many of the extant herons will not survive in the future, based on what has been lost in the recent past and considering the increasing pressure on natural resources due to a growing human population. Lastly, as the survival of heronries is also dependent on conditions in their foraging grounds, there is also a need to document problems facing these sites, as wetlands in India are under severe threats from human population pressures, especially since the past 2–3 decades (Scott 1989; Anon 1993; Lee Foote et al. 1996; Vijayan et al. 2004).

Many studies are descriptive or draw conclusions based on one location or species with limited applicability to other taxa and regions. More comparative and predictive studies are needed that are explicitly designed to test the generalisation of hypotheses. For example, many studies have found that conflict tends to increase closer to protected areas, but these observations are rarely compared to findings from other regions. Similarly, numerous studies have described and evaluated individual compensation programmes. Still, few studies have explicitly tested assumptions about factors that might influence the success of these programs by setting up experiments to control for specific variables (e.g., the amount or timing of compensation payments). A growing number of studies are moving beyond simple surveys, using increasingly sophisticated and rigorous quantitative methods, and adapting analytical approaches from other disciplines to assess Conflict (Liberg et al. 2011). One step towards the practical evaluation of population impacts is to consider the science of sampling design on population monitoring (Creel et al. 2015).

9.2 Materials and Methods

Surveys and studies were carried out on the Spot-billed Pelican during the nesting season to study the ecology of the species across breeding sites in southern India. Surveying colonial waterbirds is difficult as they are vulnerable to human intrusion and often located in remote areas with limited access. Colonial waterbirds concentrate their nesting activity in a few locations, which are highly susceptible to disease, predation, weather events and other disturbances (Sovada et al. 2005, 2013). Large fluctuations in abundance often go unnoticed for colonial waterbirds because censuses or censuses are not routinely performed, inventory methods are inconsistent, estimates are made and reliability is unknown (Hutchinson 1979). This study focused on colonised waterbirds at selected sites in southern India. Understanding coexistence and conflict at their breeding sites and the countries where waterbirds are most exposed to climate change and adaptation challenges from the human population is the principal objective of this study. However, intensive studies were carried out at the nesting colonies of Nelapattu and Telineelapuram in Andhra Pradesh, Kokkare-Bellur in Karnataka, and Koonthankulam in Tamil Nadu.

The study was confined to the states of Andhra Pradesh (12° 40′ N to 76° 45′ E), Karnataka (11° 30′ N to 74° 10′ E), and Tamil Nadu (8° 04′ N to 76° 14′ E) in southern India. The intensive study was carried out at Nelapattu (13° 51′ N to 79° 57′ E), Telineelapuram (19° 07′ N to 84° 40′ E) in Andhra Pradesh; Kokkare-Bellur (12° 13′ N to 77° 05′ E) in Karnataka; and

Koonthankulam (8° 28' N to 77° 43' E) in Tamil Nadu located in southern India (Fig. 9.1).

Nelapattu is a bird sanctuary and an old pelicanry that encompasses an area of 458.92 ha, of which 82.56 ha constitute the tank area remaining forms the reserve forest surrounding the tank. Initially, the birds were reported to be nesting in the Nelapattu village. The locals intentionally distributed the pelicans to make them shift the colony due to the problem of the bird droppings, which now continue in the tank inside the sanctuary limit (Nagulu 1983; Krishnan 1993; Narasimheulu 1995; Scott 1989). Telineelapuram Pelicanry is a village pelicanry located in the Telineelapuram village of the Srikakulam district of Andhra Pradesh. This pelicanry is of recent origin, and the pelican's nest is in the vicinity of the villagers. According to the locals and forest department, it has been there for the last 20 years. Kokkare-Bellur pelicanry is believed to be the pelicanry that T.C. Jerdon discovered in the '*Carnatic*' in the 19 century (Jerdon 1877) and was rediscovered again in the 1970s (Neginhal 1977). Pelicans used to nest in the villages of Kokkare-Bellur and Bannali villages, but the birds shifted totally to Kokkare-Bellur over the years (Kannan and Manakadan 2005). Before its discovery in the 1960s, more than 2000 birds were bred in the villages (Subramanya 1995; Subramanya and Manu 1996). Koonthankulam Pelicanry is one of the oldest pelicanry discovered during the nineteenth century (Rhenius 1907), and the pelicans built their nest in the village trees in the human vicinity. Later the pelicans moved to the nearby tank in the village (Webb-Peploe 1945; Wilkinson 1961; Mangalaraj Johnson 1971).

9.3 RESULTS AND DISCUSSION

Pelicans breeding in villages in southern India do not have thick tree cover; however, they use traditional sites and nesting trees. The survey found that all the colonies were close to the human settlements and roads (Table 9.1).

Almost 33% of the breeding trees of pelicans at sites in southern India belonged to the private growing near their house or in the backyard. The rest were protected, also the nesting trees used by Spot-billed Pelican and other colonial nesting species comprised seven species of tree, among which the *Barringtonia acutangula* and *Acacia nilotica* were the most frequently used. In southern India, the Spot-billed Pelican and other colonial nesting species are associated during breeding. Six different bird species are found breeding with pelicans of south India. The Oriental White Ibis *Threskiornis melanocephalus*, the Painted Stork *Mycteria leucocephala* and the Open-billed Storks *Anastomus oscitans* are the major associated birds found breeding with pelicans in southern India which is almost 50%. Other species include Little Cormorant *Phalacrocorax niger*, Indian Shag *Phalacrocorax fuscicollis* and Black-crowned Night Heron *Nycticorax nycticorax* found only at Nelapattu and Uppalapadu. Nesting requirements of similar colonial nesting species may require adequate nesting trees and nesting requirements, and social factors

9 COEXISTENCE AND CONFLICT—CASE STUDY ... 115

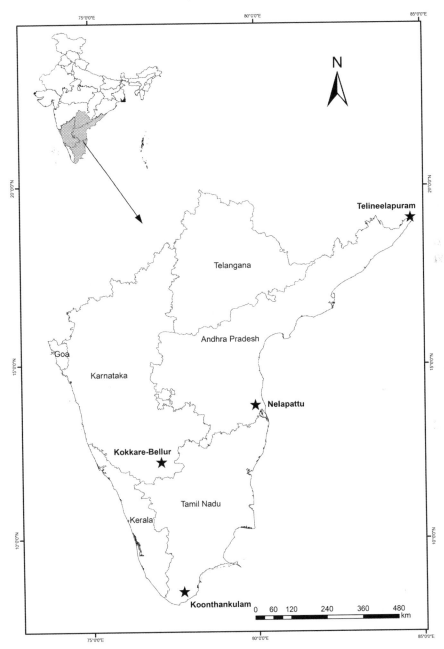

Fig. 9.1 Map showing the study sites of colonial waterbirds breeding in southern India

Table 9.1 Characteristic features of breeding colonial waterbirds at the villages in southern India

State	Colony/breeding site	Nearest house/ human settlements (km)	Nearest road (km)	Associated with other colonial breeding birds
Andhra Pradesh	Nelapattu	1.0	0.5	Open-billed Stork, Oriental White Ibis, Little Cormorant, Indian Shag, Night Heron
	Telineelapuram	0.5	1.0	Painted Stork
Karnataka	Kokkare-Bellur	0.5	1.0	Painted Stork
Tamil Nadu	Koonthankulam	1.0	1.0	Oriental White Ibis, Painted Stork

influence the selection of nesting trees and nesting (Donazar et al. 1994). For successful long-term conservation of colonial nesting species, including the Spot-billed Pelican, people's participation and raising fast-growing adequate nesting trees in the existing colonies for greater protection is warranted.

A bird colony is usually composed of a large congregation of individuals of one or more species. These species can closely nest or spend the night (known as roosting) at a particular location. So, colonies can be nesting or roosting and can be made up of single or multiple species. The presence of waterbirds can indicate a healthy aquatic ecosystem. The fact is that these species are protected by the Indian Wildlife Protection Act 1972 on hunting, and the protection of wild birds is too often overlooked in India. Colonial waterbirds like egrets, cormorants and Asian open-bill live in flocks near human settlements. Their presence is a vital sign of healthy wetlands and positive human–wildlife relationships. **Harmonious living with these waterbirds can restore the tarnished human–wildlife relationship.**

Furthermore, promoting bird villages can repair the damaged human–wildlife relationship. For a long, southern India has been a hot spot for human-induced wildlife mortality. In many of the traditional sites of south India, birds that rely on trees for nesting face difficult conditions—not least due to a lack of human acceptance. The villagers claim that bird droppings are a health hazard.

Formalising an ecological model depicting the key environmental components and the underlying cause-and-effect processes is required for successful conservation management. Over the past century, the breeding abundance of these colonial water species in the study area exhibits relationships between rainfall and breeding in frequencies and total quantities. The likelihood of breeding for these colonial waterbirds also increased. Management of complex ecosystems depends on a good understanding of the responses of organisms to the main drivers of change. Building a credible, long-term inventory of information on colonial waterbirds may help to keep these birds' populations resilient in the face of sea-level rise and human development. Research and protection efforts must engage local communities and the public through innovative outreach and social marketing to reduce human disturbance, increase awareness, and broaden public protections for birds that breed, winter, and migrate in southern India. **Establishing and promoting safe refuge for these birds in rural areas may make our conservation successes and fosters human–wildlife coexistence**.

Several studies on monitoring birds show that waterbirds are changing their distribution in response to climate change. The significant mismatch between the areas where climate change adaptation measures are most needed and those with adequate financial and technical capacity is a significant challenge that remains to be addressed. Therefore, donors supporting climate change adaptation must adopt a more integrated approach to food and water security, disaster risk reduction and biodiversity conservation, focusing on nature-based solutions that benefit both people and biodiversity. Birds are among the best-studied and best-monitored organisms, but many have been threatened with extinction due to habitat loss, over-exploitation and pollution; their population at the global level can provide us with biogenic information about species and ecosystems and its effects on climate change on species and ecosystems (Pearce-Higgins and Green 2014).

Colonial waterbirds and other birds are attracted to the rural areas and wetlands in southern India on the East Coast. With global climate change likely to affect extreme weather patterns (floods, droughts, hurricanes and tsunamis) around the world, it's crucial to understand wildlife responses and weather events can inform conservation efforts. Understanding the traditions that live outside of protected areas and perpetuating them is the key to combatting the effects of extreme weather events. Climate change is the leading cause of species extinction. The distribution of 47% of flightless terrestrial mammals and 23% of threatened bird species may have been negatively affected by climate change, as reported in the participation and inclusiveness in the Intergovernmental Science–Policy (Larigauderie and Mooney 2010).

Similar reflections in India's National Wildlife Action Plan 2017–2031 (NWAP) are that the country's protected areas were designed when climate change 'could hardly be the norm' to conserve wildlife. Furthermore, it emphasises that species will need to disperse into more suitable habitats to respond to climate change and calls for proper planning, appropriate wildlife migration, and management plans to reduce disaster risk and climate change adaptation (https://wii.gov.in/images/images/documents/national_wildlife_action_plan/NWAP_Report_hi_Res_2017_31.pdf).

Avian responses to climate change are interesting because birds are comparatively easy to identify and measure and their consequence to environmental disturbances are relatively well-known; they are valuable indicators of ecological change (Niemi and McDonald 2004). In addition, the birds are of conservation concern and challenge as their population is declining globally with a high risk of extinction (Vie et al. 2009). Birds play a vital role for humans in terms of health, well-being, pest control, sanitation, seed dispersal and pollination (Sekercioglu et al. 2004).

Bird distribution is closely related to winter and summer temperatures. Rising temperatures due to climate change could, directly and indirectly, affect birds by forcing them to use more energy to regulate heat. Their baseline condition for the energy needed to maintain their activity levels for reproduction and migration reduces suitability or health. Nevertheless, habitat and other resources may or may not be sufficient to fit the needs of birds due to the range shifting to other areas with more suitable thermal conditions over time (Devictor et al. 2008).

However, indirect effects related to climate change impact the asynchrony of bird reproduction with food sources. Many birds synchronise their nesting cycles so that their young's peak food requirements coincide with the time of peak food availability (Visser et al. 2006). Dramatic changes in global climate in the past and the current climate challenge for species and ecosystems are not just the degree of change but also the speed. Changing environmental conditions faster is likely to overwhelm many birds' ability to adapt through natural selection (Visser 2008). This has led to identifying species characteristics linked with vulnerability to climate change biological processes (Cormont et al. 2011). Reduction of anthropogenic pressures and their amplifying impact on bird populations to mitigate the effects of climate change that can interact with the resilience of habitats.

Cohabitation is defined as a dynamic but persistent state in which humans and wildlife adapt to living in a typical landscape, thus ensuring the long-term survival of wildlife populations, social legitimacy and tolerable risk (Madden and McQuinn 2014). Researchers have only recently used the term 'coexistence'; scientific focus on human–wildlife coexistence emerged much earlier in the 1970s. However, interactions can be positive or negative. On the other hand, conflicts (human–wildlife) are defined as interactions between wildlife and humans with a negative outcome (Madden 2004) in terms of damaging crops, injuring or killing domestic animals and threatening or killing people. The literature on HWCs, interactions and coexistence has grown exponentially in the last 20 years (König et al. 2020) (Fig. 9.2).

In India, conflict tends to be associated with large animals such as elephants, wild pigs, black buck and nilgai that raid crops. In contrast, big cats or elephants rarely bears those conflicted with human causing injury or death. Birds are seldom in the 'conflict' picture as pests—they can be taken care of, and few bird species are a source of the problem for humans. Birds come into conflict through direct and indirect. The immediate competition that birds compete with agriculture compete with fishermen or impacts the earnings of fish farmers, soil buildings and surroundings in urban set-ups or poses a serious threat to aircraft. The indirect conflict caused by birds are carriers of diseases transmitted to domesticated animals, pets and humans. Fish-eating birds, especially the larger *and* flocking species like pelicans, cormorants, storks, herons

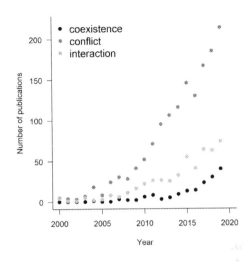

Fig. 9.2 Number of literature published on coexistence, conflicts and interaction (*Source* König et al. 2020)

and egrets, have been a threat to fisheries. For example, the piscivorous birds-fishermen conflict in the Indian subcontinent is from Pakistan in the 1980s, which was once numerous over the rivers and *jheels* of Pakistan, but by 1969 almost exterminated because they overate fish! Large birds are considered competitors for fish resources.

During the surveys under the Spot-billed Pelican project, conflict was recorded to be an issue in the fish farms at the edges of Kolleru Lake, Andhra Pradesh. It was recorded that carcasses of a few waterbirds entangled in nets placed over the fish ponds, and cormorants and pelicans attempted to forage in the fish ponds. In the Pulicat Lake area, the conflict was restricted to prawn farms. Measures adopted to deter birds were guarding and scaring; also, lethal methods were used. The case in urban areas is the urban birds that conflict with people. In southern India, the existence of birds to the protection people has on sentimental or religious grounds, and in some cases, the relationship was mutually beneficial. Traditional support for heronries in human habitation areas is on the decline. Trees that supported heronries at some sites have been felled to prevent birds from nesting. The lost pelicanry at Kolleru, Andhra Pradesh and Moondradaippu in Tamil Nadu are examples.

The loss of community support to the breeding birds attributed to the noise created, foul smell, litter or faecal matter and other wastes, fouling of water as the same water is used as a source by the locals, lop or cut trees for timber, fuel, fodder or sale and loss of revenue from fruiting trees due to nesting activities appeared to be the reason. In towns, the sharp appreciation in the price of land adds anti-sentiment for heronries in public places. Since the coexistence and conflict of birds with the human-dominated landscape is increasing, decision-makers need evidence-based information to design sustainable management plans; however, providing objective decision support can be challenging as the realities and perceptions of human–wildlife interactions vary widely between and within rural, urban and peri-urban areas. Most importantly, wildlife-related losses should be compensated, and conflict prevention should be subsidised.

9.4 Conclusion and Future Research Needs

Human–wildlife conflict and coexistence theory to ecological perspectives (animal behaviour) in wildlife-induced damages; however, human views include land use, psychology, governance, local attitudes and perceptions and costs and benefits. Therefore, developing a conceptual model with inter- and trans-disciplinary approaches and multilevel governance is sought in a changing scenario with an integrated research focus on socioeconomic, socio-ecological and essential ecological research needed to understand and overcome the challenges of HWC to transform them into coexistence.

Humans and wildlife are the subject of considerable research across all fields, including biodiversity conservation in climate change (Urban 2015). Studies of Conflict in the face of changing climate, including strategies for resilience, how climate change will stress coupled human-natural systems, and how current patterns of conflict are likely to change in the future, are few. Land use, land cover and entire ecosystems are changing because of the changing climate and the growth and scale of human enterprise. Efforts to conserve wildlife populations in the face of these changes, efforts to manage new species assemblages, or successful wildlife restoration may lead to novel challenges. Our increasingly connected planet offers abundant opportunities to consider how to expand the definition or zone of influence of what constitutes a human–wildlife conflict.

We have made significant progress in understanding the importance of human–wildlife conflict, the biological and social factors that influence conflict, and strategies to reduce friction and promote coexistence. Still, the field is in its infancy, and there is much room for further research. The following observations identify gaps and emerging opportunities for future funding related to human–wildlife conflict and coexistence. The technology available to study and mitigate conflict is changing rapidly. The emergence of inexpensive mobile phones and communication networks, digital photographs, satellite imagery, global positioning systems, lighter and longer-lasting radio collars and powerful portable tablet computers are just a few technologies already transforming the study and mitigation of human–wildlife conflict. The recent emergence of inexpensive drones and the global ubiquity of electronic social networks will almost certainly revolutionise how information is gathered and used. Therefore, engaging different disciplines in the studies of human–wildlife conflict, including anthropology, biology (including animal behaviour, conservation biology, ecology, genetics, wildlife ecology, zoology), economics, environmental studies, geography, history, natural resource management, political science and psychology, among others. This disciplinary diversity, with a rich mix of exciting studies, will likely continue to be an incubator for novel ideas in the future.

Acknowledgements I thank Professor (Dr.) Sunil Nautiyal, Head, CEENR, ISEC, Bengaluru, for inviting and supporting me to participate in the 3-Day International Workshop cum Training on 'Green Growth Strategies for Climate Resilience and DRR: Policies, Pathways and Tools'. He has been a source of inspiration to accomplish the task. I wish to record my gratitude to my guru Ranjit Manakadan (BNHS), for his advice and guidance during the study period. I am grateful to my colleague Dayesh Parmer, for helping to map the study area.

References

Anon (1993) Directory of Indian Wetlands. WWF-India and Asian Wetland Bureau, New Delhi.

Cormont A, Vos CC, van Turnhout CAM, Foppen RPB, ter Braak CJF (2011) Using life-history traits to explain bird population responses to increasing weather variability. *Climate Research* 49: 59–71.

Creel S, Becker M, Christianson D, Dröge E, Hammerschlag N, et al. (2015) Questionable policy for large carnivore hunting. *Science* 350: 1473–1475.

Devictor V, Julliard R, Couvet D, Jiguet F (2008) Birds are tracking climate warming, but not fast enough. *Proceedings of the Royal Society B* 275: 2743–2748.

Donazar JA, O Ceballos A, Travaini A, Rodriquez M, Funes, Hirald F (1994) Breeding performance in relation to the nest site substratum in Buffbacked Ibis *Theristicus caudatus* population in Patagonia. *Condor* 96: 994–1002.

Hutchinson AE (1979) Estimating numbers of colonial nesting seabirds—A comparison of techniques: Proceedings of the Colonial Waterbird Group, vol. 3, pp. 235–244.

Jerdon TC (1877) The Birds of India. Calcutta: P.S.D' Rozario and Co.

Kannan V, Manakadan R (2005) Status and distribution of the Spot-billed Pelican Pelecanus philippensis in southern India. *Forktail* 21: 9–14.

König HJ, Kiffner C, Kramer-Schadt S, Fürst C, Keuling O, Ford AT (2020) Human-wildlife coexistence in a changing world. *Conservation Biology* 34: 786–794.

Krishnan M (1993) The Aredu Pelicanry—A factual rejoinder. *Blackbuck* 9: 44–46.

Larigauderie A, Mooney HA (2010) The Intergovernmental science-policy Platform on Biodiversity and Ecosystem Services: Moving a step closer to an IPCC—Like mechanism for biodiversity. *Current Opinion in Environmental Sustainability* 2: 9–14.

Lee Foote A, Pandey S, Krogman N (1996) Process of wetland loss in India. *Environmental Conservation* 23(1): 45–54.

Liberg O, Chapron G, Wabakken P, Pedersen HC, Hobbs NT, Sand HK (2011) Shoot, shovel and shut up: Cryptic poaching slows restoration of a large carnivore in Europe. *Proceedings of the Royal Society B* 279: 910–915.

Madden F (2004) Creating coexistence between humans and wildlife: Global perspectives on local efforts to address human-wildlife conflict. *Human Dimensions of Wildlife* 9: 247–257.

Madden F, McQuinn B (2014) Conservation's blind spot: The case of conflict transformation in wildlife conservation. *Biological Conservation* 178: 97–106.

Manakadan R, Kannan V (2003) A study of the Spot-billed Pelican *Pelecanus philippensis* in southern India with special reference to its conservation. Final Report. Bombay Natural History Society, Mumbai, India.

Mangalaraj Johnson J (1971) The heronry at Koonthankulam, Tirunelveli district, Tamil Nadu. *Newsletter for Birdwatchers* 11: 1–3.

Nagulu V (1983) Feeding and breeding biology of Grey Pelican at Nelapattu Bird Sanctuary in Andhra Pradesh, India. PhD thesis, Osmania University, Hyderabad.

Narasimheulu E (1995) Studies on the ecology of selected coastal wetlands of Andhra Pradesh: Distribution of mangroves and associated flora and fauna with special reference to conservation strategies, PhD thesis, Sri Venkateswara University, Tirupati.

National Wildlife Action Plan 2017–2031 (NWAP). https://wii.gov.in/images/images/documents/national_wildlife_action_plan/NWAP_Report_hi_Res_2017_31.pdf.

Neginhal SG (1977) Discovery of a pelicanry in Karnataka. *Journal of the Bombay Natural History Society* 74: 169–170.

Niemi GJ, McDonald ME (2004) Application of Ecological Indicators. *Annual Review of Ecology, Evolution, and Systematics* 35: 89–111.

Pearce-Higgins J, Green R (2014) Effects of climate change mitigation on birds. In *Birds and Climate Change: Impacts and Conservation Responses* (Ecology, Biodiversity and Conservation, pp. 308–358). Cambridge: Cambridge University Press

Rhenius CE (1907) Pelicans breeding in India. *Journal of the Bombay Natural History Society* 17: 806–807.

Scott DA (1989) A Directory of Asian Wetlands. IUCN, The World Conservation Union.

Sekercioglu CH, Daily GC, Ehrlich PR (2004) Ecosystem consequences of bird declines. *Proceedings of the National Academy of Sciences of the United States of America* 101:18042–18047.

Sovada MA, King DT, Erickson M, Gray C (2005) Historic and current status of breeding American White Pelicans at Chase Lake National Wildlife Refuge, North Dakota: Waterbirds, vol. 28, Special Publication 1, pp. 27–34.

Sovada MA, Pietz PJ, Woodward RO, Bartos AJ, Buhl DA, Assenmacher MJ (2013) American White Pelicans breeding in the northern plains—Productivity, behaviour, movements, and migration: U.S. Geological Survey Scientific Investigations Report 2013-5105, 117 pp.

Subramanya S (1995) Save the Spot-billed Pelican. *Asian Wetland News* 8: 31

Subramanya S (1996) Distribution, status and conservation of Indian heronries. *Journal of the Bombay Natural History Society* 93: 459–486.

Subramanya S (2001) Heronries of Andhra Pradesh. *Mayura* 13: 1–27.

Subramanya S (2005a) Heronries of Tamil Nadu. *Indian Birds* 1(6): 126–140.

Subramanya S (2005b) Heronries of Kerala. *Malabar Trogon* 3(1): 2–15.

Subramanya S, Manu K (1996) Saving the Spot-billed Pelican a successful experiment. *Hornbill* 2: 2–6.

Urban MC (2015) Accelerating extinction risk from climate change. *Science* 348: 571–573.

Urfi AJ, Sen M, Kalam A, Meganathan T (2005) Counting birds in India: Methodologies and trends. *Current Science* 89(12): 1997–2003.

Vie JC, Hilton-Taylor C, Stuart SN (eds.) (2009) Wildlife in a changing world—An analysis of the 2008 IUCN red list of threatened species. Gland, Switzerland: IUCN. 180 pp.

Vijayan VS, Prasad SN, Vijayan L, Muralidharan S (2004) Inland Wetlands of India. Conservation Priorities. Salim Ali Centre for Ornithology and Natural History, Coimbatore.

Visser ME (2008) Keeping up with a warming world; assessing the rate of adaptation to climate change. *Proceedings of the Royal Society B* 275: 649–659.

Visser ME, Holleman LJM, Gienapp P (2006) Shifts in caterpillar biomass phenology due to climate change and its impact on the breeding biology of an insectivorous bird. *Oecologia* 147: 164–172.

Webb-Peploe CG (1945) Notes on a few birds from the south of Tinnevelly district. *Journal of the Bombay Natural History Society* 45: 425–426.

Wilkinson ME (1961) Pelicanry at Kundakulam, Tirunelveli. *Journal of the Bombay Natural History Society* 58: 514–515.

CHAPTER 10

Estimation of Rural Drinking Water Supply in Southern India: A Contingent Valuation Method

A. Xavier Susairaj and A. Premkumar

10.1 Introduction

Water is the most valuable precious natural resource and basic need of human life, animal consumption, and crop production good (Bakker 2007). At the global level water is an economic good and about 97% of water on earth is salt water, 69% in the form of ice caps and around 3% of water on our planet is freshly available for use (UNESCO 2003).

As world population was 7 billion in 2010 and is expected to increase to 8.6 billion by mid-2030 and 9.8 billion by 2050 (WWAP 2012). United Nations Population Division estimated that nearly 50% of the global population will be living with water shortages due to demand out stripping supply with climate change. The annual water demand has increased from 4130 km^3 in 1990 to 5190 km^3 in 2000 to 1654 km^3 in 2010. The present level of water demand is 680 km^3 in 2016. A sharp reduction in the availability of water per capita in the world (Gleick 1999).

India has only 4% of fresh water. India is a water stress country and moving towards a water scarce status to its low per capita water consumption. In 2025, the demand for water will be 170 litres per capita per day (lpcd). The rural domestic water demand was 70 lpcd in 2000 and it is estimated to be 150 lpcd in 2025. India is stricken by low water use efficiency. India gets the average

---

A. X. Susairaj (✉) · A. Premkumar
Department of Economics, Sacred Heart College (Autonomous),
Tirupattur, Tamil Nadu, India
e-mail: xsusairaj@shctpt.edu

© The Author(s), under exclusive license to Springer Nature
Singapore Pte Ltd. 2023
S. Nautiyal et al. (eds.), *Palgrave Handbook of Socio-ecological Resilience in the Face of Climate Change*,
https://doi.org/10.1007/978-981-99-2206-2_10

annual rainfall of 1.19 mm and only 1000 cubic metres water is available for consumption.

The global temperature has increased by 0.2 °C every ten years due to effect of the water crisis and climatic change. If left unchecked, this might lead to water shortage in the world (IPCC 2007). According to the Millennium Ecosystem Assessment (Harsan et al. 2005), the long-term shifts in seasonal weather patterns that define renewable freshwater was caused by the climate variations.

The impacts of climate variation in temperature and rainfall affect the natural calamities such as floods, storms, and hot waves; drought and other weather-related disasters have caused more damage to nature. These major disasters are expected to happen frequently with high intensity because of the variation in climate change. To end all forms of issues such as poverty, hunger, and inequality problems and to control the climate variation; the UN has declared the MDG and SDG goals. The 7th Goal in Millennium Development Goal and SDG 6 Sustainable Goals (UNESCO 2001) emphasize providing access to safe drinking water and sanitation by 2030.

In India the per capita supply of drinking water was 55 lpcd. Through piped and public tap connection in rural and urban areas according to 69th round of NSSO report in 2012. The average size of rural household needs to access the improved drinking water is 88.5% (NSSO 2012), the central government provided funds for an investment in water supply projects were 75% whereas only 11% of the investment by private sector (Mehta 2003). The budgetary allocation to the rural drinking water increased only 2% in the plan period. Census of India 2011 has assessed that only 6.6% of household's access to drinking water within their places and about 43.5% of household access to tap water. According to NSSO 2012, 70% of the urban households had access to piped water supply whereas only 18.7% of the households access the piped drinking water supply from the village panchayat.

Drinking water crisis has now become a common problem in every nook and corner of the state. Tamil Nadu State Environmental Report identified that the total precipitation is 32,909 mcm and ground water level and surface water level is 15,345 mcm and 17,563 mcm (Government of Tamil Nadu 2010).

Poor quality, scarcity of water, and unreliable supply of water affect the life of a human being in various ways, thus, providing safe and adequate water to the people at the right location and at the right price has become an important public policy issue.

Empirical Studies on Contingent Valuation Method

Contingent Valuation Method (CVM) is used to understand the people's preference by adopting survey questions method is used to know about the people's WTP. The advantages of CVM are that it can measure the economic benefits (or damage) of a wide assortment of beneficial effects in a way consistent with economic theory. Some of the empirical studies on CVM are as follows:

Smita and Bishwanatha (2008) studied on WTP for improved water supply in Delhi. A sample of 8000 households was studied and they found that the average WTP for improved water was Rs. 163 per month. Venkatachalam (2015) studied on water market in Chennai. He used stratified random technique to collect data from 302 households with the help of well structured interview schedule. It found that public resource are the major source of water for 97% of households in which 78.5% are public taps and 18.5 are hand pumps. In the study region, family spends 2.76% for water. The study suggests that government should improve existing policies of informal water markets for poor families. According to WHO (2009), 1.8 million deaths were due to water contamination. 2 lakh people in India died without good quality of water. Many researches have been organized in developing countries to measure the WTP for water. Therefore, the present study highlights the WTP for improved drinking water in the study area.

Research Questions

1. Are rural sample households willing to pay for improved quality water?
2. Do socio-economic and demographic factors affect the willingness to pay for quality drinking water?

Objectives

1. To estimate the cost of assessing the drinking water in the study area.
2. To estimate the Willingness to Pay in accessing improved quality of drinking water in the study area.

10.2 Materials and Methods

Selection of Study Area

In the first stage, Tamil Nadu has been selected purposively for this study and in second stage rural area of Tiruvannamalai district has been selected as the study area on the basis of predominating a rural district without big urban or trade centres. After purposively selecting the district level multi stage random sampling method was employed to choose the study area. At the first stage three blocks namely Tiruvannamalai, Thandarampet, and Arni were selected randomly from the 10 blocks of the district. As in the second stage revenue villages from each block were selected randomly. Accordingly, out of 6 revenue villages in 3 blocks, from Tiruvannamalai block two revenue villages Kolakudi

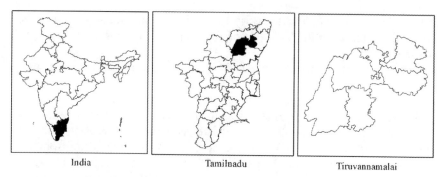

Fig. 10.1 Location of the study area

and Adaiyur were selected. Similarly, Vanapuram and Perugulathur revenue villages from Thandarampet block, Nesal and Velapadi revenue villages from Arni block were selected randomly (Fig. 10.1).

Data Collection

A questionnaire was used to collect the primary data from the households.

Sampling Design

The major occupational groups were identified to select the sample households. Accordingly, five occupational groups were found, 10% of the households in each occupational group were selected randomly. The total number of sample is 540 households from rural households.

10.3 Results and Discussion

Socio-economic Characteristics

Table 10.1 reveals the socioeconomic characteristics of respondents in the study area. It has been found that out of 540 respondents 135 (25%) were males and 405 (75%) were females. Majority of the respondents are females. Age of the respondents given below reflects that only 32 out of 540 respondents (6%) were above 60 years of age. The young respondents (below the age of 30 years) were little more than double of old people (above 60). In the overall analysis more number (33.52%) respondents were found in the 30–40 years age group and 50–60 years age group (31%). The age varied from 19 to 74 years and the mean age calculated came to 42 years. The divisions of the respondents according to Education. 70 percentage of the respondents are SSLC *and* Below, 24% respondents have studied up to Higher Secondary level and only 6 percentage of the respondents are graduates. The majority of the respondents are below 10th standard. It has also been found that there is a correlation between education and employment.

In occupational status, it reveals that out of 540 total number of sample farmers are 46 (8.15%), Agricultural labourers are the maximum of 238 (44.07%), Non-agricultural labourers are 184 (34.07%), Self-employed totally in the sample villages are 38 (7.03%), and the government employees of the above villages stand at 34 (6.29%).

In income status, it shows that 54% of sample household's annual income was below Rs. 100,000, 22% of sample household's annual income ranges between Rs. 100,001 and Rs. 150,000, 15% of sample household's annual income ranges between Rs. 150,001 and Rs. 200,000, and the remaining 9% of the households fall on the category of above Rs. 200,000 which is notably very less compared to the other income ranges.

Table 10.1 Socio-economic characteristics

Variables	Value (N = 540)	Percentage
Gender		
Male	135	25
Female	405	75
Age		
Less than 30 years	67	12.41
30–40 years	181	33.52
40–50 years	93	17.22
50–60 years	167	30.93
Above 60 years	32	5.93
Educational status		
Up to SSLC	378	70
HSC	128	24
Graduates and above	34	6
Occupational status		
Farmers	46	8.51
Agricultural labours	238	44.07
Non-agricultural labours	184	34.07
Self employment	38	7.03
Government employees	34	6.29
Income status		
Less than 100000	294	54.44
100,001–150,000	117	21.66
150,001–200,000	80	14.82
Above 200,000	49	9.08

Source Primary data

Water Supply

Table 10.2 explains the supply of water facilities. As far as source of drinking water is concerned, many households collect water from Public Water Taps (Street Taps) which are operated and maintained directly by local bodies. More than 37% of the households depend upon street public tap. Another 32.78% of the households are getting water from mini power pumps maintained and installed by the local bodies and other agencies and followed by 11.11% of households using bore well, 7.22% of the households getting water from their house-service connection. And out of the 540 households only 25 use Hand Pump as a main source for their drinking water. In water usage status, out of 540 respondents 338 (65.59%) of them get drinking water through public water taps.

The numbers of available public taps are classified into four categories. A total of 200 sample respondents collect water from the public taps. Of them, 65 sample respondents have the public water taps from one to five. This is followed by 82 respondents who avail public taps from five to ten. The number of respondents who avail the public water taps from ten to fifteen and more than 15 were found to be 32 and 21 respectively. The distance between the respondent's households and water taps from which they collect their water is 30 feet. They used to spend 1 hour to collect the water from the public tap. One person was involved in collecting water, most of them were girls and children.

Table 10.2 Water supply

Variables	Value (N = 540)	Percentage
Sources of drinking water		
Public tap	200	37.03
Private vendor	20	3.70
Bore well	61	11.29
Hand pumps	25	4.62
RO	17	3.14
Water supply	39	7.22
Mini power pumps	178	32.96
Usage		
Usage of public water taps	202	37.41
Usage of Non public water taps	338	62.59
Number of public taps (N = 200)		
1–5	65	32.5
5–10	82	41
10–15	32	16
More than 15	21	10.5

Source Primary data

Cost and Sources of Water

Table 10.3 reveals the cost for water system in the study area. Individual water supply connection is popular among the people in the study area but only limited number of the respondents were found to possess individual water connection. Out of 540 respondents in this study, there are 40 respondents who have individual water connection. A total of 28 respondents were found to be incurred less than 5000 rupees and only 12 respondents incurred more than 5000 rupees spent as total connection expenses. 25 respondents were collecting water from hand pumps. A total of 178 respondents are availing water from the mini power pumps. 17 respondents who use the reverse osmosis system, they spent Rs. 7501 to Rs. 15,000 for water purification. The households collecting water from private water vendors for drinking and cooking purpose 20. As far as cost per week for water is concerned, out of 20 respondents 14 of them are paying less than 100 for water per week and only 6 respondents pay more than 100 it may be influenced by size of the household.

There are 60 respondents who are having own bore well depending upon the necessity and size of income of the households. It is evident from the above table the minimum cost of setting up of bore well is below Rs. 100,000 in which there are 14 respondents. There are 16 respondents who spent for the cost of setting up of bore well between Rs. 100,000 to Rs. 200,000 and the majority of 30 respondents spent around more than Rs. 200,000.

Table 10.3 Cost for water system

Variables	Value N = 40	Percentage
Cost incurred for individual water connection		
Less than 5000	12	30.00
Above 5000	28	70.00
Cost for setting up of RO system (N = 540)		
Less than 7500	6	1.11
7501–10,000	3	0.55
10,001–15,000	5	0.92
Above 15,000	3	0.55
Total users	17	3.14
Total Non users	523	96.85
Cost per week for water (N = 20)		
Less than 100	14	70.00
More than 100	6	30.00
Cost for setting up Own Bore well (N = 60)		
Less than 100,000	14	23.34
100,001–200,000	16	26.66
Above 200,000	30	50.00

Source Primary data

Theoretical Framework

CVM is used to estimate the monetary values of environmental goods and services. Conceptually, WTP is the amount paid by the person for the water.
Therefore utility function as

$$U = U(x, z, Q)$$

where,

- U Utility
- x Quality of Water
- z Composite of other market goods
- Q Quantity of water

$$E\left(P_x^0, P_z^0, Q^0, u^0\right) = \text{Min}\left\{P_x^0 X + P_z^0 Z | U(X, Z, Q) = u^0\right\}$$

where P_x is the price of X and P_z is the price of Z. Then, the expenditure function is increasing in $P = P(P_x, P_z)$ and u (i.e., $E_p(P_x, P_z, Q, u) > 0$ and $E_u(P_x, P_z, Q, u) > 0$), and it is decreasing in Q (i.e., $E_Q(P_x, P_z, Q, u) < 0$).

The water quality (Q) is improved from Q^0 to Q^1, but the prices of X and Z remain constant. The position of the economy is changed from S^0 to $S^1 = (P_x^0, P_z^0, Q^1)$. Then the minimum expenditure that the individual must spend to achieve a utility level, u^0, at the new state of the economy, S^1, is $E(P_x^0, P_z^0, Q^1, u^0)$. Since $E_Q(P_x, P_z, Q, u) < 0$ by the assumption, $E(P_x^0, P_z^0, Q^1, u^0) < E(P_x^0, P_z^0, Q^0, u^0)$. Therefore, individual's welfare function for Q is determined by absolute value terms as

$$CS = \left| E\left(P_x^0, P_z^0, Q^0, u^0\right) - E\left(P_x^0, P_z^0, Q^1, u^0\right) \right|$$

The measure of Hicksian compensating surplus (CS).

WTP weighted average is an appropriate measure for computing aggregate WTP than either of other approaches (dichotomize model) in case of non-responses bias was deducted. In order to get quality of drinking water the respondents were asked to rate on the five-point scale 1-being very poor 5 for very good.

Estimation of WTP Function

$$\text{WTP}_i = X_i^1 \beta + e_i.$$

The ordinary least squares (OLS) method is used to estimate the unknown parameters, β_{OLS}. However, if the WTP data are censored negative values appear as zero, then the ordinary least squares coefficient estimates would be less desirable (Maddala 1983; Greene 1994). The proper method for estimating unknown parameters from CV data that contain a large number of zero values for WTP is a censored Tobit model. The censored Tobit model is of the form:

$$WTP = X_i^1 \beta + e_i \text{ if } WTP\, T_i > 0$$

$$WTP = 0 \text{ if } WTP\, T_i \leq 0$$

The expected value of WTP$_i$ is obtained from

$$E[WTP_i] = X_i^1 \beta_{tobit} \cdot \varphi\left(X_i^1 \beta_{tobit}/\sigma\right) + \sigma \cdot \varphi\left(X_i^1 \beta_{tobit}/\sigma\right)$$

The specific model for factors influencing household willingness to pay for improved water quality is as follows.

$$WTP = F(ARO, AWT, TASTE, HARD, COLOR,$$
$$AGE, GENDER, EDUC, INCOME, BILL)$$

Table 10.4 presents the description of explanatory variables used in willingness to pay function. It is predicted that purchased bottled water (ARO) and household water treatment device (AWT). AWT is predicted to be relatively positive to WTP. The perceptions of water quality characteristics (COLOUR and TASTE) are predicted to be a negative influence on WTP. When consumers assume that their water quality is high then there will be less willingness for improved quality of water.

Educational level (EDUC) of the respondents determines their awareness on drinking water quality. Therefore, it assumed that highly educated people may have more awareness on water quality and they may have high willingness

Table 10.4 Explanatory variables included in the WTP function

Variable	Description	Standard mean	Deviation
ARO	Purchased bottled water	0.346	0.448
AWT	Using water treatment device in home	0.414	0.464
TASTE	Taste of water	2.891	1.122
HARD	Hardship of water	2.911	1.281
COLOR	Colour of water	2.687	1.284
AGE	Age of the respondent	48.31	17.595
GENDER	Gender of the respondent	0.672	0.496
EDUC	Education of the respondent	14.059	3.541
INCOME	Income of the respondent	3.61	1.673
BILL	Water bill paid	63.21	15.001

to pay for water quality. Age is one of the factors that determine the quality of water; middle-age group may demand less WTP than the younger and older age group and they may have less willingness to take risks. Therefore, AGE is predicted to be a positive sign. The monthly water bill (BILL) considered as the price for water which predicted to be negatively influencing factor for WTP. Probably, the individual's willingness to pay may increase when their income level (INCOME) increases.

From the analysis in Table 10.5, it can be noted that variables such as purchased bottled water (ARO), gender, education, and age significantly determine the willingness to pay for water at 1% level of significance. Followed by Water colour (COLOUR), Bill and Income by 5% level of significance. Whereas, household water treatment device (AWT) at 10% level of significance. Therefore, positive coefficient of household water treatment device (AWT) variable indicated that households with this facility are willingness to pay more to reduce the water contamination in their existing water supply system than the household without water treatment device. On the other hand, Colour of water (COLOUR) variable is statistically significant with negative coefficient representing that consumers who determine their poor water quality in terms of the colour are also willing to pay. The estimated coefficient of education, age, and gender are positively significant which shows that as education of the consumers increases at the same time their concerned about drinking water quality and awareness will also increase. Women are more concerned about their health, therefore, they have more willingness to pay than the men. With addition to these, age and income are also positive significant on willingness to pay for drinking water.

Willingness to Pay in Rupees

In this process, the first step is to fix a level to which the beneficiaries are willing to adhere. That is, payment options like amount per month or per year or amount per unit of goods or services provided with improved quality is more than what is available currently. Accordingly, the options given were (Rs. Per month) 50–75, 75–100, 100–125, 125 are shown in Table 10.6. It is found that out of 540 households 527 households are willing to pay a certain amount of money for quality drinking water.

Policy Implications

For the efficient use of water, individual tap connection to every household in the villages may be initiated. In order to avoid leakages and wastages of water on the water points, to monitor these issues a monitoring committee comprising of two persons for one street may be set up. The village level water monitoring committee comprising of minimum members may be set up, the committee will bring the water-related issues to the notice of the concerned authority. In some of the sample villages many of the hand pumps remain

Table 10.5 Tobit analyses of WTP for improved quality of drinking water

Variable	Coefficient	Standard error	Z	P value
CONSTANT	5.0910	2.5347	2.008	0.044**
ARO	−2.7370	0.4651	−5.884	0.000***
AWT	1.1247	0.6003	1.874	0.610*
TASTE	−0.2110	0.2272	−0.9287	0.353
HARDSHIP	−0.0243	0.1596	−0.1525	0.878
COLOUR	−1.3492	0.5478	−2.463	0.013**
GENDER	0.3610	0.110	32.53	0.000***
EDUCATION	1.5813	0.0862	18.33	0.000***
AGE	2.9552	0.2968	9.955	0.000***
BILL	0.0017	0.008	1.989	0.046**
INCOME	0.1255	0.5292	0.2372	0.812**

Chi-Square = 360,805.0
Sigma = 2.24794
Log-likelihood = 1203.634
p-value = 0.0000

***, **, * represents 1%, 5% and 10% level of significance

Table 10.6 Willingness to pay in rupees by the households

Willingness to pay slab (Rs.)	Total Value	Percentage
50–75	278	51.48
75–100	130	24.07
100–125	71	13.15
125 and above	48	8.89
Not willing	13	2.41
Total	540	100.00

Source Primary data

unused; this may be due to depletion in the ground water level. These unused bore wells may be re-bored to the extent of lowest ground water level. So that the unused hand pumps may be brought into reuse. In some of the sample villages the ground water level is too below the normal level. Initiatives may be taken to rejuvenate the water level in these villages through proper rain harvesting methods, creating awareness about the efficient use of water and the problem they would face in future if is wasted water, etc. The demand for water rapidly has been increasing due to fast increasing population; the government may again frame policies to curb population, and to check rapid increasing population, especially in over populated countries like India.

The public taps at a particular point in a street corner may be gradually eliminated, because there is possibility of leakages of water, spill overs, and unwanted quarrels. To overcome these problems individual house tap connections may be set up. On the basis of increasing size of population the high priority may be given to rural drinking water supply. The officers of the water board should ensure all the complaints with regard to water distribution and technical side issues have to be solved within a stipulated time period. The water releasing time should be uniform in all the villages of the district. If there is any change in the time should be informed in advance to the households, through local media.

10.4 Conclusion

Global climate change and the growing population make water management difficult in developing countries like India. Rainfall, land changes, and severe weather conditions due to climate change have a serious impact on water resources that results in the shortage of drinking water and inadequate availability of safe drinking water, especially in the rural areas, as well as conflicts among agriculture, industry, and domestic users. Effective policies and programmes should be implemented to overcome the issues in water resource management.

Acknowledgements Authors acknowledge the ICSSR for the financial support.

Declaration There is no conflict of interest.

References

Bakker Karan (2007) The common and versus the commodity, Alter globalization Anti Privatisation and the Human Right to water in the Global South. *Antipode* 39(3): 430–455.
Gleick Peter H (1999) Introduction: Studies from the water sector of the National Assessment. *Journal of the American Water Resources Association* 35(6): 1297–1300.
Government of Tamil Nadu (2010). http://cgwb.gov.in/gw_profiles/st_TN.htm.
Greene WH (1994) Accounting for excess zeros and sample selection in Poisson and negative binomial regression models.
Harsan Rashid, Scholes Robert, Ash Neville (2005) Millennium Ecosystem Assessment, Ecosystem and Human Well- Being: Current State and Trends, Volume 1, Island Press.
Intergovernmental Panel on Climate Change (2007) Climate Change 2007—Mitigation of Climate Change: Working Group III contribution to the Fourth Assessment Report of the IPCC. Cambridge: Cambridge University Press. https://doi.org/10.1017/CBO9780511546013.
Maddala GS (1983) Methods of estimation for models of markets with bounded price variation. *International Economic Review* 361–378.
Mehta Meera (2003) Meeting the Financing challenge for water supply and sanitation. Investment to promote reforms behaviors resources and importance Washington World Bank.

NSSO (2012) Drinking water sanitation and Hygienic and housing condition in India, Functionary July 2014.
Smita Misra, Bishwanath Golder (2008) Likely impact of reforming water supply and sewerage services in Delhi. *Economic Political Weekly* 43 (41): 57–66.
UNESCO (2001) Millennium Development Goals, World Water Assessment Programme.
UNESCO (2003) World Water Assessment Programme. The United Nations World Water Development Report: Water for People Water for Life. UNESCO, Paris.
Venkatachalam L (2015) Informal water markets and willingness to pay for water: A case study of the urban poor in Chennai City, India. *International Journal of Water Resources Development* 31(1): 134–145.
WHO (2009) Global Water Supply and Sanitation Assessment, Joint Monitoring Report.
WWAP (2012) World Water Assessment Programme: The United Nations World Water Development Report 4: Managing Water under Uncertainty and Risk.

PART II

Adaptation—Approaches and Mechanisms

CHAPTER 11

Assessment of Climatic Risk and Adaptation Measures in Rural India: A Case Study of Vanvasi Village, Maharashtra

Sujit Chauhan and Madhuri Pal

11.1 Introduction

The Intergovernmental Panel on Climate Change (IPCC) has reported a voluminous amount of evidence on current and projected climate change, indicating a very high impact of climate change on the natural system (IPCC 2021). Nowhere is the impact more pronounced than on agriculture and its overlapping concomitant sectors, i.e., food, water, and energy. The intensifying rise in temperature and declining precipitation rates impact agriculture and entrench an overwhelming risk for sustainable food production (IPCC 2007). In addition, climate change causes the degradation of land resources, loss of biodiversity and ecosystem services, and soil fertility decline. Developing countries are the worst sufferers of climate change, as they lack the enabling mechanisms to confront extreme events and severe weather conditions (IPCC 2021). Within a country, smallholder farmers and indigenous communities are the most vulnerable groups to climate change and extreme events. Studies have shown that they heavily rely on climate-sensitive sectors, and the availability of

S. Chauhan (✉)
Department of Economics, Indian Institute of Technology Bombay, Mumbai, India
e-mail: sujit.chauhan@iitb.ac.in

Present Address:
M. Pal
Energy Economics, Graduate School of Energy Science, Kyoto University, Kyoto-Shi, Japan
e-mail: madhuri.pal.62f@st.kyoto-u.ac.jp

© The Author(s), under exclusive license to Springer Nature Singapore Pte Ltd. 2023
S. Nautiyal et al. (eds.), *Palgrave Handbook of Socio-ecological Resilience in the Face of Climate Change*,
https://doi.org/10.1007/978-981-99-2206-2_11

resources (forests, agriculture, and water) greatly influences their livelihood and sustenance (Morton 2007).

The long-term changes in agricultural practices include changes in the period when land is prepared for cultivation, crops are sown, and harvesting is done. Further, the selection of the various crops for production is affected by rainfall variability and dry spells (Aryal et al. 2020). For instance, dry spells interfere with crops' growth rate and development stages. Moreover, it alters the nutrient–water relations, photosynthesis, and assimilation process, leading to reduced absorption rate of necessary radiations. These issues are directly linked with the declining productivity of small farming lands and the low profits of smallholder farmers.

United Nations reports that Sustainable Development Goals (SDGs), such as eradication of poverty (SDG 1), elimination of hunger (SDG 2), economic growth (SDG 8), and reduction of inequalities (SDG 10), critically depend on the achievement of climate action goals (SDG 13) (UN 2015). In the context of agriculture, improvements in productivity and yields are critical to tackling food security and poverty concerns. Apart from sustaining economic growth, climate change has far-reaching implications for the achievement of several other SDGs, such as ensuring decent health and well-being (SDG 3), provisioning of clean water (SDG 6), productive employment and decent working conditions (SDG 8), responsible consumption and productions (SDG 12), and the conservation and sustainability of water and terrestrial resources (SDG 14 & SDG 15, respectively). As many SDGs are dependent on climate action, such as mitigation of greenhouse gases and adaptation, the failure to overcome the issues of agricultural productivity and the welfare of the farmers directly undermines the achievement of SDGs (UN 2015).

In order to understand the sustainability of agricultural practices, studies are increasingly focusing on rural agricultural systems. It provides a suitable ground for researchers to delve into the intricacies of climate change and how their interaction with political, socioeconomic, and institutional factors creates conditions of vulnerability for smallholder farmers. IPCC describes the vulnerability to climate change as "the degree to which a system is susceptible to, or unable to cope with adverse effects of climate change, including climate variability and extremes" (IPCC 2014). Climatic risk and vulnerability emanate from political, institutional, and socioeconomic conditions (Blaikie et al. 1994). In particular, the economic, demographic, and political processes give rise to vulnerability by distorting the allocation and distribution of resources between the stakeholders. Against this background, there is a need to systematically analyze and understand the intrinsic relationship between the drivers of climate change, the political and socioeconomic factors that accelerate climate change, the impact on agricultural outcomes, and adaptation options and decisions.

11.2 Conceptual Framework

The study addresses three distinct relationships. Firstly, it examines how the drivers of climate change directly impact agricultural yield and productivity. It uses rainfall data to analyze the impact on agriculture. Secondly, the study argues that climate change entrenches several climatic risks and vulnerability concerns for smallholder farmers. The traditional farmers depend on the monsoon's vagaries and do not possess the modern technologies to adapt to the frequent onset of extreme events. Thirdly, the study proposes adaptation mechanisms as traditional agricultural practices and community-based adaptation (CBA). Farmers diversify crops, use fertilizers and pesticides to offset declining soil fertility, and engage in learning by doing from the community. Figure 11.1 describes the conceptual framework adopted for the study.

Climate change and extreme events have wide-ranging implications for agriculture and food security (UNFCC 2007). Researchers and policymakers are interested in understanding the extent of the risks and uncertainties involved in the agricultural sector as well as the underlying mechanisms that increase the farmers' adaptive capacities. Berger et al. (2014) suggest that adaptive capacity can be enhanced by

1. strengthening the ability of the people, organizations, and networks,
2. promoting regular learning and reflection,
3. ensuring the flexibility of the policies and practices to tackle climate risk, and
4. using tools and methods to plan for uncertainty and unexpected events.

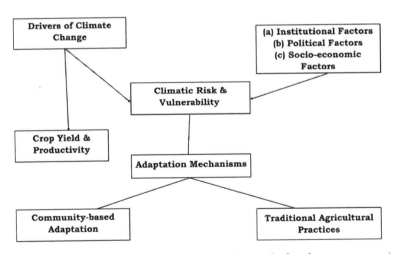

Fig. 11.1 Conceptual framework of the study (*Source* Authors' own construct)

In our framework, we understood two routes to adaptation mechanisms: (a) sustainable agricultural practices (foster learning and preservation of the existing traditional farming practices) and (b) community-based adaptation to climate change.

11.3 LITERATURE REVIEW

Much research has been undertaken to understand climate change in conjunction with agriculture. This section provides a synoptic view of the literature focusing on developing countries, particularly India. Essentially, the focus is on describing how climate change impacts crop yield and productivity, the decision-making related to the diversification of crops, and the need for agronomic practices and community-based adaptation.

Climate Change and Smallholder Farming

Studies find that while climate variability is responsible for 60% variability of crop yield, climate change affects decisive parameters of agriculture, such as the length of the season and the availability of water resources (Aryal et al. 2020). At the micro-level, agricultural output and the revenue generation of farmers, especially in Southern Asia, are declining. At the macro-level, food insecurity and energy shortages are becoming more common than ever (IPCC 2007).

The Indian subcontinent is subject to extreme climate change and variability (Cruz et al. 2007). The study by Khan et al. (2009) evaluates the increase in atmospheric carbon dioxide and temperature in relation to climate change and climate variability. The study reveals that depending on current ambient levels, an increase in temperature, solar heat, and precipitation rate severely impacts crop productivity, livestock agriculture, and food security. Moreover, changes in the amount of rainfall affect surface moisture availability in rainfed areas, an essential requirement for germination.

Morton (2007) discusses the impact of climate change and climate variability on the subsistence or smallholder farmers in developing countries by foregrounding farmers' limited adaptive capacity. It also discusses farmers' resilience factors, such as the use of family labor, indigenous knowledge, and diversification of crops. The income of the smallholder farmers depends on the sale of agricultural harvests. In the absence of non-farm income opportunities, income from harvest sales tends to be very low when climatic conditions are unfavorable. Studies find that the relationship between adaptive capacity and income is positively related to smallholders. The income of the smallholder farmers depends on the sale of agricultural harvests. In the absence of non-farm income opportunities, income from harvest sales tends to be very low when climatic conditions are unfavorable. Studies find that the relationship between adaptive capacity and income is positively related to smallholders.

The need for maintaining policy attention on smallholder farmers is highlighted in Bukchin-Peles and Fishman (2021). The authors drew attention to the relevance of smallholder farmers from a long-term policy perspective. The study pursued *whether focusing on smallholder productivity is a justified policy focus* given that the future projections reveal that the prevalence of smallholder farming will fall. The main finding of the study suggests that the global share of land under smallholder farming would remain stable (after accounting for migration and structural transformation in the rural regions). However, the share of smallholders will decrease by 2050. Therefore, the policy focuses on smallholders of land in the long term remains valuable due to its significant potential to tackle the twin problems of the food security crisis and the incidence of poverty.

Sustainable Agricultural Practices

Crop diversity is an indispensable agricultural strategy. Food and Agriculture Organization (FAO) recognizes the significance of sustainable storage practices, especially the conservation of "good" seeds in gene banks to maintain crop diversity (FAO 2013). Furthermore, the practice of crop diversity, especially in gene banks, helps researchers and scientists create other varieties of crops that can be effective against the environmental changes associated with global climate change. The study by Hazra (2001) further considers crop diversity a valuable tool to combat the various risks of climate change and overcome the issues of pests and diseases in crops.

Akinnagbe and Irohibe (2014) reviewed various agriculture adaptation mechanisms applied by farmers in various regions of Africa. The study found diverse crops, including drought-resistant tillage methods, changing cropping pattern dynamics, and crop rotation, to conserve soil fertility. Similarly, building robust irrigation infrastructures and harvesting rainwater reduces the monsoon's dependency. Moreover, it encourages agroforestry practices and afforestation that help to arrest adverse climate change impacts.

The age-old traditional and sustainable agriculture draws widespread attention to building resilience against changing climate (Singh and Singh 2017). The existing sustainable agricultural practices in India include organic manures and pastures, soil management techniques, crop diversification, water-saving irrigation techniques, cultivation of native plant species, and agroforestry practices. Sustainable farming practices facilitate a climate-smart approach to climate vulnerability, which helps the farming system to adapt, improve crop productivity, and restore biodiversity without relying on chemical and harmful fertilizers and technology. Concerning the sustainable production of food grains, Attri and Rathore (2003) suggest an "adjustment in sowing dates, breeding of plants that are more resilient to the variability of climate, and improvement in agronomic practices." By implementing such changes, the harmful effects of climate change arising from increased temperature and carbon dioxide can be reduced in the development stages of crops (Khan et al. 2009).

Community-Based Adaptation

Community-based adaptation (CBA) to climate change is "a community-led process, based on communities' priorities, needs, knowledge, and capacities, which should empower people to plan for and cope with the impacts of climate change" (Reid et al. 2009). At the community level, adaptation is an intrinsic part of any development interventions undertaken by self-help groups (SHGs), women development centers (WDCs), or external civil society organizations (CSOs) which engage in community-based governance (Ludi et al. 2014; Vedeld et al. 2014).

CBA is an adaptive mechanism that aims to adjust to climatic risk by reducing the burden on the most impoverished communities (Forsyth 2013). As cited in Wright et al. (2014), there is emerging evidence from Mozambique, Bangladesh, Uganda, and India, that the direct involvement of communities in implementing CBA at local levels proved effective in enabling farmers to overcome specific social and technological barriers to climate change adaptation. Mall et al. (2006) suggest identifying vulnerable agroclimatic regions as well as adopting altered cropping systems to maintain soil fertility. The study also recommends shifting toward community-based technologies (e.g., tractor use on a rotation basis) by foregrounding adaptation and risk-reduction activities in community agricultural practices.

11.4 Research Problem and the Main Objectives

Indian agrarian crisis concerning the environmental and climatic problems is deep-rooted and is not amenable to a swift solution. It is characterized by smaller landholdings (inequitable land distribution), lack of advanced technologies, shortage of financial resources, poor public infrastructure, weak research institutions, and unsound educational background of farmers. The literature review section recognizes how climate change affects the agriculture system, briefly touching upon the implications on food security. In particular, smallholder farmers of developing nations bear the highest brunt of climate change as they often depend on subsistence farming. Further, smallholder farmers respond ineffectively to unpredictable climatic conditions due to the various challenges in adopting appropriate adaptation strategies.

Past literature discusses adaptation interventions in several agrarian regions that design strategies that create rural livelihoods, increase resilient factors, and foster sustainable agricultural practices. Adaptation is often necessitated to address emerging issues of climate poverty and food security that are accentuated by declining productivity as well as the destruction of crops by droughts and floods (Aryal et al. 2020). In view of the above, the current study aims to sketch a broad outline of the impacts of climate change on agriculture in Vanvasi village by analyzing the precipitation rates and the traditional and sustainable agricultural practices of the village. In particular, the main objectives of the study can be described in the following manner:

1. To analyze the impact of climate change on crop yield and productivity in traditional agriculture. In other words, the objective is to detail the rainfall data and farmers' experiences in the past two decades to discuss the possible impacts on agricultural outcomes.
2. To delineate an understanding of village-level "adaptation" against climatic risk through traditional agricultural practices and CBA to climate change. In other words, we seek to understand (a) cropping pattern (types of crops grown in different seasons), (b) the pattern of crop rotation, (c) the decision to use fertilizers and pesticides, (d) storage mechanisms, and (e) the use of irrigation facilities.

This chapter is organized in the following manner. Section 11.5 outlines the area chosen for the case study and describes the methodology and data sources used in this study. Section 11.6 presents the main findings and discussions. The final section of the chapter contains the conclusions, policy implications, and directions for future research.

11.5 Materials and Methods

Study Site

The required data for the study was collected by undertaking a primary survey in Vanvasi village. It is a small tribal village in the taluka of Jawhar, situated in the Palghar district of Maharashtra, India (see, Fig. 11.2). It mainly comprises households from three communities: Kokna, Warli, and Thakar. Our field survey reveals that since the early 1980s, the village has been facing an acute water shortage. It has led farmers to engage in sand mining and settle for a low-income job. Many farmers from the village have also migrated to urban cities within Maharashtra and other states of India. They have either taken up daily wage labor or set up food/drinks stalls at the roadside.

Field Survey

We have conducted transect walks and household visits in the village to study village composition, existing institutions and physical infrastructure, and natural resources (e.g., water, land, and forest). In November 2018, we visited the village on a daily basis to conduct in-depth interviews with the Warli and Kokna communities of the village. The primary data was collected using the participatory statistics method, in which interviews were guided through a semi-structured questionnaire. The objective was to collect information on farmers' perceptions of the change in crop yield, productivity of the soil, farmers' decision to diversify crops, and the adoption of adaptation strategies to adjust to climate change. We used the snowball sampling technique because of the difficulty finding respondents as the villagers always leave for

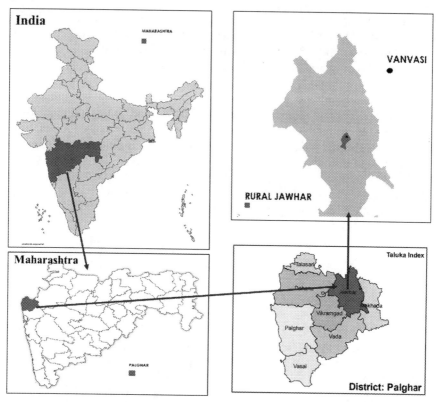

Fig. 11.2 Location of study area

their daily activities in their farmlands. To understand the pattern of precipitation in the region, the study also employed empirical data on rainfall. This data was collected from the Taluka Krishi Adhikari Office, Jawhar.

11.6 Results and Discussion

We first describe the agricultural timeline of crops. Secondly, we discuss the steps taken up by the farmers to do cultivation. It is followed by describing the impact of rainfall change on agricultural outcomes. The rest of this section discusses cropping patterns, crop rotation practices, irrigation practices, use of fertilizers and pesticides, and storage mechanisms in the village.

Table 11.1 Cropping pattern in Vanvasi village

Dam construction	Kharif (June–November)	Rabi (December–May)	Year-round
Pre-construction of dam	Rice and *nachni*	Forest leaves, e.g., *kundu* leaves and *maad* leaves	NA
Post-construction of dam	Rice, *nachni*, *varai* (a variety of millet), urid (black gram), toor (pigeon pea), and *khursani* (niger seeds)	Vegetables and Cash crops: bitter gourd, pumpkin, *ambadi* (roselle plant), spinach, fenugreek, *gawar ki falli* (cluster beans), ladyfinger, coriander, maize, bottle guard, drumstick, brinjal, green leafy vegetables, tomato, brinjal, potatoes, and grams	Cash crops: mango, cashew, jasmine, turmeric, ginger, papaya, sapodilla fruit, guava, black pepper, banana, and coconut

Source Authors' field survey, 2018

The Agricultural Timeline

Almost every family in the village cultivated food crops such as rice, *nachni* (ragi or finger millet), pulses, *urid* (black gram), *khursani* (niger seeds), and vegetables such as drumstick, brinjal, tomato, spinach, and sweet potatoes. A local NGO intervention helped the farmers set up various plantations, such as mango, cashew, and jasmine. With the help of the NGO, a seed bank and grain storage were also set up in the village.

Table 11.1 describes the crops grown and the cropping pattern by focusing on two agricultural seasons: Rabi and Kharif. The table also depicts the critical timelines for the village: before and after the construction of the dam. The interviews with the farmers revealed that after the dam construction, the number and variety of crops grown in the village increased manifold. It resulted in increased income for the farmers, which also provided the opportunity to engage in cash crop farming.

Steps in Agriculture

The first step in crop cultivation is the preparation of *Raab*. It is a traditional agricultural practice of clearing the land with naturally available biomass, including slashing and burning vegetation. This technique is applicable only to rice seedlings. Both women and men undertake the process of *raab*. In the next step, bulls are used to prepare the entire arable land for plantation. In the proceeding stage, known as *Ropni*, the sowing of saplings is done. It also involves farmers undertaking *tann nikalana*, i.e., removing all the unnecessary weeds from the crops. The following procedure is *katai*, i.e., harvesting the crops, followed by threshing, which removes the husk from the crops. In the last phase of the cycle, the winnowing of crops using *soop* (traditional basket)

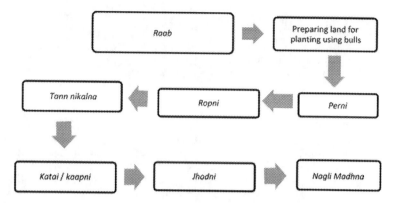

Fig. 11.3 The timeline of agricultural preparation and harvest in Vanvasi village (*Source* Authors' illustration [based on field survey, 2018])

to remove unwanted materials is undertaken. Figure 11.3 describes the entire process of farming in the village.

Impact of Climate Change on Agricultural Outcomes

Impact of Rainfall

Every household responded unanimously regarding the amount of rainfall the village has received for the past two decades. Respondents conveyed that it continuously used to rain for two to three months in the previous years. It has now been reduced to 10–12 days only. Figure 11.4 depicts the rainfall patterns in the Jawhar region for the 2012–2018 period.

The survey reveals a definite decline and irregularity in the rain, especially in the past four–five years. This has contributed to the destruction of crops that were heavily dependent on rainfall. The graph suggests that the highest rainfall of 3663.5 mm was recorded in 2017. The lowest rainfall of 1720 mm was observed in 2015. Many respondents insisted that the primary cause behind the destruction of crops, such as *nachni*, *urid*, and *toor* was the scanty rainfall.

Apart from the impact of the rainfall, the farmers also revealed that *tambera*, a crop disease found in the *nachni* crop, has led to the declining productivity of the crop. Interestingly, the farmers continue to cultivate irrespective of the challenges posed by climate change, such as irregular rainfall. During heavy rainfall, the soil from uphill descends to the plain area making it difficult to cultivate on hills. Farmers shift their cultivation to plain areas whenever such a situation arises.

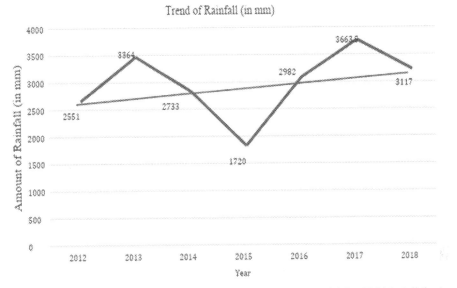

Fig. 11.4 The rainfall trend during 2012–2018 (*Source* Taluka Krishi Adhikari Office, Jawhar)

Crop Yield and Cropping Pattern
A large variety of crops are cultivated in Vanvasi village. As shown in Table 11.1, before the dam's construction in the town, the farmers were engaged in subsistence farming and mainly produced rice and *nachni*. They heavily relied on the edible plants from the forests for their nutritional needs. With the construction of the dam, farmers started diversifying, switching to other crops and vegetables. Some farmers also grew cash crops for self-consumption and commercial sales as they could afford better irrigation facilities. However, some households only produced rice because they did not have access to the water supply from the dam.

The yield from cultivation depends on three factors: ownership of land, the usage of fertilizer, and access to water resources. We find that the yield of some farmers has reduced due to the infertility of the soil. For example, one respondent used to produce 20–30 quintals of rice annually, which has now been reduced to 1.5 quintals. The respondent revealed that they had accumulated 11 acres of barren land, causing the declining yield. Some farmers rely on ration shops for household needs due to the declining yield of their grown crops.

Table 11.2 The pattern of crop rotation in Vanvasi village

Number of days	Amount of rainfall
90	Less
120	High
130	Highest

Source Authors' field survey, 2018

Crop Rotation

The village has a distinctive crop rotation pattern. First, only one variety of rice is cultivated for the first three years, and from the fourth year onward, the second variety of rice is introduced. Secondly, the cultivation of rice varieties is based on the rainfall the village receives. The crop rotation of rice is determined by the number of days and rainfall amount (see Table 11.2).

The farmers follow crop rotation when the yield is sufficiently low. Besides practicing crop rotation of rice, the farmers also practice rotation of other crops. For instance, in the first year, *nachni* is cultivated; in the second year, *varai* is grown; and in the third year, *urid* is sown.

Crop Calendar

Further, we used a crop calendar to analyze farm planning as well as to identify the cropping pattern among the farming households (see Appendix I, Table 11.4). FAO defines a crop calendar as a "tool that provides timely information about seeds to promote local crop production" (FAO 2020a). The calendar contains all the information on the agriculture timeline and practices ranging from sowing to harvesting crops that are adopted in particular regions. The calendar is beneficial in deciding the cultivation period of crops based on their corresponding agro-ecological conditions. The calendar also effectively responds to climatic disasters by providing rehabilitation measures for the agriculture system.

Traditional Agricultural Practices

Fertilizers and Pesticides

The farmers use chemical and organic fertilizers to enhance the productivity and yield of the crops they grow. The natively produced organic fertilizers include cow dung and dry leaves. These are used on all the crops. Chemical fertilizers, such as urea and Di-ammonium Phosphate (DAP), are used only on rice and *nachni*. On the other hand, Suphala and Ujjwala are applied to cashew and mango.

Table 11.3 The practice of fertilizers and pesticides in Vanvasi village

Fertilizers/pesticides (chemical/organic)		Rice and nachni	Cash crops
Chemical	Fertilizer	Urea, DAP 15.5	Suphala, Ujjwala
	Pesticides	NA	Endosulfan
Organic	Fertilizer	Cow dung, dry Leaves	
	Pesticides	Ash, *bhawarni*, *nirguri* leaves	

Source Authors' field survey, 2018

Organic pesticides and insecticides, such as ash, *bhawarni*, and *nirguri* leaves, are used in various crops. The survey finds that only a few farmers are aware of the appropriate dosages of fertilizers. But the farmers are aware of the adverse effects of chemical fertilizers, especially the impact of urea on human health was well understood by the farmers. Some respondents detailed their adverse effects after applying urea to their crops. It includes declining soil fertility, falling life expectancy, and the emergence of new crop diseases. Table 11.3 documents the fertilizers and pesticides used by the farmers in the village.

The field survey further reveals that the farmers use minimal urea and pesticides on vegetables as these crops are produced primarily for self-consumption. Further, the farmers discontinued wheat cultivation due to the increasing presence of rodents in the cultivated area. Despite having significant rat and other insect issues on the crop fields, the farmers rarely resort to chemical pesticides and rely primarily on the naturally available organic pesticides.

Storage Mechanisms and Cultivation Tools
Each cultivating household has a very adaptive storage system that facilitates the storage of harvests longer than three to four years. Without getting perished, *nachni* can be stored for at least two years in the storage facilities. The farmers have two kinds of storage. The first kind is made of plastic, in which they keep rice (before threshing). The second one is bamboo-made, which is used to keep the rice grains (after threshing) (see Appendix II, Fig. 11.6). To keep the pests and rodents away from the storage, they put *nirguri* leaves and spread a layer of cow dung over the stored item. In this way, the storage system is used by the farmers for the temporary storage of grains as well as it act as a means of food security. The survey also finds that some households keep their grains in a *tataki* or *pitara*, which can save harvests for 4–5 years. *Neem* leaves and salt are applied over this stored item.

Most farmers use wooden *nangar* (plow) and bullocks for cultivation (see Appendix II, Fig. 11.5). Only a fraction of the village owns tractors. Those who do not own tractors borrow from the community. There is cooperation among the farmers as they generally share their resources whenever and wherever required. The survey finds that the reasons for the non-replacement of traditional and primitive tools by modern technologies are the lack of financial ability to purchase the tools. Farmers also revealed that they do not want manual work to be replaced by machines.

Irrigation Facilities

Before the construction of the dam, farmers relied on small canals and wells for irrigation (see Appendix II, Fig. 11.7, for sources of irrigation). Even after the dam's construction, many farmers could not use the dam due to the cost involved with the diesel-run motor pumps, which are required to pump out the water from the well to farmlands. Farmers are waiting for assistance from organizations or groups, enabling them to receive the necessary water supply and reduce their financial burden. Most farmers depend on rainfall for rice production, while motor pumps are required to produce *toor, gawar ki falli*, and jasmine. Only a few households reap the benefits of vegetable cultivation as it requires a continuous water supply. Without rain and continuous water supply, vegetable crops fail to yield. The study found only one household in the village that used drip irrigation to cultivate cash crops.

11.7 Conclusions and Policy Implications

Climate change risks are a living reality. The growing intensity of climate risks and concerns about food security issues necessitates the adoption of transformational strategies in rural regions in order to achieve SDGs. The field visit to Vanvasi village revealed many important issues engulfing villages in forested and hilly locations populated by tribal farmers. Anecdotal evidence of the study indicated unplanned expansion of agriculture, land use changes, and soil degradation in the village. We find that the agrarian crisis in the village intensified in recent years due to a decrease in rainfall levels. It is also observed that the agricultural activities in Vanvasi village have made significant progress in sustainable farming practices owing to the timely interventions by local NGOs.

The study recognizes a growing need to go beyond piecemeal solutions by considering the broader perspective of sustainable development with a thrust on sustainable management of natural resources, diversification of farming systems, and building social capital and cohesion in the rural regions of developing countries. In particular, to address the climatic risks, stakeholders need to accept adaptation measures, such as strategizing crop diversity, revamping irrigation facilities, and improving the storage mechanisms of perishable foods. The results from the field survey indicate that although most of the farming practices in the village are sustainable, there is a need to adapt to growing extreme events and climate variability. By removing the market inefficiencies, such as the provisioning of credit and saving facilities and linking the input and output markets, the farmers can be encouraged to take up modern technologies that can increase yield and profitability, and offset many of the uncertainties associated with extreme events.

It is observed that the grafting[1] of cash crops, such as mango, jasmine, and cashew, can provide opportunities for the diversification of crops. Addressing the apprehensions regarding the cultivation and sale of commercial crops, particularly jasmine and mango, can inspire farmers to practice crop diversity. Adopting crop diversification is another important strategy that can adapt to low rainfall and droughts. Due to the lack of continuous water supply and irrigation facilities, the cultivation of vegetables is infeasible, resulting in the dominance of rice and *nachni* cultivation. In this context, crop diversification provides broader choices to farmers. Secondly, the traditional storage system cannot store the harvest of *nachni* for more than two years. It gives rise to the need for more adaptive storage mechanisms for such crops.

The present study is limited on account of considering only rainfall data and missing out on several climatic and non-climatic data. By collecting data on temperature, frequency of extreme events, availability of groundwater level and agricultural resources, terrestrial ecosystems, and biodiversity, future studies can improve our understanding of the impacts of climate change on traditional agriculture practices.

Acknowledgements The study area was chosen as a part of the field study undertaken to fulfill the degree requirements of M.Phil. Program at the Indian Institute of Technology, Bombay. The authors wish to place their sincere gratitude to the class of M.Phil. (2018–2020), Department of Humanities and Social Sciences (HSS), for their support in the fieldwork. The fieldwork was financially supported by the Department of Humanities and Social Sciences (HSS), Indian Institute of Technology, Bombay.

Declaration The authors declare that they do not have any known competing financial interests or personal relationships that could have influenced the work reported in this chapter. The contents are solely the authors' responsibility. They do not necessarily represent the views of the Department of HSS, IITB.

Appendix I

See Table 11.4.

---

[1] Grafting is the "connection of two pieces of living plant tissue in such a manner that they will unite and subsequently grow and develop as one plant" (FAO 2020b).

Table 11.4 Crop Calendar of Vanvasi village

Seasons/Crops	December	January	February	March	April	May	June	July	August	September	October	November
Toor								▬▬▬	▬▬▬	▬▬▬	▬▬▬	▬▬▬
Nachini								▬▬▬	▬▬▬	▬▬▬	▬▬▬	
Varai								▬▬▬	▬▬▬	▬▬▬		
Rice									▬▬▬	▬▬▬	▬▬▬	
Urid								▬▬▬	▬▬▬	▬▬▬		
Khursani								▬▬▬	▬▬▬	▬▬▬		
Vegetables	▬▬▬	▬▬▬	▬▬▬	▬▬▬	▬▬▬							
Cash crops (e.g., mango, cashew, black Pepper, coconut)	▬▬▬	▬▬▬	▬▬▬	▬▬▬	▬▬▬	▬▬▬	▬▬▬					

Source Authors' own construction (based on field survey, 2018)

APPENDIX II

See Figs. 11.5, 11.6, and 11.7.

Fig. 11.5 Tools used in Agriculture

11 ASSESSMENT OF CLIMATIC RISK AND ADAPTATION … 157

Fig. 11.6 Storage system of the village (*Source* Authors' field survey, 2018)

Fig. 11.7 The sources of irrigation (*Source* Authors' field survey, 2018)

References

Akinnagbe O, Irohibe I (2014) Agricultural adaptation strategies to climate change impacts in Africa: a review. Bangladesh Journal of Agricultural Research 39: 407–418. https://doi.org/10.3329/bjar.v39i3.21984.

Aryal JP, Sapkota TB, Khurana R, Khatri-Chhetri A, Rahut DB, Jat ML (2020) Climate change and agriculture in South Asia: adaptation options in smallholder production systems. Environment, Development and Sustainability 22: 5045–5075. https://doi.org/10.1007/s10668-019-00414-4.

Attri SD, Rathore LS (2003) Simulation of impact of projected climate change on wheat in India. International Journal of Climatology 23: 693–705. https://doi.org/10.1002/joc.896.

Berger R, Ensor J, Wilson K, Phukan I, Dasgupta S (2014) Adaptive capacity. In: Schipper EL, Ayers J, Reid H, Huq S, Rahman A (eds) Community-based adaptation to climate change: scaling it up. Routledge, Oxon, pp. 22–35.

Blaikie P, Cannon Y, Davis I, Wisner B (1994) At-Risk: Natural hazards, people's vulnerability, and disasters. Routledge, London.

Bukchin-Peles S, Fishman R (2021) Should long-term climate change adaptation be focused on smallholders? Environmental Research Letters 16: 114011. https://doi.org/10.1088/1748-9326/AC2699.

Cruz RV, Harasawa H, Lal M, Wu S, Anokhin Y, Punsalmaa B, Honda Y, Jafari M, Li M, Huu Ninh N (2007) Climate change 2007: impacts, adaptation and vulnerability. In: Parry ML, Canziani OF, Palutikof JP, Van der Linden PJ, Hanson CE (eds) Contribution of Working Group II to the fourth assessment report of the Intergovernmental Panel on Climate Change. Cambridge University Press, Cambridge, pp. 469–506.

FAO (2013) Genebank standards for plant genetic resources for food and agriculture. Rome.

FAO (2020a) Crop calendar—an information tool for seed security. http://www.fao.org/agriculture/seed/cropcalendar/welcome.do. Accessed 2 Oct 2022.

FAO. (2020b). Definitions and terminology. http://www.fao.org/3/ad224e/AD224E02.htm. Accessed 2 Oct 2022.

Forsyth T (2013) Community-based adaptation: a review of past and future challenges. Wiley Interdisciplinary Reviews: Climate Change 4: 439–446. https://doi.org/10.1002/wcc.231.

Hazra CR (2001) Crop diversification in India. In: Papademetriou MK, Dent FJ (eds) Crop diversification in the Asia-Pacific Region. Food and Agriculture Organization, Bangkok, pp. 32–50.

IPCC (2007) Climate Change 2007: Impacts, Adaptation and Vulnerability. Contribution of Working Group II to the Fourth Assessment Report of the Intergovernmental Panel on Climate Change. Cambridge University Press, Cambridge.

IPCC (2014) Climate change 2014: synthesis report. Contribution of Working Groups I, II and III to the fifth assessment report of the Intergovernmental Panel on Climate Change. Intergovernmental Panel on Climate Change (IPCC), Geneva.

IPCC (2021) Summary for Policymakers. Climate Change 2021: The Physical Science Basis. Contribution of Working Group I to the Sixth Assessment Report of the Intergovernmental Panel on Climate Change. Cambridge University Press, Cambridge and New York.

Khan SA, Kumar S, Hussain MZ, Kalra N (2009) Climate change, climate variability, and Indian agriculture: impacts vulnerability and adaptation strategies. In: Singh SN (eds) Climate change and crops. Springer, Berlin, Heidelberg, pp. 19–38.

Ludi E, Wiggins S, Jones L, Lofthouse J, Levine S (2014) Adapting development: how wider development interventions can support adaptive capacity at the community level. In: Schipper EL, Ayers J, Reid H, Huq S, Rahman A (eds) Community-based adaptation to climate change: scaling it up. Routledge, Oxon, pp. 36–51.

Mall R, Singh R, Gupta A, Srinivasan G, Rathore L (2006) Impact of climate change on Indian agriculture: a review. Climatic Change 78: 445–478. https://doi.org/10.1007/s10584-005-9042-x.

Morton FJ (2007) The impact of climate change on smallholder and subsistence agriculture. Proceedings of the national academy of sciences (PNAS) 104: 19680–19685. https://doi.org/10.1073/pnas.0701855104.

Reid H, Alam M, Berger R, Cannon T, Huq S, Milligan A (2009) PLA 60: Community-based adaptation to climate change. International Institute for Environment and Development (IIED), London.

Singh, R, Singh GS (2017) Traditional agriculture: a climate-smart approach for sustainable food production. Energy, Ecology, and Environment 2: 296–316. https://doi.org/10.1007/s40974-017-0074-7.

UNFCC (2007) Climate change: impacts, vulnerabilities, and adaption in developing countries. United Nations Framework Convention on Climate Change (UNFCCC), Bonn.

UN (2015) Transforming our world: the 2030 agenda for sustainable development. https://sdgs.un.org/2030agenda. Accessed 2 Oct 2022.

Vedeld T, Salunke SG, Aandahl G, Lanjekar P (2014) Governing extreme climate events in Maharashtra, India. Final report on WP3.2: Extreme Risks, Vulnerabilities and Community-based Adaptation in India (EVA): A Pilot Study. CIENS-TERI, TERI Press, New Delhi.

Wright H, Vermeulen S, Laganda G, Olupot M, Ampaire E, Jat M (2014) Farmers, food, and climate change: ensuring community-based adaptation is mainstreamed into agricultural programs. Climate and Development 6: 318–328. https://doi.org/10.1080/17565529.2014.965654.

CHAPTER 12

Climate-Smart Agricultural Practices and Technologies in India and South Africa: Implications for Climate Change Adaption and Sustainable Livelihoods

Juliet Angom and P. K. Viswanathan

12.1 INTRODUCTION

Gross projections by the *"Food and Agricultural Organisation of the United Nations"* indicate the urgency to step up food production and supply to sustain an estimated 9 billion global population by 2050 (FAO 2014; Barasa et al. 2021; Totin et al. 2018). Achieving *"zero hunger"* is part of the aspirational global goals, yet agricultural productivity continues to be ravaged by natural or anthropomorphic stressors, including climate change (Altieri and Nicholls 2017; Ghosh 2019; Wekesa et al. 2018; Kishore et al. 2018).

Climate change (CC) is a universal phenomenon that grossly alludes to variations in climatic capricious (*"viz humidity and rainfall"*) overtime, which may be due to natural variability or anthropogenic perturbations (Praveen and Sharma 2019; Naumann et al. 2021; Kraaijenbrink et al. 2021). The *"Intergovernmental Panel on Climate Change (IPCC)"* cites CC as *"alterations within climatic conditions identified (for instance via econometric testing) by changes in the atmosphere that lasts for extended periods, predominately decades or longer"* (IPCC 2012; Hegerl et al. 2007). Climate change threatens sustainable development through the so-called *"nonlinear events"*, in

---

J. Angom (✉)
Amrita School for Sustainable Development, Amrita Vishwa Vidyapeetham, Amritapuri, India

P. K. Viswanathan
Department of Management, Amrita Vishwa Vidyapeetham, Amritapuri, India
e-mail: viswanathanpk@am.amrita.edu

© The Author(s), under exclusive license to Springer Nature Singapore Pte Ltd. 2023
S. Nautiyal et al. (eds.), *Palgrave Handbook of Socio-ecological Resilience in the Face of Climate Change*,
https://doi.org/10.1007/978-981-99-2206-2_12

which an ecosystem fundamentally shifts after rapidly and irreversibly passing a specific environmental threshold (Hoegh-Guldberg et al. 2018).

The threat posed by CC to food security and the improvement of management strategies to achieve the same remains cardinal to strengthen the self-adaptivity of agrarians in the face of a changing climate and simultaneously build resilience to the climate emergency (FAO 2014).

India and Africa are epicentres of CC and its collocated effects (Okoba 2018; Picciariello et al. 2021; IPCC 2021). While food production should be scaled up, this still needs to be achieved with minimal negative ramifications on the environment (Totin et al. 2018; Godfray et al. 2010). The World Bank (2018), estimates that the detrimental repercussions of CC have the potential to force at least 45 million Indians into poverty if not smartly addressed by following a low-carbon growth path. Consequently, India has been earmarked as one of the global pioneers in promoting policies and interventions to address climate change (The World Bank 2018; Picciariello et al. 2021). To this end, the country has adopted some, "*Climate Resilient Agriculture (CRA)*" or "*Climate Smart Agriculture*" innovative technologies.

"*Climate Smart Agriculture (CSA)*" applies an integration of innovative technologies in agricultural systems to buttress production, investment atmosphere and policy environment sustainably to advance agricultural transformation, food security and productive livelihood development while ensuring farmers adapt to a changing climate (Rao et al. 2016; FAO 2014; Zougmoré et al. 2014). It is based on three pertinent objectives embedded in CRA, a subset of CSA. These include: (1) accumulating agricultural productivity and sustainable income generation, (2) enhancing resilience capacities to climate change and (3) minimising and eliminating green-gas emissions, thereby facilitating carbon sequestration (FAO 2014). The three contexts ("*food security, adaptation and mitigation*") have received characterisation as the "*triple win*" unison of CRA (Rao et al. 2016; CSA Guide 2020).

Taken together, CSA and CRA embody practices aiming at realising both interim and future agricultural improvement priorities amidst climate change, and as such, serving as a bridge to other developmental initiatives (Rao et al. 2016). The adoption of CSA implies that there should be a gradual transition to efficient and sustainable utilisation of natural resources and application of good agricultural practices. This shift will necessitate crafting and amending state and decentralised governance, policy processes and financial systems to navigate potential synergies, trade-offs and optimal profits (FAO 2014; Kanitkar et al. 2019). Economic, cultural and social barriers are known to hinder communities from adopting CSA practices and technologies (Tankha et al. 2020; Vera et al. 2017; Tanko and Ismail 2021; Davies et al. 2019). Unfortunately, the appropriate policy mechanisms to reduce or eliminate such barricades remain insufficient. Therefore, a probe to offer an insight into the barriers, practices and technologies, as well as their impacts on climate change adaption cannot be overemphasised (Barnard et al. 2015). Given these premises, this study explores some of the CSA practices and technologies that have been adopted in India and South Africa, and the feasibility of out-scaling the practices and related technologies.

The chapter is assembled as follows; (1) brief introduction of CSA, and why the approach is inevitable in addressing the ongoing climate emergency and realisation of food security, (2) the sustainable development theoretical framework in the milieu of India and South Africa, (3) the literature search strategy (methodology) adopted, (4) results and discussion of the best CSA practices in India and South Africa, feminisation of agriculture under CC in India and South Africa, policies and programme interventions to address CC in India and South Africa and lastly (5) the conclusions and recommendations.

12.2 Theoretical Framework (Sustainable Development Framework)

The basic tenet of the contemporary concept of sustainable development derives from the 1987 Brundtland Report in which the; *"United Nations General Assembly"* saw resolution *"72/279"* promote the centrepiece of its reform process (*"the United Nations Development Assistance Framework, currently known as the United Nations Sustainable Development Cooperation Framework"*) as *"the most important instrument for planning and implementation of the United Nations development activities at country level in support of the implementation of the 2030 Agenda for Sustainable Development"* (UN Sustainable Development Group 2019, 2020). With the multipronged nature of the 2030 Agenda and its timeline urgency, the current framework resolution is an essential shift for guiding the integral programme sphere, planning, implementation, monitoring, evaluating and collaborative report development towards the realisation the *"Global Goals"*. Moreover, the framework ascertains and reflects the United Nations development system's contributions to any country. In other words, it symbolises the nationalisation of the sustainable development goals (Government of India and the United Nations 2018).

India released *"The Government of India and United Nations Sustainable Development Framework (UNSDF)"*, spanning *"2018–2022"* as a national fabric of alliance, results and policy strategies channelled at contributing to the realisation of various developmental priorities (Government of India and the United Nations 2018). Some of these national priorities are also articulated in the National Institution for Transforming, *"India's (NITI Aayog's) Three-year Action Agenda (2017–2020)"* and other policy documents viz; *"A New India by 2022"* and *"Transformation of Aspirational Districts programme"*, all of which are developed from excerpts of the Sustainable Development Goals. The framework aims at:

1. purveying technical assistance based on comparable worldwide experiences to meet new difficulties,

2. configuring and executing high-impact, adaptable and replicable programmes to support national issues like affordable housing, vaccines, excellent education, health and nutrition, renewable energy and the environment, as well as skill development for young people, particularly women,
3. promoting enhanced planning, implementation and monitoring of the Global Goals at the national and local levels, including within and outside of national institutions,
4. supporting the government's participation in multilateral forums on Sustainable Development Goals, exceptionally on regional and international concerns such as renewable energy sources, urbanisation and resilience to climate change.

The projected outcomes of the framework are to:

1. promote equitable economic transformation and alleviate the complex forms of poverty experienced by vulnerable communities,
2. establish sustainable as well as original approaches to major issues, notably involving collaborations with more than just the commercial sector,
3. reinforce accountability mechanisms and aid in the provision of equitable, high-quality social welfare programmes,
4. sponsor the government's initiatives to have a bigger voice in international forums on issues like economic growth, climate change and contingency planning, notably.

On the other hand, the South African government and the United Nations entity drafted the *"United Nations Sustainable Development Cooperation Framework"* for the period of, "2020–2025". It is the initial Intergovernmental Framework created in accordance with the new principles published in 2019 (United Nations South Africa 2021). This framework is closely aligned with the 2030 Agenda, prioritising poverty elimination, reducing inequality and promoting an inclusive economy's growth. The framework is also aligned with the African Union Agenda 2063. Enshrined and central to the sustainable development frameworks in India and South Africa is the emphasis on mitigating or enhancing climate change adaptability. The current national priorities of the countries may differ, but all are aligned to realise Sustainable Development Goals.

12.3 Methodology

Brief Description of the Study Areas

India

India is a South Asian sovereign state with approximately 1.395 billion people in 2021 and the world's second-most populated nation after China (Macrotrends 2021; Wolpert 2021). It occupies an area of 1,147,955 square miles (Fig. 12.1) and shares common frontiers with sovereign nations such as Pakistan, Sri Lanka, Nepal, Bangladesh, Bhutan, Myanmar and China (Singh 2021). The country is also bordered between the Arabian Sea and the Indian Ocean all of which influence its climate (Patra 2016).

India is subdivided into various climatic zones due to its distinguished geography and geology, especially in the Himalayas and the Thar Desert. The Tropic of Cancer traverses India, explaining why its climate is considered tropical (Patra 2016). Thus, the country hosts two distinct climatic subtypes namely: tropical monsoon and tropical wet and dry climates (New World Encyclopedia 2017). It possesses quatern seasons: "*winter (December to February)*", "*pre-monsoon season/summer (March to May)*", "*rainy season/monsoon (June to September)*" and "*autumn season/post-monsoon (October to November)*".

India is ranked the third-largest economy in the world, though many Indians still wallow in abject poverty with limited access to resources (land, water and energy) or subsist on rain-fed agriculture and pastoralism which are climatically sensitive livelihood systems (Bisht et al. 2020; Patra 2016). The agricultural performance of India is influenced largely by the summer monsoon which accounts for nearly 75% of its annual precipitation (Gupta et al. 2020). However, other hydrometeorological events such as cyclones, droughts and hailstorms affect the country's agricultural output to varying degrees (Rao et al. 2016). For example, between 2002 and 2014, India registered severe droughts, accompanied by occasional cyclones and hailstorms. Eastern India on the other hand was ravaged several times by seawater intrusion (Rao et al. 2016).

The major agri-exports of Asian nations viz; India, Bangladesh, Indonesia and Vietnam are cereals principally rice and wheat, herbs and spices, assorted edible nuts, tobacco, tea and low and maritime merchandise (FAO 2015). Agriculture makes at least a 15% contribution to the overall economic output of India (Gupta et al. 2020). Obstructive climate change ramifications on this sector of the economy lies between "*0.7% - 1.35%*" of its annual economic growth (Rao et al. 2016). Up to 80% of Indian farmers are subsistence (smallholder) agriculturalists who practise monsoon-dependent rain-fed agriculture (Rao et al. 2016; Bhatia et al. 2010). An estimated 194.6 million Indians are undernourished amidst the persistent climatic changes or variability (Rao et al. 2016; Praveen and Sharma 2019). For this reason, there is a need to strategise, launch and maintain CSA initiatives to ensure environmental sustainability and food security.

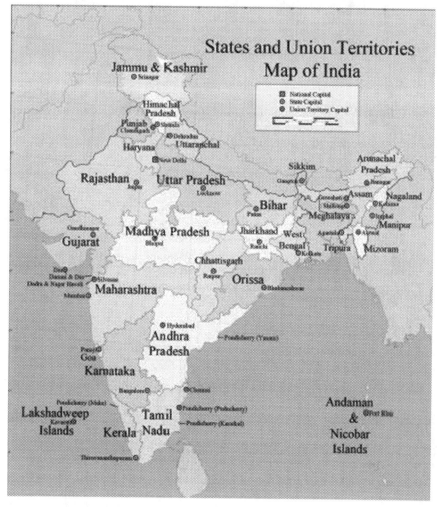

Fig. 12.1 Map of Asia showing the location of India

South Africa
South Africa is an African state with about, "*59 million people*" (UNDP 2020). It has a total area of 1,221,037 square kilometers, bordered by the South Atlantic and Indian Ocean which influences its climate (Fig. 12.2). South Africa shares common frontiers with Mozambique, Zimbabwe, Botswana, Namibia and Swaziland. Inside it is the enclaved country of Lesotho (Sartorius et al. 2011). The climate of South Africa is generally temperate, with a desert climate in southern Namibia, lush greenery subtropical climate towards the southeast along the Indian Ocean and Mozambique borders. The extreme southwest is predominantly Mediterranean in climate with hot, dry summers

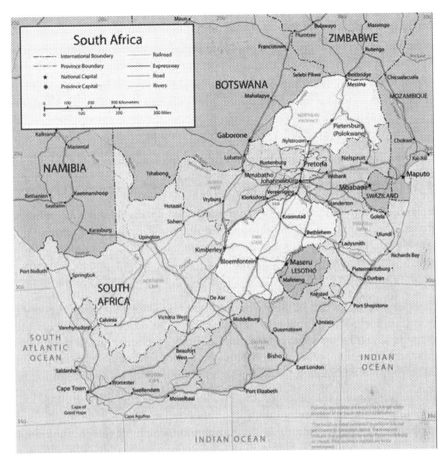

Fig. 12.2 Map of Africa showing the location of South Africa

and damp winters (South Africa Gateway 2017). Just like India, the economy of South Africa depends on agriculture and other climate-sensitive activities that can readily be impacted by climate change (Anseeuw and Boche 2015).

South Africa possesses a market-oriented agricultural economy, with the agricultural sector reported to have contributed about 10% of the country's overall export revenue in 2019. The highest-value exports were citrus, liquors, wheat and wool. In addition, South Africa exports a variety of goods, including almonds, artificial sweeteners, wool, apples and apricots (ITA 2020).

Literature Search Strategy

The current study is a non-systematic study that examined peer-reviewed reports and articles on strategies for adapting to climate change and technologies in India and South Africa dated until December 2021. The reports under consideration were sourced from multidisciplinary databases namely: Springer

Link, Wiley Online Library, Scopus, Google Scholar and Taylor & Francis Online. The Google search engine was then used to conduct a more extensive search to find papers and reports from international organisations, regional, national, and subnational authorities. The documents analysed were examined for key phrases using their titles, abstracts and keywords, *"climate smart agriculture"*, *"climate resilient agriculture"*, *"climate change"*, *"adaptation"*, *"climate change policies"* and *"climate change programs"*. Information about the climatic variability was collected from the reports and agricultural development perspectives in India and South Africa were extracted. The study further examined the current climate-smart agriculture and climate resilient strategies being employed in India and South Africa and the trends in feminisation of agriculture. The last section of the review gives an overview of the ongoing interventions in policies and programmes to address climate change or promote agriculture that is climate-smart in India and South Africa.

12.4 Results and Discussion

Agriculture remains the backbone of most countries worldwide. The sector furnishes raw materials for agro-processing industries as well as food, livelihood (employment) and income for households and nations alike (Willy et al. 2020). The Indian subcontinent and Africa are particularly vulnerable agroclimatic regions in this regard, especially in light of rapid population growth rates, excessive pressure on available natural resources, climate vagaries and the recent coronavirus pandemic (Willy et al. 2020; Varshney et al. 2020; Rao et al. 2016; Kishore et al. 2018). India and Africa have many similarities in their agroecologies, economies and civilisations (Reno 2019).

Climate-Smart Agriculture Practices in India and South Africa

Numerous research has shown that climate change is exerting adverse impacts on Indian agriculture (Tripathi and Mishra 2017; Picciariello et al. 2021). A number of approaches and innovations within the seven CSA entrance points, including soil management (Shah and Wu 2019), crop management (GoK 2014), water management (Vora and Parikh 1996), livestock management (Olarinmoye et al. 2011), forestry, fisheries and aquaculture (Viswanathan et al. 2020) and energy management (DSUUSM 2019) have been recommended by previous authors to avert the looming threat (Patra 2016). These are discussed comparatively citing India and South Africa in the penultimate subsections. The technologies (Fig. 12.3) may be incorporating new elements into both new and obsolete procedures such as high-yielding, drought-water tolerant crop varieties, hardy pest and disease animal breeds (Zougmoré et al. 2014). Most entry points involve interventions at the farm levels.

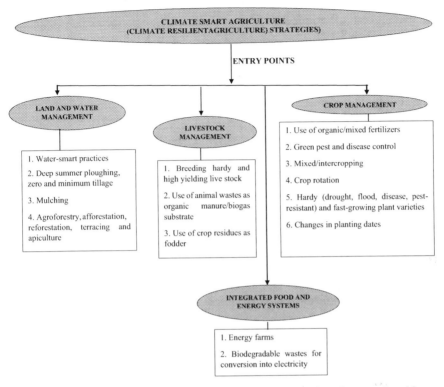

Fig. 12.3 Synopsis of climate-smart and resilient agricultural strategies (*Source* Viswanathan et al. 2020; Angom et al. 2021; Andrieu et al. 2019; Ali et al. 2014; Abegunde et al. 2020; Adger et al. 2013; Aggarwal et al. 2018)

Crop Management

Crop production must adapt to the drawbacks posed by climate change and become resilient to both climate changes' irregularities and severity (Aryal et al. 2016). Some of the CSA used in the management of crops in India and South Africa include.

Improved Fertiliser Management Practices (Application of Organic Fertilisers)

There are many types of biofertilisers available at present in India with *Rhizobium* being the relatively more effective and widely used type (Rao and Reddy 2005). Others such as + "*Azotobacter*" (results from its application in non-legume crops have been unclear). According to reports, only a few traditional paddy regions can benefit from blue-green algae and azolla of India. It is yet unknown how organisms like Azospirillum and Azotobacter affect the subduing of soil-borne pathogenic illnesses in crops (Rao and Reddy 2005). Exhilarating growth factors, which encourage germination and plant life, are key functions of bio fertilisers. The Government of Karnataka, for instance took application of organic fertilisers as a measure to enforce organic farming (Kumar and Indira 2017).

According to FAO (2005), South Africa has a decadal history of enriching manures (mostly chicken manure) with inorganic fertilisers. This practice confers the advantages of inorganic fertilisers with enhanced plant nutrient concentrations when mixed with manure.

Green Pest and Disease Management Approaches

Global warming has given rise to pests and diseases that lower crop yields and destroy agricultural investments. Investments have been made to plant pest and disease-resistant seedlings, employ manual weeding as much as possible, noting noxious weeds while soft weeds that can replenish the soil and prevent erosion of nutrient-rich topsoil are kept intact. In the "*jhum*" (mixed cropping) system, most Indian farmers employ plant and animal parts/products as components of indigenous efforts in the regulation of plant diseases and pests (Chhetry and Belbahri 2009). For instance, in many parts of southern India, farmers are applying paddy husk for five months to contain the blast of rice for effective control of blast disease (Chhetry and Belbahri 2009).

Analogously, plant decoctions such as chilli, garlic and stinging nettle sprays, wood ash and are similarly used in South Africa as repellents for aphids, beetles, snails, cabbage flies and to treat plant diseases (Schoeman 2020). Other approaches have been the introduction of herbicide-tolerant crop varieties. For example, the "NK603" maize variety is tolerant to the glyphosate herbicide, Roundup (Iversen et al. 2014).

Planting Shade Trees and Vegetative Barriers

Growing the right number, species and canopy of shade trees is important. These can protect planted crops and farms from excessive heat and act as live fences, providing an appropriate overall shade-tree system to curb harsh winds and heavy rains while providing support to other creeping plants. Trials using Vetiver ("*khus*") trees at Kabbalanala Watershed in Karnataka (India) indicated soil moisture retention and higher crop yields for plots with live hedges (Bandyopadhyay et al. 2015). Henderson (1983), identified that nearly 428 naturalised, invader and exotic plant species have a record of use as vegetative barriers in South Africa. Such diversity has enhanced the restoration of South African natural evergreen forests, soil fertility and agricultural productivity (Geldenhuys et al. 2017).

Crop Rotation

Indian farmers have had to deal with numerous irregular monsoon-related challenges and have adopted planting different crops on the same parcel of land alternately (Gupta et al. 2020; Jat et al. 2019). This farming system affords an improved cost-effective use of soil fertility, improved weed, pests and disease management and a reduction in the amount of land needed for crop production. Along with wheat, bajra, sorghum and radish, grain legume crops are grown in India (Naaiik et al. 2018), this has often improved harvest yields. As opposed to monoculture cropping, crop rotation has been practised for a long time in South Africa (Maas and Kotzé 1990; Nel 2005). A recent study (Nel 2005) indicated that this old cropping system improved South African maize and wheat grain yields in seven trials when compared with monoculture.

Mixed/Intercropping
This involves the cultivation of two or more crops, such as rice and soybeans, being grown simultaneously on the same parcel of land. It may also refer to an agricultural technique of concurrently planting two crops in the same land (Das et al. 2019). This old form of polyculture (additive or replacement) farming employs companion planting principles that confer various agronomic benefits capable of mitigating the spread of pests and disease epidemics, increasing productivity, providing as well water and soil conservation, hazards insurance against complete crop failure in the event of anomalous climatic conditions.

In the case of compatible crops, it encourages biodiversity through the provision of a habitat for a homogeneous crop ecosystem, which makes it impossible for numerous insects and soil organisms to exist (Das et al. 2019). Intercropping systems involving cereals and legumes are practised in India (Das et al. 2019; Tripathi and Mishra 2017).

Legume-cereal intercropping is also a widespread agricultural practice in South Africa (Mzezewa and Gwata 2016; Odhiambo 2011; Tsubo et al. 2003). It has been indicated to have superiority over sole cropping systems. For instance, trials of legume-maize with or without fertilisers generated approximately; "*19 -to- 58 percent*" more maize grain yield (Odhiambo 2011). Green manure legumes: sun hemp ("*Crotalaria juncea*") and lab-lab ("*Lablab purpureus*") were reported to be the most suitable cover crops when planted in South Africa's early summer season (Odhiambo 2011; Nel 2005).

Breeding and Planting Drought-Tolerant Crop Varieties
Breeding stress-tolerant crop varieties has been a revered strategy for achieving CSA. Recent biotechnological advances have afforded breeders diverse tools to improve phenotypic screening. In India, these techniques such as genetic engineering, scientific breeding and marker-assisted screening of essential qualities have been reported to be an attractive alternative with up to 30% increase in crop yields (Pray et al. 2011; Gautam 2009; Birthal et al. 2014). Another assessment of the benefits of growing drought-tolerant groundnuts (variety ICGV 91,114) indicated that its adoption could minimise agricultural food production risks by 46% in the drought-prone Anantapur district of India, when compared to other local varieties (Birthal et al. 2012). For local cultivars of cereals and pulses, the registration of these farmer's varieties for Plant Variety Protection have been performed for self-pollinated crops which passed

the distinctiveness, uniformity and stability assessment for their protection and conservation through a legislative mechanism in India (Singh and Agrawal 2020).

In this regard, the improvement that provides farmers with finance and crop insurance packages could also increase people's confidence in and capacity to implement these mitigation and adaptation biotechnological measures. In South Africa, drought-tolerant crop varieties such as "*ZM521*" maize variety, stem borer resistant Bt-maize ("MON810"), wild watermelon, mustard and bambara groundnuts have been encouraged in opposition to other indigenous susceptible cultivars (Tirado and Cotter 2010; Modi and Mabhaudhi 2013; Iversen et al. 2014; GIZ 2017).

Planting Fast-Growing Crop Varieties
Arid zone Indian farmers have adopted cultivation of vigorous-growing varieties of pearl millet, cotton, cluster and moth beans that not only require less water but also utilise rainwater available in short spells for which it rains. These plants are also highly responsive to the application of outside resources, including certain fertilisers and pesticides (Gupta et al. 2020). Some fast-growing plant varieties (cereals e.g. wheat) have been encouraged in South Africa (GIZ 2017).

Changes in Planting, Weeding and Harvesting Dates
Changes in farming calendar have been recognised to be instrumental in the mitigation of pest and disease infestations (Jalota et al. 2012), which ultimately increase yields. Planting earlier than expected due to wheat damage spurred on by climate was indicated to have a potential to minimise up to 65–75% of agriculture barriers in India (Birthal et al. 2014). Another study predicted that under changed climatic conditions, yields in the central Indian rice–wheat farming scheme, precisely in the Punjab region subjected to shifting transplanting dates from 171 to 178 for rice (Dar et al. 2020) and 309 to 324 for wheat (Pedraza et al. 2015) would result in the least reduction in crop yields. The authors suggested that this could be a practicable adaptation strategy for promoting yields of these cereals in India (Jalota et al. 2012). Other studies in India has also attested to yield improvement with alteration of planting dates (Mahajan et al. 2009; Rajwade et al. 2015). Concurrently, these practices have been encouraged as well in South Africa (Moeletsi 2017), with improved maize yields recorded in one study (Akinnuoye-Adelabu and Modi 2017). In South Africa, winter wheat is harvested when it is dry in the spring, after the bulk of the winter rainfall to reduce the risk of crop losses (GIZ 2017).

Broad-Bed and Conservation Furrow System
On some Indian agricultural lands, waterlogging and water scarcity are endemic within the cropping season. Thus, myriad "in situ" soil and water conservation measures as well as drainage technologies have been adopted. Broad-bed and furrow systems for instance have been verified to be a feasible

practice; it recorded a 21% yield advantage for chick peas and 35% yield increase of soybeans in Central India (Bandyopadhyay et al. 2015).

In comparison, the fallow-land-based method of producing maize also led to increased maize grain yields in South Africa though it had drawbacks exemplified by losses from excessive soil erosion and sedimentation (Haarhoff et al. 2020).

Livestock Management

Livestock can create oversized contributions to climate-sensitive food offer systems. Choices to condense greenhouse emissions are obtainable across value-chain operations and subject to feed and manure regulation alongside celiac fermentation. The common practices under this are:

Breeding Hardy and High Yielding Livestock

Animal breeding has a long history in India. Excessive demand for accessible land and a constantly growing population have culminated in the breeding of pest and disease-resistant, high yielding and fast-growing poultry, goat, sheep, buffalo and cattle breeds, particularly to increase dairy and beef production (Gowane et al. 2019). The Indian government has invested in progeny testing with success recorded in cross-bred cattle and buffaloes. Malpura sheep breeding projects have also been demonstrated in Rajasthan's semi-arid regions with at least 60 breeders in 21 villages and up to 5000 animals (Gowane et al. 2019). Howbeit, this area is growing at a slow pace, due to the diverse agroecological climates in India, different ethnic groups as well as the systems of governance.

Though breeding of animals has been done in South Africa (Iversen et al. 2014), smallholder farmers prefer land race cultivators of different crops and raise indigenous livestock breeds like Nguni cattle and free-range chicken due to their tolerances to biotic and abiotic stresses (FANRPAN 2017).

Use of Animal Manure and Wastes in the Production of Energy

This theme has been explored to supplement the electricity, thermal and nuclear potential of India. Livestock biomass has been harnessed in electricity production in biogas in accordance with the government's priority towards rural development (Kaur et al. 2017). This has curtailed greenhouse emissions from indiscriminate ignition of non-renewable crude oils to match the energy demands of the Indian population. This resource if tapped sustainably has the potential of supplying nearly 477 terawatt/hour of electricity annually to the Indian energy basket (Kaur et al. 2017). In South Africa, the first industrial anaerobic digester in the world utilising animal manure had been constructed in 1957 but this sector did not develop any further (Laks 2017). Since then, over 700 biogas installations have been made in the country (Mutungwazi et al. 2018). However, most of them are not primarily producing electricity for use (Mukumba et al. 2016). Recently, a Bronkhorstspruit Biogas plant running on manure from over 25,000 cattle and other organic wastes was installed and is currently reducing both the country's electricity deficit and volumes of landfill waste (Sunref 2013; Norfund 2012). The turbines potentially generate, "4.4 MW" of energy, which is boosting green industrial growth. Just like India, biogas from organic wastes is an under-tapped energy potential that is still in its infancy in South Africa (Mukumba et al. 2016; ESI Africa 2016; Uhunamure et al. 2020; Laks 2017).

Feeding Livestock on Crop Residues as Fodder

India is an agrarian nation that generates a considerable volume of agricultural residues. In order to avert the negative effects of burning crop harvests that usually follow harvests, various Indian communities have resorted to using part of these residues in preparation for dry fodder for livestock (Rathod et al. 2017). This practice has been reported to increase dry fodder utilisation, reduce handling costs, paramount to improvement in the quantity and quality of milk (Rathod et al. 2017). Similarly, an important source of dry season nourishment is leaves (stover) and maize stalks that many South African farmers precisely livestock farmers rely on to feed their livestock (Mthembu et al. 2018; Haarhoff et al. 2020).

Application of Animal Wastes as Organic Manure

Other than exploring the use of animal wastes in biogas production, some Indian farmers have utilised them as a cheap source of organic fertilisers to improve soil fertility. This reduces the emission of greenhouse gases. Cow dung for instance is heaped for 1–2 years to convert it into compost manure or vermin composting is done using earthworms. The manure is thereafter applied in agricultural fields or vegetable gardens (Singh 2016). The compost is also revered as a nutrient boost for fodder crops such as Lucerne ("*Medicago sativa*") and Berseem ("*Trifolium alexandrium*") in rural India. Animal manure usually contains nitrogen, phosphorous and potassium required for plant growth (Singh 2016).

An elaborate print by Mokgolo et al. (2019) indicated that the use of poultry and cattle manure practised in South Africa resulted in an increase in the second season's grain yield compared to the former season. In the second cropping season, the use of organic manure severely impacted sunflower returns, plant height and stem girth at varying growth stages, with poultry manure generating the highest results (Mokgolo et al. 2019). Additionally, vermiculture technology, which uses earthworm composting in vermibins to produce high-quality vermicompost from organic agriculture and residential waste, is being encouraged (Mnkeni and Mutengwa 2014).

Land and Water Management

Within CSA, land and water are important focal points. Changes in the manner land and water are governed will be necessary for productive and sustainable agriculture to ensure that they are utilised effectively (Dell'Angelo et al., 2017). Accelerated groundwater extraction for irrigation has increased crop output and food security in India, but excessive groundwater mining has endangered agriculture's long-term sustainability (Archana et al. 2017).

Practices to sustainably manage land include soil carbon sequestration as well as the reclamation of peat bogs and degraded lands. Some adopted climate-resilient techniques and approaches that can be scaled up under the "*National Action Plan for Climate Change*" cited in the "*National Mission on Sustainable Agriculture*" of India include.

Staggered Community Paddy Nursery

This is an adaptation approach to offset the effects of rainfall shortage in some seasons of lowlands in India (Prabhakar and Reddy 2010). On, "*565 hectares*", an estimated, "*1274 farmers*" from different villages in eleven Indian states were shown how to operate it. Adoption of this strategy has offered promising yield advantages of 9.4–80.2% with a benefit–cost ratio of, "*1.4–5.1*" vis-à-vis the orthodox farming techniques previously adored in South Africa (Ghosh 2019).

Water-Smart Practices

Some of the water-smart approaches in CSA include; recharging aquifers, collecting rainwater, managing water resources in communities, levelling soil with lasers and minuscule (drip and sprinkler) irrigation (Suresh and Manoj 2020), raised-bed planting, use of solar pumps for taping irrigation water (Bhatia et al. 2010; Jayan 2018; Hartung and Pluschke 2018; ICID 2019; Suman 2018) and the subsurface control of surface floods with help from pipes (*holiyas*) (Naireeta Services 2018; Pavelic 2020; Bunsen and Rathod 2016).

Though water management technologies and practices, as well as CSA are known, their practical applications are low and limited. Case in point, drip irrigation which directs water to plant roots and averts flooding the entire field which serves to reduce both water and electricity wastage is well known (Tankha et al. 2020). Unfortunately, its usage is not well spread to achieve the intended aims, with flood irrigation still dominating (nearly 97%) as compared to sprinklers (2%) and localised drip irrigation (1%) (Gupta et al. 2020; Birkenholtz 2017; Namara et al. 2007; Fishman et al. 2014; Bahinipati and Viswanathan 2019). Earlier reports indicated that micro-irrigation augmented by watershed management and insurance cover have the potential to avert up to 70% of the avoidable losses induced by droughts (ECA 2009; International Finance Bank 2015). Some of these water-conservation techniques have improved crop production and as such have been and continue to be encouraged in South Africa (Mnkeni and Mutengwa 2014).

Deep Summer Ploughing
In India, this practice has been previously advocated prior to the commencement of the summer monsoon season as an adaptable technique for capturing rainwater in the surface soils (Gupta et al. 2020). However, tillage effects are usually short-lived and may exacerbate costs as there may expound a need for more tillage. In addition, a study undertaken in South Africa shows that in other areas summer deep tillage practices may lead to accelerated erosion and nutrient loss from fertile surface soil, along with poor soil accumulated stability, which may be counterproductive (Gupta et al. 2020).

Mulching
This involves covering the topsoil with plant materials. For the Indian monsoonal climate, the application of crop cultivation and field mulching done before the rains arrive reduces soil erosion and evaporation while enhancing efficient utilisation of the rainwater and boosting soil fertility. Thus, "*in situ*" rainfall conservation can be considerably aided by preventing residual combustion and preserving the agricultural soils covered (Gupta et al. 2020; Bhattacharyya et al. 2016). Mulched crops can meet their water requirements for longer durations without stress because mulched soils preserve more moisture than typical till systems.

Studies in South Africa on crop productivity have shown that improved assemblage, carbon sequestration, increased ability for water to permeate and decreased erosion have all been linked to increased water storage in soils (GIZ 2017).

Zero and Minimum Tillage
Indian land and water resources can be preserved by the use of laser land levelling and zero tillage (Jat et al. 2019). The zero tillage approach in India dates as far as 1960. It is predominant in the Indo-Gangetic farmlands where rice and wheat are grown as crops. In India however, zero tillage is more practical for direct-seeded rice, maize, pigeon peas, soybean, cotton and wheat (Laxmi et al. 2007; Keil et al. 2017).

Agroforestry

Climate change jeopardises the conveyance of goods and ecosystem services (such as food, fuel, carbon reduction, water, biodiversity). However, forestry and agroforestry contributes to global efforts in mitigating climate change (Shashidharan 2012). Farmland forests and trees have the capacity to absorb more carbon sink (GoF 2005). Agriculture continues to be the main factor causing deforestation and is the main contributor to carbon emissions from the lumber industry (Longobardi et al. 2016). Some of the good CSA practices in agroforestry include:-

Afforestation
Soil carbon sequestration is enhanced through the planting of trees and India has since launched such compensatory afforestation policies. This practice translates into an increased generation of organic manures that replenishes the soil. Increasing soil organic carbon has been referred to as a significant mitigating strategy in agricultural systems by the IPCC (Altieri and Nicholls 2017).

Since 1980, India has been undertaking extensive afforestation under various agroforestry programmes, including farm forestry, social forestry and collaborative sustainable forest management. This may reduce pressure on forests which triggers climate change (Ravindranath et al. 2008). Gupta et al. (2020) suggested that saline Mewat area of India could be diverted for afforestation or agroforestry with planting of woody-salt-tolerant-tree varieties such as, "*Acacia nilotica*" and "*Prosopis juliflora*". Similar to this, extensive areas of Indian wastelands have undergone afforestation (Balooni 2003).

In South Africa, commercial and small-scale farmers in the Western Cape Province are considering the cultivation of carob ("*Ceratonia silliqua*"), an annual and drought-resistant tree (Mnkeni and Mutengwa 2014).

Reforestation
This refers to the regrowth of forests utilising regionally native tree species after they had previously been cleared of woods (Ravindranath et al. 2008). India and South Africa have both put in efforts to replant trees since 1980 as a measure of mitigating climate change.

These resurgent forests will enhance the ecosystem, protect threatened species, replenish precious resources and encourage carbon sequestration (Ravindranath et al. 2008; Shashidharan 2012).

Terracing
Terrace farming involves the creation of steps ("*terraces*") on hill slopes to conduct farming operations. In India, states such as, "Punjab, "Haryana", "Himachal Pradesh" and "Uttaranchal" to mention a few practice terrace cultivation (Mishra and Rai 2011). This farming method was reported to be economically profitable as a climate conservation measure in Papung-Ben Khola watershed of Sikkim Himalaya (Mishra and Rai 2011).

In the case of South Africa, each terracing stage involves growing a variety of crops, which has the benefit that rainfall cannot completely dissipate the nutrients since they are driven to the lower levels instead (Mishra and Rai 2011). Additionally, by taking these precautions, a free-flowing avalanche of water is prevented from destroying all the crops on the farmland.

Apiculture (Apisilviculture)

India's economy boasts of large bee farms in Punjab with over 8000 bee colonies, mostly of *"Apis mellifera"* and *"Apiscerana"*. The planting of preselected woody species with flowers that produce nectar and pollen that bees value has been used to improve wax and honey production in India and the case is similar in South Africa (Tej et al. 2017).

Integrated Food and Energy Systems

The Government of India (GoI) wants to increase the proportion of food and renewable energy due to concerns about climate change, food security and energy access through application of mixed ecosystems (Jain et al. 2020). Indeed, every process of the agri-food system, including the pre-production of raw materials, agricultural production, marine, livestock and forested products, depends heavily on energy, storage, processing, transport and distribution. Many such ecosystems need explicit energy, which includes fuels like electricity, solid, liquid and gaseous fuels, or intermediate energy, which includes energy associated with producing inputs like; agricultural equipment, fertilisers and pesticides (Jain et al. 2020). FAO (2020) expounds that the major focus of food-energy systems as, integration, intensification and increase by converting waste products from one system into feedstocks for another in the process of food production and energy generation (Bogdanski et al. 2010).

In India, the idea of, *"energy farm"* facility constructed to generate energy, typically for conventional transmission to far-off metropolitan markets was introduced. In this case, the biogas plant has been connected to the urban toilet systems, generating heated water and power for street lighting simultaneously easing the sewage treatment issue. Similarly, livestock biomass has been harnessed for power generation in biogas reactor plants in rural areas (Kaur et al. 2017). Another initiative has been biomass production from mustard crop residues (Bogdanski et al. 2010). *"The Indian Ministry of New and Renewable Energy"*, estimates that the country generates approximately, *"500 million tonnes"* of agricultural-crop waste annually. Howbeit only 12.2% of this is harnessed into energy production through combustion as solid fuel, leaving most of the wastes (with nearly 80%) biomass remains untapped. This untapped waste biomass has a potential to be bio transformed into biofuels such as biogas.

The *"Municipal Solid Waste Management Directive"* (2000), mandated source segregation of biodegradable wastes for conversion into electricity. However, it has received little success partly owing to poor implementation and reporting mechanism, and low levels of awareness. Further, Indian urban waste is frequently co-mingled with rubble, construction and demolition wastes, which render them unsuitable for harnessing into energy (Goyal 2011).

Comparatively, since 2016 waste-to-energy projects have been started in South Africa by the, *"United Nations Industrial Development Organization (UNIDO)"* and the *"Global Environment Facility (GEF)"*, to create awareness and capacity for harnessing agro-processing organic wastes into biogas (GEF 2020). Biogas production from human excreta has been piloted in, *"KwaZulu-Natal"*, *"Mpumalanga"*, *"Limpopo"*, *"Free State"*, *"Gauteng Province"*, *"Northern, Western* and *Eastern Cape"*. These sewage-dependent biogas are however unutilised, possibly as a result of South Africa's cost-effective electricity prices (Mukumba et al. 2016).

12.5 Policies and Programmes Intervention to Address Climate Change in India and South Africa

The 2014 United Nations climate summit launched pillar efforts towards addressing climate change, including, *"Global Alliance for Climate-Smart Agriculture (GACSA)"*. This policy was to enhance exchange of knowledge and inter-regional collaboration on CSA amongst members-states including non-governmental organisations, academia, governments, farmer organisations and the private sector (Dinesh et al. 2017).

India has made progressive strides notably seen through the piloted and scaled-up climate change initiatives, regulations, alliances and investments domestically. On June 30, 2008, the first, *"National Action Plan on Climate Change (NAPCC)"* was made public (Government of India 2018; Jörgensen et al. 2015) with eight primary objectives outlined (Dinesh et al. 2017; Tripathi and Mishra 2017).

These include nationwide initiatives for a *"Green India"*, sustainable habitat, water conservation, renewable power, improved energy efficiency, preserving the Himalayan ecology, sustainable agriculture and a strategic knowledge repository for climatic changes (Government of India 2018). The *"National Mission for Sustainable Agriculture (NMSA)"*, is specially geared towards increasing agricultural productivity, particularly in the weather-dependent agricultural areas by integrating farming, efficient water usage, soil health management and ecological conservation.

South Africa's 2012, *"National Development Plan (NDP)"*, provides a, *"2030 Vision"* that guides its sustainable development trajectory (FANRPAN 2017). This is also enumerated in its climate action, through the; *"National Climate Change Response Policy"*, sector-specific plans that are climate-compatible, and the *"National Sustainable Development Strategy"*. Some sectoral strategies for South Africa's power and integrated energy, economic reform action plan and sustainable growth route have all made progress toward implementation (FANRPAN 2017).

The National Innovations on Climate Resilient Agriculture

Over 150 climate-resilient villages and climate-smart villages have been established thanks to the *"National Innovations on Climate Resilient Agriculture (NICRA)"*, a national programme that facilitates research (CRVs/CSVs) in, *"Punjab-Haryana"*, *"Bihar"* and *"Andhra Pradesh states"* (Rao et al. 2016; Kishore et al. 2018). The agriculture analysis for development (*"ar4d"*) agenda may heavily rely on the CSV approach to address the issues posed by changes in temperature for food and nutrition security (Kishore et al. 2018). Launched in NICRA's Technology Demonstration Component in at least 151 vulnerable districts to counter climatic vagaries viz, salinity, extreme temperatures, droughts and floods (ICAR-CRIDA 2016). The CSV approach has been launched in Uganda and Kenya, among other African and Asian countries (Bonilla-Findji et al. 2017; Tripathi and Mishra 2017). Following the 21st Conference of Parties (COP 21) under the *"United Nations Framework Convention on Climate Change (UNFCCC)"*, in Paris, India's response to climate change was expanded with the introduction of four new missions in 2014. These included, *"Health Mission"*, *"National Mission"*, on *"Waste to Energy Generation"*, *"National Mission on India's Coastal areas"* and *"National Wind Mission"*.

Of these, *"Mission on Health"* was to utilise a multifaceted strategy to mitigate the health effects of climate change, including the creation of an integrated primary prevention monitoring system, state and local government-specific response to emergency plans and instating climate-resilient health facilities (Government of India 2018).

In India, an initiative to educate farmers and provide agro-meteorology advice has been commenced (Tripathi and Mishra 2017). The *"Indian Meteorological Department"*, established the former, an outdated programme in 1945 and was able to extend its services to the states level in 1976. This programme did not fully meet the expectations of most farming communities which culminated in its integration into a multi-channel transmission system at the federal, state and local levels that integrates agro-met assistance programmes. Together with other stakeholders, it has launched various farmer awareness programmes in different parts of India. Its objective is to disseminate and capacitate smallholder farmers' knowledge on weather and climate information alongside farm management (Moinina et al. 2018). Between 2009 to 2011, Indian agricultural institutions of higher learning and other organisations are hosting sixty Agro-met field sections and took over the programme, and has since trained over 7000 farmers (Tripathi and Mishra 2017). So far, agricultural extension services and capacity development initiatives were identified as crucial for promoting resilience and adaptation to climate change in the agriculture sector (Tripathi and Mishra 2017). In South Africa, Climate Change Programmes are being implemented. These involve initiatives to raise awareness, formulate policies, implement sector mitigation and adaptation strategies, undertake nationwide vulnerability assessments, find and organise climate-related research and development projects and enhance sector capacity (Mnkeni and Mutengwa 2014).

Policy on Organic Farming

The National Project on Organic Farming initiative was launched by the Indian centralised administration during the tenth plan period to encourage organic agriculture in India (Kumar and Indira 2017). Several Indian states have adopted similar organic agriculture promotion strategies (Yadav et al. 2013). For instance, *"Karnataka State Organic Farming Policy"* in 2004, established an initiative to encourage the use of organic fertilisers by the Government of Karnataka. Furthermore, the state introduced schemes such as the organic village programme (2006–2011), *Savayava Bhagya Yojane*, and issuance of certifications for organic products, as well as creating public awareness on organic farming (Kumar and Indira 2017).

South Africa has encouraged the application of organic–inorganic enriched manure and fertilisers as well as the use and planting of genetically modified organisms, GMOs (Iversen et al. 2014; GIZ 2017). The primary mechanism for regulating the importing and exporting of GMOs from and into South Africa is the, *"South African Genetically Modified Organism Act"* of 1997 and its related policy (Department of Agriculture Forestry and Fisheries 1997). Among other restrictions, the end-users of GMOs are necessitated to enter into technology licensing contracts with permit holders, in which such parties attest to their adherence to the terms of the permits (Iversen et al. 2014).

Renewable Energy Policy and Programmes

For more than 50 years, energy security issues, renewable energy policies and sustainable energy have been areas-of-interest in India (Jörgensen et al. 2015). The majority of state policies implemented in India provide for their own regulations and site requirements for investments in renewable energy, making them crucial to the attainment of the country's sustainable energy policy aspirations. Karnataka, Maharashtra, Rajasthan and Gujarat states with some solar and wind sectors have shown location benefits and some success stories in this context (Jörgensen et al. 2015). In the past two decades, India has as well initiated a renewable energy programme under her Ministry of Non-Conventional Energy Sources.

According to Panigrahi (2020), *"3–4 cubic meters of biogas per day"*, are being utilised in nearly "3.3 million home biomass gasification systems". This has furnished energy for cooking applications in rural settings (Keerthana et al. 2020). The government has also engaged public investments to construct the natural gas transmission network for both protracted and local delivery (Panigrahi 2020). For instance, approximately, *"4–5 billion cubic meters of gas"* from offshore exaction are transported to northern Delhi along a *"1500-kilometers"* elevated gas pipeline close to Mumbai. Gas now accounts for 8%

of total power generation capacity, up from 2% previously recorded about a decade ago.

Liquefied petroleum gas has also been used as coal and kerosene substitutes in urban households (Panigrahi 2020). Additionally the mitigation-focused initiative under the *"Jawaharlal Nehru National Solar Mission"* (Akoijam and Krishna 2017), has been, executed in three phases: *"phase-I (2010–2013)"*, *"phase-II (2013–2017)"*, and *"phase-III (2017–2022)"*. This was initiated to establish India as a global leader in solar energy and meant to see the installation of 2.970 MW of grid-connected solar generation capacity, installed 364 MW of off-grid solar generation capacity and 8.42 million square meters of solar thermal collectors (Patra 2016).

Similarly, South Africa has taken steps in enhancing the adoption of renewable energy. Initiatives such as conducive economic biogas feed-in tariffs and numerous current environmental policies in favour of biogas plant developments such as, *"National Environmental Management Act"*, *"National Environmental Management"*: *"Waste Act"*, *"National Environmental Management"*: *"Air Quality Act"*, *"National Waste Management Strategy"* and the *"Income Tax Act Amendment 12L (Cleaner Development Mechanism)"*, are indicative of the country's commitment to cleaner energy technologies (ESI Africa 2016). It formed the *"Southern Africa Biogas Industry Association (SABIA)"*, which accentuates its prospect of a biogas industry expansion (ESI Africa 2016).

Water Policy and Conservation and Development Programmes

Under this adaption-focused strategy, India through its national water mission revised its, *"National Water Policy (2012)"* which led to the creation of nearly, *"1 1,082 new groundwater"* monitoring reservoirs constructed, geared towards water conversation (Patra 2016). Additionally techniques and technologies for generating viz, *"Integrated Watershed Development Project"*, *"Watershed Development in Shifting Cultivation Areas"*, *"Desert Development Programme, Drought-Prone Area Programme"* and *"National Watershed Development Project for Rainfed Areas"* (Bhattacharyya et al. 2016). These projects were launched in three steps; the first-generation watershed projects targeted soil conservation, the second aimed at the conservation of degraded land and the third was meant to emphasise participatory strategies (Bhattacharyya et al. 2016).

Research cited by Bhattacharyya et al. (2016) performed a meta-analysis (at a macro-level a total of 636 micro-watersheds lying on 100–1000 hectares of land) which indicated that watershed programmes improved income, generated employment for the rural population, accelerated crop yields and intensity, reduced poverty, soil degradation and runoff water. Other innovative responses in water management focusing on addressing climate change risks in India have resulted in the introduction of technologies such as desalination, reusing, harvesting rainwater and recharging aquifers (Kumar and Gautam 2014).

These have been addressed by the Central Indian government initiatives notably subsidies for micro-irrigation (to optimise water use in agriculture) and artificial recharge to groundwater through wells drilled in hard rock regions, as well as programmes to improve rural water systems using the drainage network method (Kumar and Gautam 2014). In South Africa, *"the National Conservation Agriculture Task Force (NCATF)"* established in 2009 constitutes Land Care's contribution towards the advancement of climate-smart agriculture. It is one of the governmental agencies taking part in the, *"Conservation Agriculture Regional Working Group"* (Mnkeni and Mutengwa 2014).

National Mission for Enhanced Energy Efficiency

This initiative aims at achieving transformation by identifying solutions that are both economical and energy-efficient, which is socio-ecologically sustainable, the Indian government initiated the *"National Mission for Enhanced Energy Efficiency"*. This programme distributed nearly 2 million light-emitting diode bulbs to the Indian population (Patra 2016), such technology development measures in the energy sector has also been launched and has directly favoured reduction in the emission of greenhouse gases. For instance, minimisation of gas leaking during the power generation of fossil fuels, improvement in biomass cookstoves and the introduction of modern renewable energy systems. However, these measures have not received full exploitation, indicating that there is still untapped potential to minimise emissions growth in the country (Panigrahi 2020). Other enhancements in coal treatment have been made, and government regulations ban the shipping of dirty coal to below 1000 kilometres.

Customers have also been encouraged to minimise the ash concentration in an effort to improve performance, decrease pollution and cut freight charges. Novel coal to lessen the effects of combustion processes on the environment and subsequently climate change, combustible technologies including coal-fired power stations that operate at extreme temperatures and coal-bed biogas encapsulation are being advocated (Panigrahi 2020).

In South Africa, a sustainable economic paradigm was advanced. This is a "system of economic growth involving the production, dissemination and consumption of commodities and services that enhances human well-being over lengthy periods without endangering the environment or contributing to ecological shortfalls for subsequent generations" (Patra 2016).

South Africa has made the transition to a green economy a priority and has started a multitude of enabling measures to make it easier, with 9 major areas of emphasis: (1) *"Green buildings and the built environment"*, (2) *"Sustainable transport and infrastructure"*, (3) *"Clean energy and energy efficiency"*, (4) *"Resource conservation and management"*, (5) *"Sustainable waste management practices"*, (6) *"Agriculture, food production and forestry"*, (7)

"*Water management*", (8) "*Sustainable consumption and production*", (9) "*Environmental sustainability*" (Mnkeni and Mutengwa 2014).

National Policy for Disaster Management

First, put forward in 2009, the policy encourages the integration of approaches and as well as promotion of techniques for coping with climate change and decreasing vulnerability to unprecedented disasters. Further, "*the National Disaster Management Authority*", assembles a team of technocrats who can identify research priorities for reducing catastrophic events' risk, focusing on climate variability (Patra 2016).

Comparatively, the South African National Agricultural Disaster Risk Management Committee was instituted to provide consultation and strategic policy direction on matters pertaining to agricultural disaster risk and mitigation management. The government parastatal routinely addresses dangers including, "*veld fires*", "*floods*", "*droughts*" and "*pest and disease outbreaks*". They provide material and financial increments to farmers who would have lost income due to infrastructure-related damages, notably those characterised by irrigation, soil conservation systems and water reservoirs (Mnkeni and Mutengwa 2014).

12.6 Conclusions and Recommendations

This review attempted to establish the most effective practices in CSA which could be replicated by smallholder farming communities in India and South Africa alike. It is only prudent to appreciate that India and South Africa pose avenues to expand the profits of agricultural chain players in the face of a dynamic climate. The creation of a national distinct verification archive is imperative to synchronise a pool of interventions applicable to the needs, challenges being envisaged and resources of rural farmers that are the foremost susceptible to climate change impacts.

To scale up the adoption of the best CSA practices to fulfil the growing food demand, agriculture should undergo a transformation that handles simultaneously the dual obstacles of food insecurity and climate change. There are multiple CSA innovations that have been engineered, tried, field tested and commended under five entry points viz; crop management, land and water management, agroforestry and integrated food-energy systems that may be replicated and upscaled across India and South Africa.

The existence of tangled operation of, "*Political, Economic, Social and Technological, Legal and Environmental (PESTLE)*" barricades to the holistic adoption of productive and promising CSA practices by smallholder farmers needs to be exercised so as to improve agricultural productivity, enhance resilience to the present and unforeseen climate shocks and where attainable mitigate pollutant emissions (Reno 2019). Building a cohesive partnership between agricultural worth chain actors such as smallholder farming communities, agro-enterprises and national research institutions will establish imperative pathways for sustainable adoption of best CSA practices geared towards attaining food security, productivity, resilience and sustainable livelihoods. Formulating favourable policies and legislation that addresses CSA synchronisation and mainstreaming across different socioeconomic sectors so as to inclusively propel agricultural transformation and advancement at the state (micro), regionalistic (meso) and international (macro) spheres is also warranted.

Furthermore, elevating CSA best practices among assorted stages of the agricultural value chains requires the application of innovative mechanisms to funding as a result of the various structures, operations and administration within agro chains. Subsidisation may be a compound hiccup to agricultural transformation in India and South Africa's agricultural sectors, specifically as regards smallholder farmers. Bank interest rates are on the upper side and often smallholders marginalised usually because they lack collaterals viz; land and housing. This has additionally pronounced gender inequality, as the majority of the financially incapable a girls and youth who hardly own as well as have complete access to farmlands.

Thus, motivating the adoption of CSA techniques consequently necessitates public–private partnership integration in agricultural supply chain systems subsidising through providing packages (such as individual and micro-credit access and insurance to protect crops and livestock). Bearing in mind the foregoing recommendations, it remains imperative to acknowledge that India and South Africa already possess operational policy structures and legislative processes in sectors of agriculture, infrastructure and surroundings among others that buttress the development, implementation and adaptation of CSA innovations.

Hence, it remains desiring to develop holistic goals that encourage the coordination of policy structures and processes across the agricultural value chains as it is essential in the propellant adaptation of CSA practices. In view of the above, cooperative synergies developed at local, national and regional levels between India and South Africa may be applied as a buffer towards engineering political and socioeconomic cooperation between the nations and further engagement towards knowledge-data exchange, research and technological development so as to cohesively address world food and nutritional insecurity alongside vulnerable livelihoods within the paradigms of a dynamical climate.

Acknowledgements This project has been funded by the E4LIFE International Fellowship programme offered by Amrita Vishwa Vidyapeetham and JA also extends her gratitude to the Live-in-Labs Academic programme for the support provided.

REFERENCES

Abegunde VO, Sibanda M and Obi A. (2020) Determinants of the Adoption of Climate-Smart Agricultural Practices by Small-Scale Farming Households in King Cetshwayo District Municipality, South Africa. *Sustainability* 12: 195.

Adger N, Barnett J, Brown K, et al. (2013) Cultural Dimensions of Climate Change Impacts and Adaptation. *Nature Climate Change* 3: 112–117.

Aggarwal PK, Jarvis A, Campbell RB, et al. (2018) The Climate-Smart Village Approach: Framework of an Integrative Strategy for Scaling Up Adaptation Options in Agriculture. *Ecology and Society* 23: 14.

Akinnuoye-Adelabu DB and Modi AT. (2017) Planting Dates and Harvesting Stages Influence on Maize Yield under Rain-Fed Conditions. *Journal of Agricultural Science* 9: 43–55.

Akoijam AS and Krishna VV. (2017) Exploring the Jawaharlal Nehru National Solar Mission (JNNSM): Impact on Innovation Ecosystem in India. *African Journal of Science, Technology, Innovation and Development* 9: 573–585.

Ali A, Riaz S and Iqbal S. (2014) Deforestation and Its Impacts on Climate Change an Overview of Pakistan. *Papers on Global Change* 21: 51–60.

Archana, P. R., Aleena, J., Pragna, P., Vidya, M. K., Niyas, A. P. A., Bagath, M., ... & Bhatta, R. (2017). Role of heat shock proteins in livestock adaptation to heat stress. *J Dairy Vet Anim Res*, 5(1), 00127.

Altieri MA and Nicholls CI. (2017) The Adaptation and Mitigation Potential of Traditional Agriculture in a Changing Climate. *Climatic Change* 140: 33–45.

Andrieu N, Howland F, Acosta-Alba I, et al. (2019) Co-designing Climate-Smart Farming Systems With Local Stakeholders: A Methodological Framework for Achieving Large-Scale Change. *Frontiers in Sustainable Food Systems* 3: 37.

Angom J, Viswanathan PK and Ramesh MV. (2021) The Dynamics of Climate Change Adaptation in India: A Review of Climate Smart Agricultural Practices among Smallholder Farmers in Aravalli District, Gujarat, India. *Current Research in Environmental Sustainability* 3: 100039.

Anseeuw W and Boche M. (2015) South Africa in African Agriculture: Investment Models and Their Dynamics Towards a Structured Conquest. *Autrepart* 76: 49–66.

Aryal JP, Sapkota TB, Stirling CM, et al. (2016) Conservation Agriculture-Based Wheat Production Better Copes with Extreme Climate Events Than Conventional Tillage-Based Systems: A Case of Untimely Excess Rainfall in Haryana, India. *Agriculture, Ecosystems & Environment* 233: 325–335.

Bahinipati CS and Viswanathan PK. (2019) Can Micro-Irrigation Technologies Resolve India's Groundwater Crisis? Reflections from Dark-Regions in Gujarat. *International Journal of the Commons* 13: 848–858.

Balooni K. (2003) Economics of Wastelands Afforestation in India: A Review. *New Forests* 26: 101–136.

Bandyopadhyay KK, Sahoo RN, Singh R, et al. (2015) Characterisation and Crop Planning of Rabi Fallows Using Remote Sensing and GIS. *Current Science* 108: 2051–2062.

Barasa PM, Botai CM, Botai JO, et al. (2021) A Review of Climate-Smart Agriculture Research and Applications in Africa. *Agronomy* 11: 1255.

Barnard J, Manyire H, Tambi E, et al. (2015) FARA. Barriers to Scaling Up/Out Climate Smart Agriculture and Strategies to Enhance Adoption in Africa Forum for Agricultural Research in Africa, Accra, Ghana, 80.

Bhatia A, Pathak H, Aggarwal PK, et al. (2010) Trade-Off Between Productivity Enhancement and Global Warming Potential of Rice and Wheat in India. *Nutrient Cycling and Agroecosystems* 86: 413–424.

Bhattacharyya R, Ghosh BN, Dogra P, et al. (2016) Soil Conservation Issues in India. *Sustainability* 8: 565.

Birkenholtz T. (2017) Assessing India's Drip-Irrigation Boom: Efficiency, Climate Change and Groundwater Policy. *Water International* 42: 663–677.

Birthal PS, Khan MT, Negi DS, et al. (2014) Impact of Climate Change on Yields of Major Food Crops in India: Implications for Food Security. *Agricultural Economics Research Review* 27: 145–155.

Birthal PS, Nigam SN, Narayanan AV, et al. (2012) Potential Economic Benefits from Adoption of Improved Drought-Tolerant Groundnut in India. *Agricultural Economics Research Review* 25: 1–14.

Bisht IS, Rana JC and Ahlawat SP. (2020) The Future of Smallholder Farming in India: Some Sustainability Considerations. *Sustainability* 12: 3751.

Bogdanski A, Dubois O, Jamieson C, et al. (2010) Making Integrated Food-Energy Systems Work for People and Climate. An Overview. *Food and Agriculture Organization of the United Nations, Rome*.

Bonilla-Findji O, Recha J, Radeny M, et al. (2017) East Africa Climate-Smart Villages AR4D Sites: 2016 Inventory. Wageningen: CGIAR Research Program on Climate Change, Agriculture and Food Security (CCAFS).

Bunsen J and Rathod R. (2016) Pipe Assisted Underground Taming of Surface Floods: The Experience with Holiyas in North Gujarat. IWMI-TATA Water Policy Research Highlight 2, International Water Management Institute, Anand, India.

Chhetry GKN and Belbahri L. (2009) Indigenous Pest and disease Management Practices in Traditional Farming Systems in North East India: A Review. *Journal of Plant Breeding and Crop Science* 3: 028–038.

CSA Guide. (2020) What Is Climate-Smart Agriculture? Retrieved from https://csa.guide/csa/what-is-climate-smart-agriculture#tid-497.

Dar MH, Waza SA, Shukla S, et al. (2020) Drought Tolerant Rice for Ensuring Food Security in Eastern India. *Sustainability* 12: 2214.

Das A, Layek J, Babu S, et al. (2019) Intercropping for Climate Resilient Agriculture in NEH Region of India. *Technical bulletin No 1 (Online). ICAR Research Complex for NEH Region, Umiam - 793 103, Meghalaya*.

Davies J, Spear D, Chappel A, et al. (2019) Considering Religion and Tradition in Climate Smart Agriculture: Insights from Namibia. In: Rosenstock T., Nowak A., Girvetz E. (eds) *The Climate-Smart Agriculture Papers*. Springer, Cham, pp. 187–197.

Dell'Angelo, J., D'Odorico, P., & Rulli, M. C. (2017). Threats to sustainable development posed by land and water grabbing. *Current Opinion in Environmental Sustainability*, 26, 120–128.

Department of Agriculture Forestry and Fisheries. (1997) Genetically Modified Organisms Act (GMO Act) No.15 of 1997, as amended. Retrieved from http://cer.org.za/virtual-library/genetically-modified-organisms-act-15-of-1997.

Dinesh D, Aggarwal P, Khatri-Chhetri A, et al. (2017) The rise in Climate-Smart Agriculture Strategies, Policies, Partnerships and Investments Across the Globe. *Agriculture for Development* 30: 4–9.

DSUUSM. (2019) Dhundi Solar Energy Producers' Cooperative Society: Tri-annual Report, 2015–18. Colombo, Sri Lanka: International Water Management Institute

(IWMI). 28p. Tri-Annual Report, 2015–18. Available http://www.iwmi.cgiar.org/iwmi-tata/PDFs/dhundi_solar_energy_producers_cooperative_society-tri-annual_report-2015-18.pdf.

ECA. (2009) Economics of Climate Adaptation. Shaping Climate-Resilient Development: A Framework for Decision-making.

ESI Africa. (2016) Biogas—South Africa's Great Untapped Potential. Retrieved from https://www.esi-africa.com/magazine-article/biogas-south-africas-great-untapped-potential/.

FANRPAN. (2017) Climate-Smart Agriculture in South Africa. Retrieved from https://media.africaportal.org/documents/Policy_Brief_Issue_15.2017_CSA_South_Africa_-_Final_Draft02.pdf.

FAO. (2005) Fertilizer USE by Crop in South Africa. Land and Plant Nutrition Management Service. Land and Water Development Division, Food and Agriculture Organization of the United Nations, Rome. Retrieved from http://www.fao.org/tempref/agl/agll/docs/fertusesouthafrica.pdf.

FAO. (2014) FAO Success Stories on Climate-Smart Agriculture. *Rome: Food and Agriculture Organization of the United Nations (FAO)*: 1–28.

FAO. (2015) Fertilizer Use by Crop in India. *Food and Agriculture Organization of the United Nations, Rome*.

FAO. (2020) Integrated Food-Energy Systems. Retrieved from http://www.fao.org/energy/bioenergy/ifes/en/.

Fishman R, Gulati S and Li S. (2014) Should Resource Efficient Technologies Be Subsidized? Evidence from the Diffusion of Drip Irrigation in Gujarat. Paper presented at the ISI, Delhi.

Gautam A. (2009) Impact Evaluation of Drought Tolerant Rice Technologies through Participatory Approaches in Eastern India. *Masters' dissertation. The State University of New Jersey, Rutgers*.

GEF. (2020) South Africa Biogas Project Continues Under New COVID-19 Guidelines. Retrieved from https://www.thegef.org/news/south-africa-biogas-project-continues-under-new-covid-19-guidelines.

Geldenhuys CJ, Atsame-Edda A and Mugure MW. (2017) Facilitating the recovery of natural evergreen forests in South Africa via invader plant stands. *Forest Ecosystems* 4: 21.

Ghosh M. (2019) Climate-smart Agriculture, Productivity and Food Security in India. *Journal of Development Policy and Practice*: 1–22.

GIZ. (2017) Agricultural Adaptation: Six Categories of Good Practices and Technologies in Africa. https://reliefweb.int/sites/reliefweb.int/files/resources/Agricultural-Adaptation-Report-Digital-low-res.pdf.

Godfray HCJ, Beddington JR, Crute IR, et al. (2010) Food security: The Challenge of Feeding 9 Billion People. *Science* 327: 812–818.

GoF. (2005) Climate Change The Fiji Islands Response: Fiji's First National Communication Under the Framework Convention on Climate Change. Pacific Islands Climate Change Assistance Programme (PICCAP) & Fiji Country Team. Fiji's Initial National Communication, pp. 1–80.

GoK. (2014) El Niño Contingency Plan. In: Ministry of Environment WaNR (ed). Nairobi, Kenya: The Government of Kenya and Humanitarian Partners.

Government of India. (2018) National Action Plan For Climate Change & Human Health. Ministry of Health & Family Welfare.

Government of India and the United Nations. (2018) Sustainable Development Framework 2018–2022.

Gowane GR, Kumar A and Nimbkar C. (2019) Challenges and Opportunities to Livestock Breeding Programmes in India. *Journal of Animal Breeding and Genetics* 136: 5.

Goyal S. (2011) Food Waste to Energy Conversion—Indian Perspectives. Retrieved from https://www.altenergymag.com/article/2011/02/food-waste-to-energy-conversionindian-perspectives/840.

Gupta R, Tyagi NK and Abrol I. (2020) Rainwater Management and Indian Agriculture: A Call for a Shift in Focus from Blue to Green Water. *Agricultural Research* https://doi.org/10.1007/s40003-020-00467-2.

Haarhoff SJ, Kotzé NT and Swanepoel PA. (2020) A Prospectus for Sustainability of Rainfed Maize Production Systems in South Africa. *Crop Science* 60: 14–28.

Hartung H and Pluschke L. (2018) The Benefits and Risks of Solar-Powered Irrigation—A Global Overview. Food and Agriculture Organization of the United Nations and Deutsche Gesellschaft für Internationale Zusammenarbeit. http://www.fao.org/3/I9047EN/i9047en.pdf.

Hegerl GC, Zwiers FW, Braconnot P, et al. (2007) Understanding and Attributing Climate Change. In: Climate Change 2007: The Physical Science Basis. Contribution of Working Group I to the Fourth Assessment Report of the Intergovernmental Panel on Climate Change [Solomon S, Qin D, Manning M, Chen Z, Marquis M, Averyt KB, Tignor M, and Miller HL (eds.)]. Cambridge, United Kingdom and New York, NY, USA: Cambridge University Press.

Henderson L. (1983) Barrier plants in South Africa. *Bothalia* 14: 635–639.

Hoegh-Guldberg, O., Jacob, D., Bindi, M., Brown, S., Camilloni, I., Diedhiou, A., ... & Zougmoré, R. B. (2018). Impacts of 1.5 C global warming on natural and human systems. *Global warming of 1.5° C.*

ICAR-CRIDA. (2016) Research Highlights 2015-16. *National Innovations in Climate Resilient Agriculture (NICRA), ICAR-Central Research Institute for Dryland Agriculture.* Retrieved from http://www.nicraicar.in/nicrarevised/images/publications/NICRA%20Res%20Highlights%202015-16.pdf.

ICID. (2019) Solar Powered Irrigation Systems in India: Lessons for Africa Through a FAO Study Tour Draft Report. *International Commission on Irrigation and Drainage.* Retrieved from https://www.icid.org/FAO-SPIS-Report.pdf.

International Finance Bank. (2015) Impact of Efficient Irrigation Technology on Small Farmers. https://www.ifc.org/wps/wcm/connect/1f630d98-dabc-41e4-9650-b8809d620664/Impact+of+Efficient+Irrigation+Technology+on+Small+Farmers+-+IFC+Brochure.pdf?MOD=AJPERES&CVID=lKbEzwG.

IPCC. (2012) Intergovernmental Panel on Climate Change. Glossary of Terms Oxford Dictionaries. Retrieved from https://www.ipcc.ch/sr15/chapter/glossary/#:~:text=Climate%20change%20refers%20to%20a,period%2C%20typically%20decades%20or%20longer.

IPCC. (2021) Climate Change 2021: The Physical Science Basis. Contribution of Working Group I to the Sixth Assessment Report of the Intergovernmental Panel on Climate Change [Masson-Delmotte V, Zhai P, Pirani A, Connors, Péan C, Berger S, Caud N, Chen Y, Goldfarb L, Gomis MI, Huang M, Leitzell K, Lonnoy E, Matthews JBR, Maycock TK, Waterfield T, Yelekçi O, Yu R, and Zhou B (eds.)]. Cambridge University Press.

ITA. (2020) South Africa—Country Commercial Guide. Retrieved from https://www.trade.gov/knowledge-product/south-africa-agricultural-sector#:~:text=The%20agricultural%20sector%20contributed%20around,name%20just%20a%20few%20products.

Iversen M, Grønsberg IM, van den Berg J, et al. (2014) Detection of Transgenes in Local Maize Varieties of Small-Scale Farmers in Eastern Cape, South Africa. *PLoS ONE* 9: e116147.

Jain A, Das P, Yamujala S, et al. (2020) Resource Potential and Variability Assessment of Solar and Wind Energy in India. *Energy* 211: 118993.

Jalota SK, Kaur H, Ray SS, et al. (2012) Mitigating Future Climate Change Effects by Shifting Planting Dates of Crops in Rice–Wheat Cropping System. *Regional Environmental Change* 12: 913–922.

Jat RK, Singh RG, Kumar M, et al. (2019) Ten Years of Conservation Agriculture in a Rice-maize Rotation of Eastern Gangetic Plains of India: Yield Trends, Water Productivity and Economic Profitability. *Field Crops Research* 232: 1–10.

Jayan TV. (2018) Solar Pumps: A Nondescript Village in Gujarat Shows the Way. Retrieved from https://www.thehindubusinessline.com/news/solar-pumps-a-nondescript-village-in-gujarat-shows-the-way/article22694612.ece.

Jörgensen K, Mishra A and Sarangi GK. (2015) Multilevel Climate Governance in India: The Role of the States in Climate Action Planning and Renewable Energies. *Journal of Integrative Environmental Sciences* 12: 267–283.

Kanitkar T, Banerjee R and Jayaraman T. (2019) An Integrated Modeling Framework for Energy Economy and Emissions Modeling: A Case for India. *Energy* 167: 670–679.

Kaur G, Brar GS and Kothari DP. (2017) Potential of Livestock Generated Biomass: Untapped Energy Source in India. *Energies* 10: 847.

Keerthana KM, Arjun C, Krishnan JR, et al. (2020) Technology Assisted Rural Futures in the Village of Moti Borvai. *Lecture Notes in Electrical Engineering*: 1654–1661.

Keil A, D'souza A and McDonald A. (2017) Zero-Tillage Is a Proven Technology for Sustainable Wheat Intensification in the Eastern Indo-Gangetic Plains: What Determines Farmer Awareness and Adoption? *Food Security* 9: 723–743.

Kishore A, Pal BD, Joshi K, et al. (2018) Unfolding Government Policies Towards the Development of Climate Smart Agriculture in India. *Agricultural Economics Research Review* 31: 123–137.

Kraaijenbrink PDA, Stigter EE, Yao T, et al. (2021) Climate Change Decisive for Asia's Snow Meltwater Supply. *Nature Climate Change* 11: 591–597.

Kumar LMP and Indira M. (2017) Trends in Fertilizer Consumption and Foodgrain Production in India: A Co-integration Analysis. *SDMIMD Journal of Management* 8: 45–50.

Kumar R and Gautam HR. (2014) Climate Change and Its Impact on Agricultural Productivity in India. *Journal of Climatology and Weather Forecasting* 2: 109.

Laks R. (2017) The Potential for Electricity Generation from Biogas in South Africa. A potential study as part of the BAPEPSA project. Retrieved from https://publicaties.ecn.nl/PdfFetch.aspx?nr=ECN-E--17-001.

Laxmi V, Erenstein O and Gupta RK. (2007) *Impact of Zero Tillage in India's Rice-Wheat Systems*. Mexico, D.F.: CIMMYT.

Longobardi P, Montenegro A, Beltrami H, et al. (2016) Deforestation Induced Climate Change: Effects of Spatial Scale. *PLoS ONE* 11: e015335.

Maas EMC and Kotzé JM. (1990) Crop Rotation and Take-All of Wheat in South Africa. *Soil Biology and Biochemistry* 22: 489–494.

Macrotrends. (2021) India Population Growth Rate 1950–2021. https://www.macrotrends.net/countries/IND/india/population-growth-rate.

Mahajan G, Bharat TS and Tasmina TS. (2009) Yield and Water Productivity of Rice as Affected by Time of Transplanting in Punjab, India. *Agriculture and Water Management* 96: 525–532.

Mishra PK and Rai SC. (2011) Cost-Benefit Analysis of Terrace Cultivation in Sikkim Himalaya, India. *The Indian Geographical Journal* 86: 29–37.

Mnkeni P and Mutengwa C. (2014) Report. *A Comprehensive Scoping And Assessment Study Of Climate Smart Agriculture Policies In South Africa. Food, Agriculture and Natural Resources Policy Analysis Network*.

Modi AT and Mabhaudhi T. (2013) Water-Use And Drought Tolerance Of Selected Traditional Crops. *Report to the Water Research Commission*. WRC Report No. 1771/1/13. Retrieved from http://www.wrc.org.za/wp-content/uploads/mdocs/1771-1-131.pdf.

Moeletsi ME. (2017) Mapping of Maize Growing Period over the Free State Province of South Africa: Heat Units Approach. *Advances in Meteorology* 2017: 7164068.

Moinina A, Lahlali R, MacLean D, et al. (2018) Farmers' Knowledge, Perception and Practices in Apple Pest Management and Climate Change in the Fes-Meknes Region, Morocco. *Horticulturae* 4: 42.

Mokgolo MJ, Mzezewa J and Odhiambo JJO. (2019) Poultry and Cattle Manure Effects on Sunflower Performance, Grain Yield And Selected Soil Properties in Limpopo Province, South Africa. *South African Journal of Science* 115: 1–7.

Mthembu BE, Everson TM and Everson CS. (2018) Intercropping maize (Zea mays L.) with lablab (Lablab purpureus L.) for Sustainable Fodder Production and Quality in Smallholder Rural Farming Systems in South Africa. *Agroecology and Sustainable Food Systems* 42: 362–382.

Mukumba P, Makaka G and Mamphweli S. (2016) Biogas Technology in South Africa, Problems, Challenges and Solutions. *International Journal of Sustainable Energy and Environmental Research* 5: 58–69.

Mutungwazi A, Mukumba P and Makaka G. (2018) Biogas Digester Types Installed in South Africa: A Review. *Renewable and Sustainable Energy Reviews* 81: 172–180.

Mzezewa J and Gwata ET. (2016) Analysis of Soil Profile Water Storage under Sunflower × Cowpea Intercrop in the Limpopo Province of South Africa. In: Alternative Crops and Cropping Systems, IntechOpen, pp. 132–143.

Naaiik RVTB, Madhavi A and Mishra JS. (2018) Effect of Preceding Legumes, Nitrogen Levels and Irrigation Schedules on Productivity of Sorghum and Soil Biological Activity. *Journal of Pharmacognosy and Phytochemistry* 7: 2453–2459.

Naireeta Services. (2018) Bhungroo. Available from www.naireetaservices.com/projects/.

Namara R, Nagar R and Upadhyay B. (2007) Economics, Adoption Determinants and Impacts of Micro-irrigation Technologies: Empirical Results from India. *Irrigation Science* 25: 283–297.

Naumann G, Cammalleri C, Mentaschi L. et al. (2021) Increased Economic Drought Impacts in Europe with Anthropogenic Warming. *Nature Climate Change* 11: 485–491

Nel AA. (2005) Crop Rotation in the Summer Rainfall Area of South Africa. *South African Journal of Plant and Soil* 22: 274–278.

New World Encyclopedia. (2017) The Climate of India. Retrieved from https://www.newworldencyclopedia.org/entry/Climate_of_India.

Norfund. (2012) Electricity From Organic Waste. Retrieved from https://www.norfund.no/app/uploads/2020/01/Bio2Watt-case-study.pdf.

Odhiambo JJO. (2011) Potential Use of Green Manure Legume Cover Crops in Smallholder Maize Production Systems in Limpopo province, South Africa. *African Journal of Agricultural Research* 4: 107–112.

Okoba BO. (2018) FAO, Ministry of Agriculture, Livestock and Fisheries. Climate Smart Agriculture—Training Manual for Extension Agents in Kenya.

Olarinmoye AO, Tayo OG and Akinsoyinu AO. (2011) An Overview of Poultry and Livestock Waste Management Practices in Ogun State, Nigeria. *Journal of Food, Agriculture & Environment* 9: 643–645.

Panigrahi SK. (2020) Climate Change and Development in Indian Context, 1–7. Retrieved from https://unfccc.int/cop8/se/kiosk/cd4.pdf.

Patra J. (2016) Review of Current and Planned Adaptation Action in India. CARIAA Working Paper No. 10. International Development Research Centre, Ottawa, Canada and UK Aid, London, United Kingdom.

Pavelic P. (2020) Mitigating Floods for Managing Droughts through Aquifer Storage: An Examination of Two Complementary Approaches. Washington, D.C.: World Bank, 16 pages.

Pedraza V, Perea F, Saavedra M, et al. (2015) Winter Cover Crops as Sustainable Alternative to Soil Management System of a Traditional Durum Wheat-Sunflower Rotation in Southern Spain. *Procedia Environmental Sciences* 29: 95–96.

Picciariello A, Colenbrander S, Bazaz A, et al. (2021) The Costs of Climate Change in India: A Review of the Climate-Related Risks Facing India, and Their Economic and Social Costs. ODI Literature Review. London: ODI. www.odi.org/en/publications/the-costs-of-climate-change-in-india-a-review-of-the-climate-related-risks-facing-india-and-their-economic-and-social-costs.

Prabhakar SVRK and Reddy SN. (2010) Effect of Different Dates of Dry Seeding and Staggered Nursery Sowing on Growth and Yield of Kharif Rice. *Nature Precedings*. https://doi.org/10.1038/npre.2010.5399.1.

Praveen B and Sharma P. (2019) Climate Change and Its Impacts on Indian Agriculture: An Econometric Analysis. *Journal of Public Affairs* 20: e1972.

Pray C, Nagarajan L, Li L, et al. (2011) Potential Impact of Biotechnology on Adaption of Agriculture to Climate Change: The Case of Drought Tolerant Rice Breeding in Asia. *Sustainability* 3: 1723–1741.

Rajwade YA, Swain DK, Tiwari KN, et al. (2015) Evaluation of Field Level Adaptation Measures Under the Climate Change Scenarios in Rice Based Cropping System in India. *Environmental Processes* 2: 669–687.

Rao AS and Reddy KS. (2005) Integrated Nutrient Management vis-à-vis cRop Production/Productivity, Nutrient Balance, Farmer Livelihood and Environment: India. Paper Number 2. Retrieved from http://www.fao.org/3/AG120E09.htm.

Rao CS, Gopinath KA, Prasad JV, et al. (2016) Climate Resilient Villages for Sustainable Food Security in Tropical India: Concept, Process, Technologies, Institutions, and Impacts. *Advances in Agronomy* 140: 101–214.

Rathod P, Veeranna KC, Ramachandra B, et al. (2017) Utilization of Crop Residues for Livestock Feeding: A Field Experience. *XXVI Annual Conference of Society of Animal Physiologists of India (SAPI)*, Veterinary College, Bidar, Karnataka, 21–22 December, 2017: 66–74.

Ravindranath NH, Chaturvedi RK and Murthy IK. (2008) Forest Conservation, Afforestation and Reforestation in India: Implications for Forest Carbon Stocks. *Current Science* 95: 216–222.

Reno C. (2019) From Theory to Practice: Perspectives on Climate-Smart Agriculture in India and Africa. ORF Issue Brief No. 290, April 2019, Observer Research Foundation.

Sartorius BK, Sartorius K, Chirwa TF, et al. (2011) Infant Mortality in South Africa-Distribution, Associations and Policy Implications, 2007: An Ecological Spatial Analysis. *International Journal of Health Geographics* 10: 61.

Schoeman A. (2020) Recipes for Natural Pest Control. Retrieved from http://southafrica.co.za/recipes-for-natural-pest-control.html.

Shah F and Wu W. (2019) Soil and Crop Management Strategies to Ensure Higher Crop Productivity Within Sustainable Environments. *Sustainability* 11: 1485.

Shashidharan N. (2012) Analysis of Low Cost Reforestation Techniques in Dry Deciduous Forests in South India: A Case Study: Lokkere Reserve Forest, Karnataka. *Technical Report*, Junglescapes NGO, Karnataka.

Singh A. (2016) The LifeCycle of Cow Dung in India. Retrieved from https://ypard.net/2016-march-27/lifecycle-cow-dung-india.

Singh H. (2021) List of India's neighbouring countries. https://www.jagranjosh.com/general-knowledge/list-of-indias-neighbouring-countries-1400669307-1#:~:text=countries%20of%20India%3F-,India%20shares%20its%20border%20with%20seven%20countries%2D%20Afghanistan%20and%20Pakistan,two%20countries%20with%20water%20borders.

Singh RP and Agrawal RC. (2020) Farmers' Varieties and Ecosystem Services with Reference to Eastern India. In: Bauddh K, Kumar S, Singh R, Korstad J. (eds) *Ecological and Practical Applications for Sustainable Agriculture*. Singapore: Springer, pp. 421–443.

South Africa Gateway. (2017) South Africa's Weather and Climate. Retrieved from https://web.archive.org/web/20171201031244/https://southafrica-info.com/south-africa-weather-climate/.

Suman S. (2018) Evaluation and Impact Assessment of the Solar Irrigation Pumps Program in Andhra Pradesh and Chhattisgarh. Report prepared for Shri Shakti Alternative Energy Limited.

Sunref. (2013) South Africa's first industrial biogas plant leads way for waste to energy developments. Retrieved from https://www.sunref.org/en/projet/south-africas-first-industrial-biogas-plant-leads-way-for-waste-to-energy-developments/.

Suresh A and Manoj PS. (2020) Micro-irrigation Development in India: Challenges and Strategies. *Current Science* 118: 1163–1168.

Tankha S, Fernandes D and Narayanan NC. (2020) Overcoming Barriers to Climate Smart agriculture in India. *International Journal of Climate Change Strategies and Management* 12: 108–127.

Tanko M and Ismail S. (2021) How Culture and Religion Influence the Agriculture Technology Gap in Northern Ghana. *World Development Perspectives* 22: 100301.

Tej MK, Aruna R, Mishra G, et al. (2017) Beekeeping in India. In: Omkar (Ed.), *Industrial Entomology*, pp. 34–65. Springer Nature Singapore Pte Ltd.

The World Bank. (2018) Country Partnership Framework for India. Climate Smart Engagement. Retrieved from https://www.worldbank.org/en/cpf/india/cross-cutting-themes/climate-smart-engagement.

Tirado R and Cotter J. (2010) Ecological Farming: Agriculture Drought-Resistant Agriculture. *Research Laboratories Technical Note 02/2010*. Retrieved from https://www.greenpeace.to/publications/Drought_Resistant_Agriculture.pdf.

Totin E, Segnon AC, Schut M, et al. (2018) Institutional Perspectives of Climate-Smart Agriculture: A Systematic Literature Review. *Sustainability* 10: 1990.

Tripathi A and Mishra AK. (2017) Knowledge and Passive Adaptation to Climate Change: An Example from Indian Farmers. *Climate Risk Management* 16: 195–207.

Tsubo M, Mukhala E, Ogindo HO, et al. (2003) Productivity of Maize-Bean Intercropping in a Semi-Arid Region of South Africa. *Water SA* 29: 381–388.

Uhunamure SE, Nethengwe NS and Tinarwo D. (2020) Evaluating Biogas Technology in South Africa: Awareness and Perceptions towards Adoption at Household Level in Limpopo Province, pp. 1–16. In: *Renewable Energy—Resources, Challenges and Applications*. IntechOpen.

UN Sustainable Development Group. (2019) United Nations Sustainable Development Cooperation Framework Guidance. https://unsdg.un.org/resources/united-nations-sustainable-development-cooperation-framework-guidance.

UN Sustainable Development Group. (2020) In: Brief: United Nations Sustainable Development Cooperation. https://unsdg.un.org/sites/default/files/2020-01/In-Brief-UN-Sustainable-Development-Cooperation.pdf.

UNDP. (2020) South Africa. Retrieved from https://www.adaptation-undp.org/explore/africa/south-africa.

United Nations South Africa. (2021) How the UN Is Supporting the Sustainable Development Goals in South Africa. https://southafrica.un.org/en/sdgs.

Varshney V, Roy D and Meenakshi JV. (2020) Impact of COVID-19 on Agricultural Markets: Assessing the Roles of Commodity Characteristics, Disease Caseload and Market Reforms. *Indian Economic Review*. https://doi.org/10.1007/s41775-41020-00095-41771.

Vera TS, Wiliams CE and Justin CO. (2017) Understanding the Factors Affecting Adoption of Subpackages of CSA in Southern Malawi. *International Journal of Agricultural Economics and Extension* 5: 259–265.

Viswanathan PK, Kavya K and Bahinipati CS. (2020) Global Patterns of Climate-resilient Agriculture: A Review of Studies and Imperatives for Empirical Research in India. *Review of Development and Change*: 1–24.

Vora A and Parikh D. (1996) Water Management to Combat Drought. *Journal of Sardar Patel Institute of Economic and Social Research. Ahmedabad* 26: 33.

Wekesa BM, Ayuya OI and Lagat JB. (2018) Effect of Climate-Smart Agricultural Practices on Household Food Security in Smallholder Production Systems: Micro-Level Evidence from Kenya. *Agriculture & Food Security* 7: 80.

Willy DK, Diallo Y, Affognon H, et al. (2020) COVID-19 Pandemic in Africa: Impacts on Agriculture and Emerging Policy Responses for Adaptation and Resilience Building. *TAAT Policy Compact Working Paper No. WP01/2020*, 15 pages.

Wolpert SA. (2021) India. https://www.britannica.com/place/India.

Yadav SK, Babu S, Yadav MK, et al. (2013) A Review of Organic Farming for Sustainable Agriculture in Northern India. *International Journal of Agronomy* 2013: 718145.

Zougmoré R, Jalloh A and Tioro A. (2014) Climate-Smart Soil Water and Nutrient Management Options in Semiarid West Africa: A Review of Evidence and Analysis of Stone Bunds and Zaï Techniques. *Agriculture & Food Security* 3: 16.

CHAPTER 13

Flood Resilience of Pokkali Rice: A Study of the 2018 Flood in Kerala

Sweta Baidya, Parvathi Radhakrishnan, and Anil Kumar Gupta

13.1 INTRODUCTION

We are increasingly becoming vulnerable to climate change-related hazards and hydro-meteorological extremes. The frequency of the Climate change-induced disasters and the aftermath is increasing day by day. The Centre for Research on Epidemiology of Disasters (2008), evidenced that the rate of disasters has increased worldwide by the first and second half of the twentieth century. In recent years, the consequences of climate change and associated issues, along with the intensity of extreme events, are increasing in the country (World Meteorological Organization, 2021). This has an adverse impact on agricultural productivity, as Indian agriculture is mostly dependent on climatic conditions. Climate variability influences the production of crops and livestock, hydrological balances, input supplies, and other factors of the agronomy system (Adams et al., 1998). The long-term variability of climate change

---

Present Address:
S. Baidya (✉) · P. Radhakrishnan · A. K. Gupta
National Institute of Disaster Management (Ministry of Home affairs), Rohini, India
e-mail: cons-cyclone@ndma.gov.in

S. Baidya
Mitigation Division, National Disaster Management Authority (Ministry of Home Affairs), Delhi, India

P. Radhakrishnan
Kerala Institute of Local Administration, Thrissur, India

© The Author(s), under exclusive license to Springer Nature Singapore Pte Ltd. 2023
S. Nautiyal et al. (eds.), *Palgrave Handbook of Socio-ecological Resilience in the Face of Climate Change*,
https://doi.org/10.1007/978-981-99-2206-2_13

results in a change in conditions such as drought, pest outbreaks on crops and livestock, changes in soil characteristics, sea-level rise, and other disaster events. This, in turn, increases the vulnerability of the agriculture-dependent regions by causing a major loss in productivity and a reduction in yield (Kurukulasuriya and Rosenthal, 2003).

Importance of Pokkali Cultivation

Pokkali is an incredible flood and salinity-resistant crop that is a cultivable, eco-friendly, organic, and traditional way of rice cultivation, practiced only in Kerala. As the name suggests, Pokkali in native language means 'tall that can stay' and is internationally acclaimed as a sustainable system of farming and accepted as a gene donor for salt tolerance in rice (Deepa, 2018). With a rich 3000 years of cultivation history, Pokkali is grown without using any chemical fertilizers or other inorganic chemicals. They are not cultivated throughout the state even though they are only seen in Kerala. The large coastline tract (580 km) of the state provides hydromorphic saline soil, which is the perfect condition for Pokkali cultivation (Chandramohanan and Mohanan, 2011). Pokkali cultivation is majorly practiced in the districts of Ernakulam, Alappuzha, and Thrissur and most of the farmlands lie between the Arabian Sea coast and Vembanad Lake. Ernakulam has a vast range of crop cultivation. Pokkali has high market value and numerous medicinal benefits along with other properties and was awarded the status of Registered Geographical Indication in 2008. In fact, Pokkali fields could be used alternatively for crops and aquaculture (Image 13.1), such as Pokkali and Prawn culture according to the season, and the Pokkali crop is primarily depending on the monsoon shower. It is the world's oldest and most organic way of crop farming, normally done during May to October. Shrimp cultivation is carried out for the next six months after the Pokkali harvest during the post-monsoon time. The crop is grown in the coastal belt, where the fields are susceptible to high tide; therefore, this crop has the ability to tolerate abiotic stress like the salinity of the soil to some extent. Most of the time, the fields are submerged in saline water and thus, the tidal amplitude causes a direct impact on the salinity and water level in the field (Government of Kerala, 2004). The salinity of the Pokkali field decreases at the time of paddy cultivation and increases at the time of shrimp culture; hence the area is under convergence with saline and freshwater with the salinity ranging from 0 to 31 ppt or above (Sudhan et al., 2016). The ability to withstand salinity and flooding is the foremost and major feature of the crop (Ranjith et al., 2018). Pokkali is a cost-effective method, even though there are many problems associated with this cultivation. The encountering problems of Pokkali involve the shortage of land for cultivation, reclamation of land for other commercial purposes, and construction of infrastructures, etc. Now in Kerala, 30 lakh tons of rice is produced each year, and it is estimated as one-third of the total requirements of the state; thus, it is recognized that the cultivation of Pokkali in Kerala is decreasing day by

Image 13.1 A Picture of natural growth of mangroves in Pokkali wetland at Kumbalangi study region. B Pokkali field at the time of prawn culture in the North Paravur study area. C At Kumbalangi study area during seasonal shrimp farming in Pokkali field. D Picture of Pokkali wetland at the time of shrimp culture at North Paravur

day due to the decreasing rate of production. The large-scale conservation of paddy fields for other purposes is the main reason for the reduction in Pokkali rice farmlands (Chandramohanan and Mohanan, 2011). Thus, the area under Pokkali cultivation has also changed from 25,000 ha to 5500 ha in the past few decades.

Properties of Pokkali Rice

The specialties of Pokkali rice cultivation include saline rice farming and its historical way of organic farming that is tolerant to abiotic stress. It is medium bold shape rice having 7.5%–8.57% of average protein content, intermediate amylose content, good cooking quality, and rich taste. These special features are the reason for the predominant organic farming of Pokkali in Kerala. Along with these properties, rice has a good presence of iron, zinc, potassium, and antioxidants (oryzanol, tocopherol and tocotrienol, etc.) and possesses great medicinal properties (Shamna and Vasantha, 2017). Since it contributes to a sustainable way of agriculture, the rice crop has a vital role in achieving a green revolution (Sudhan et al. 2016). Pokkali has been registered for the status of Geographical Indication (GI) in the year of 2008–2009. Geographic indications are the indication of a specific origin of a product and provide qualities and reputation by the origin (Anson and Pavithran, 2014). The flood and salinity-tolerant rice variety of Pokkali acts as a goal to achieve food security.

Traditional Cultivation of Pokkali

The seasonal cycle of Pokkali cultivation system consists of an array of steps and each step of farming from sowing to harvest are different. The process of farming starts in the mid of May. At end of April, when the shrimp harvesting is completed, the soil of the Pokkali lands is dried. After that mounds of half meter to one meter width are constructed to remove salinity of the soil. According to the season and water inflow, Pokkali grows in low to medium-saline condition and cannot withstand high salinity (Image 13.1). The salinity of the soil is tested for proper crop management, and the required amount and depth of water is filled to the field for this rice cultivation.

The soaked seeds are sown in the field at the rate of 100 kg/ha and need to maintain the water level after sowing to protect the seeds from tidal inflow. After 45 days, the germinated seeds reaching a height of 40–45 cm are pulled apart and dispersed around the mounded land uniformly, and the water level is increased to maintain the level once the seedlings are rooted in the soil. June is favorable for the Pokkali cultivation and the draining of the fields is done in that same month. The crop can withstand the high tide of this season due to the tall nature of the plant. Weed management is the further step for farming after it becomes mature. Since, there is no perfect weed management technology, it is done by skilled laborers. Harvesting is the final step of Pokkali farming; after three to four months of growing, Pokkali is harvested in the month of October. At the time of harvest, the crop grows up to a height of one to two feet. Traditionally the yield of Pokkali ranges from 1 to 1.5 tons per hectare (Ranga et al., 2006; Sudhan et al., 2016).

Overview of 2018 Kerala Flood

Kerala is a relatively small state situated in the southwest part of the Indian peninsula, having an area of 38,863 km^2 (Vishnu et al., 2019) and are vulnerable to natural disasters such as cyclone, tsunami, storm surge, coastal erosion, drought, landslide, etc. due to long coastline area and the steep slope along the Western Ghats.

The massive monsoon in 2018 in Kerala caused a sudden rise in the water level among the water bodies leading to severe damage throughout the state. And it was the worst flood that was experienced in over a century since 1924 the great flood. It was declared a calamity of severe nature or a level 3 calamity by the Government of India. According to IMD data, the state has experienced 23,463.3 mm of rainfall against the normal rainfall of 1649.85 mm from June to August 2018. Alappuzha, Idukki, Ernakulam, Kottayam, Pathanamthitta, Malappuram, and Wayanad are the most extremely flood-affected districts in Kerala (Vineesh, 2019). The spontaneous event of heavy rainfall occurred from June to August in the catchment area of the Western Ghats in Kerala, increasing the water level in the reservoirs, and the release of water through floodgates resulted in inundation by the excess water in most of the districts.

The floodwater inundated huge built areas and small cities and distracted thousands of life (Vishnu et al., 2019).

A total of 7.5 lakh people were potentially affected, and overall 1.75 lakhs of buildings were completely or somewhat damaged. Due to the torrent of floodwater in the region, about 14 lakhs of people were relocated to relief camps. The main difficulties during and after the flood were the contamination of water due to the inundation of 95,000 household latrines were resulted in the pollution of groundwater, water bodies, wetlands, and other water resources. The aftermath of a flood created many major distractions in all sectors, including productive sectors, infrastructure, cross-cutting sectors, and social sectors, etc. (PDNA). Agriculture is the major sector that faces widespread damage after a flood. A total of 11% of the GDP is contributed by the agriculture sector in the state, which has been negatively impacted by the flood due to heavy rainfall. Paddy is one of the agricultural products that has been adversely affected by the floods which caused the decline of paddy fields (Vineesh, 2019). Damage to the agricultural sector, infrastructure, and property after the floods led to a temporary decline in agricultural-related incomes and an increase in government agricultural reform spending after the disaster. Thus the production losses had last for several months and lead to a depression in household incomes. Among these, 15% of the loss is in the productive sector including agriculture, fisheries, and livestock (PDNA). In 2017–2018, 52.7% of the total area of Kerala is agricultural land, and is about 20, 48, 109 ha. In this agricultural land, among all types of crops, 5214.3 ha is used for paddy cultivation in the state. The devastating flood of Kerala demolished the agricultural system and produce of the state especially rice. After the flood in 2018, the Ernakulum district experienced a crop loss that extends up to 1296.66 ha (Report of Kerala Flood, 2018a). By the analysis of the agriculture department, a total of around 56,844.44 ha of agriculture was affected by the flood and had a loss of 1355.68 crores. Rubber, tea, banana, paddy, tapioca, poultry, cattle, and cardamom are the major various agricultural produce, that was rigorously affected by the flood. Among all, Paddy had a huge loss in the agriculture sector and a loss of 26,106 ha of crops after the flood (Rajan and Anjana, 2019).

Cause of the 2018 Flood in Kerala

Floods are a widespread natural and climate-related disaster around the world. It is generally defined as the flow of water into an arid area. It can occur as a result of heavy rainfall or in other ways unrelated to the weather (Doswell, 2003).

There was a notable range of heavy rainfall that occurred during 1924 throughout Kerala. Similarly in 2018 also significant amounts of rainfall occurred throughout the state in August, at the time of monsoon. Among the several other reasons for havoc rainfall, the presence of Western Ghats and the Arabian Sea are one of the main causes of heavy rainfalls and floods in Kerala.

Western Ghats is the rain pocket of the country during monsoon season. The heavy rainfall in Kerala is due to the orographic rainfall in Western Ghats which leads to the way to an extreme event called flooding. The small and medium storage size of the reservoirs resulted in an increase in the amount of water in the reservoirs at the time of heavy rainfall. The surplus amount of rainfall of about 40–50% occurred during the first week of May to August in Kerala. This resulted in an increase in the water level at the earliest and the continuous rainfall that occurred during August 14 to 16 led to the release of a substantial amount of water in a short period of time. The combination of heavy rainfall and reservoir release at the same time might be one of the foremost reasons for the severity of 2018 Kerala flood. The heavy rainfall and limited capacity of the reservoirs in Kerala played a triggering role in worsening the flood situation in the state (Mishra & Shah, 2018; Mishra et al., 2018).

The discharge of toxic waste and other toxic gases, including greenhouse gas, into the atmosphere is another set of reasons for climate change that lead the heavy rainfall and flooding in Kerala (Agarwal, 2018). Other than climate change, man-made factors such as changes in land use and land cover, changes in the reservoirs, antecedent hydrological conditions, the encroachment of floodplain, and other factors are the human-induced reason for the Kerala flood as these factors have locally played an important role on the conversion of massive rainfall to flooding condition (Mishra & Shah, 2018; Mishra et al., 2018). According to the study conducted by Axel Bronstert (2003) the origin and overflow of a flood is an extremely nonlinear system. The topography, meteorology, groundwater, drainage channel, soil characteristics, vegetation, and climate, are the temporal or natural variability that is susceptible when a flood occurs. The factors such as meteorology and hydrological conditions, hydraulic characteristics of rivers, and land use conditions are to be considered and assessed for identifying the flood risk. Among these, the meteorological factors are significant regarding climate change. Also, the soil condition in the catchment area and vegetation can be affected by climate change. These were the determining reasons for the water retention and evaporation process, which had an impact on the development of floods in Kerala.

13.2 Method and Study Design

The methodology adopted for the study includes a review of related literature, sampling procedure, data collection, and data analysis. The results presented here are the consolidated outcome of data collection. Ernakulam is the district that was purposely chosen for the study due to the unique system of Pokkali farming. The district comprises of 4050 ha of Pokkali field out of 5700 ha of land all over in state (Jayan and Sathyanathan, 2010). Out of 14 blocks in Ernakulam, Palluruthy and North Paravur are the two blocks selected for the study. Kumbalangi (KL) Panchayath in Palluruthy block and Ezhikkara (EZ) Panchayat in North Pravur block are affected by the Kerala flood and are the selected area near Kochi city for the survey. From each Panchayath,

25 Pokkali farmers were randomly selected for the study and a total of 50 respondents were taken for the survey.

The primary data was collected through interviews with selected respondents through a questionnaire developed for the study. A series of questions consisting of three parts were prepared for the interviews. The first part of the questionnaire consisted of primary information about the respondents' including names and details about the family. The second part of the questionnaire comprised general details about Pokkali farming for the past 20–30 years. And the final part of the questionnaire incorporated information regarding cultivation after a flood. The household survey was conducted by selecting the respondents randomly, and each selected respondent was personally communicated with and interviewed with the help of the previously prepared questionnaire. The behaviors, emotions, feelings, and background of the respondents were correspondingly observed during the time of interview for future information. Data including field visits and relevant photographs of the study areas were also taken along with the survey for identifying the current condition of the field. Another set of the questioner was also made for the secondary data collection. Here the questionnaire also comprised general information of crop and the details after the flood. And the data was collected from administrations such as Panchayath, Village office and Krishi Bhawan. The relevant variables of Pokkali cultivation are taken from the reviews of related literature including the research papers, articles, documentations etc. The information taken from the literature are included in the study and the survey for more information.

The dependent variables and qualitative data such as the documentation of the status of Pokkali rice farming before and after the flood and the farmers' perception of attributes of Pokkali rice are subjected to percentage analysis. Calculation of the data included identification of a total area of land, identified the benefit and loss of Pokkali cultivation and changes after the flood in the field, the reason for the loss after the flood, etc. The details and prevalence were calculated using percentages and represented through suitable illustrations.

13.3 Result and Discussion

As per the field survey data regarding the Pokkali cultivation, the following observations have been made.

Damaged Farmland

From the total area of 435 acres in KL, only 33% (144 acres) of the field is surveyed, and in that, 24% of the farmland is affected by the Kerala flood. From the survey, it could be noted that the percentage of the affected areas is comparatively greater than the non-affected areas. Similarly in EZ, 31% of the total 32% (118.7) of surveyed land, is affected by the flood. Therefore, it is evident that the 2018 Kerala flood critically affected most of the agriculture fields of this region. But the inundated flood water ranges in the fields are different in both study areas. This is because, in the Kerala floods of 2018, EZ was severely affected compared to KL.

Normal Crop Production

The production of the Pokkali crop is unsteady in the study areas. Each landowner has a different area of land, in which each one has been getting a different rate of production i.e. majority has got low yield from their farming (Table 13.1). It is understood that production does not depend on the size of the land because owners holding the large and small fields show a major loss in production. So the production of Pokkali may change by the influence of some factors. The probable reason for the same is due to the loss of interest in farming. The other set of issues is about the pest attack and drainage problems rather than other issues in the farming. The major pest attack in the field is the Western swamp hen, it destroys the maturing panicle of the paddy by sucking the juice of ripened rice seeds and also they build their nest by using the growing paddy. Since the attack of this pest decreases the yield of Pokkali very drastically. Drainage problem is associated with the Pokkali field in both regions. The irresponsible and untimely measures taken by the organization every year for the draining out of the water from the farmland make the production decrease. Along with these issues, problems associated with laborers is another important reason for the losses in yield including high labor charge, unavailability of laborers, and irresponsible workers, etc. The reason might be the loss of interest among farmers due to a small area of cultivable land and less amount of production. So overall these are the issues that are impacting the production of Pokkali in both areas.

Profit and Loss of Farming

From the Table 13.1, it is evident that, the net profit of farming in the study areas have both profit and loss and that are depending upon the production and income. So the net incomes of cultivation are not the same for all the farmers. The expenditure and income are based on the method of cultivation, and the investment in farming is based on personal interest. It is seen that the investment in farming is high among the respondents with more interest in farming. Due to the profitless farming, most of them are not interested in cultivation and some of them have stopped this Pokkali cultivation practice due to the losses. The major reason for the vast amount of loss is the higher expenses in farming.

Table 13.1 Dataset used for the study, represents the set of data on Pokkali cultivation including production, income, expenditure, and the overall net profit from Pokkali farming in Kumbalangi (KL) and Ezhikara (EZ)

KL

Land size (acre)	Production (kg)	Income	Expenditure	Net profit
2.75	1272	8909	6290	2619
0.35	7	0	0	0
6	500	5500	12,000	-6500
2.75	909	5090	14,545	-9455
2.25	600	8000	21,777	-13,777
3.5	571	7000	5714	1286
0.56	26	26,785	7142	19,643
2.56	1679	11,757	7200	4557
1.5	1666	14,000	5666	8334
2.35	1489	10,425	7659	2766
1.5	2000	14,000	6466	7534
2.37	1687	11,814	5400	6414
1.36	14	29,411	34,558	-5147
30	466	3266	1320	1946
5	1000	4400	12,000	-7600
2	750	4000	7500	-3500
22	454	318	909	-588
5	1000	4000	5000	-1000
0.5	8	0	0	0
4.5	333	0	1000	-1000
2.25	120	15,000	20,000	-5000

EZ

Land size (acre)	Production (kg)	Income	Expenditure	Net profit
5	900	8000	10,000	-2000
3.5	571	5714	10,857	-5143
7	785	5714	7142	-1428
4.75	736	6315	8510	-2195
2.5	600	9600	12,000	-2400
0.36	8	8333	0	8333
6.75	889	4444	7384	-2940
5	900	6000	8000	-2000
4	750	4500	7500	-3000
6	1000	10,000	5000	5000
6	750	6666	8000	-1334
4	500	5000	10,000	-5000
3	750	8333	13,333	-5000
4	875	6250	7500	-1250
1.25	800	12,000	16,000	-4000
0.21	8	6857	5000	1857
7	502	11,428	10,000	1428
5.38	464	3234	9000	-5766
2.5	600	3000	5200	-2200
2.5	750	14,000	20,000	-6000
3.5	560	11,428	18,571	-7143

(continued)

Table 13.1 (continued)

KL

Land size (acre)	Production (kg)	Income	Expenditure	Net profit
2.5	280	19,600	11,200	8400
2.75	220	20,740	13,818	6922
36.28	385	2701	1267	1434
1.56	288	20,192	18,589	1603

EZ

Land size (acre)	Production (kg)	Income	Expenditure	Net profit
6.5	800	15,384	15,384	0
18	222	4444	12,000	−7556
7	1000	14,285	14,285	0
3	350	20,000	6666	13,334

The representation of a negative sign in the net profit shows a loss in production

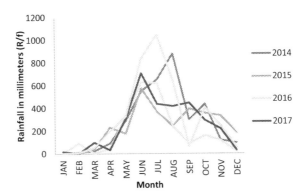

Fig. 13.1 Showing the annual rainfall data of Ernakulam (2014–2018) (*Source Indian Meteorological Department*)

Weather Condition

The amount of annual rainfall in the monsoon season (June to September) has increased in each consequent year (Fig. 13.1). It is also found that there is a change in the monsoon season, in the case of 2018. The monsoon is highly affected in the month of July and it shows a decrease in monsoon rain at the start of the monsoon. But in previous years, the enormous monsoon rain started in the month of May itself (Fig. 13.1). Since it is clear that a drastic change in the monsoon rainfall system in Kerala was the reason for the huge devastating flood of 2018 in the region.

Change in Weather and Productivity

The above-mentioned graph illustrates the total production of paddy crops and average rainfall in the Ernakulam district (Fig. 13.2). Here, the above graph shows a drastic reduction of annual rainfall in 2016 in the district along with a great amount of crop production in the same year. From the five-year data, 2016 only shows the least amount of rainfall and a higher amount of crop production. When considering crop production, it shows a sudden variation in 2016 and 2018. In 2018 there is a drastic increase in the occurrence of precipitation in the district having a range of more than 300 mm, and consequently, there is also a massive decrease in the amount of production of the crop in the region (Fig. 13.2). 2018 is the year of the flooded condition in Kerala which has caused a severe loss in all sectors. Other than these two years, the rainfall and crop production of the region does not show any considerable changes. It could be concluded that the production of the crop constantly depends on the average rainfall in the region. Therefore, the reason for the decrease in the drastic reduction in the agriculture sector after the flood is due to the heavy rainfall.

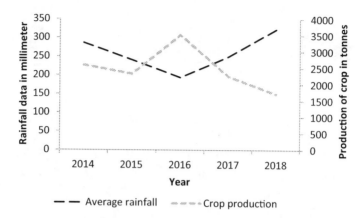

Fig. 13.2 Showing crop production and average rainfall in Ernakulam district (*Source* Department of Economics and Statistics, Government of Kerala and Indian Meteriological Department)

Survival of Crop from Flood

By comparing the pie charts, Pokkali survived more in EZ than KL, i.e., 5% of Pokkali survived in EZ after the flood in the field when compared to KL (Fig. 13.3). The major reason for the difference in survival quantity of the crop is due to the difference in the inundation rate in the farmland. Because in EZ the field of Pokkali is inundated only for 2–3 days but in KL the inundation was for about one week. The level of inundation of floodwater is almost the same in both the study regions. So the change in the inundation period is the primary reason for the difference in the survival rate of crop. And also in both instance, the yield production from the survived Pokkali is very low and the yield was only 1% of the total in KL (Fig. 13.3).

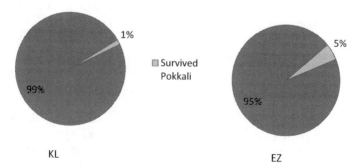

Fig. 13.3 Showing total survived Pokkali after a flood from the survey in Kumbalangi

Reason for Crop Damage

The causes of damage are different in the two study areas. In the case of KL, the reason for the crop loss is primarily due to the long period of water inundation and the level of inundated water in the field (Fig. 13.5). In the area, inundation was about one week and the water level was over 1 meter. So, logically, the cause of the damage in this area is the severe flooding of the farm associated with enormous rainfall. But in EZ the polluted water was the reason responded by majorities than other issues, so the reason in the area was due to the sewage waste from the industries combined with flooding caused damage to crops in the province (Fig. 13.5). From this understanding, it is clear that Pokkali can withstand flood under several conditions. If necessary actions are taken at the right time, it could be better for increasing the survival rate and also thus provides comparably better yield and outcomes.

Changes in Farmland After the Flood

From Fig. 13.4, it is clear that the changes after the flood in the field are not the same. Because one of the foremost problems in the region is associated with pollution, other than this problem most of the other issues are almost the same in both regions (Fig. 13.5). But compared to KL, EZ has seen more changes in the farmland after the flood. The percentages of crop disease and salinity problems in the regions are very less after the flood. The major issues are high sedimentation, change in water levels, pollution, and the pest attacks (Fig. 13.5). It can therefore be concluded that the change in salt water and crop diseases in the study area does not occur seriously after the floods but the pollution and effluent from industries after the floods are a serious threat in the farms in EZ (Fig. 13.5). This is due to the large presence of nearby chemical industries on the farmland. This is because Ernakulam has a large number of industries as compared to other districts in the state. Factories have a significant impact on the agricultural sector in the region as the study areas are close to the city limits of Kochi.

13.4 CONCLUSION

Pokkali is aggro-ecological farming and it is an incredible crop traditionally known for its resilient capacity toward salinity and flood. Although in present scenarios, in comparison to other technologically advanced cultivation practices, Pokkali cultivation is not profitable, it is seen that it can get the desired result if properly maintained. In 2018, the overall farmlands of the state were submerged by the devastating rainfall and flood. The study got an undesirable outcome where most of the Pokkali in the region did not survive the deluge. But, the reason for this result was due to the presence of extreme conditions which were never seen before. The Prolonged period and level of inundation

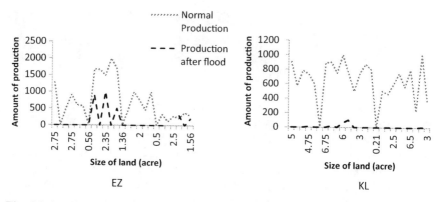

Fig. 13.4 Comparison of the overall rice production before and after the Kerala Flood 2018

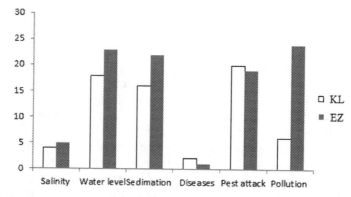

Fig. 13.5 Showing the comparison of farmland changes according to the number of respondents after flood in the two study regions

in addition to chemically polluted floodwater. Had it not been for this unusual condition, the crop would certainly have provided a large amount of survival rate and food security in the event of a normal flood disaster. As Kerala is a coastal district and most of the paddy varieties are non-resilient to salinity and flood, and also being a tall crop, as the name suggests, it will last longer than any other crop normally would. Therefore, it would be more beneficial to cultivate Pokkali in the coastal areas as now it is only cultivated in a few districts. Thus, Pokkali is a nature-based solution toward flood resilience and can be implemented in other coastal villages and farmlands of India and other countries that have similar agro-climatic regions. Due to all these attributes, with the exception of loss in the 2018 flood, it can be said that Pokkali is an ideal organic farming practice for the coastal areas.

REFERENCES

State Disaster Management Authority, Kerala. (2018a). *Kerala Floods—2018: 1st August to 30th August 2018*. State Relief Commissioner, Disaster Management. Government of Kerala.

State Disaster Management Authority, Kerala. (2018b). *Kerala Post Disaster Needs Assesment: Flood and Landslides August 2018*. Government of Kerala.

Adams, R. M., Hurd, B. H., Lenhart, S., and Leary, N. (1998). Effects of global climate change on agriculture: An interpretative review. *Climate Research, 11*: 19–30.

Agarwal, R. (2018). Lesson learned from killer floods in Kerala: Time for retrospection (B. A. Iqbal, Ed.). *Management and Economics Research Journal, 4*(S2).

Anson, C., and Pavithran, K. (2014, January). Pokkali rice production under geographical indication protection: The attitude of farmers. *Journal of Intellectual Property Rights, 19*.

Bronstert, A. (2003, May). Floods and climate change: Interactions and impacts. *Risk Analysis: An International Journal, 23*(3).

Chandramohanan, K. T., and Mohanan, K. V. (2011). *Rice cultivation in the saline wetlands of Kerala—An overview*. University of Calicut, Genetics and Plant Breeding Division, Department of Botany. Gregor Mendel Foundation Proceedings.

Deepa, K. M. (2018). *Seasonal Variation in Avifauna with Respect to Habitat Changes in the Pokkali Fields of Ernakulam District, Kerala*.

Doswell, C. A., III. (2003). Flooding. University of Oklahoma, Norman, OK.

Jayan, P. R., and Sathyanathan, N. (2010, December). Overview of farming practices in the water-logged areas of Kerala, India. *International Journal of Agricultural and Biological Engineering, 3*.

Kurukulasuriya, P., and Rosenthal, S. (2003). *Climate change and agriculture: A review impact of adaptation*. Agriculture and Rural Development Department.

Mishra, V., and Shah, H. L. (2018, November). Hydroclimatological perspective of the Kerala Flood of 2018. *Journal of Geological Society of India*.

Mishra, V., Aadhar, S., Shah, H., Kumar, R., Pattanaik, D. R., and Tiwar, A. D. (2018). The Kerala flood of 2018: Combined impact of extreme rainfall and reservoir storage. *Hydrology and Earth System Science Discussions*.

Ranga, M. R. (2006). *Transformation of Coastal Wetland Agriculture and Livelihoods in Kerala, India*. Natural Resources Institute, University of Manitoba.

Ranjith, P., Karunakaran, K., and Sekhar, C. (2018). Economic and environmental aspects of Pokkali rice prawn production system in central Kerala. *International Journal of Fisheries and Aquatic Studies*.

Remmiya Rajan, P., and Anjana, P. (2019, November). The darken phase of flood on economy: A study on the influence on various sectors in Kerala. *A Journal of Composition Theory, 2*(11).

Shamna, R., and Vasantha, R. (2017, October). A study on farmers perception on prospects and problems of Pokkali rice farming in the state of Kerala. *Indian Research Journal for Extention Education*.

Sudhan, C., Mogalekar, H. S., Ranjithkumar, K., and Sureshbhai, P. D. (2016). Paddy cum prawn farming (Pokkali fields) of Kerala. *International Journal for Innovative Research in Multidisciplinary Field, 2*(7). ISSN–2455-0620.

Vineesh, O. K. (2019). Impact assesment of Kerala Flood 2018 & 2019. *A Journal of Composition Theory, 12*(11).

Vishnu, C., Sajinkumar, K., Oommen, T., Coffman, R., Thrivikramji, K., Rani, V., et al. (2019). Satellite-based assessment of the August 2018 flood in parts of Kerala, India. *Geomatics, Natural Hazards and Risk, 10*.

World Meteorological Organization. (2021). Weather-related disasters increase over past 50 years, causing more damage but fewer deaths. https://public.wmo.int/en/media/press-release/weather-related-disasters-increase-overpast-50-years-causing-more-damage-fewer

CHAPTER 14

The Role of Gharbari Model to Support Climate-Smart Agriculture

Kumudini Mishra

14.1 Introduction

Kalahandi district is one of the most drought-prone areas in India, occupying the South Western portion of Odisha. The drought in Kalahandi in 1993 was a severe environmental and humanitarian crisis that affected the Kalahandi district in the state of Odisha, India. The region experienced a prolonged period of drought, resulting in widespread crop failure, food shortages, and a devastating impact on the local population (Economic and Political weekly, May 29, 1993). As per the 2011 census, 92.26% of the population of Kalahandi district live in rural areas. The district is primarily an Agriculture-based economy. About 86% of the total working population is engaged in Agriculture and allied activities, comprising mainly marginal farming households, which are highly food insecure (Kalahandi District Statistical Handbook, 2018). In Kalahandi, like in many rural areas, gender plays a significant role in agriculture. Women are actively involved in various agricultural activities, from sowing seeds and transplanting crops to weeding, harvesting, and post-harvest activities. They contribute significantly to the agricultural workforce and are essential for the success of agricultural production in the region. However, despite their significant contributions, women in agriculture often face various challenges and inequalities. They may have limited access to resources and

K. Mishra (✉)
Planning and Convergence Department, Government of Odisha, Micro Planning and Livelihood Expert, Bhubaneswar, India
e-mail: mishra.kumudini@gmail.com

© The Author(s), under exclusive license to Springer Nature Singapore Pte Ltd. 2023
S. Nautiyal et al. (eds.), *Palgrave Handbook of Socio-ecological Resilience in the Face of Climate Change*,
https://doi.org/10.1007/978-981-99-2206-2_14

inputs, including land, credit, and agricultural extension services. Gender-based norms and discrimination can also restrict women's decision-making power, access to markets, and control over income generated from agricultural activities (Mary Bage & Mishra, 2019; Mishra, 2021). The climate of the district is of an extreme type. It is dry except during monsoon. The people cultivate rain-fed staples (mainly paddy), and any vegetables that are rarely found, are only grown in the rainy season. For the rest of the year, the economically vulnerable households depend on the local market for their food and vegetable requirements.

Since the area is rain-dependent, drought poses a severe threat at regular intervals when there is an event of monsoon failure. In such a scenario, the farmers need to intelligently adapt to the changing climate in order to sustain crop yields and farm income. Moreover, enhancing the resilience of agriculture to climate risk is of paramount importance for protecting the livelihood of small and marginal farmers. In such a scenario, the current study aims to bring out the effective implementation of the CRAFT-K Project (Climate Resilience Adaptive farming in Tribal Communities in Kalahandi) through its innovative model known as **"Gharbari"** in Karlamunda block of Kalahandi district.

Understanding CRAFT-K Project

The CRAFT-K project has been operating in Karlamunda Block of Kalahandi District by Indo-Global Social Service Society (IGSSS), an International NGO, and has helped the tribal people, in particular, the women, to sustain their livelihood also resulted in supporting climate-smart agriculture. The project aims at engaging in resilience building of small and marginal farmers against drought and other climate variabilities by adopting drought-resilient climate-smart practices. It also aims to promote and enable the community to undertake lead mapping, planning and implementing measures for drought and climate-induced vulnerabilities. It focuses on promoting demonstration, replication and upscaling innovative models for combating drought and climate change vulnerabilities in Karlamunda Block of Kalahandi District.

FAO's work on gender and climate-smart agriculture focuses on addressing gender disparities, promoting women's empowerment, and ensuring that women are equal partners in efforts to build climate-resilient and sustainable agricultural systems (FAO, 2002, 2010, 2013b, 2015). The women of twenty villages in Karlamunda Block were mostly unaware of the debates and discussions on climate-smart agriculture and its impact. They also needed to be made aware on the impact of climate change in the area. The IPCC recognizes that women, particularly in rural areas, often play critical roles in agriculture, natural resource management, and food security. They also emphasize that gender inequalities can exacerbate the vulnerabilities of women to climate change impacts and hinder their ability to adapt and mitigate effectively (IPCC, 2001, 2007). In such a scenario, it was crucial to involve the women farmers and sensitize them on the CRAFT-K Project and generate awareness among them on the amount of annual household vegetable requirement for the number of family members, annual expenditure on vegetable purchasing, cultivation of vegetables in the Gharbari as per interest and choice and systematic management of Gharbari. In this regard, an assessment was conducted in 2018–2019 to study the knowledge, skill and attitude of the tribal communities in the study area. Further, it was revealed that the women were cultivating vegetables only in the rainy season on 6.3 acres of Gharbari land of 200 families. Only season-based vegetable cultivation of three to six types was seen during this.

Thus, there was a need to train the women in particular and training sessions were organized for the women, youth and lead farmers on the benefits of food, and season-specific multiple vegetable cultivation in a small piece of Gharbari land around the year. Prior to this intervention, the common view was that it was impossible to cultivate multiple crops in the traditional Gharbari with no source of irrigation and infertile land.

Methodology and Coverage

A participatory approach and methodology were followed during all stages of the study, including Focused Group Discussions (FGDs), Case studies, village meetings, and discussions with key informants. In addition, Participatory Learning Exercises were conducted during field visits. Discussions and interactions were made with the farm families, farm women, local Village agricultural workers, members of the Women Self-Help Groups and the line department officials to collect information about the villages, the farm families, various projects, programmes, and crop and non-crop enterprises. Besides respondents, representatives of Panchayati Raj Institution, Field level officers of Agriculture Department, district statistical officer and experienced senior staff of Block were also consulted during the primary data collection process.

Understanding Gharbari Model

Gharbari (Ghar means House and Bari means a small piece of land situated in the backside of the home or in-front of the home). If managed systematically, this small piece of land can provide a household with clean, green, fresh and nutritious vegetables around the year with very little expenditure. The traditional type of Gharbari is now being cultivated systematically. Along with catering to the household nutritious food requirement, women participate in reducing dependency on the market, monetary saving, seed conservation, protecting agro-biodiversity, preparing household crop plans and making farmers self-reliant. Gharbari cultivation has not only helped improve the nutritional status of rural households but also generated a small and constant source of income for women.

Gharbari is a small low-cost initiative to produce clean, green, safe and nutritious vegetables round the year. The Gharbari cultivation contributes to the nine goals adopted by the United Nations-Sustainable Development Goals 2030.

Objectives of the Gharbari Model

- To initiate climate adaptive and resilient cropping initiatives
- To introduce and facilitate Climate-smart agricultural practices
- To produce safe and fresh vegetables through organic methods without using any chemical pesticides.
- To produce multiple varieties of seasonal nutritious vegetables in a little piece of land.
- To produce vegetables in summer and winter seasons using water coming out from domestic works like kitchen and washing utensils.
- To conserve seeds from Gharbari (home yard) cultivation and reduce the dependency on the market for purchasing seeds.
- To conserve farm-friendly insects cultivating in the Gharbari (home yard) through the organic method.
- To involve women and youths in household crop selection, seed conservation and farming.

14.2 GHARBARI MODEL—UNDERSTANDING THE STAGES

1. **Need assessment through farmers-led planning process**—In the Gharbari Model, there are different stages and each stage is important for the success of the model. In the first stage, a need assessment is conducted involving the farm households. Importance is given to mobilizing the members of the community for efficient and sustainable use of land and water resources. This is mainly done involving the women members of the households. Formulation of Village Development Council—Village development committee is crucial as the female farmers mainly the women Self-Help Group members are part of the Village Development Council. This encourages better participation and results in better outcome. The female members are trained on the vegetable production assessment through the Village Development Committee and it is finally signed by the group members. Weekly meetings, discussions and debates are encouraged with the active involvement of youths on the nutrition value of seasonal vegetables, calendar of climate change and seed conservation and protection. Training and knowledge sharing meetings on season-wise vegetable production is done. Preparation of harvesting and cultivation of vegetable calendar as per household requirement and interest of the family members is done.

2. **Preparation of Gharbari land**—The second stage becomes crucial as it involves land preparation, seed collection and seed exchange. Preparation of different saplings like peas, gram, papaya, drumsticks, tomato, etc. along with sapling distribution and preparation of Gharbari as per the geographical location of the land is done in the second stage.
3. **Learning and exchange of experiences**—Exchange and sharing of experiences by the females in the field. The training consists of activities on Ecological agriculture skills (growing vegetables organically, soil health management, climate stress tolerant seeds collection and exchange in Village Development Committee (VDC) & Self-Help Group (SHG) meetings. This is a crucial stage where efforts are made for mobilizing joint support from local Agriculture and Horticulture Department officials for sharing of technical knowledge. The participants learn as well as share their knowledge on organic way of pest control. Thus, is through all these initiatives, the success of the Gharbari Model is ensured.
4. **Use of technology and knowledge**—This is an important step. Distribution of climate stress-tolerant seeds, saplings, low-cost vermin compost tanks, organic fertilizer and bio-pesticide preparation inputs, low-cost drip irrigation kits are the important aspects of this stage in the Gharbari Model. Planning for harvesting of vegetables even in the summer season and winter season is chalked out with a focus on diversity of crops and ensuring at least fifty varieties of vegetable plants are grown in the project area.
5. **Monitoring of production, consumption and sharing of experiences**—This is a very crucial stage and it involves active participation and coordination of all stakeholders. Meetings are organized at the village level. There is sharing of experiences on Climate Smart Agricultural practices with a focus on the production of vegetables with a little piece of land and little amount of water. Knowledge exchange workshops in the villages on the production, and utilization of local resources along with visits to the farms are done in the final stage. Participation in various fairs for the exchange of knowledge and showcasing of best practices forms the focus of the last stage.

Thus, the above five major steps of Gharbari Model have their own relevance, and each and every step is important for the sustainability of the Project. These innovative steps focus on Land and water Management, season-wise nutritional sapling preparation, maintenance of genetics of the village community members, diversity of seeds, sharing of traditional knowledge and lastly producing safe and nutritious vegetables. The various steps discussed are reflected in Fig. 14.1.

Fig. 14.1 Stages in Gharbari model (*Source* Field visit)

14.3 Impact of the Gharbari Model

The Gharbari Model contributes towards achieving the Nine Goals of the United Nations SDGs-2030. Table 14.1 explains the same in detail.

Gharbari Model has not only helped the community to become self-sustaining but also they are able to scale up the practice of soil management, and storage of use of draught tolerant seeds, they follow the season calendar and have adopted an integrated approach. Thus, scaling up the practice of Climate Smart Agriculture in their villages.

In the initial stage, there were only a few households but slowly each and every household from the village joined, currently more than 200 women farmers are producing around eight to twenty-four varieties of vegetables organically in 6.3 acres of land i.e., the Gharbari land. Every household has joined hands and this has scaled up resulting in providing around 500 grams to 1.5 kgs of clean, green, fresh and safe vegetables. The tribal household is able to consume vegetables that are rich in carbohydrates, protein, minerals and multivitamins and have conducted seed festivals last year and this year during April.

During the Kharif season the tribal communities produced 132 tons of vegetables and during the Rabi season, they produced 88 tons of vegetables from their Gharbari or kitchen gardens. Moreover, the income source has improved. Earlier they were only in paddy cultivation but slowly they started vegetable cultivation. Out of 200 women farmers, 80 farmers are earning Rs. 300 to Rs. 500 per week by selling vegetables in own villages and nearby weekly markets. Besides consumption by self the family members, women are

Table 14.1 Impact of the Gharbari model

1	End poverty in all its forms everywhere	Gharbari Model provides multiple vegetables in lean period. Provides instant cash through sale of vegetables. Kitchen garden provides additional income of Rs. 8000 to Rs. 10,000 annually
2	End hunger, achieve food security and improved nutrition and promote sustainable agriculture	Round the year production of seasonal nutritious safe, clean, green and fresh vegetables. Saving of money and reducing market dependency. Increasing farmer's capacity on development of climate-resilient cropping patterns, conservation of climate stress tolerant seeds
3	Ensure healthy lives and promote well-being for all at all ages	Gharbari model provides nutrition food at hand as per the need of every member of the household. Pregnant women, lactating mothers, children and old age persons are easily availing safe and nutrition food from their Gharbari
4	Ensure inclusive and equitable quality education and promote lifelong learning opportunities for all	Members of every family engaged in Gharbari model are involved in production of nutrition-rich vegetable and fruits. Experienced persons share their knowledge on production of seasonal vegetables and health benefits of fresh vegetable to youth and children
5	Achieve gender equality and empower all women and girls	Women are gradually getting involved in agricultural decision-making activities like preparation of crop plans, selection of crops, seed conservation, organic farming and cultivation of food crops instead of more cash crops
8	Promote sustained, inclusive and sustainable economic growth, full and productive employment and decent work for all	Gharbari Model reduces market dependency and provides ample of opportunities for the development of micro-enterprises based on local resources like organic vegetable markets, mushroom cultivation, marketing of organic compost and other inputs
11	Make cities and human settlements inclusive, safe, resilient and sustainable	Gharbari model produces safe food, conserves indigenous climate stress-tolerant seeds, effective use of water and help to conservation of farmer's friendly insects hence farmers are not using any kinds of chemical fertilizers and pesticides. Gharbari model is free from polluted elements of environments. It supports to build safe and resilient environment

(continued)

Table 14.1 (continued)

12	Ensure sustainable consumption and production pattern	Gharbari model produces poison-free nutritious vegetables and fruits. It fulfils household nutrition food requirements
13	Take urgent action to combat climate change and its impact	Regular knowledge exchange and judicious use of land, water, seeds and fertilizers strengthen adaptive capacity of household

sharing vegetables with guests and neighbours in lean time. Every household has been able to save Rs. 8000 to Rs. 10,000 towards purchasing vegetables annually. This is 7% of the annual earning of one household. Every family is applying organic fertilizers and pesticides instead of chemical fertilizers and pesticides used in vegetable cultivation. Due to this, every family has been able to save Rs. 2000 and getting poison-free vegetables to eat along with checking environmental pollution and biodiversity loss.

In addition to the production of vegetables in the Gharbari various leafy vegetables, ladies' finger, ridge gourd, bitter gourd, pumpkin, bottle gourd, etc. vegetable seeds have been possible to be collected. The women and youths are jointly preparing a vegetable production calendar as per the food choice of the family. During the lockdown period, all the kitchen garden holders consumed 50% of vegetables as per their family needs and reserved (for seed & food during the scarcity period) ripe vegetables like pumpkin and other nuts, seeds and legumes related vegetables. All the kitchen garden holders have sold very few amounts of vegetables because they have shared vegetables with their neighbours and relatives during COVID-19-led lockdown scarcity periods.

Thus, the overall observation is that the case of Gharbari model of agriculture supports climate-smart agricultural practices. The farm women are producing climate-smart crops in the unused land with waste water and growing vegetables organically. They are comparatively paid more, decide more, supplement more income and finally empowered more than their male counterpart. Secondly, their role and contribution are better than the women of the irrigated tract in the block as well as the district (Fig. 14.2).

The above pictures reflect that small initiatives can bring great changes like the sowing of seed balls in common land and forest. It is a low-cost initiative where community members identify Rare Endangered Threatened (RET) species, and their habitats and take steps of regeneration through natural processes. The community members of the twenty villages of Karlamunda block have collected thirteen varieties of wild species and prepared 12,000 seed balls which were showcased during World Environment Day this year. The demonstration of different climate resilient models by the community leaders of Karlamunda block was very much appreciated. The Gharbari model has also helped the beneficiaries to understand the importance of seed conservation. They are helping in building up a climate-smart village and help reduce hunger and poverty in the face of climate change.

Fig. 14.2 Demonstration of climate-resilient farming gharbari models (*Source* Field visit)

14.4 Gharbari—Supporting a Society to Be a Resilient One Along with the Scalability of This Practice

Learning from the Field

An Adivasi female farmer named Mamata Majhi of Kansil Village in Karlamunda Block demonstrated how climate degradation and climate change affect food and nutritional security. She also explained how the Gharbari model has helped to understand the effects of frequent changes in local climatic conditions and its effects on pulses production and uncultivated foods. She says, "Uncertain rain, drought, flash flood, fog, soil erosion, insect, pest and diseases are inevitable since 2010".

Jayanti Bhoi, an Adivasi woman of Kansil village of Gajabahal Gram Panchayat was a keen participant in the Gharbari Model project training and actively participated in the discussion on climate resilient nutritious food crops, production of round-the-year nutritious food giving safe, clean, green and varieties of nutritious food. As an impact of various critical reflections on agriculture-related issues and knowledge sharing, she began a field demonstration unit in her own home garden. Her husband Jagannath Bhoi supported her in this initiative.

Jayanti's homestead land with her own house is about 5000 sq ft. Three years ago, she had planted some fruit plants in the home yard. She also grew brinjal, tomatoes and leafy vegetables only in raining season. Her husband Jagannath grew only paddy on their two acres of farmland. Faced with the rising costs of paddy cultivation, he discussed cotton cultivation with local agents.

Both Jayanti and Jagannath continued to participate in knowledge exchange meetings held for critical reflection on agriculture-related issues, Jayanti learned to estimate her family's vegetable requirements and the production to meet them. She then planned to cultivate vegetables in her own home yard around the year She shared the estimation of the assessment of family vegetable requirements and money expenditure on every weekly market day.

Jayanti family's daily intake was 800–1500-gram vegetables which costs Rs. 50 to Rs. 60 as per market price or a weekly expenditure of Rs. 350–400. Since vegetables were scarce in her village, she had to buy them from the market. Further, she did not have storage facilities so these vegetables wouldn't last for seven days so she had to buy small quantities only. With the learning from project training, Jayanti prepared a crop cultivation plan of mixed cropping to produce more crops with little land and water, and climate-resilient farming practices (Fig. 14.3).

Jayanti and Jagannath after assessment of their food requirement-initiated vegetable cultivating for round-the-year production. In the Kharif cultivation season of 2020, Jayanti received support for seed and a vermin bed. With this, she started the campaign of producing nutritious vegetables. Jayanti

Size of Kitchen Garden (SqFt)	5000 SqFt
Types of Species	21
➢ Fruit Vegetables	11
➢ Beans & Legumes	2
➢ Leafy vegetables	2
➢ Leafy with fruit vegetables	2
➢ Roots & Tubers	2
➢ Herbs & Spices	2
Total Vegetable Production in Kgs	1064
➢ Total Vegetable Production Kharif Season	414
➢ Total Vegetable Production Rabi Season	341
➢ Total Vegetable Production Summer Season	314
Total Money Savings (Rs)	30428

Fig. 14.3 Lessons from the field, the Case study of Jayanti Bhoi (*Source* Field visit)

always gives importance to women for taking decisions in the process of crop selection and poison-free cultivation. She had involved women members of the Self-Help Group (SHG) for vegetable sapling production, preparation of bio-pesticides and round-the-year mixed cropping.

Now Jayanti is cultivating 21 types of vegetables and 7 types of flowers jointly with her husband Jagannath. They have prepared 8 types of bio-pesticide, 3 types of hormones and 3 types of composts and having applied in own field, they have become able to share their experience on their use and benefit among farmers. In the past year, Jayanti stopped buying vegetables from the market. Jagannath began earning Rs. 300 to Rs 400 from the sale of vegetables produced in their Gharbari through the organic method. Jayanti now has 8 types of indigenous seeds. Her husband Jagannath has dropped the idea of cotton cultivation and now has cultivated green gram, black gram, and beans after harvesting of paddy in addition to Arhar cultivation in the farm bund. By dint of the endeavour of 18 months, Jayanti has succeeded to make her Gharbari a farmer's field school. 15 farmers of the nearby villages have learned the experiences and knowledge of Jayanti and Jagannath (Fig. 14.4).

In 18 months, Jayanti has transformed her Gharbari into a farmer's field school. 15 farmers of the nearby villages have learnt the experiences and knowledge of Jayanti and Jagannath. Jayanti says that every farmer should produce and conserve seeds, organic compost and natural bio-pesticides in Gharbari to become self-reliant. Unlike Jayanti and Jagannath, a greater

224   K. MISHRA

Fig. 14.4 Innovative practices in Gharbari model

number of households have started Gharbari Model in their own spaces. Currently, they have formed Producer Groups and scaled up the practice in their villages. The practice of Climate stress-tolerant seeds (focusing on seed exchange not procurement) is one of the best take away from Gharbari Model. It has also resulted in innovative practices of land and water management, preparation of season-wise sapling preparation, resulting inorganic farming and producing clean, green and fresh vegetables around the year. The figure given below represents the innovative practices adopted in the different villages.

14.5 Suggestions

However, for the replication and wider dissemination of the best practices through the Gharbari model, there are some suggestions based on field visits and discussions with the farm communities in the study area. These have been mentioned in the points given below:

1. Establishment of Rural Haats or weekly markets in the Gram Panchayat. This would help the farmers get the appropriate price for their organic vegetables if an organic vegetable weekly market could be set up by.
2. Initiatives such as the collection of indigenous seeds of all vegetables and training farmers on seed production techniques should be encouraged in the tribal villages. Sharing of Best Practices/Success Stories at the Panchayat level should be encouraged.
3. Gharbari Model can be introduced as model kitchen garden in Anganwadi Centres and Schools by the School Management Committees and Gaon Kalyan Samities, and this could be appreciated at Government level.
4. Capacity Building Programmes on Climate Smart Agriculture involving Women Self-Help Groups, particularly tribal women should be encouraged so that they are not only able to revive traditional agriculture practices but also adopt Climate-smart practices at grassroots level.
5. Awareness Programmes on Climate Smart Agriculture Practices in Local Language by District Administration involving the Government, NGO and CSR should be there at the grassroots level.

14.6 Conclusion

Tribal communities in Kalahandi District of Odisha have revived the traditional practice of growing food without the help of chemical fertilizers and made it viable economically by making pragmatic changes. The Gharbari model is the best example of Climate Smart Agriculture Practices.

In the study area, more than 200 women have been benefiting from the Gharbari model. They are now recognized as change makers and part of the campaign for eating and producing poison-free safe vegetables. In the drought-prone area like Kalahandi where the farmers cultivate vegetables only in the Kharif season due to lack of water and eat vegetables bought from the market for eight to ten months, this small initiative of Gharbarimodel of cultivation has brought faith in many farmers, in particular the women farmers. There was a mind-set prevalent that cultivation cannot be in water scarcity, cultivation cannot be without chemical fertilizer and pesticides and the women and youth did not know farming technique. The impact of Gharbari farming has altered this mind-set.

Home-grown vegetables are organic, low cost and could be totally free from chemicals and pesticides. These gardens have an established tradition and great potential for improving household food security and alleviating micronutrient deficiencies. If cultivation is done in a small land in a planned way, this can help to save eight to ten thousand every year and sustain their livelihood most importantly; it gives direct access to diverse nutritionally rich vegetables. It also increases the purchasing power through savings on food bills. This small initiative of the tribal women who have adopted such an easy and simple method has incorporated various tasty vegetables into the daily food plate of every member of the family. Adoption of this model at the household level can be promoted for replication in similar ecological and social conditions.

Strengthening resilience through climate-smart village and climate adapting agriculture practices in every aspect of life is most important. We should focus to build community capacity on Weather Smart, Soil Smart, Water Smart, Carbon Smart, Energy Smart, Knowledge Smart and Market Smart strategy. All the flagship programmes, schemes, projects must be driven to focus on above said themes. All government departments, private and corporate should think and act to enable vulnerable people to protect existing livelihood systems, diversify their sources of income, and change their livelihood strategies on a priority basis.

Acknowledgements I sincerely thank Mr. Amar Kumar, Program Officer, Indo Global Social Service Society, Kalahandi for the information provided on the Gharbari Project operational in Karlamunda Block and his team for the support extended during field visits conducted.

Declaration There is no conflict of interest.

References

Drought in Kalahandi. (1993, May 29). The real story, Reproduction of an article published in the Economic and Political Weekly, XXVIII(22), 14–16.

FAO. (2002). From farmer field school to community IPM—Ten years of IPM training in Asia. Bangkok, Thailand. Food and Agriculture Organisation of the United Nations (FAO), Rome, Italy.

FAO. (2010) "Climate-Smart" Agriculture: Policies, Practices and Financing for Food Security, Adaptation and Mitigation, Food and Agriculture Organisation of the United Nations, Rome.

FAO. (2013a). Sourcebook on Climate Smart Agriculture, Forestry and Fisheries, Food and Agriculture Organisation of the United Nations (FAO), Rome, Italy. Retrieved from http://www.fao.org/climatechange/37491-0c425f2caa2f5e6f3b9162d39c8507fa3.pdf.

FAO. (2013b). Climate Smart Agriculture: Sourcebook. Food and Agriculture Organisation of the United Nations, Rome, Italy.

FAO. (2015). Building resilient agricultural systems through farmer field schools – Integrated Production and Pest Management Programme (IPPM). Food and Agriculture Organisation of the United Nations, Rome, Italy.

IPCC. (2001) Climate change 2001: The scientific basis. Contribution of Working Group I to the Third Assessment Report of the Intergovernmental Panel on Climate Change (Houghton, J. T., Ding, Y., Griggs, D. J., Noguer, M., Van der Linden, P. J., Dai, X., Maskell, K. and Johnson, C. A. (Eds.) Cambridge University Press, Cambridge and New York.

IPCC. (2007). Climate Impact s, Adaptation and Vulnerability.Working Group II to the Intergovernmental Panel on Climate Change Fourth Assessment Report, DRAFT technical summary 2006, Intergovernmental Panel on Climate Change, Geneva.

Kalahandi District Statistical Handbook. (2018).

Mishra, K. (2021). Gender Bias in Agricultural Wage Employment A Case study of Rural Kalahandi, Mahila Pratishtha, 6(3), 138–154.

Mary Bage, and Mishra, K. (2019). A study on the Impact of Mahila Kisan Sashaktikaran Pariyojana (MKSP) in empowering the Tribal Women of M.Rampur Block of Kalahandi District, Mahila Pratishtha, 5(1), 150–166.

SDG Webpage. https://sustainabledevelopment.un.org/. Accessed 20 September 2019.

CHAPTER 15

Understanding the Mental Models that Promote Water Sharing for Agriculture Through Group Micro-Irrigation Models in Maharashtra, India

Upasana Koli, Arun Bhagat, and Marcella D'Souza

15.1 Introduction

Water security is considered as a tolerable level of water-related risk to society (Grey et al., 2012). David Grey and Claudia W. Sadoff (2007) defined water security as 'the availability of an acceptable quantity and quality of water for health, livelihoods, ecosystems and production (agriculture, industry, energy and transport), coupled with an acceptable level of water-related risks to people, environments and economies'. The concept of tolerable and acceptable levels refer to the threshold (Falkenmark et al., 1989; Alcamo et al., 2000; Sullivan, 2002; Young et al., 2019) below which the water security elements of availability, accessibility, safety and affordability (Agarwal et al., 2000; Young et al., 2021) are compromised leading to crisis in the domains of health, livelihood, ecosystems and production. Several factors affect water security such as an increasing population, urbanization, economic development, lifestyle changes, water pollution, etc. (Falconer & Norton, 2012), however in the domain of agricultural production, that our study focuses on, the most important factors affecting water security are climate change and excessive abstraction of freshwater (surface and groundwater). Climate change further aggravates the excessive abstraction of surface and groundwater.

U. Koli (✉) · A. Bhagat · M. D'Souza
WOTR-Centre for Resilience Studies (W-CReS), Watershed Organization Trust, Pune, India
e-mail: upasana.koli@wotr.org.in

© The Author(s), under exclusive license to Springer Nature Singapore Pte Ltd. 2023
S. Nautiyal et al. (eds.), *Palgrave Handbook of Socio-ecological Resilience in the Face of Climate Change*,
https://doi.org/10.1007/978-981-99-2206-2_15

The Intergovernmental Panel on Climate Change (IPCC) report (Bates et al., 2008; Field et al., 2012) pointed out that climate change would have a greater impact on the freshwater resources resulting from higher climatic and hydrological variability with consequences reaching out to the society and water security (Habiba et al., 2014). In the Indian context, studies have found that the average temperature has increased by about 0.7 °C during the period 1901–2018. It has mainly been caused by anthropogenic aerosols and changes in land use land change. Rainfall data trends comparison between 1951–1980 and 1981–2011 shows a shift towards more frequent dry spells (27% higher). During the last 6 to 7 decades, the frequency of drought conditions has significantly been increasing. The data climatology and weather data for 1951–2016 showed at least 2 droughts per decade occurred in the regions of the southwest coast, southern peninsula, central India, and north-eastern India. Climate model projections indicate an increased frequency (>2 drought in each decade) in the coming future caused by increased monsoon variability and a warmer atmosphere (Krishnan et al., 2020; Maharana et al., 2021). This reducing trend of the only source, i.e.; rainwater, to support most of the surface and groundwater affect the recharge rate of groundwater in India (Bhanja et al., 2019; Kumar et al., 2021). Simultaneously, the dependence of humans on freshwater for domestic, irrigation and industrial use is increasing exponentially. The increasing demand and the low recharge rate of freshwater put pressure on this natural resource. It severely affects groundwater quantity and quality (Kulkarni & Shankar, 2014; Dangar et al., 2021; Chindarkar & Grafton, 2018). The 2011 Ministry of Water Resources report of the Government of India says, freshwater reserves have reduced from 5177 m^3 in 1951 to 1820 m^3 in 2001, reducing to more than half i.e. 1545 m^3 in 2011 (Kumar et al., 2021). Groundwater supplies for 85% of rural domestic needs and 62% of the agricultural production needs. In 2013, where 70% of irrigation was dependent on groundwater, it drastically increased to 90% in 2018 (Saha et al., 2018; Joshi et al., 2021).

Agricultural production and farmers livelihood are sensitive to climate variability. The temporal and spatial variability of climate change holds the potential to impede food production and supply (Tirkey et al., 2018; Bewketa & Conwayb, 2006). Crop development has a systematic regime, compatible with a certain type of climate, adequate water supply and inputs, to follow (Challinor & Wheeler, 2008). Even the slightest variation in these elements creates a yield gap, causing reduction in production output and its quality (Agarwal, 2007; Rötter & Geijn, 1999; Mora et al., 2015). The quantity and quality of crop yields have a bearing on the market compensation farmer receive as an income (Kawasaki & Uchida, 2006). Since climate change cannot be regulated at the local level, the natural response of farmers to sustain production, to maintain if not increase the level of income they have been receiving, is to over-indulge in the extraction of groundwater. The extraction of groundwater is happening at a much faster rate than it can naturally replenish, majorly from monsoon rainfall (Famiglietti, 2014). As groundwater levels drop, wells are dug to deeper levels which have implications on the wealth of the farming community. This over-indulgent behaviour has far-reaching consequences at the farmer's level and the environmental and ecosystem levels. At the farmers' level, subsidies for and accessibility of technology and electricity have led to an explosion in groundwater extraction (Srinivasana & Kulkarni, 2014; Janakarajan & Moench, 2006). These benefits are reaped by wealthy farmers, however, farmers living on subsistence face the problem of inequity of access to the distribution of water, which widens the economic and social divide (Cuadrado-Quesada & Joy, 2021; Vaidyanathan, 1996). According to the 2015–2016 Agricultural Census of the Government of India, about 86.08% of the farmers are small and marginal farmers operating an area of 46.94% and 13.35% are medium farmers operating an area of 43.99% (GoI, 2019), of these, many are economically weaker and are unable to make huge individual investment. As a result, these are the farmers subject to problems of inequity and are compelled to take up agricultural practices that are unsustainable in nature. At the environmental and ecosystem level, mismanagement in groundwater extraction causing soil subsidence, reduction in vegetation area, disturbing the biota essential for crop development, etc. (Carrillo-Rivera et al., 2008; Danielopol et al., 2003) are constraining agricultural out-turn. These issues have been prevailing for long and are seen to be widening with time as farmers behaviour becomes more aggressive towards resources.

Irrigation as a concerted effort was adopted as an adaptation strategy to address water security. Various irrigation-related schemes, technology, practices and management approaches and adaptation strategies have been implemented to manage water use, besides watershed development that improves water security. For instance, In India, the Pradhan Mantri Krishi Sinchai Yojana (PMKSY) project by the Department of Agriculture and Farmers Welfare, Govt. of India, promotes end-to-end solutions in irrigation supply, micro-irrigation technologies such as drips and sprinklers, etc. Precision irrigation management systems, a sustainable water security adaptation strategy, ensure precise supply of water to crops at precise locations at precise time however uniformly distributed across the irrigated area (Smith & Baillie, 2009; Zhang et al., 2021). However, in the backdrop of the magnitude of farmers with economic and resource constraints, the uptake of these technology and management practices individually and for a sustained period becomes difficult. To support such farmers, the Watershed Organisation Trust, an NGO based in Pune, India, established an irrigation management system to be implemented through a group, called the Group Micro-irrigation Model (GMI). This system entails sharing water as a common pool resource along with the application of climate-resilient agricultural practices. Our study focuses on the behavioural aspects related to the adoption of the system, sharing of water resources and cooperative management by the group. Through the mental model method, we draw a mental structure of the group of farmers who use this approach. It covers the interaction between the external resources and their perspectives, beliefs and attitudes to understand their point of view for adopting a sustainable practice of sharing water, from their experiences of joining and operating in the group. This study aims to highlight the importance of considering the drivers of people's behaviour as an independent component, which has been given limited attention in the Indian context, and in framing sustainability and adaptation frameworks and studies. Besides, it also has relevance for the practitioners and policymakers, as factoring these aspects will enhance their ability to formulate effective water-sharing policies, and/or other sustainable and adaptation interventions.

15.2 Study Area

The state of Maharashtra, characterized by varied geography, topography and climatic conditions has led to regional differences in environmental conditions, thus allowing to group into different agro-climatic zones (FAO, 1996; Gajbhiye & Mandal, 2000). This diversity in the environmental conditions has divided the state into 4 meteorological regions and 9 agro-climatic zones. They can be differentiated based on soil conditions, precipitation, weather, and physiography and crop suitability. The Very High Rainfall with Lateritic Soils, Very High Rainfall with Non-Lateritic Soils and the Ghat agro-climatic zone forms the Konkan meteorological region; the Transition Zone I, Transition Zone II and part of the Scarcity Zone form the Madhya Maharashtra region; the rest of the Scarcity Zone and the Assured Rainfall Zone form the Marathwada region and the Moderate Rainfall Zone and High Rainfall Zone with Soils from Mixed Parent Material form the Vidharbha region. The Madhya Maharashtra and the Marathwada region being located in the interior of the state and with the Western Ghat range of the elevation of about 1200 m above mean sea level obstructing monsoon flow, has rendered the regions to be semi-arid (Ratna, 2012; Kelkar et al., 2020).

The Group Micro-Irrigation models were chosen to be established in the semi-arid region of Marathwada and Madhya Maharashtra which needs water-related adaptation strategies the most as compared to other regions. In the Marathwada region, the GMI model was installed in the village of Tigalkheda in the Bhokardhan block of Jalna district. And in Madhya Maharashtra, two models were installed in the villages of Bhangadewadi and Ranmala hamlet belonging to the Dhawalpuri Panchayat in the Parner block of Ahmednagar district. These two villages are to a distance of approximately 5 km from each other. Figure 15.1 illustrate the location of these three models:

Jalna: GMI Group I—Tigalkheda

The district of Jalna is located in the central part of Maharashtra state and is a part of the scarcity zone in the Madhya Maharashtra region (Ratna, 2012; Kelkar et al., 2020). Its district boundaries stretch for about 7687.39 sq.km.

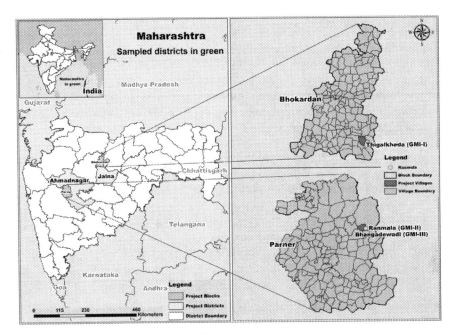

Fig. 15.1 Location map of the study area

Jalna district is 400–500 km away from the coastline of the state towards the east and about 160–170 km east of the Ahmednagar district where our other two GMI group models are installed. It belongs to the Marathwada region. The physiography of the region are of 04 types: Ajantha and Satmala hills (part of the Western Ghats), undulating (wave-like structure) plateau, denudation (reduction in elevation caused by either weathering, erosion, deposition or transportation) slope and older flood plain. Most of the central and southern part of the district comprise of undulating and denudation slopes. Jalna is divided into 4 revenue sub-divisions, of which Bhokardhan block is one. The GMI group I established falls in the Bhokardhan sub-division. Bhokardhan lies on the northernmost side of the Jalna district. The block is mostly of denudation and older floodplain physiography. The normal annual rainfall in this region varies between 400 and 600 mm on average with, sometimes, even less than 37 rainfall days as compared to the overall 122 rainfall days of monsoons of the state. The unconfined aquifer, that is the primary level of groundwater, is of weathered/fractured basalt form at a depth of 5 m to 30 m, while the second aquifer of jointed/fractured basalt form goes 35 m to 145 m deep (CGWB, 2016).

GMI Group I—Tigalkheda Model Details

Tigalkheda village lies on the southeastern side of the Bhokardhan block. The distance of the village between the Jalna district headquarters located in the Jalna main town is approx. 34 km and located to its north and approx. 29 km between Bhokardhan town located to south-east. The nearest water body to the village is the Khadakpurna dam which is approx. 40 km to its east. It has no evidence of groundwater influence on the village. The model in the village area is located at 20.118005 latitudes and 75.866839 longitude. It was establishment commenced in the year 2017 between April and May. After an aquifer delineation was conducted an area of 32.45 acres of area was selected. This area had 03 wells tapping the same aquifer. The model restricts the extraction of water from multiple resources (dug-wells) for agricultural use and facilitates sharing water from one well only. An automation and fertigation system of water distribution was installed next to one of these dug-wells with drips attached to it and spread across the 32.45 acres of adjacent farmlands. Before the installation, the selected dug-well was de-silted to increase the depth by 35.38 ft. in order to increase the storage capacity to 50 ft. The 32.34 acres of the area belong to 14 farmers who belong to the same familial lineage. Of these, 3 farmers belong to the small (2.6 to 5 acres landholding), 6 farmers to medium (5.1 to 10 acres landholding), 1 to marginal (>2.5 acres landholding) and 4 to large (<10 acres landholding) land class categories. The area allocated for the GMI model range between 0.45 to 5 acres. This group was engaged with rain-fed farming only.

Ahmednagar: GMI Group II—Ranmala and GMI Group III—Bhangadwewadi

The district of Ahmednagar is the largest by area coverage in Maharashtra state, covering 17,196 sq.km of the total state area of 307,713 sq. km. It lies about 200–250 km to the east of the western coastline of India, in the state. The district's physiography has four major landforms i.e.; Ghats and hills (7.6%), foothills (19.4% area), plateau (3.71% area) and plains (69.30% area). The Ghats and the majority of the hills fall on the western side of the district (part of the western mountain range [Western Ghats]), shifting the landform to foothills, plateau and plains towards the eastern side. Due to the orographic effect caused by the Western Ghats, the majority of the district area that fall on the leeward side/rain-shadow side, receives scant rainfall of about on an average of 574 mm annually spread over 47–59 rainfall days as compared to 3200 mm average rainfall in 95–110 days in its neighbouring district of Raigad that lies on the west coast and to the windward side of the Western Ghats. Ahmednagar district is one of the drought-prone districts in Madhya Maharashtra (Ratna, 2012; Kelkar et al., 2020). The district consists of 7 revenue divisions, with each division having 2 blocks under its administration. The two GMI models in Ahmednagar were established in the Parner block belonging to the Shrigonda revenue division. This block is located on the southwestern side of the district. It is part of the plateau region of the district, with hillocks at certain places. The unconfined aquifer, the primary source of groundwater replenished by rainfall water, in this region, is to a depth of 20–40 mbgl. Therefore, the scant rainfall conditions with a low groundwater holding capacity of the region compel farmers to carry out rainfed farming or sustain on irrigation.

GMI Group II: Ranmala Model Details

Ranmala village is located 28 km west of the Ahmednagar district headquarters. It lies on the plain region of the Parner block. The nearest water body from the village is the Bhalwani Lake to its east, with no evidence of groundwater influence in the model area. The Ranmala GMI model is located at 19.15054 latitude and 74.534989 longitudes. It was inaugurated in May 2020. This group consists of 06 farmers, of which 04 farmers belong to the medium land class category (5.1 to 10 acres landholding) and 02 farmers belong to the small land class category (2.6 to 5 acres landholding). In this group, only two farmers i.e.; 01 small and 01 medium farmer land class category, own a water resource. Each of the farmers allocated 1 acre each, adjacent to each other, i.e.; 6 acres total, for the model. Water for this model is distributed from one dug-well owned by one of the group farmers. The dug-well earlier which was in a dilapidated condition, was reconstructed. The automation and fertigation system of water distribution was connected after due approval was obtained from the well owner. Drips pipelines attached to the system then were distributed to each 1 acre of land of the model.

GMI Group III: Bhangadewadi Model Details

Bhangadewadi village is situated approximately 29 km to the northeast of Parner town and 30 km west of Ahmednagar district headquarters. The nearest water body from the village is the Kalu dam, Dhoki, to its west at about 9–10 km distance. Water to this GMI group farms is supplied from this dam. A check dam (weir) is constructed downstream of the Kalu dam for the purpose of obstructing excess water flow. This surface water is lifted through a pipeline and transferred for about 9 km to a farm pond situated midst the group farmland. An automation and fertigation system of water distribution is connected to this farm pond from which water is distributed to all farms. The Bhangadewadi GMI model is located at 19.150983 latitudes and 74.512865 longitudes. It started its operation from April–May 2020. There are a total of 47 farmers who have allocated 65.5 acres of their adjacent lands for the model. Of these 47 farmers, 19 are medium (5.1 to 10 acres landholding), 24 are small (2.6 to 5 acres landholding), 2 are marginal (>2.5 acres landholding) and 2 large (<10 acres landholding) land category farmers. They allocated land in the range of

Table 15.1 GMI Tigalkheda, Ranmala and Bhangadewadi group details

GMI model	District	Location	Year of establishment	Type of farmers	No. of farmers	Total land owned (acres)	Area under GMI (Acre)	Water source
GMI-I	Jalna	Tigalkheda, Bhokardan Block	April–May, 2017	Marginal (0.1–2.5 acres)	1	2	1.25	1 Shared Dug-well
				Small (2.6–5 acres)	3	12	3.7	
				Medium (5.1–10 acres)	6	49.5	17.25	
				Large (>10 acres)	4	52	10.25	
				Total	14	115.5	32.45	
GMI-II	Ahmednagar	Ranmala, Parner Block	May, 2020	Marginal (0.1–2.5 acres)	–	–	–	1 Shared Dug-well
				Small (2.6–5 acres)	2	9	2	
				Medium (5.1–10 acres)	4	36	4	
				Large (>10 acres)	–	–	–	
				Total	6	45	6	

GMI model	District	Location	Year of establishment	Type of farmers	No. of farmers	Total land owned (acres)	Area under GMI (Acre)	Water source
GMI-III		Bhangadewadi, Parner Block	April–May, 2020	Marginal (0.1–2.5 acres)	2	4	3	Surface water (lifted from Kalu dam excess water stored in check dam downstream) to Farm pond
				Small (2.6–5 acres)	24	91	27.5	
				Medium (5.1–10 acres)	19	152	31.5	
				Large (>10 acres)	2	28	3.5	
				Total	47	275	65.5	

3 to 0.5 acres for the model. In this group, 22 farmers have water resources while the rest have been doing rainfed farming (Table 15.1).

15.3 Group Micro-Irrigation Intervention

The initial stage of the group formation was about promoting the GMI concept among the farmers through group meetings. Key people of the villages were approached to discuss the potential of the model, who further disseminated the information to fellow farmers. After back-and-forth discussions, we finally arrived at the aforementioned groups to begin with the establishment of the models. Water sharing from common resources and application of the Climate Resilience Agriculture (CRA) practices were the crux of the model to which the group agreed to comply. The two approaches are explained below.

GMI Operations

The Group Micro-Irrigation (GMI) approach considers water as a common good rather than privately owned. This approach intends to promote managing scarce water resources in a judicious and equitable way. The GMI approach has four integral components: (1) Groundwater management by supporting both the supply and demand side needs, (2) Application of Climate-Resilient Agriculture (CRA) practices, and (3) Supporting farmer's market linkages, and (4) Providing tools and techniques to support agricultural operations through applied research.

The primary component comprises taking water-related measures such as harvesting rainwater and construction of soil and water conservation structures to recharge groundwater. These measures support the supply-side needs. While, collectivization of private groundwater resources, sharing water from the same aquifer and equitably distributing of water through a common micro-irrigation system to the farm area of the group, supports the demand side needs. This entails the development and maintenance of the collectivized common groundwater resources, pumping house and water distribution pipe network that is spread across the field of the farmers who are part of the group. An automation system is installed to eliminate the manual work of supply water and send precisely equitable distribution of water to each farm. Groundwater collectivization is a sustainable water security solution for the conservation, efficient use and equitable distribution of groundwater by considering the resource as a common resource. Key principles of this component are common sharing, social regulations, technical support and gaining a scientific understanding of the operations. The second component suggests adopting CRA practices to boost soil health and plant resilience to ensure a good harvest in the face of weather and environmental challenges. The third component involves market linkages support for better market prices by forming new and engaging with existing FPOs. And the last component comprises integrating applied research to generate tools and techniques to support farmers in evaluating their agricultural performance. This will help them make informed agricultural decisions for next seasons. Assessments are also conducted to provide research-based evidence on the impact of various measures undertaken. This method considers creating simple tools and methods like maintaining field books by farmers, crop water budgeting and planning and assessing groundwater availability by testing well water depth and pump discharge.

GMI approach provides a robust solution with the adoption of both micro-irrigation and climate-resilient farming practices, which are otherwise stymied due to financial and institutional constraints. Besides, it also has social benefits as it creates an attitude of cooperation rather than competition and strengthens interpersonal relationships through constant and effective coordination.

Climate-Resilient Agriculture Practices

Climate change and variabilities like erratic rainfall, frequent interchanging dry–wet spells, intensive short-duration rainfall, and other extreme weather events affect agricultural production negatively at the local and global levels. Additionally, the excess and injudicious use of chemical inputs, excessive water application, and faulty agricultural practices, as an outcome of these externalities, have triggered the severe loss of soil health. These multidimensional impacts of climate change have made farmers realize the importance of following sustainable ways to build resilience to manage agriculture (Chaubey et al., 2018; Patra et al., 2016; Nayak & Solanki, 2021). In this regard, to combat the situation Climate-Resilient Agriculture practices were devised to sustainably improve agricultural production. CRA is an approach that uses natural resources existing in the surrounding, rather than synthesized, to achieve continuing higher productivity in the context of climate variability (Lorenz & Lal, 2018).

The Watershed Organisation Trust (WOTR) consults a package of CRA practices for a variety of crops including indigenous and climate stress-tolerant crop varieties. The package of practices involves applying techniques of water conservation like in-situ moisture conservation, use of micro-irrigation, mulching, and water harvesting structures for protective irrigation. Apart from water application, this package also suggests methods of conservation agriculture like minimum tillage, contour cropping, intercropping, mixed cropping, and crop rotation. To manage the sudden climate changes information use of weather-based location and crop-specific advisories are provided. WOTR has developed an android mobile software to disseminate this information.

CRA also provides Integrated Nutrient Management (INM) and Integrated Pest Management (IPM) solutions. Integrated Nutrient Management (INM) helps to efficiently and in a balanced way, use organic and synthetic fertilizers based on the existing soil health status. This deals with the application of organic and inorganic fertilizers, in addition to farmyard manure, vermin compost, legumes in rotation and crop residue for sustaining soil health for the long term. And, Integrated Pest Management (IPM) is an approach for pest and disease management that includes bio-pest management practices such as the use of insect traps, trap crops and bio-pesticides. The application of these approaches is flexible and applied as per the needs of the agricultural conditions. The IPM approach use a wide range of information related to the life cycles of pests and how they interact with the environment. This information, combined with the pest control methods, is applied by the best economical means to manage pest damage. It is ensured there is low damage to people, property and the environment.

15.4 Research Approach and Methodology

Mental Model Approach

Mental Models can be defined as a cognitive representation of the real world system. The real world system image in the mind is drawn from the individual's selected concepts and relationships which are the basis for the formation of perceptions and experiences that he or she further uses to make decisions (Doyle & Ford, 1998; Johnson-Laird, 1983). This approach is useful in a qualitative study (Desthieux et al., 2010) enquiring the cognitive representation of the environmental system of a person. It gives us an understanding of how an individual structure the environmental issues in his/her mind, how it changes over time and how these changes might influence the behaviour and actions in the future (Lynam & Brown, 2012). Mental models are also one of the important tools useful in exploring the different understanding of concepts one holds about a particular issue, integrating perspectives from various stakeholders and assisting in the decision-making processes of resources related to complex systems, while also learning about the social system (Pahl-Wostl & Hare, 2004). In the context of water resource management, this has been used to understand the perceptions and dynamics of water-related issues and the impact of climate change on it (Kolkman et al., 2007; Levy et al., 2018). M. J. Kolkman's frames and mental model's framework is mentioned below to demonstrate how a mental model is structured. The mental process entails uncovering and segregating systematically hidden information and identifying feedback and delay that enable or inhibit the sustainable function of natural resources (Jones et al., 2011). It basically helps in understanding the cause and effects as an interaction between the individual's mind and the external world (Doyle & Ford, 1998). The process of representing this information in a mental model further involves converting the data into a visual representation in the form of diagrams or models. The concepts and relations are connected and highlighted to show the flow of perceptions of the individual or the group. In our study, we have used the indirect elicitation technique of constructing mental models, that is to draw a conceptual and relational model extracted from interviews or verbal texts (Jones et al., 2011; Wood et al., 2012). In this, we have created models that represent the group's experience, beliefs, values

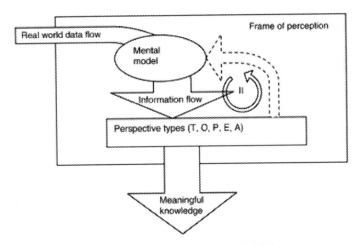

Fig. 15.2 Theoretical framework by Kolkman et al. (2007)

and understanding of sustainable groundwater resources from the information obtained through interviews received in a narrative form (Fig. 15.2).

The agriculture system is a composite of environmental resources that are managed by humans to produce food and food-related resources. The action an individual takes is the outcome or the behaviour one arrives at based on the quantum of resources one has, the influence of the external factors on these resources and the psychological attitude one possesses that is shaped by their experiences and perceptions about the system. Here, we have classified the behaviour influencing factors into three groups and drawn a mental model elaborating on the consequences of these factors:

1. **Behavioural aspects**—Behavioural aspects in a human are responsible for one's behaviour after considering the endogenous (socio-economics) and exogenous (outside forces) factors. These aspects are perceptions, beliefs, moral norms, habits, attitudes, feelings & emotions, thoughts, knowledge, etc. based on the cognition of a being, from the senses and thought processes one develops from external experiences.
2. **Endogenous** (Socio-economics/Dependent variables)—The factors are the tangible resources farmers physically deal with while having them in their possession. Farm characteristics such as farm size, soil quality, mechanization, number of plots, labour, distance between market and farm, membership with the institutions, etc. The factors are receptors of the individuals' behaviour from the decisions made.
3. **Exogenous** (outside forces/independent variables)—These factors are those that are not in the farmer's control. These are outside forces that have an effect, in the form of support or pressure on the resources the farmers deal with. The factors involve market functioning (price), climatic

factors, market, institutional support and assistance, national and local policy and planning, accessibility of information and inputs, etc. These have a direct or indirect impact on the socio-economic factors. The intensity of these factors on the dependent variables has a bearing on the decision-making process of an individual.

Behaviour is an outcome of the psychological processes of perception, attention, memory, language, motivation, emotions, etc. when encountered with external events which require a response and reaction to (Heidbreder, 1945; Resick et al., 2010). In the water management context, understanding the psychological functioning is imperative as it determines the coping, adaptive and/or sustainable behavior of an individual towards surface and groundwater resources (Blackstock et al., 2010; Waldman et al., 2020).

Data Collection

For the study, a stratified sampling technique was applied to select members from the 03 GMI group for the interviews. Since each group is a composition of farmers belonging to different class categories, namely; marginal, small, medium, and large land owing class categories, the list was bifurcated accordingly and selections were made proportional to the composition in each category. The purpose to follow this technique was to cover the perspectives of farmers coming from different land class backgrounds. There were about 26 respondents selected for the interviews. Table 15.2 shows the details of this bifurcation. They were approached to with the help of WOTR's local staff or Jal Shevak/Wasundhara Sevak.

A semi-structured open-ended questionnaire was prepared pertaining to questions related to water, agriculture, climate, market, GMI group formation and functioning, relation with group members and related points, etc. The questions were framed in a way to capture the extensive narrative each member had about the aforementioned topics. The duration of interviews ranged between 30 minutes to 1 hr 15 minutes. The interviews were conducted in Marathi, the local language of the model villages. They were recorded in an audio recorder. The member were asked open-ended questions including, but not limited to, the following questions:

1. What was the agricultural, water, and market sales condition before joining the GMI group? What challenges did you face then?
2. What were the encouraging/supporting factors that led you to join the GMI group?
3. What changes do you see in the agricultural, water, and market sales condition after joining the group?
4. How has your relationship been with the group members before and after joining the group?

5. Are there any suggestions you would like to give for better functioning of the model and group?

Data Analysis

Interview Translation

As mentioned above, the 26 interviews were recorded in Marathi language. They were translated and transcribed in English for the purpose of initiating the analysis. It is pertinent to be familiar with the local context, intonnation and terminology to capture the essence of the interviews and translate the same in English. Hence, the audio was divided between two researchers who were familiar with all these aspects. Once the interviews were translated, there was one round of data scrutiny done by the other two researchers who conducted the interviews. The scrutiny involved listening to the audio interviews and reviewing the translation. It was done to ensure data accuracy and capture those that were missing. After this process, the translated files were assembled to proceed with the next step of coding the interviews.

Interview Coding

The output of the translated interviews was in a narrative form transcribed on word files. The next step required separating the narrative into its respective themes. Themes are narratives having a common reference point and different ideas around it have an association to (Vaismoradi et al., 2016; Graneheim & Lundman, 2004). The Dedoose qualitative analysis software was used for this process of separating and recording the themes. All 26 interviews word

Table 15.2 Interviewed group details

Group micro-irrigation group details		Jalna	Ahmednagar	
		bhokardhan block	Parner block	
		Tigalkheda—GMI group I	Ranmala—GMI group II	Bhagadewadi–GMI group III
GMI Area (acre)		32	6	65
Total. No. of Farmers		14	6	47
Farmers Composition	Marginal	1	–	2
	Small	3	2	24
	Medium	6	4	19
	Large	4	–	2
Interviews conducted	Marginal	1	–	1
	Small	2	2	6
	Medium	2	2	7
	Large	3	–	1

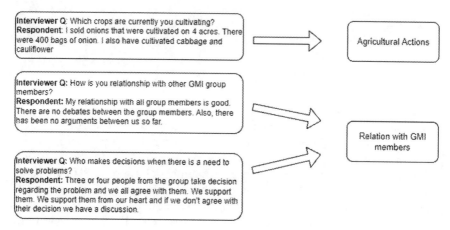

Fig. 15.3 Thematic segregation of narrative data

files were uploaded on the software. After this, there were codes created for each theme (thematic code) such as water actions, water challenges, climatic conditions, reasons for joining the GMI group, etc. The codes were created considering the objective of the analysis i.e.; to identify the factors responsible for the farmers' behaviour (output as an outcome of inputs (factors)) for adopting sustainable agricultural practices or technology to adapting from climate-related challenges and vice-versa. Each word file was scanned, narratives were highlighted and added to their respective thematic code (Fig. 15.3). These steps are illustrated below:

Analysis

Once the narratives were added to their respective thematic codes, the next step was to start with the analysis of the segregated data. The data was downloaded in the form of an Excel sheet. The codes were clustered and added under the psychological, endogenous, and exogenous categories as where they belonged to. Within each category, the codes were added under the titles 'enablers' and 'disablers'. Enabler in the psychological aspects means positive perception, in the endogenous aspects means positive behavioural outcomes and in the exogenous aspects means external factors supporting the endogenous aspects. And, the disablers are the existing challenges highlighted in the perceptions, endogenous and exogenous aspects that are dissuading factors to GMI adoption. These are the factors, according to the farmers, that suppress the optimal capability of the GMI model of water sharing. Following this step of separation, the mental models for each GMI models were created. The positive variables are coded in *black* and the disablers are highlighted in *red*. The results of these models are elaborated below.

15.5 Results

This results section explains the mental models of the farmers of the three diverse features in terms of GMI model structure and locations they belong to.

Mental Model Flow of GMI Group Farmers in Tigalkheda

In the Tigalkheda GMI model, the farmers mentioned that the perception of better, clear and timely, guaranteed and systematic agricultural, water supply and transportation operations were the encouraging factors to join the GMI group. They believe group unity formed from their father's and grandfather's times and faith in a key person have made it easier, mentally, to join the group. They have been witnessing faster growth in agriculture, an increase in knowledge due to the operation as a group. Earlier, they were pessimistic about the success of the model, however, discussions, enquiries and clarification, and support from family members changed their view about the model. The resultant (actions/behaviour) of these positive aspects was adopting the micro-irrigation model. They talked about cultivating food and vegetable crops for commercial sales. They started using upgraded technology and transitioning gradually from complete chemical to organic fertilizers. They shared about the use of water from their ancestral well, refurbished for the model, working efficiently through drip irrigation. The systematic functioning of the GMI has contributed to an increase in income. From the external sources they received assistance and guidance from key village persons, WOTR and the Agricultural departments' officers for water efficient use and agriculture. Due to better quality of farm produce they received good market rates. Merchants have started visiting their farms to make purchases of production, and they also sell the produce to NAFED-FPO who give them better prices. Since merchants have been purchasing directly from their farms, they have been saving most of their transportation costs. There are some who continue transporting production for sales individually or in small groups.

Besides these positive aspects, they reported negative perceptions and challenges they are currently facing. They reported that during the formation of the group some farmers were discouraged by other farmers to join. Moreover, as people have different perspectives and goals, it takes time to unite the farmers for participating in the intervention. They mentioned climate change including increasing instances of dust storms, erratic and irregular rainfall and fog occurrences to be the major deterrent factor for the challenges they face in agriculture. They face crop damage due to erratic and irregular rainfall, therefore low production. Water in the GMI well is available only till February or March, therefore cultivation is restricted to these months and hence they are unable to reach the optimal functioning of the GMI. As water availability is still a concern, they cannot take high-yielding crops. The cost of cultivation and transportation has increased due to climate change. Inconsistent market rates are the cause they think doing single-farming or taking different crops is more beneficial. And continuous electricity supply is an issue also contributing to low production (Fig. 15.4).

Mental Model Flow of GMI Group Farmers in Ranmala

In the Ranmala GMI group, their primary motive to opt for the GMI model was to make water available and have technology for efficient use of water. The participating farmers intended to receive guaranteed, systematic, organized planning and distribution of water supply, production and transportation of sale. They believed group farming would reduce transportation costs that they pay individually. They believed these aspects for them was believed to bring prosperity to farming in the future. These farmers too were pessimistic about the success of the model, however with being battered by the water crisis, after discussions they accepted the idea of forming a group. Sustained belief in key persons who help operate the group also has helped them maintain the change in their perspective. Talking about the inter-relational aspect, they mentioned that as members belong to the same village or relatives, therefore there is better coordination and no conflicts. They think that increasing memberships in the existing group or forming a new group require farmers to have a good level of compatibility with other member farmers, like-mindedness and trust in each other for handling water distribution. This positive change in perception and uptake of the model has encouraged them to cultivate commercial crops such as food and vegetable crops (cereals, cauliflower, cabbage, etc.). Earlier, the land was uneven, but later land leveling was done to start the cultivation. Land levelling work was a part of the GMI model essentials. They mentioned about improvement in crop production after upgrading to new technology, transitioning to organic fertilizers and efficient use of water. Better quality crop production has given them better market rates. Also, they started selling the farm produce in groups which have reduced their transport costs.

On the other side, they shared disappointment about the GMI operations which led one member to withdraw from the model. Here too, changing weather conditions is a major factor in the disturbance of the agricultural operations in this group. Water through GMI is available till February or March. Weather-varying conditions are causing low productivity, increasing instances of pests and insect attack, and crop damage due to low water supply, therefore causing an increase in chemical fertilizer usage. As their financial status is comparatively lower, they are unable to make capital investments. And even if they wish to make investments, low rainfall and groundwater availability dissuade them. Shortage of labour during the harvest season is another reason for the low production. They have issues with bulk production as they receive insufficient market rates when the supply is more and demand is less (Fig. 15.5).

15 UNDERSTANDING THE MENTAL MODELS THAT PROMOTE … 247

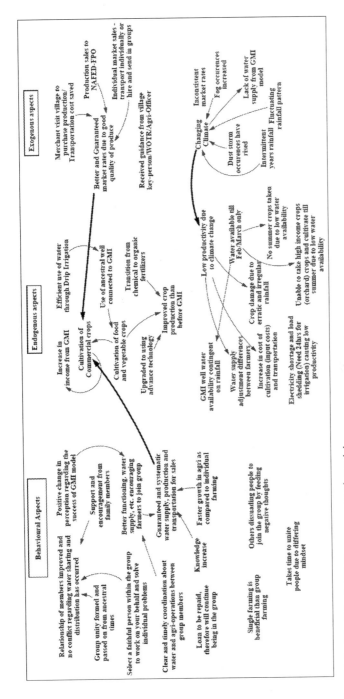

Fig. 15.4 Tigalkheda GMI group mental model

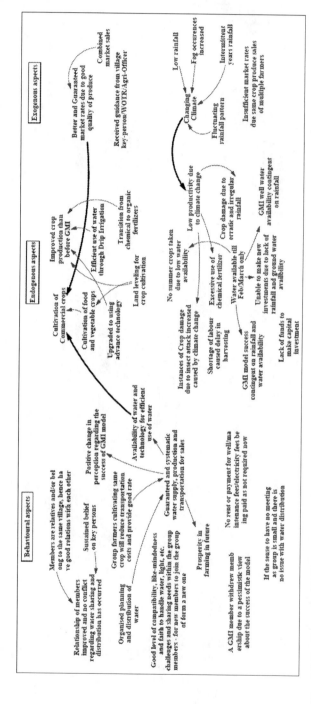

Fig. 15.5 Ranmala GMI group mental model

Mental Model Flow of GMI Group Farmers in Bhangadewadi

In the Bhangadewadi group, members mentioned the perception of guaranteed, systematic, clear and timely coordinated water supply, production and transport led them to join the group. The other factors that encouraged group decision-making was engaging in joint farming, easy access to schemes and grants. The absence of water resources led them to change their pessimistic view of the model and joined the GMI group. Exposure visits to the model village of Tigalkheda, Jalna, helped increase their knowledge about the model and also helped realize the importance of micro-irrigation over flood irrigation. They believe they are receiving water as per their requirement and would feel the shock of water scarcity only in famine situations. They have good relations with the group members because they belong to the same village or are relatives. According to them, good management of water within the GMI group would attract new memberships. These positive feedbacks from their perceptions have led them to join the group and start cultivating commercial crops such as food, vegetables and horticulture crops. There were land levelling work done, the farmers transitioned to using organic fertilizers, upgraded to using advanced technology. They have been applying fertilizers through a drips system. They did not have to wait for the entire duration of irrigation as before as it is automated. There is efficient use of water being done through drip irrigation, their land fertility has improved and have water until summer at times. There are smaller groups formed within the group taking up same crops to manage for good market rates. And their productions has increased they reported that they have been receiving a good return on their production that has increased their income and further reduced their debts. Merchants have been visiting villages and as the sale is done as a group their cost of transportation has been reduced.

Climate change has been a major factor causing low production. Apart from these, they mentioned inconsistent and insufficient market rates as problems. They also consider bulk production reducing the chances of getting a better price. Some have mentioned government officers and their schemes and support do not reach general farmers. Climate change has caused an increase in borrowed funds to agricultural operations as cost of inputs has increased, and excessive chemical fertilizer application. They mentioned most of the market fertilizers they receive are adulterated. Shortage of labour causes delays in harvesting and cultivating unsustainable crops. They were unable to automatize the system further due to a lack of funds. Some are dissuaded from taking up organic crops as they have to delay results with regard to an increase in productivity and therefore income. Regarding the perceptions, they believe a smaller group size with a comparatively lesser land area would have been better for better management, adequate supply of water and get the good market price. Electricity problems, according to them, dissuade new membership as water-related issues may arise later. Uniting people for the formation of groups is tedious (Fig. 15.6).

250 U. KOLI ET AL.

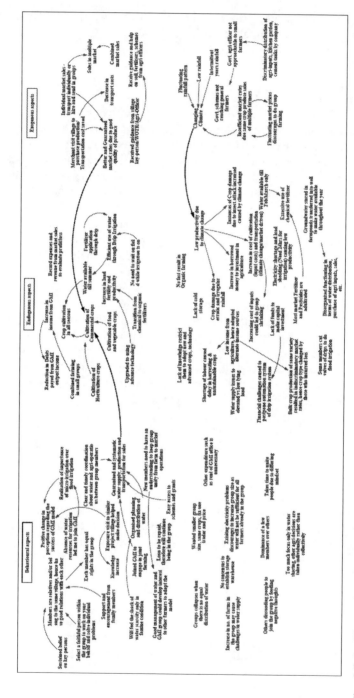

Fig. 15.6 Bhangadewadi GMI group mental model

15.6 Discussion and Conclusion

Groundwater is considered a reliable source of water; acting as a key parameter for the sustenance of economic activities. Any decrease in this, increases the level of vulnerability of the users against external forces such as climate change (Burke et al., 1999). P. M. Kelly and W. N. Adger defined vulnerability with the support of an analogy of a 'wounded soldier', with the perspective of drawing attention to the constraints a soldier possesses that limit his capacity to respond to stress i.e.; cope, recover or adapt to, effectively (Kelly & Adger, 2000). An individual or a social group's adaptive capacity is not only determined by the socio-environmental-economic factors such as five capitals (natural, physical, financial, social and human), access to knowledge, institutions, etc. in their surrounding but also the socio-cognitive factors perceptions, cognitive biases, etc. Studies conducted on the socio-cognitive factors for adaptation have found that adaptation are contingent on social values, perceptions and intentions of individuals and social groups, and that it is a more effective measure to assess adaptive capacity than the traditional socio-economic factors which are substitutes for measuring adaptation (Grothmann & Patt, 2005; Truelove et al., 2014; Mortreux & Barnett, 2017).

The socio-cognitive-behavioural aspect can be assessed from different dimensions of cognition that when combined becomes bases for a behavioural response. For instance, the risk-taking perspective and willingness play an important role in determining the level of involvement with adaptation and reduction in disaster risk. The absence of hope in oneself caused by the lack of the five capitals, to tackle water or climate setbacks, can be a barrier to one's adaptation capacity (Grothmann & Patt, 2005). An individual dependent on the psychological aspect of cognitive biases could be highly low in adaptation capacity and display unsustainable behaviour. Cognitive biases can be explained as a process of thinking deviating from rationality or creating a subjective reality based on decision heuristics, or cognitive shortcuts (attributes that comes easily to mind), social or cultural pressure or emotions (Waldman et al., 2019). Building adaptation capacity needs a well thought plan that should have far-reaching effects in the future. Motivation for adoption has majority of the time been a catalyst for an individual to overcome or safeguard from current and future stress. Practitioners and policymakers play an eminent role in stimulating this part of the cognition process. Such as in our study areas, farmers were motivated to adopt the GMI approach by providing the idea of having water available for multiple seasons by starting with sharing water through micro irrigation. Also, they were provided with financial assistance to begin with.

These factors as discussed are only a few of the long list of cognitive-behavioural aspects that need traction to study the adaptation process. In the Indian context, it is more of an urgency than just an exploratory phenomenon, considering the level of vulnerability to hazards emanating from climate change. India is a country of diverse beliefs, social practices, customs and knowledge, besides having diversity in climatic, topographical and ecological aspects. A thorough assessment is needed to understand the diverse perceptions, beliefs and conceptual understanding of the environmental factors that affect people and in particular farmers in India. There should be a gradual shift from a mere descriptive approach to a predictive approach to inform policy. Conducting such studies have the potential to generate salient approaches that when integrated will benefit policy and practices, raising it from sub-optimal levels to more impactful ones. To begin with adopting the already existing approaches such as the Theory of Planned Behaviour (TPB), the Value-Belief-Norm (VBN), Trans-theoretical Model (TTM), and other such approaches that explain or predict behavioural actions from different mitigation related dimensions, talk about motivating pro-environmental behaviour (Whitmarsh et al., 2021). In our context of water sharing, these approaches would be valuable in finding ways to accept the idea of common pool resources and encourage the adoption of group micro-irrigation with healthy agriculture practices as a community.

Limitation: Climate change and water stress produce dynamic effects. Therefore, the feedback process to the mental model will be fluctuating as some may come early and some may delay. In this uncertainty and complexity, the mental model captured for one particular time may or may not be replicable or scalable for some other place and time. Hence, multiple ways of addressing the problem are to be applied to cover a major extent of the issue. The mental model should be revisited to adjust the dynamic feedback. Not doing so will result in misinterpreting the reality and incorrectly informing policy recommendations or interventions that may lead to adverse effects. Therefore, caution needs to be exercised in capturing information and be more elaborate as possible.

REFERENCES

Agarwal, A. et al. 2000. *Integrated water resources management—TAC background papers 4*. Denmark: Global Water Partnership Technical Advisory Committee (TAC).

Agarwal, P. 2007. Climate change: Implications for Indian agriculture. *Jalvigyan Sameeksha*, 22.

Alcamo, J., Henrichs, T. & Rösch, T. 2000. *Global modeling and scenario analysis for the world commission on water for the 21st century—Report A0002*. University of Kassel, Kurt Wolters Strasse 3, 34109 Kassel: Center for Environmental Systems Research.

Bates, B., Kundzewicz, Z. W., Wu, S. & Palutikof, J. 2008. *Climate change and water. Technical paper of the Intergovernmental Panel on Climate Change, IPCC Secretariat, Geneva*. Geneva: IPCC.

Bewketa, W. & Conwayb, D. 2006. A note on the temporal and spatial variability of rainfall in the drought-prone Amhara region of Ethiopia. *International Journal of Climatology*, 27, pp. 1467–1477.

Bhanja, S. N. et al. 2019. Long-term groundwater recharge rates across India by in situ measurements. *Hydrology and Earth System Sciences*, 23(2), pp. 711–722.

Blackstock, K. et al. 2010. Understanding and influencing behaviour change by farmers to improve. *Science of the Total Environment*, pp. 5631–5638.

Burke, J. J., Sauveplane, C. & Moench, M. 1999. Groundwater management and socio-economic responses. *Natural Resources Forum*, 23, pp. 303–313.

Carrillo-Rivera, J. J., Cardona, A., Huizar-Alvarez, R. & Graniel, E. 2008. Response of the interaction between groundwater and other components of the environment in Mexico. *Environmental Geology*, 55, pp. 303–319.

Challinor, A. J. & Wheeler, T. R. 2008. Crop yield reduction in the tropics under climate change: Processes and uncertainties. *Agricultural and Forest Meteorology*, 148(3), pp. 343–356.

Chaubey, D., Prakash, V., Patel, A. B. & Yadav, T. C. 2018. Role of agro-meteological advisory services on risk mitigation. *Indian Journal of Pure & Applied Biosciences* 6(1), pp. 27–32.

Chindarkar, N. & Grafton, R. Q. 2018. India's depleting groundwater: When science meets policy. *Asia & the Pacific Policy Studies*, 6(1), pp. 108–124.

Cuadrado-Quesada, G. & Joy, K. 2021. The need for co-evolution of groundwater law and community practices for groundwater justice and sustainability: Insights from Maharashtra, India. *Water Alternatives*, 14(3), pp. 717–733.

Dangar, S., Asoka, A. & Mishra, V. 2021. Causes and implications of groundwater depletion in India: A review. *Journal of Hydrology*, 596.

Danielopol, D. L., Griebler, C., Gunatilaka, A. & Notenboom, J. 2003. Present state and future prospects for groundwater ecosystems. *Environmental Conservation*, 30(2), pp. 104–130.

Desthieux, G., Joerin, F. & Lebreton, M. 2010. Ulysse: A qualitative tool for eliciting mental models of complex systems—Methodological approach and application to regional development in Atlantic Canada. *System Dynamics Review*, 26(2), pp. 163–192.

Doyle, J. K. & Ford, D. N. 1998. Mental models concepts for system dynamics research. *System Dynamics Review*, 14(1), pp. 3–29.

Falconer, R. A. & Norton, M. R. 2012. Global water security: Engineering the future. In: *National security and human health implications of climate change, NATO science for peace and security series C: Environmental security*. Springer Science+Business Media B.V. pp. 261–269.

Falkenmark, M., Lundqvist, J. & Widstrand, C. 1989. Macro-scale water scarcity requires micro-scale approaches—Aspects of vulnerability in semi-arid development. *Natural Resources Forum: A United Nations Sustainable Development Journal*, 13(4), pp. 258–267.

Famiglietti, J. S. 2014. The global groundwater crisis. *Nature Climate Change*, 4, pp. 945–948.

FAO. 1996. *Agro-ecological zones, their soil resource and cropping systems*. Rome: Soil Resources, Management and Conservation Service FAO, Land and Water Development Division.

Field, C. B. et al. 2012. Managing the risks of extreme events and disasters to advance climate change. In: *IPCC 2012—A special report of working groups I and II of the intergovernmental panel on climate*. Cambridge, UK: Cambridge University Press, p. 582.

Gajbhiye, K. & Mandal, C. 2000. *Agro-ecological zones, their soil resource and cropping systems*. Nagpur: National Bureau of Soil Survey and Land Use Planning.

GoI. 2019. *All India report on number and area of operational holdings—Agriculture census.* Agriculture Census Division Department Of Agriculture, Co-Operation & Farmers Welfare Ministry Of Agriculture & Farmers Welfare Government Of India.

Graneheim, U. & Lundman, B. 2004. Qualitative content analysis in nursing research: Concepts, procedures and measures to achieve trustworthiness. *Nurse Education Today*, 24(2), pp. 105–112.

Grey, D. et al. 2012. Water security in one blue planet: Twenty-first century policy challenges for science. *Philosophical Transactions of the Royal Society A*, 317.

Grey, D. & Sadoff, C. W. 2007. Sink or Swim? Water security for growth and development. *Water Policy*, 9, pp. 545–571.

Grothmann, T. & Patt, A. 2005. Adaptive capacity and human cognition: The process of individual adaptation to climate change. *Global Environmental Change*, 15(3), pp. 199–213.

Habiba, U., Abedin, M. A. & Shaw, R. 2014. Defining water insecurity. *Water Insecurity: A Social Dilemma. Community, Environment and Disaster Risk Management*, 13, pp. 3–20.

Heidbreder, E. 1945. Toward a dynamic psychology of cognition. *The Psychological Review*, 52(1).

Janakarajan, S. & Moench, M. 2006. Are wells a potential threat to farmers' wellbeing? Case of deteriorating groundwater irrigation in Tamil Nadu. *Economic and Political Weekly*, 41(37), pp. 3977–3987.

Johnson-Laird, P. N. 1983. *Mental model.* Cambridge, UK: Cambridge University Press.

Jones, N. A. et al. 2011. Mental models: An interdisciplinary synthesis of theory and methods. *Ecology and Society*, 16(1).

Joshi, S. K. et al. 2021. Strongly heterogeneous patterns of groundwater depletion in Northwestern India. *Journal of Hydrology*, 598.

Kawasaki, K. & Uchida, S. 2006. Quality Matters more than quantity: Asymmetric temperature effects on crop yield and quality grade. *American Journal of Agricltural Economics*, 98(4), pp. 1195–1209.

Kelkar, S. M., Kulkarni, A. & Rao, K. K. 2020. Impact of climate variability and change on crop production in Maharashtra, India. *Current Science*, 118(18).

Kelly, P. M. & Adger, W. N. 2000. Theory and practice in assessing vulnerability to climate change and facilitating adaptation. *Climatic Change*, 47, pp. 325–352.

Kolkman, M., Veen, A. d. & Geurts, P. 2007. Controversies in water management: Frames and mental models. *Environmental Impact Assessment Review*, 27, pp. 685–706.

Krishnan, R. et al. 2020. *Assessment of climate change over the Indian region—A report of the Ministry of Earth Sciences (MoES), Government of India.* Pune: Springer.

Kulkarni, H. & Shankar, P. V. 2014. Groundwater resources in India: An arena for diverse competition. *Local Environment: The International Journal of Justice and Sustainability*, 19(9), pp. 990–1011.

Kumar, P. J. S., Schneider, M. & Elango, L. 2021. The State-of-the-art estimation of groundwater recharge and water balance with a special emphasis on India: A critical review. *Sustainability Journal*, 14(340).

Levy, M. A., Lubell, M. N. & McRoberts, N. 2018. The structure of mental models of sustainable agriculture. *Nature Sustainability*, 1, pp. 413–420.

Lorenz, K. & Lal, R. 2018. Importance of soils of agroecosystems for climate change policy. *Carbon Sequestration in Agricultural Ecosystems*, pp. 357–386.

Lynam, T. & Brown, K. 2012. Mental models in human–environment interactions: Theory, policy implications, and methodological explorations. *Ecology and Society*, 17(3), p. 3.

Maharana, P., Agnihotri, R. & Dimri, A. P. 2021. Changing Indian monsoon rainfall patterns under the recent warming. *Climate Dynamics*, 57, pp. 2581–2593.

Mora, C. et al. 2015. Suitable days for plant growth disappear under projected climate change: Potential human and biotic vulnerability. *PLoS Biology*, 13(6).

Mortreux, C. & Barnett, J. 2017. Adaptive capacity: Exploring the research frontier. *Wires Climate Change*, 8(4).

Nayak, P. & Solanki, H. 2021. Pesticides and Indian agriculture—A review. 9(5), pp. 250–263. https://doi.org/10.29121/granthaalayah.v9.i5.2021.3930

Pahl-Wostl, C. & Hare, M. 2004. Processes of social learning in integrated resources management. *Community & Applied Social Psychology*, 14(3), pp. 193–206.

Patra, S., Mishra, P., Mahapatra, S. C. & Mithun, S. K. 2016. Modelling impacts of chemical fertilizer on agricultural production: A case study on Hooghly district, West Bengal, India. *Modeling Earth Systems and Environment*, 2, pp. 1–11.

Ratna, S. B. 2012. Summer monsoon rainfall variability over Maharashtra, India. *Pure and Applied Geophysics*, 169, pp. 259–273.

Resick, C. J. et al. 2010. Team composition, cognition, and effectiveness: Examining mental model similarity and accuracy. *American Psychological Association*, 14(2), pp. 174–191.

Rötter, R. & Geijn, S. V. D. 1999. Climate change effects on plant growth, crop yield and livestock. *Climatic Change*, 43, pp. 651–681.

Saha, D., Marwaha, S. & Mukherjee, A. 2018. Groundwater resources and sustainable management issues in India. In: *Clean and Sustainable Groundwater in India*. Singapore: Springer, pp. 1–11.

Smith, R. & Baillie, J. 2009. *Defining precision irrigation: A new approach to irrigation management*. Swan Hill, Australia, Irrigation Australia 2009: Irrigation Australia Irrigation and Drainage Conference: Irrigation Today—Meeting the Challenge.

Srinivasana, V. & Kulkarni, S. 2014. Examining the emerging role of groundwater in water inequity in India. *Water International*, 39(2), pp. 172–186.

Sullivan, C. 2002. Calculating a water poverty index. *World Development*, 30(7), pp. 1195–1210.

Tirkey, A. S., Ghosh, M., Pandey, A. & Shekhar, S. 2018. Assessment of climate extremes and its long term spatial variability over the Jharkhand state of India. *The Egyptian Journal of Remote Sensing and Space Sciences*, 21(1), pp. 49–63.

Truelove, H. B. et al. 2014. Positive and negative spillover of pro-environmental behavior: An integrative review and theoretical framework. *Global Environmental Change*, 29, pp. 127–138.

Vaidyanathan, A. 1996. Depletion of groundwater: Some issues. *Indian Journal of Agricultural Economics*, 51(1).

Vaismoradi, M., Jones, J., Turunen, H. & Snelgrove, S. 2016. Theme development in qualitative content analysis and thematic analysis. *Journal of Nursing Education and Practice*, 6(5).

Waldman, K. B. et al. 2019. Cognitive biases about climate variability in smallholder farming. *Weather, Climate, and Society*, 11(2), pp. 369–383.

Waldman, K. B. et al. 2020. Agricultural decision making and climate uncertainty in developing countries. *Environmental Research Letters*, 15.

Whitmarsh, L., Poortinga, W. & Capstick, S. 2021. Behaviour change to address climate change. *Current Opinion in Psychology*, 42, pp. 76–81.

Wood, M. D. et al. 2012. A moment of mental model clarity: Response to Jones et al. 2011. *Ecology and Society*, 17(4), p. 3.

Young, S. L. et al. 2019. The household water insecurity experiences (HWISE) scale: Development and validation of a household water insecurity measure for low-income and middle income countries. *BMJ Global Health*, 4.

Young, S. L. et al. 2021. Perspective: The importance of water security for ensuring food security, good nutrition, and well-being. *Advances in Nutrition*, 12(4), pp. 1058–1073.

Zhang, J. et al. 2021. Challenges and opportunities in precision irrigation decision-support systems for center pivots. *Environmental Research Letters*, 16.

CHAPTER 16

Water Budgeting for Sustainable Development: A Micro Level Study on Climate-Resilient Agriculture

Gireesan K

16.1 Introduction

Climate change refers to the change in temperature, rainfall and other environmental factors/conditions that affect agriculture and other sectors in varying levels of frequency and intensity. It is a fact that 'Climate change affects agricultural productivity through physiological changes in crops' (Chakraborty et al. 2000). In view of global warming and its impact on the environment across the globe, the agricultural sector is treated as the most vulnerable to climate change all over the world. The prospects of agriculture in India are no exception. Any negative variation in the agriculture sector in India could affect the country's food security and nutrition security programmes, primarily aiming at the targeted population, who are either below the poverty line or living in conditions like 'destitute'.

According to Agriculture Census 2015–2016, about 86% of operational holdings belong to small and marginal farmers (GoI 2019). It is noted that 'the average size of land holding in India 1.15 ha and will decrease even more in coming years due to land fragmentation, industrialisation and urbanisation which are forcing farmers to quit farming and look for more viable livelihood options' (Bhattacharya et al. 2018). A large segment of small and marginal

---

Gireesan K (✉)
MIT School of Government, Dr. Vishwanath Karad MIT World Peace University, Pune, India
e-mail: gireesan.k@mitwpu.edu.in

© The Author(s), under exclusive license to Springer Nature Singapore Pte Ltd. 2023
S. Nautiyal et al. (eds.), *Palgrave Handbook of Socio-ecological Resilience in the Face of Climate Change*,
https://doi.org/10.1007/978-981-99-2206-2_16

farmers has a low literacy level, large family size, poor financial stability, inadequate technical guidance and less scope for credit and financial support. They are also affected by unpredictable and varying rainfall, very high dependence on chemical fertilisers and pesticides, degenerating soil quality, reluctance to diversity crops, high dependence on informal sources of finance, poor storage facilities, very low bargaining capacity, etc. Hence, it is not easy for cultivators to adapt quickly to the impacts of climate change.

In this context, it is pertinent to analyse how far the farmers are taking up agricultural practices to adapt and manage the effects of climate change 'Whether farmers adapt agricultural practices to climate change depends on whether they perceive it or not. However, this is not always enough for adaptation. Therefore, it is important how a farmer perceives the risks associated with climate change' (Tripathi and Mishra 2017).

Evidence suggests a clear indication of climate change happening in the past. Global average surface temperatures have been on the rise over the last few decades, and the increase in temperature has resulted in changes in the weather patterns, wide-spread melting of snow and ice and a rise in global average sea level. Moreover, this rise in temperature is not uniform across the earth, which also results in an increase in sea level and decrease in snow and ice. 'The global average sea level rose at an average rate of 1.8 mm per year over the period 1961–2003, and the rise is more intense at the rate of 3.1 mm per year between 1993 and 2003' (Mahajan and Joshi 2013).

FAO report on greenhouse gas (GHG) data shows that 'emissions from agriculture, forestry and fisheries have nearly doubled over the past fifty years and could increase an additional 30% by 2050' (FAO 2014). This is the first time FAO has released its global estimates of GHG, emissions from agriculture, forestry and land use patterns, while contributing to the Fifth Assessment Report of the IPCC.[1]

In addition to the aspects mentioned above, the shrinking of arctic sea ice and changes in weather patterns were also noted. Over the period from 1900 to 2005, significant change in precipitation trend was also noted across the world. The frequency of heavy precipitation events has increased in several areas, causing water-borne natural disasters in some areas and increased drought, tropical cyclone activity, etc., in other parts. 'Sea levels are rising as polar glaciers, and ice sheets melt away. Communities around the world are living with the consequences and working together on solutions' (WWF 2020).

Research studies have shown that the planet earth is suffering from the consequences of a profoundly unbalanced relationship between people and the planet, with deepening climate change and a spiralling loss of nature, coupled

---

[1] The Intergovernmental Panel on Climate Change (IPCC) is the United Nations body for assessing the science related to climate change. IPCC was established by the United Nations Environment Programme (UNEP) and the World Meteorological Organization (WMO) in 1988.

with deep inequalities in the consumption of resources. It was reported that 'we must tackle the deforestation and environmental degradation that leads to risky interactions between humans and wildlife' (Lambertini 2020).

Considering the rise in population of the country, it is known that the per capita availability of water for the country as a whole has decreased from 5177 km^3 per year in 1951 to 1654 km^3 per year in 2007. It is significant to note that owing to spatial variations in the rainfall, the per capita water availability varies from basin to basin, where 'the average per capita water availability in Brahmaputra and Barak basin was about 14,057 m^3 per year whereas it was 308 m^3 per year in Sabarmati basin in the year 2000' (GoI 2008a).

The main water resources of India consist of the precipitation on the Indian territory, which is estimated to be around 4000 km^3 per year. Out of the total precipitation, including snowfall, surface water and refillable groundwater availability is estimated as 1869 km^3. However, due to various constraints of topography, uneven distribution of resources over space and time, it has been estimated that only about 1123 km^3, including 690 km^3 from surface water and 433 km^3 from groundwater resources, can be put to beneficial use (GoI 2008a).

Precipitation over a large part of India is concentrated in the monsoon season during the period from June to September/October. Precipitation in different parts of the country varies from 100 mm in the Western parts of India (in certain parts of Rajasthan) to over 11,000 mm in the North-Eastern parts of the country (in certain parts of Meghalaya). Extreme conditions also exist in the country as in some places where floods may be followed by droughts and vice versa.

The critical environmental challenges in India have been sharper in the last few decades. Climate change is impacting the natural ecosystems and is expected to have substantial adverse effects in the country, mainly on agriculture, in which the majority of the population still depends for their livelihood. It needs no emphasis that climate change will significantly impact the country's water security, food security and nutrition security. As climate change impacts are felt at multiple levels, from global to local, responses to climate change shall be analysed at multiple levels. It looks for strategic interventions at each level in a suitable manner.

At the global level, India's contribution to multilateral negotiations in the UNFCCC has been significant, whereas the country continues to 'advocate for effective, cooperative, and equitable global approaches based on the principle of common but differentiated responsibilities and respective capabilities' (GoTN 2014).

At the national level, India developed the National Action Plan on Climate Change (NAPCC) in 2008, which consists of eight National Missions, of which the National Water Mission and National Mission for Sustainable Agriculture (NMSA) are pretty significant. National Water Mission sets a goal of 20% improvement in water use efficiency through various measures.

NMSA aims to support climate adaptation in agriculture by developing climate-resilient crops and appropriate agricultural practices (GoI 2008b).

Along with the rise in temperature, total annual rainfall in the State of Tamil Nadu is decreasing during the monsoon seasons, South-West and North-East. There is a distinct inter-annual and spatial variability in rainfall across the State, with the southern and western parts receiving maximum rainfall. It is expected to be a decrease in rainfall in the 2070s by about 1 to 9% with reference to the baseline data of 1970–2000. The maximum temperature is likely to increase in the area by about 3.1 °C. In view of very high spatial variability, catching water where it falls and transferring it to water-scarce areas will be a big challenge (GoTN 2014).

In this context, a research study was carried out in Sembedu Village Panchayat (VP) of Tamil Nadu to ascertain the demand and supply of water through a systematic approach and scientific process of 'Water Budgeting'. The paper analyses the influence of climate-resilient agricultural practices in the study area by examining the demand and supply of water on a longitudinal pattern. It puts forward relevant and practical suggestions for the preservation of water in the area, consumption patterns, cropping patterns, adaptation of new varieties of crops, etc.

Water Budgeting has been viewed here as an essential tool for the diversification of farm-based initiatives and sustainable development, where the changes in the demand and supply of water in a specified area could be analysed on a longitudinal pattern. Here, attempts were made to study the availability of water, and its usage over a specific period, examine its effects in the study area in 2011 and 2018, compare the changes in the agricultural patterns and practices in the area and analyse its reasoning and possible impacts.

Water budgeting has its own manifestations at the micro and macro levels aiming towards sustainable development. By taking up a micro level analysis of farm-based initiatives in a given area, an effort was made to study the impact of climate change on the lives and livelihood of the local farmers. That will enable the farmers to modify the agricultural practices, if needed and to effectively tide over the situation. Moreover, it enables the farmers to be aware of the vagaries of climate change and adopt climate-resilient agriculture using their traditional wisdom.

At the macro level, it could contribute to the policy discourses on disaster management and other cross-cutting domains to mitigate the hardships of climate change. In addition, it could provide valuable input to the policy-makers, administrators and scientific community to take appropriate measures to tackle the ill-effects of climate change effectively.

Participatory tools and techniques, such as Resource Mapping, Social Mapping, Mobility Mapping, Historical Transect, Seasonal Calendar, Before and After Analysis and Venn Diagram, were used for the collection of useful inputs from the field. In addition, Focus Group Discussion (FGD) was held with the young farmers and community members (male and female) separately to capture their views and experiences.

Detailed interviews were held with the key informants of the area, such as experienced farmers and senior citizens of the area to gather primary data from the field. For the collection of data and its analysis, a longitudinal approach was adopted, which includes comparing and contrasting with the relevant data gathered from the same area about seven years back. In addition, relevant secondary data was obtained from the Village Panchayat (VP), relevant departments of the State Government and the Institute of Sustainable Development (ISD).[2]

A Brief Profile of the Study Area

The research study on water budgeting was carried out in Sembedu VP of Ellapuram Panchayat Union, Tiruvallur District, Tamil Nadu. It would be essential to capture a brief profile of the State, District, Panchayat Union and the Village Panchayat before the analysis of 'water budgeting' is taken up.

Topographically, the State of Tamil Nadu could be divided broadly into three natural divisions—coastal plains at the East, hilly areas in the West and the plains in between. Mean annual temperature in Tamil Nadu is 28.2 °C in the plain areas and 15.2 °C in hills. Generally, temperature is very low in December with 24.7 °C and maximum in May with 37.3 °C. Soil temperature data available used to be in the range of 30.7 °C to 32.3 °C in plains and around 14.4 °C in hills (GoTN 2014).

Usually, the State receives rainfall in three seasons such as South-West monsoon, North-East monsoon and pre-monsoon. Normally, the State gets an annual rainfall of 958.4 mm. About half of the total annual rainfall is received during North-East monsoon, slightly less than one-third of rainfall during South-West monsoon and the balance gets scattered during the remaining period. Interior districts of the State receive about 40–50% of rains whereas the coastal districts receive about 65–75% of annual rainfall during the season (GoTN 2014).

The North-Western, Western and Southern parts of the State are hilly and are very rich in vegetation. The Western Ghats block much of the rain-bearing clouds of the South-West monsoon. The eastern parts of the State are fertile coastal plains, but the northern parts are a mix of hills and plains. The central and the south-central regions are arid plains and receive less rainfall when compared to other regions. As the State is highly dependent on monsoon rains for recharging its water resources, any failure in monsoon lead to acute water scarcity and drought.

---

[2] Institute of Sustainable Development (ISD) was formed in the year 2005 with a vision to emerge as a full-fledged multidisciplinary resource centre to think globally and act locally in the arena of sustainable development. For more details, please contact at: info@isdindia.org.

On 1 January 1997, Tiruvallur District was bifurcated from the erstwhile Chengalpattu district. The district is located in the North-Eastern part of Tamil Nadu between 12°15′ and 13°15′ North and 79°15′ and 80°20′ East. The district is surrounded by Kancheepuram District in the South, Vellore District in the West, Bay of Bengal in the East and Andhra Pradesh in the North. The district is spread into an area of 3422 sq. km.

The climate of the district is moderate. The Summer (between April and June) is generally very hot with temperatures going up to an average of 37.9 °C. At the same time, heat is considerably influenced in the coastal areas by the sea breeze. During winter (between December and February), the average temperature is 18.5 °C. The average normal rainfall of the district is 1104 mm, of which about 52% occurs during the north-east monsoon and 41% during south-west monsoon. The district completely depends upon monsoon rains and faces distress conditions in the event of any lapse of monsoon.

Apart from seasonal rivers like Kosasthalaiyar, Araniar, Nandi, Kallar, Coovum and Buckingham Canal, there is no perennial river flowing through the district. As seasonal rivers in the District are not sufficient to meet the demands, irrigation through tanks, tube wells and open wells are very much imperative for agriculture and other activities.

Tiruvallur District has 14 Panchayat Unions (Block Panchayat) and Ellapuram is one of them. Ellapuram Panchayat Union consists of 53 Village panchayats. As per the census 2011, the total population of Ellapuram Panchayat Union is 120,509 (59,727 male and 60,782 female) which includes 44,541 Scheduled Castes (22,108 male and 22,433 female) and 4264 Scheduled Tribes (2131 male and 2133 female).

Sembedu VP is one of the constituents of Ellapuram Panchayat Union. The VP has 941.64 Hectare areas and has 752 households. As per Census 2011, Sembedu VP has the population of 2586 and incorporating the details obtained from the Anganawadi workers of the GP during Dec 2018, the population was calculated as 3108 that includes Total 1541 male and 1567 female). As no election to the Panchayats in the State was held for more than two years, the day-to-day affairs of the VP used to be handled by a designated individual who is on a short-term contract. Significantly, this temporary and untrained official does not have any well-defined roles, functions and responsibilities. In many VPs in the State, this post has been a perfect example of 'absentee staff' with no decent pay package attached which ultimately makes him/her vulnerable on many counts. During several visits made to the village, the VP office was found closed most of the times, which is a sad reflection of the seriousness of the administrative system. Interactions with the community members in the neighbourhood confirmed the doubts about the punctuality, efficacy and performance of the system at the Grassroots.

As part of the study, efforts were made to go around the village, identify the water bodies, interact with the villagers and the local leadership towards exploring the initiatives and interventions made in the area towards protection and preservation of water by taking up relevant programmes and activities. During the interactions with the elected leaders and officials of the VP, it was noted that though several activities were organised, incorporation of works under Mahatma Gandhi National Rural Employment Guarantee Scheme (MGNREGS) and Village Panchayat Development Plan (VPDP) were quite insignificant. However, no significant intervention in this direction was noted to be implemented in the area, considering the views of the community leaders from the locality, officials in different Govt. Departments and a cross-section of the society.

16.2 Water Budgeting—What and Why?

Budget is one of the main instruments of the Government/institution/organisation/agency to achieve its goals and objectives. 'It represents the preferences and priorities of the Department/Institution/Agency. The process of arriving at a budget involves important political dimensions regarding how resources should be raised and allocated over a specified period' (Gireesan 2015). The term 'Budgeting' generally refers to financial aspects like anticipated revenues and proposed expenses of a Department/Institution/Agency within a specific time period. However, the term 'water budgeting' has been used here with a different connotation without any bend towards the economic dimensions. It is visualised more like a 'social budgeting', where the requirement (demand) and availability (supply) of water in a locality over a specific time period are documented and analysed.

Water budgeting is the process of judiciously identifying the use of water for human being, livestock, agriculture, industrial purpose, etc. with a view to optimise benefits in the context of climate change. Here, the climate change may be manifested in terms of climate variability, off-seasonal rainfall, erratic rains, excess rains, drought, etc. And it can also be viewed as 'an important tool for diversification of farm-based initiatives, as the changes that occur in the demand and supply of water in a specific area can be analysed for long-term water security' (Gireesan 2021).

Water budgeting is 'geared towards ensuring optimum and most efficient use of water. It involves an understanding of water availability, community's existing needs and requirements of water, crop planning based on water availability, optimizing irrigation, equitable sharing of excess water, and considered decisions on ground water withdrawals' (WOTR, n.d.).

The process of Water budgeting demands calculation, analysis and documentation of the potential water resources and estimated use of water in a specific geographical area for a specified period. Towards carrying out water budgeting, it is necessary to gather the basic data regarding population, details of livestock in the area with their types and number, average rainfall in the area, total area, storage area of water bodies in the locality, cultivable area used for agriculture, types of crops cultivated, demand for industrial consumption, etc. It can be visualised as a tool to motivate the communities to protect, preserve and optimally use the water by taking informed, calculated and considered decisions about the use of water, selection of crops, patterns and practices.

Availability of water in the area was ascertained by gathering the rainfall data, calculating the extent of surface water, estimating the storage in water bodies and measuring the quantum of recharge of ground water in the area. The quantum of water usage can be calculated by examining the quantity of water for human, livestock, agriculture, requirement of water for industrial units, etc. This process enables to analyse how far the 'water budgeting' could be factored into the agricultural practices in a given area.

16.3 THE PROCESS AND INPUTS FOR CALCULATION OF WATER BUDGETING

The following data inputs were necessary for the purpose of carrying out Water Budgeting exercise in a given locality.

a. Geographical Area (in Hectare)
b. Annual Rainfall (in mm)
c. Area of water bodies such as ponds, tanks, etc. (in Hectare)
d. Population—Total number of persons staying in the locality (including floating population such as migrants).
e. Number of Livestock—Disaggregated data of buffaloes, cows, goats, sheeps, etc.
f. Area under cultivation—Disaggregated data of different crops such as paddy, wheat, pulses, vegetables, fruits, flowers, etc.
g. Water for industrial requirements in the area—Based on the need for operation of village/cottage industries; Micro, Small and Medium Enterprises; and Heavy Industries in the area.

For the purpose of calculating the demand of water during this exercise, the following standard consumption pattern was considered.

For human-being (including children)	:	70 litres per capita per day (lpcd)
For Livestock		
Cows/Bulls/Buffaloes, etc. (Big animals)	:	60 lpcd
Goats/Sheeps/Dogs (Small animals)	:	25 lpcd
For agriculture		
Paddy	:	1200 mm per Hectare
Groundnut/Pulses	:	450 mm per Hectare
Vegetables/Fruits/Flowers	:	400 mm per Hectare

Note The data on water consumption for human being, livestock and agriculture was obtained from the Tamil Nadu Water Supply Board.

Water Budgeting Carried Out in Sembedu VP in the Year 2018

This section provides the details of data inputs gathered from the locality during the year 2018 to carry out the calculations, total water availability in the area from rainfall, water storage in the area in terms of Ground Water (GW) and Surface Water (SW), and water demand in the area.

As discussed in the previous section, the following data was gathered to enable the calculations.

Total population of Sembedu VP : 3108 (as on Dec 2018)
   Male : 1541
   Female : 1567

Note These figures were arrived at considering the data from Census 2011 and incorporating the additional details gathered by the Anganawadi workers of Sembedu VP during Nov–Dec 2018.

Total No. of Livestock : 1462 (as in Dec 2018)
   Cows, Bulls, Buffaloes, etc : 682
   Goats/Sheeps : 780

Note The data on livestock was obtained from the Livestock Hospital, Perandur, Tiruvallur District in Nov–Dec 2018.

Details of crops cultivated:

Paddy : 94.72 Hectare
Pulses/Groundnut : 159.5 Hectare
Flowers/Vegetables/Fruits : 74.37 Hectare

Note The details regarding crops and cultivable area were obtained from the Village Administrative Officer, Sembedu in Nov–Dec 2018.

Considering the above data, the water budgeting exercise was carried out as given below.

In line with the calculations carried out for analysis of water budgeting in Sembedu VP, the details such as availability of water in the village, storage of water in terms of GW and SW in the locality, and the demand of water in the area were collected.

For the calculation of total availability of water, geographical area and annual rainfall received in the area were considered. It was noted that the VP has an area of 941.64 Hectares and the average rain fall received during the year 2018 was 984 mm. Hence, the total availability of water in the area was arrived at 92,657 lakh litres. Considering that the Ground water (GW) will be about 10% of the total availability of water in the area, the figures of GW during the year came to be approximately 9266 lakh litres. Surface water (SW) availability in ponds, tanks and other water bodies in the area was calculated as 20% of the total availability of water during the period, thus the figures came to be about 18,531 lakh litres. In view of the estimation that only 60% of the SW will be available during the year taking into account the evaporation, the figures of SW were arrived at about 6102 lakh litres. Subsequently, the contribution of SW into GW was arrived at 2486 lakh litres. Considering the Ground water availability, surface water availability and contribution of surface water to ground water during the period, the Net Water Storage (NWS) in Sembedu VP during 2018 came as 17,853 lakh litres.

Regarding the demand of water in the area, 794 lakh litres was needed for human consumption, considering the population and per capita consumption of water. Towards the consumption of water by livestock, the figures came to be at 221 lakh litres. And for agriculture activities (for various crops), it was 21,520 lakh litres in which the figure for paddy has 11,367 lakh litres, followed by Ground nut and Pulses as 7178 lakh litres. In addition, the demand of water supply for flowers and vegetables in the area came to be about 2975 lakh litres. Considering the water for human beings, livestock and agriculture, the total demand for water in the area in the year 2018 was arrived at 22,534.63 lakh litres. Considering the availability of water and its demand in the area during 2018, it shows a deficit of 4682 lakh litres.

Water Budgeting Carried Out in Sembedu VP in the year 2011

During the year 2011, water budgeting in Sembedu VP was carried out by ISD. The details gathered from them are compiled in Table 16.2.

From Table 16.2, it is noted that the total water storage in the area was 18,856 lakh litres whereas the water requirements for the area was 41,533 lakh litres, which is less than half (45.4%) indicating a water deficit of 22,622 lakh litres. This was a very disturbing trend for an area which is mostly agrarian and in which the livelihood activities depend completely on ground water and surface water.

Comparative Analysis of Water Budgeting Carried Out During 2011–2018

A comparison of different key aspects of data arrived at 2011 and 2018 is indicated in Table 16.3.

In Sembedu VP, when compared to 2011, rainfall in 2018 was lesser by 96 mm and hence the total water availability was also decreased. The total water availability in the area came down by 9040 lakh litres in 2011 and 2018. Subsequently, the availability of GW and SW storage in the study area was decreased by 904 lakh litres and 1809 lakh litres respectively. On similar lines, recharging of GW from SW also showed a declining trend from 2848 lakh litres (2011) to 2486 lakh litres (2018) with a difference of 362 lakh litres.

Regarding the variations in demand for water in the study area during the period 2011–2018, it was noted that there is a rise in population figures from 2536 (2011) to 3108 (2018), resulting in higher demand for water for human consumption.

Difference in the number and type of livestock available in the village was also noted during the period. As per the livestock data gathered from the

Table 16.1 Water budgeting analysis in Sembedu VP for the year 2018

Qty. of Water Received from Rainfall		
Total Area of the village	:	941.64 Hectare
Annual Rainfall	:	984.00 mm
Water availability in the area (941.64 Hectare × 0.984 m × 10,000)	:	92,657.00 lakh litres
Total Qty. of Water received from Rainfall	:	92,657.00 lakh litres
Water Storage in the Area		
Ground Water (GW)—10% of total water received in the area from rainfall	:	9265.70 lakh litres
Surface Water (SW)—20% of total water received in the area from rainfall	:	18,531.40 lakh litres
SW Storage* in Tanks, Ponds and other water bodies		
Tanks—60% of the total storage capacity (29.26 Hectare × 2.5 m × 60)	:	4389.00 lakh litres
Ponds—60% of the total storage capacity (7.21 Hectare × 2 m × 60)	:	865.20 lakh litres
Other water bodies—60% of the total storage capacity (5.65 Hectare × 2.5 m × 60)	:	847.50 lakh litres
Surface Water Storage (Tanks + Ponds + Other water bodies)	:	6101.70 lakh litres
Recharging of GW from SW (SW received from rainfall—SW storage × 20%)	:	2485.80 lakh litres
Net Water Storage (NWS) (GW + SW + Recharging of GW from SW)	:	17,853.20 lakh litres
Projected Requirements/Demand for Water in the Area		
Requirement of water for human being**		
For human being (3108 persons × 70 lpcd × 365 days)	:	794.09 lakh litres
Requirement of water for human being	:	794.09 lakh litres
Requirement of water for livestock		
Cows, Bulls, Buffaloes, etc. (Big animals) (682 Nos. × 60 lpcd × 365 days)	:	149.36 lakh litres
Goats, Sheeps, etc. (Small animals) (780 Nos. × 25 lpcd × 365 days)	:	71.18 lakh litres
Requirement of water for livestock	:	220.54 lakh litres
Requirement of water for agriculture		
Paddy (94.72 Hectare × 1.2 m × 10,000)	:	11,367.00 lakh litres
Pulses/Ground nut (159.5 Hectare × 0.45 m × 10,000)	:	7178.00 lakh litres

(continued)

Table 16.1 (continued)

Vegetables/Flowers/Fruits (74.37 Hectare × 0.40 m × 10,000)	:	2975.00 lakh litres
Requirement of water for agriculture	:	21,520.00 lakh litres
Total requirement of water in the Area (For human being + Livestock + Agriculture)	:	22,534.63 lakh litres
Surplus/**Deficit** of water in the area during the year (Net water Storage − Total Requirement of water)	:	(−) 4681.63 lakh litres

Source Primary Data collected from Sembedu VP area

*From the records of Village Administrative Officer, Sembedu, no change in the measurement of water bodies in the locality during the period 2011 and 2018 was noted. Else, the fresh measurements of water bodies should be considered

**Here, average consumption of 70 lpcd was considered for the entire population. The difference in consumption owing to age, gender, profession and health condition of the individuals were not accounted. And differential consumption of water as per the economic status and use of domestic gadgets by the families was also not counted

nearest Veterinary centre, it was noted that the number of cows/buffaloes/bulls (Big animals) increased from 327 (2011) to 682 (2018) and the number of goats/sheeps (Small animals) showed a minor reduction from 802 (2011) to 780 (2018). The rise in cows/buffaloes in the area may be attributed to a special scheme to support the rural community by the Animal Husbandry Department, Govt. of Tami Nadu.

During the FGD, it was known that the number of persons from the village received benefits owing to the scheme during the period. In addition, there was an increase in the number of big animals in the area as the figures included calves as well during the period. But no specific reason for a slight dip in the number of goats in the area was cited as several farmers in the village were benefited by the State Sponsored Scheme (SSS) to enhance animal wealth in the area. During the interactions with the community members, the changes in food patterns and preferences in the area were also noted by the researcher. And, during the interactions with the community members, it was noted that there was an increased consumption of meat by them that was more evident during weekends, School/College holiday period, village festivals and other social gatherings.

Regarding the requirements of water for agriculture, it was noted that there was a drastic reduction in the size of cultivable area for paddy, considerable reduction in the cultivable area for Pulses/Groundnut and significant rise in the land used for cultivable area for flowers/vegetables during the period 2011–2018. Cultivable area for paddy has shown a sharp decline of 61% during the period, registering a fall from 246 Hectares (2011) to 94.72 Hectares (2018). Looking into the patterns of cultivation of pulses/groundnut during the period, it was noted that there is a considerable reduction in the cultivable area from 212 Hectare (2011) to 159.5 Hectare (2018) registering a decline

Table 16.2 Water budgeting analysis in Sembedu VP for the year 2011

Water Received in the Area from rainfall		
Total Area of the village	:	941.64 Hectare
Annual Rainfall	:	1080.00 mm
Water availability in the area (941.64 Hectare × 0.984 m × 10,000)	:	1,01,697.12 lakh litres
Total Qty. of Water received from Rainfall	:	1,01,697.12 lakh litres
Water Storage in the Area		
Ground Water (GW)—10% of total water received in the area from rainfall	:	10,170.00 lakh litres
Surface Water (SW)—20% of total water received in the area from rainfall	:	20,340.00 lakh litres
SW Storage in Tanks, Ponds and other Water bodies		
Tanks—60% of the total storage capacity (29.26 Hectare × 2.5 m × 60)	:	4389.00 lakh litres
Ponds—60% of the total storage capacity (7.21 Hectare × 2 m × 60)	:	865.20 lakh litres
Other water bodies—60% of the total storage capacity (5.65 Hectare × 2.5 m × 60)	:	847.50 lakh litres
Total Surface Water storage (Tanks + Ponds + Other water bodies)	:	6101.70 lakh litres
Recharging of GW from SW (SW received from rainfall—SW storage in Tanks, Ponds and other water bodies) × 20%	:	2847.60 lakh litres
Net Water Storage (NWS) in the Area (GW + SW + Recharging of GW from SW)	:	18,856.30 lakh litres
Projected Requirements/Demand for Water in the Area		
Requirement of water for human being		
For Human beings (2536 persons × 70 lpcd × 365 days)	:	647.94 lakh litres
Requirement of water for human being	:	647.94 lakh litres
Requirement of water for livestock		
Cows, Bulls, Buffaloes, etc. (Big animals) (327 Nos. × 60 lpcd × 365 days)	:	71.61 lakh litres
Goats, Sheeps, etc. (Small animals) (802 Nos. × 25 lpcd × 365 days)	:	73.18 lakh litres
Requirement of water for livestock	:	144.79 Lakh litres
Requirement of water for Agriculture		
Paddy (246 Hectare × 1.2 m × 10,000)	:	29,520.00 lakh litres

(continued)

Table 16.2 (continued)

Pulses/Ground nut (212 Hectare × 0.45 m × 10,000)	:	9540.00 lakh litres
Vegetables/Flowers/Fruits (42 Hectare × 0.40 m × 10,000)	:	1680.00 lakh litres
Requirement of water for agriculture	:	40,740.00 lakh litres
Total requirement of water in the Area (For human being + Livestock + Agriculture)	:	41,533.00 lakh litres
Surplus/**Deficit of water** in the area during the year (Availability of water − Total Requirement of water)	:	22,622.00 lakh litres

Source: ISD's records

Table 16.3 Comparative analysis of water budgeting in Sembedu VP

	2011	2018	Difference Between 2011 & 2018
Rainfall during the year (in mm)	1080	984	96.00
Population as per Census 2011 (No. of persons including Children)	2536	3108*	572
Total Qty. of Water received from Rainfall (in Lakh litres)	1,01,697	92,657	9040
Net Water Storage (NWS) (in Lakh litres)	18,856	17,853	1003
Requirement of water for human being (in Lakh litres)	648	794	146
Requirement of water for livestock (in Lakh litres)	145	221	76
Requirement of water for agriculture (in Lakh litres)	40,740	21,520	19,220
Total Projected Requirement of Water in the Area (In Lakh litres)	41,533	22,535	18,998
Deficit of Water in the Area (in Lakh litres) (Total Projected Requirement − NWS)	22,622	4682	17,940

Source Primary Data collected from Sembedu VP area
*Population in the VP as per Census 2011 data updated with the field data gathered from Anganawadi workers during Nov–Dec 2018

by about 25%. However, in the case of flowers/vegetables, there was a significant increase in the cultivable area by about 77%, where land under cultivation was expanded from 42 Hectare (2011) to 74.37 Hectare (2018).

The variation in the cultivable areas for different agriculture crops cannot be viewed as an accidental phenomenon in the locality during the period. The variation in the cultivable areas of paddy, pulses/ground nut and flowers/vegetables have its reasoning in the differential demand for water for these categories. It was noted that cultivation of paddy demand almost three times of additional water, when compared to pulses/groundnut as well as flowers/vegetables, the details of which was indicated in the section 'The Process and Inputs for Calculation of Water Budgeting'. It was also noted that when compared to paddy as well as pulses/groundnut, flowers/vegetables demand less water for cultivation in terms of number of days of irrigation as well as quantity.

Variations in the cultivable area for different crops along with diversification of crops in the study area during the period were the result of informed, calculated and considered decisions made by the farmers. Reduction in rainfall coupled with reduced water availability, reduced irrigation capacity and desire to advance the harvesting period for their products, forced them to think of diversification of crops in the area. In addition, easy as well as quick returns from flowers were also influential factors.

'Going by conventional cultural practices in Tamil Nadu, every woman loves to wear a bunch of jasmine in her hair; almost every believer wishes to hang garlands around idols at home, in religious places and on the photographs placed in their vehicles' (Gireesan 2021) on a day-to-day basis, the demand for jasmine and other flowers is high in the State. It is significantly very high when compared to other neighbouring States such as Andhra Pradesh, Karnataka and Kerala except Puducherry, where the cultural ethos and practices are quite similar to Tamil Nadu.

There was a preference for the provision of financial support from banks and other financial institutions for the cultivation of flowers/vegetables when compared to paddy and other crops. Lesser dependency on extra labour (non-family members) which further may have increased the input cost of the products and the availability of an every-ready market for flowers/vegetables, also may have encouraged the farmers to diversify the cropping areas, patterns and preferences.

While analysing the variations in the animal wealth vis-à-vis agriculture in the study area, it was noted that increase in the number of cows/buffaloes by the farmers in the area may be the manifestation of their increased dependency on 'milch animals rather than draught animals'[3] to compensate the lesser and delayed returns from agriculture and such other perennial crops. Probably, the farmers may have been forced to bring animal fodder from distant places, when the paddy cultivation has come down at Sembedu. The extent of reduction in animal fodder (hay of paddy) due to reduction in paddy cultivation and its economic impact on farmers in the village who own cows/buffaloes needs a detailed analysis.

It was also noted that there was an increase in demand for more value-added products of milk such as curd, paneer and cheese owing to the steady increase in the number of migrant labourers from Eastern, Northern and

---

[3] Milch animals are those who are kept for the purpose of milk production whereas draught animals are those whose males are used for ploughing and carriage purpose.

North-Eastern regions of India to South Indian states, especially to Tamil Nadu. It was noted that they were mostly working in different infrastructure development projects such as Chennai Metro Rail works and in the industrial belts of Ambattur, Avadi, Oragadom, Perambur, Perungudi, Sirussery, Sriperumbudur, etc. in the satellite areas of Chennai. Expansion of IT corridors in the peri-urban areas of Chennai and enhanced scope for labour in the industrial sector in cities like Coimbatore, Erode, Madurai, Salem, Tirupur, etc. have also attracted lot of migrants to the state. It is inferred that the farmers in Sembedu VP were quite aware of the increase in market demand for value-added milk products, though they may not aware about 'migration statistics'.[4]

Even though the farmers may not have made formal water budgeting in the area, they are very rich in traditional knowledge, practical wisdom and application of common sense. They have fair understanding about declining rainfall, decreased water availability and varying market demands. In addition, they are very keen to reduce the input cost for agriculture by relying more on crop diversification where family labour could be gainfully engaged with lesser dependency on non-family labour force. And the scope for quick returns from diversification of agricultural crops for improved patterns also may have influenced them. All these factors enabled the cultivators in Sembedu VP to tackle the ill-effects of climate change with hope and confidence by diversifying the income sources and to manage the pressures of a 'shoe-string family budget'.[5] Their preference to big animals rather than small animals could be noted as their manifestation of practical wisdom and perfect understanding about market fluctuations with increased demand for milk, owing to the rise of in-migration to their locality and neighbourhood.

Diversification of farm-based initiatives in the study area that include experiments with agricultural crops, patterns and practices, increased preference for livestock, and continued faith on sustainable development practices with thrust on agriculture and allied sectors rather than switching over to industrial options manifest the presence of climate-resilient agriculture at the study area which has its roots in the traditional knowledge, practical wisdom, common sense, belief in the social capital and unrelenting faith in nature.

---

[4] Here, 'migration statistics' refer to the limited dimensions such as trends, patterns, quantity, scale of expansion and its spill-over influence in the socio-economic matters of a locality.

[5] 'Shoe-string family budget' indicates the absence of any robust mechanism to enhance income or to reduce expenses on a domestic budgeting exercise. It refers to the forced need of maintaining the balance between the reduced income by limiting/containing the expenses for a family, even at the cost of cutting down the expenses on food, clothes, entertainment, travel, etc.

16.4 Suggestions

In view of the field research in Sembedu VP, the following suggestions are put forward. The suggestions are made to enhance the extent, reach and utility of the exercise in the larger interests of sustainable development.

Farmers may be encouraged, guided and supported to adopt the 'System of Rice Intensification (SRI)' method for paddy cultivation in the area, which consumes less water per hectare (Roughly 800 mm per hectare in SRI method when compared to 1200 mm in conventional method for paddy cultivation). Appropriate orientation, guidance and support shall be provided to the farmers to encourage them to take up the SRI method. To begin with, selected farmers in a specified area may earmark a piece of farmland for the pilot intervention.

Farmers who are cultivating flowers, vegetables, fruits, coconuts, etc. shall be encouraged and supported to incorporate Micro irrigation system such as Drip irrigation and Sprinkler irrigation with the requisite Government support under various Centrally Sponsored Schemes (CSS) and State Sponsored Schemes (SSS). Necessary awareness about the CSS and SSS be imparted to the farmers, especially to the young farmers.

There is a need to preserve and protect the traditional variety of seeds by incorporating conventional techniques which are quite simple and cost-effective. It has been proved that the traditional variety of seeds have better coping pattern in climate change and could withstand pest attacks more effectively.

Spreading 'Water Literacy' shall be a major theme to be disseminated through the schools and other academic institutions as through children, it will be easy to take the message to their families. Quiz, debate, skit and other programmes shall be organised for the promotion of water literacy.

Students in academic institutions shall be motivated to develop 'Model Agro Parks' in their institutions with the guidance of teachers and officials of agriculture department officials and with the support of School Management/Parent Teacher Association, Panchayati Raj Institutions (PRIs) and NGOs. They shall be encouraged to conduct 'Energy Audit' at home, school and other public institutions in the locality.

During the preparation of Annual Action plan for MGNREGS activities in the GPs, the cultivators in the area shall be invited to the discussions. The scheme in its present form itself provides a lot of scope to take up watershed-based activities and to boost the agricultural sector. However, there is a need to ensure proper integration of activities of agriculture with the MGNREGS activities by developing a Seasonal Calendar.[6] This exercise could be taken up

---

[6] Seasonal Calendar is one of the participatory tools which can be used to portray the particular activities in the agricultural process in a specified area in the annual cycle with thrust on the duration for each agricultural activity and the number of persons to be engaged. This information can be gainfully used to plan the MGNREGS activities during the lean period when no demand for labourers for agriculture. Formulation of Seasonal Calendar specific to a locality will enable to properly integrate the human resource

by involving the key stakeholders such as cultivators, functionaries of farmers' collectives, elected members of PRIs and officials from relevant Govt. Departments.

A 'Sub-Committee on Climate-resilient Agriculture' may be constituted at different spheres[7] from Gram Panchayat to District Panchayat. The Sub Committee shall consist of representatives of farmers especially from youth and women, one or two elected members of PRIs having interest in this domain, officials from relevant Govt. Departments, teachers/researchers in the domain, functionaries of development organisations/NGOs, etc. The Sub Committee can be entrusted with the responsibility to suggest measures for the promotion of sustainable agriculture practices, protection and preservation of water bodies and management of other common property resources in the area. Constitution of the Sub Committee is well within the provisions of Panchayati Raj Act and could bring in lot of horizontal and vertical sharing about various development and welfare schemes by the Governments along with ensuring better co-ordination among different stakeholders.

Each GP may take appropriate steps for the creation of 'Bio-Diversity register' at the village. This exercise shall be carried out in a participatory manner by involving all the key stakeholders including school students. Specific efforts may be made to carry out the exercise leading to the formation of a 'People's Bio-Diversity Register' at the local level. The register needs to be updated at least once in two years through a participatory process.

Each State Government may think of formulating a policy on 'Climate-resilient agriculture and sustainable development' with specific roles, responsibilities, functions and actionable points to the different spheres of Governments. This policy shall be formulated through a participatory process involving all key stakeholders by convening detailed discussions at different parts of the State and virtual meetings for non-resident citizens.

16.5 Conclusion

The exercise of 'Water Budgeting' carried out in the study area on a longitudinal pattern is an effective tool as well as strategy to identify the availability and demand for water considering several factors into account, such as rainfall, population, livestock, cultivable area, type of crops, productivity, etc. The paper analyses the different ways practised by small and marginal farmers in the study area to adapt with the climate-resilient agriculture and allied activities. It also portrays the significance of a comprehensive and integrated approach in

---

requirements of agriculture sector and ensure proper planning for the MGNREGS activities that ultimately benefit the farmers as well as agricultural labourers.

[7] Instead of the conventional words such as Levels, Tiers or Strata, the word 'Sphere' has been used here, to indicate different Local Governments from Village Panchayat to District Panchayat. 'Spherical Autonomy' is one of the core principles of decentralisation. It denotes 'sufficient autonomy' to the Local Governments in their own domain, irrespective of their size, position and range of functions.

agriculture factoring climate changes, crop diversification, livelihood aspects, consumption patterns, etc.

Being a unique initiative, the experiences of water budgeting carried out in a small locality has its own significance towards enabling the farmers to effectively address climate change and adapt climate-resilient agriculture using their traditional knowledge, practical wisdom, understanding about market needs and common sense. The 'methodology and procedures of water budgeting' may be replicated at different parts of the country by making relevant contextual variations. The exercise shall be repeated at periodical intervals, involving cultivators (including women farmers and agricultural labourers), elected members and officials of PRIs, officials from agriculture and irrigation departments, and other key stakeholders. The results of the micro level study and intervention could be effectively used as a strategy to scientifically diversify the farm-based initiatives as well as to mitigate the hardships of climate change to a certain extent, leading to climate-resilient agriculture practices. It has the potentials to nurture, promote and diversify nutrition-sensitive agriculture also in a conscious, informed and considerate manner in line with the preferences and patterns of consumption. Such initiatives and interventions have the latent potentials to contribute significantly to achieve sustainable development goals in a participatory manner by ensuring active ownership of key stakeholders at the grassroots.

Acknowledgements Author acknowledges the inputs received from Mr. Prabagaran M, alumnus of RGNIYD, Sriperumbudur and Dr. Rema Saraswathy, Director, ISD, Chennai.

References

Bates BC, Kundzewicz ZW, Wu S and Palutikof JP (Eds) (2008) *Climate Change and Water. Technical Paper of the Intergovernmental Panel on Climate Change*, IPCC Secretariat, Geneva, 210.

Bhattacharya S, Burman RR, Sharma JP, Padaria RN, Paul S and Singh AK (2018). Model villages led rural development: A review of conceptual framework and development indicators. *Journal of Community Mobilisation and Sustainable Development* 13(3): 513–526.

Chakraborty S, et al. (2000) Climate change: Potential impact on plant diseases. *Environmental Pollution* 108(3): 317–326.

Directorate of Census 2011 Operations, Tamil Nadu (2017). http://www.census.tn.nic.in. Accessed 2 April 2017.

Food and Agricultural Organisation (FAO). (2014). *Agriculture's Greenhouse Gas Emissions on the Rise*. http://www.fao.org/news/story/en/item/216137/icode.

Gireesan K (2015) Youth budgeting at the local level. *KILA Journal of Local Governance* 2(2): 19–27.

———. (2020) Youth in agriculture: Opportunities, challenges and the way forward in the scenario of COVID-19 pandemic. *IAHRW International Journal of Social Sciences Review* 8(7–9): 321–329.

———. (2021) Water wise. *Down to Earth* 29(18): 54–55.
Government of India (GoI). (2008a). *National Water Mission under National Action Plan on Climate Change*. New Delhi: Ministry of Water Resources.
GoI. (2008b). *National Water Mission*. http://wrmin.nic.in/writereaddata/nwm287 56944786.pdf. Accessed 3 April 2017.
GoI. (2019). *Agricultural Statistics—At a Glance 2019*. New Delhi: Directorate of Economics and Statistics, Ministry of Agriculture and Farmers' Welfare, 315.
Government of Tamil Nadu (GoTN). (2014). *Tamil Nadu State Action Plan for Climate Change*. Chennai: Government of Tamil Nadu. http://www.environment.tn.nic.in/doc/TNSAPCC%20PDF/Chapter%204%20%20Observed%20Climate%20.pdf. Accessed 5 April 2017.
GoTN. (2017). Rural development and Panchayati Raj department. http://www.tn.gov.in. Accessed 3 April 2017.
Gupta A et al. (2019). Roadmap of resilient agriculture in India. Thematic paper released on International Symposium on Disaster Resilience and Green Growth for Sustainable Development organised by Centre for Excellence on Climate Change, National Institute of Disaster Management, New Delhi, 26–27 Sep 2019, 18.
Intergovernmental Panel on Climate Change (IPCC). 2001. *Climate Change 2001: Synthesis Report. A Contribution of Working Groups I, II, and III to the Third Assessment Report of the Integovernmental Panel on Climate Change* [Watson, R.T. and the Core Writing Team (eds.)]. Cambridge University Press, Cambridge, United Kingdom, and New York, NY, USA, 368.
IPCC. (2007). *Climate Change 2007: Mitigation. Contribution of Working Group III to the Fourth Assessment Report of the Intergovernmental Panel on Climate Change* [B. Metz, O.R. Davidson, P.R. Bosch, R. Dave, L.A. Meyer (eds)], Cambridge University Press, Cambridge, United Kingdom and New York, NY, USA.
IPCC. (2014). *Climate Change 2014: Mitigation of Climate Change. Contribution of Working Group III to the Fifth Assessment Report of the Intergovernmental Panel on Climate Change,* [Edenhofer, O., R. Pichs-Madruga, Y. Sokona, E. Farahani, S. Kadner, K. Seyboth, A. Adler, I. Baum, S. Brunner, P. Eickemeier, B. Kriemann, J. Savolainen, S. Schlömer, C. von Stechow, T. Zwickel and J.C. Minx (eds.)]. Cambridge University Press, Cambridge, United Kingdom and New York, NY, USA.
Lambertini, Marco. (2020). *A Crucial Opportunity for Change Following the COVID-19 Pandemic*. https://arcticwwf.org/work/climate. Accessed 12 May 2020.
Mahajan, Keshav Lall and Joshi, Niraj Prakash. (2013). *Climate Change, Agriculture and Rural Livelihoods in Developing Countries*. Tokyo: Springer.
Tiruvallur District. (2017). http://www.tiruvallur.tn.nic.in. Accessed 5 April 2017.
Tripathi A and Mishra A (2017) Farmers need more help to adapt to climate change. *Economic and Political Weekly* LII (24): 53–59.
Watershed Organisation Trust (WOTR). (n.d.). Brochure of WOTR: Water budgeting—Tool for improving water governance at local level. Pune: WOTR. www.wotr.org. Accessed 9 May 2020.
World Wildlife Fund (WWF). (2020). *Arctic Melt Matters*. Washington, DC: WWF. https://arcticwwf.org/work/climate. Accessed 12 May 2020.

CHAPTER 17

Making of a Climate Smart Village: A Study on Meenangadi Gram Panchayat in Kerala (India)

Jos Chathukulam and Manasi Joseph

17.1 INTRODUCTION

As per the 2021 Intergovernmental Panel on Climate Change (IPCC), human activities are 100 per cent responsible for the global warming. It was found that earth's global surface temperature warmed by 1.09 degree Celsius compared to the pre-industrial period of 1850–1900 (IPCC 2021). The previous IPCC reports have also pointed out that the alarming increase in anthropogenic (human-made) greenhouse gas emissions has contributed to a rise in global temperature (IPCC 2013) and such changes are causing a negative impact on biological and physical systems across the globe. The 2020 Human Development Report (HDR 2020) titled *Human Development Report, The Next Frontier: Human Development and Anthropocene* states that humanity and planet have entered into a geological epoch known as "Anthropocene or Age of Humans" and warns that countries should redesign their paths to development by keeping in mind the "dangerous pressures humans put on planet and dismantle the gross imbalances of power to prevent

---

J. Chathukulam (✉)
Centre for Rural Management (CRM), Kottayam, India
e-mail: joschathukulam@gmail.com

Sri. Ramakrishna Hegde Chair On Decentralization and Development, Institute of Social and Economic Change (ISEC), Bengaluru, India

M. Joseph
Centre for Rural Management, Kottayam, India

© The Author(s), under exclusive license to Springer Nature Singapore Pte Ltd. 2023
S. Nautiyal et al. (eds.), *Palgrave Handbook of Socio-ecological Resilience in the Face of Climate Change*,
https://doi.org/10.1007/978-981-99-2206-2_17

that change" (HDR 2020). The 2021 HDR report titled *Uncertain Times, Uncertain Lives: Shaping our Future in a Transforming World* cautions that "planetary-level and human-induced changes of the Anthropocene would be sufficient to inject frightening uncertainties for the mankind" (HDR 2021). Though previous climate conferences from Stockholm (1972) to Kyoto (1997) to Cancun agreement (2010) have been demanding to keep the global temperature below 2 °C, it was the 2015 Paris Agreement that turned out to be a historic one in convincing European Union and 192 countries including India regarding the need and urgency to adopt carbon-neutrality. In fact, the term carbon-neutrality got widespread acceptance during the Paris Climate agreement. The term carbon-neutrality in general refers to *"achieving zero carbon emissions by balancing a measured amount of carbon released with an equivalent amount sequestered or offset or buying enough carbon credits to make up the difference,"* (Brown 2018). The Paris Agreement aims to *"substantially reduce global greenhouse gas emissions to limit global temperature increase in this century to 2 degree Celsius,"* (Paris Agreement Article 2 (a)).

After the 2015 Paris agreement, the irreversible changes in the climate and its repercussions on the environment triggered discussions to substantially reduce the emission of global greenhouse gases and the need to embrace carbon-neutrality gained momentum. To overcome the challenges posed by the climate change, the Left Democratic Front (LDF) government in Kerala launched a carbon–neutral project on June 5, 2016 in Meenangadi Gram Panchayat. The idea was conceived in 2015 during a discussion with the then Finance Minister T.M. Thomas Issac and the then Gram Panchayat President Ms. Vijayan[1] and other elected functionaries of the Panchayat. They informed the Finance Minister (who also happened to attend the Paris Climate Agreement) regarding the activities already implemented by the Panchayat to preserve and conserve the environment. During the discussion, it was decided to introduce a carbon–neutral project in Meenangadi on a pilot basis. Though Meenangadi has been involved in environment-friendly initiatives, the concept of carbon-neutrality was a challenging one. The Panchayat first conducted a carbon audit in 9000 households to identify the key sectors and sources that lead to maximum amount of carbon emission and carbon sequestration in Meenangadi. The carbon-auditing held in 2016–2017 found that the total carbon emission of Meenangadi was estimated to be at 33,375 tonnes of CO_2 Eq while carbon sequestration was 21,962.53 tonnes of CO_2 Eq (Raghunath 2021). It was also found that there was an excess emission of 11,412.57 tonnes of CO_2 Eq. To bring down the carbon emission and to increase the sequestration, the Panchayat initiated a lot of schemes including Tree Banking

---

[1] The authors of this paper interviewed Ms. Vijayan on February 21 and 22, 2021 and October 17, 2022. She said that climate change had reduced the yield from agriculture in Meenangadi and Wayanad district. Hence, the Panchayat was mulling to change its approach towards development. So, when the state government wanted local governments to take up projects to tackle climate change, Meenangadi took the first initiative to make the village carbon–neutral.

Scheme in which people were incentivised for planting trees to climate literacy programmes to promoting the use of solar powered street lights and LED lights in homes to decentralized waste management among a few notable initiatives. In terms of fund allocation, for implementing the carbon–neutral project, the Panchayat in 2017 earmarked Rs. 11 crores (US $1.52 million) and state government allotted Rs. 10 crores (US $1.38 million). While the initial target was to make Meenangadi carbon–neutral by 2020, the Covid 19 pandemic and the change of panchayat administration owing to 2020 local government elections in Kerala posed hurdles in accomplishing it. At present, the Panchayat is aiming to achieve the carbon–neutral tag by 2025. The ongoing pandemic along with Russia—Ukraine war is like to delay the accomplishment of targets outlined in 2015 Paris Agreement. Though the pandemic had its positives and negatives on environment, it has diverted the attention from the issue of climate change and the ongoing escalations between Russia and Ukraine will certainly undo the progress the world has made in preserving our mother earth.

Another major issue when it comes to achieving the carbon–neutral tag is the concept of carbon trade. Carbon trade operates under the notion that if a country (developed) is promoting an industry that causes high carbon emissions it can compensate for those harmful emissions by planting trees as well as establishing environment-friendly programmes in developing and third world countries or within the rural areas of the country. Experts are of the opinion that capitalism itself needs to be transformed if we were to decarbonize the "global" (Bohm et al. 2012). The problem with rural Meenangadi as well as capitalistic countries in general has sought refuge in carbon trade as they have placed their mission to become carbon–neutral within a growth-centred development paradigm and it does more harm to the environment (Newell 2012). In the recent times, in a capitalistic and growth-centred economy, "embracing carbon markets by financial and political elites constitute a possible first step towards turning capitalist ventures into a new form of greener and sustainable climate capitalism" (Newell 2012; Newell and Patterson 2010). While a growing number of countries are making commitments to achieve carbon-neutrality within the next few decades, the major problem is that they are operating within a capitalistic—materialistic—growth-centred paradigm and going carbon—not only for Meenangadi but for the rest of the world. Here comes the significance of sustainable and simplistic model of Gandhi-Kumarappa Framework in the race towards carbon-neutrality and the philosophy of degrowth.

This paper looks into carbon-neutrality in India with special emphasis to Meenangadi in Wayanad, Kerala. The first part begins with a general introduction followed by a brief discussion on Gandhi- Kumarappa framework. The second part gives s a brief overview regarding carbon-neutrality in India. The third part offers the profile and the political economy of ecology in Meenangadi in Wayand district. The fourth part of the paper highlights the relevance of carbon-neutrality in Meenangadi and its significance in preserving the ecologically fragile Western Ghats. The fifth part of the paper looks into the steps undertaken to assess the carbon emissions in Meenangadi. The sixth part is followed by mitigation strategies adopted by the Meenangadi and the result followed by conclusion.

Gandhi-Kumarappa Framework and Carbon-Neutrality

Mahatma Gandhi in his seminal work *Hind Swaraj* (1909) wrote that earth has enough for "everyone's need" but not for "everyone's greed" and it is evident from these words to understand his views on environment and the need to embrace sustainable model of development. Long before, the Stockholm Conference of 1972 (First Earth Summit) there was Gandhi, the visionary, who raised concerns on environment (Tiwari 2019). Gandhi had cautioned the world about the hidden dangers posed by large-scale industrialization and rapid urbanization and how it will eventually lead to destruction of environment (Hind Swaraj 1909). Gandhi gave emphasis to "production by the masses" as it can result in the development of an economic system that minimizes destruction to environment and lead to a more sustainable model of development and he suggested *Swaraj* as a means to accomplish it. It is a widely accepted fact that Gandhi led a zero-carbon footprint lifestyle and Khadi would be the perfect example here. Though Gandhi encouraged the use of Khadi to make India self-reliant and to generate self-employment at the grassroots level, it is also one of the most eco-friendly fabric, a fact many failed to notice even during the time of Gandhi. Not many know that Khadi is a "minimum carbon-footprint fabric" since spinning and weaving of khadi are largely done by hand (with minimal use of water, electricity or fuel). Gandhi used the spinning wheel (chakra) to spin thread and make clothes. Traditionally there are two types of chakras—Bardoli Chakra (Box type) and Yerwada Chakra (round wheel) and these days people use Amber Chakra (Solar Chakra) which make use of solar energy to operate the wheel. There is minimum transportation and packaging involved. The Khadi and Village Industries Commission supports the production, research and sale of Khadi in India. At a time when we are attempting to mend our ways to save the earth, Gandhian way of life can be helpful not in reducing carbon emissions but also in generating self-employment opportunities[2] J. C. Kumarappa, an economist trained at Columbia University and a close associate of Mahatma Gandhi, has suggested in his book *Economy of Permanence*, that a total non-violence in production and consumption is required for transitioning towards sustainable society to sustainable economy. Kumarappa argues that only through an "economy of permanence" (sustainable social order) in which human beings

---

[2] The first author of this paper interacted with Ms. Asha Buch, a retired bilingual teacher and reputed social worker on February 22, 2021. Asha Buch grew up in a family of activists supporting India's freedom movement under the guidance of Gandhi. She learnt to spin at the age of about seven or eight, a skill that she has honed over the years and enjoys passing it on.

collaborate with nature to meet their needs without disrupting the natural patterns of growth and renewal can lead to a sustainable economy (Lindley 2007; Nair and Moolakkattu 2018). During the field visit to Meenangadi, it was evident that neither the local citizens nor the politicians, panchayat functionaries, civil society organizations were aware of the Gandhi-Kumarappa perspectives on preserving the environment. They are not ready to move away from the growth-centric development and what they prefer is to place carbon-neutrality within growth-centric or development-centric paradigm.

In general terms, being carbon–neutral means balancing out carbon emissions by reducing them elsewhere. For instance, a country can plant as many as trees to suck up the carbon put out by their industrial establishments. This concept of carbon-neutrality is in favour of growth-centred development paradigm. Since Meenangadi is trying to balance between growth-centred development and carbon-neutrality, there are concerns that even if the village earns the tag of carbon-neutrality whether it will remain sustainable in a long run. The concept of solidarity economics also enjoys huge significance in the context of carbon-neutrality. Solidarity economy seeks to *"resist the colonizing power of the individualistic, competitive, and exploitative Economy of Empire"* (Nair 2020). Some of the features of solidarity economy are reciprocity, unity in diversity, shared power, autonomy, horizontal communication, cooperation and mutual aid and local rootedness. Solidarity Economics *"is a process which identifies, connects, strengthens and creates grassroots, life centered alternatives to capitalist globalization, or the Economics of Empire"* (Nair 2020).

India and Carbon-Neutrality

As per India Energy Outlook 2021, India is the world's third-largest primary energy consumer and not surprisingly it is also the third-largest carbon emitter. India has signed the declaration of 2030 Agenda for Sustainable Development Goals (SDGs) of United Nations. The Goal 13 of the SDGs focuses on urgent action required to combat climate change and to strengthen resilience and adaptive capacity to climate hazards and to integrate climate change measures into policies and strategies. In 2016, India ratified Paris Agreement and under India's Intended Nationally Determined Contributions (INDCS), the country is committed to creating a cumulative carbon sink of 2.5–3 billion tonnes of carbon dioxide through additional forest and tree cover by 2030. To achieve the targets, India has to adopt a carbon–neutral trajectory. India has already started ways to reduce carbon emissions and Phayeng in Manipur, Majuli in Assam and Union Territory of Ladakh and Meenangadi in Wayanad, Kerala are a few examples in this regard.

Phayeng, a village surrounded by forested hillocks in Imphal West district in Manipur is considered to be the first village in India to earn a "carbon positive[3]

---

[3] The general consensus is that a village is given the tag of carbon positive settlement if it sequesters more carbon than it emits.

eco-model village" tag in the country. The inhabitants in the village belong to the Chakpa (a scheduled caste) community. Phayeng is one among the five villages in Phayeng Gram Panchayat. As per the 2011 Census, there are 890 households with a total population of 3835 out of it 1874 are male and 1961 female. The transition of Phayeng to a carbon-positive village is a remarkable one. In the 1970s and 80s, deforestation was rampant in forest area in this village. In the absence of green cover, Phayeng had an extreme warm climate and scanty rainfall and it led to severe shortage of water. Occasional heavy rains led to flash flooding and it affected the soil moisture which made agriculture and farming activities difficult. In the 1970s, the once dense forest in this village which was spread across 7678.78 hectares was just 579.38 hectares in 2016 (Anand 2017). As the situation worsened villagers decided to do something to save their place and they approached Directorate of Forest and Climate Change, Government of Manipur. They jointly started reforestation of the forest land in the Phayeng village. In addition to that, Ministry of Environment, Forest, and Climate Change, Government of India also provided assistance under National Adaptation Fund for Climate Change (NAFCC) to develop Phayeng as a carbon-positive eco-village (Nandi 2020). National Bank for Agriculture and Rural Development (NABARD) was the implementing agency for this Rs.10 crore (US $1.38 million) project. The project duration was from 2016 to 2020.[4] As part of the project, villagers even formed a forest protection committee (*umang kanba*) and also came up with stringent rules including a complete ban on hunting, restriction on entry of outsiders into forest without permission, patrolling in the forest area and monitoring forest fires. The forest developed by villagers in Phayeng have now more than 95 tree species with medicinal value (Nandi 2019). Though the villagers in Phayeng are not explicitly following the Gandhi-Kumarappa framework, the measures they have undertaken to preserve their environment are simplistic and affordable as envisioned by Gandhi-Kumarappa. In short, directly, or indirectly, the foundation of Phayeng community in saving their environment lies in Gandhi and Kumarappa. It is also interesting to note that Phayeng is not involved in any sort of carbon trade within or outside their jurisdiction.

Majuli, the river island district in Assam is also on its way to become a carbon–neutral region. An island in Brahmaputra river, it became a district in 2016. Majuli, which was previously part of Johrat district in Assam has a population of around 167,304. Flooding and soil erosion are the major issues faced by the people of Majuli as Brahmaputra has devoured half of the island in the last 40 years (Singh 2021). It was in this context, the Department of Environment and Forest, Assam launched a project titled Sustainable Action for Climate Resilient Development (SACReD) in 2016 to combat climate change by reducing greenhouse gas emissions and eventually turning Majuli into a

---

[4] The first author of this paper interacted with Phayeng native Shri. Angom Gojendro Singh, a recipient of Panchayat Sashaktikaran Puraskar 2015, on March 3 and 4, 2021. He is the Sarpanch of Phayeng Gram Panchayat.

carbon–neutral district. In addition, the state government is also carrying out forestry activities including bamboo cultivation and biodiversity conservation under Mahatma Gandhi National Rural Employment Guarantee Scheme (MGNREGS).[5] Like Phayeng, Majuli is also by default operating within the Gandhi- Kumarappa framework. There is no deliberate attempt from the part of the residents and officials to follow this perspective.

The Union Territory of Ladakh, one of the least populated UTs in India, is also going the carbon–neutral way. The decision to develop Ladakh, Leh and Kargil into carbon–neutral regions was announced by Prime Minister Narendra Modi on August 16, 2020 (Nandi 2020). The construction of a 7500 MW solar park is underway in Ladakh. Meanwhile Ladakh also functions as a natural carbon sink as the agroforestry sector in Ladakh including the Nubra Valley filled with 575,000 plantations of willow and poplar trees have the potential to sequester 75,000 tonnes of carbon (Kumar et al. 2009) and along with that the locals have been engaged in the cultivation of climate-resilient crops for years.

Meenangadi in Kerala aiming to become carbon–neutral by 2025. Though initially the plan was to achieve the carbon–neutral tag by 2020, the Covid 19 pandemic and local government elections resulted in an inordinate delay. Since 2016, Meenangadi has been in the race to become carbon–neutral. Prior to embarking on the project, a carbon auditing was conducted to assess the total amount of carbon emission and carbon sequestration in Meenangadi. It was found that in 2016–2017, the total carbon emission of Meenangadi was estimated to be at 33,375 tonnes of CO_2 Eq while carbon sequestration was 21,962.53 tonnes of CO_2 Eq (Raghunath 2021). It was found that there was an excess emission of 11,412.57 tonnes of CO_2 Eq. To bring down the carbon emission and to increase the sequestration, the Meenangadi Gram Panchayat initiated a lot of schemes including Tree Banking Scheme in which people were incentivized for planting trees to climate literacy programmes to promoting the use of solar powered street lights and LED lights in homes to decentralized waste management among a few notable initiatives. Meanwhile before delving into the strategies and procedures adopted by Meenangadi to achieve carbon-neutrality, it is equally important to understand more about Meenangadi and its political economy of ecology.

17.2 Profile and the Political Economy of Ecology in Meenangadi in Wayanad District

Meenangadi is one among the 25 Gram Panchayats in Wayanad district in Kerala. Wayanad, a highland region lies in the Western Ghats. Wayanad district which was created on November 1, 1980, is located on the south-western tip of Deccan Plateau, at an altitude of 700 m above sea level, extends over an

---

[5] The first author of this paper interviewed Dr. Rakesh Chetry, Assistant Commissioner, Manjuli district, Assam on March 16, 2021.

area of 2125 sq.km. Meenangadi is part of Sulthan Bathery constituency and Wayanad Parliamentary constituency. Around 70 per cent is table land and around 20 per cent is fertile plains. Around 3 per cent of the total area in Meenangadi is covered by forest. Nearly 94 per cent of the land (excluding the forest area) is used for agricultural purpose. Some parts of Meenangadi fall under shadow region. As per the 2011 Census, Meenangadi has a total population of 34,601 across 8199 households. The Panchayat is divided into 19 wards and the Panchayat falls in two revenue villages-Krishnagiri and Purakkadi.

Meenangadi and Wayand had been a hub of "new wave of social movements" (Chathukulam and Moolakkattu 2006). Even before carbon-neutrality, green politics and movements to protect environment and land were frequent in Wayanad and Meenangadi, especially several land struggles between tribals also known as *Adivasis*, one of the most marginalized communities in the state constantly fighting for land reforms and for their own land. Muthanga agitation in Wayanad district in 2003 is the biggest example in this regard. The failure of the successive state governments in Kerala to allot one acre of cultivable land to each tribal family, (an assurance that was given to them by the state governments following the displacement of the Adivasis/tribal dwellers in the 1960s and 1980s to make way for eucalyptus plantations in Wayanad) led to one of the landmark agitations by the tribal community in Kerala. As per the Kerala Scheduled Tribes (Restriction of Transfer of Land and Restoration of Alienated Land), Act 1957, the state must transfer cultivable land to all Adivasis in the state but the governments over the years refused to implement this in letter and spirit (Bachan 2019). Considering the betrayal of the governments in the state over the years, around 617 tribal families set up their tenets in Muthanga forest as of protest. The government used police machinery to forcefully evict the protestors from the forest and it turned into a violent clash (Ameerudheen 2016). The socio-economic profile of Adivasis has always been outside of the *Kerala Model of Development* (Chathukulam et al. 2013). According to the 2011 census, adivasis constitute 18.5 per cent of the total population in Wayanad. Paniyas are the largest tribe and they constitute nearly 45.6 per cent of the tribal population, followed by Kurichiyas at 16.6 per cent, Kurumas at 13.8 per cent and Kattunayakas at 11.2 per cent (Census of India 2011; Government of India, Chathukulam et al. 2013). People in Meenangadi, which is home to 23 per cent of tribals[6] have been very vocal about protecting and preserving their land and environment. In Meenangadi itself, nearly seven years ago, following the mass eviction of people from various tribal settlements, nearly 68 families occupied a 70-acre plot belonging to Kerala Forest Department in Meenangadi (Ameerudheen 2016). The adivasis made makeshift huts inside the vast take plantation and lived there for months withstanding eviction threats from

---

[6] Kurumar, Paniyar, Kaattunaykkar, Vettakkurumar and Kurichyar are the tribal communities of Meenangadi are the major tribes found in Meenangadi.

government. The adivasis occupied the land until the state government agreed to distribute one acre of land each to all families. The new settlement in Appad in Meenangadi came to be known as Panchami Colony, named after the first baby, born during the struggle (Ameerudheen 2016). The tribal communities are highly vulnerable to climate change as their livelihood is entirely dependent on climate-sensitive crops and minor forest produces. According to the 2011 Census, the literacy rate of Meenangadi is 81 per cent and it is below the overall literacy rate of Kerala (93.91 per cent). Tribals in Kerala, like any other marginalized communities are a crucial vote bank for political parties. In fact, mainstream and local political parties will be competing to woo these innocent people with poll sops and promises during the election season but none of them are ready to offer their full-fledged support to the cause of indigenous people.

Over the years both LDF led by Communist Party of India (Marxist) and United Democratic Front (UDF) led by Indian National Congress have ignored the demands of the landless adivasis. (Ameerudheen 2016). As per the Biodiversity Register[7] prepared by the Panchayat, wetlands, forests plantations and homesteads are the major ecosystem. A total of 240 species of plants and trees have been identified out of which 140 species are herbs, 55 tree species, 14 species each of water plants, 12 species each of wild plants and wild ornamental plants. The faunal diversity includes 88 species of birds, 20 species of mammals, 16 species of fishes, 15 species of reptiles, 12 species of aquatic animals, five species of frogs, 40 species of butterflies, five species of dragonflies and damselflies and 38 species of insects are also documented in the Biodiversity register (Jayakumar et al. 2018).

When the carbon-neutrality project was launched in Meenangadi way back in 2016, LDF was ruling the Meenangadi Gram Panchayat. The major phases towards the implementation of carbon–neutral Meenangadi took place when the LDF was ruling the Meenangadi Panchayat. While the target was to make Meenangadi carbon–neutral by 2020, the Covid 19 pandemic and local government elections in 2020 completely disrupted the plan. In the recent local government elections, LDF lost power to UDF. One of the major reasons for the setback was the failure of LDF to internalize the carbon-neutrality project. The in-party fighting's and factions and adversarial politics led to the

---

[7] Meenangadi Gram Panchayat prepared the first Biodiversity register on 31–03–2013, it was revised in 2016 and on 8–03–2018 it was further updated. The first author of this paper conducted an interview with Shri.O V Pavithran Master, Convenor of Meenangadi Biodiversity Management Committee (BMC) on March 5 and 6, 2021. The above-mentioned details were provided by Shri. Pavithran to the author. The first author conducted an interview with O V Pavithran on October 18, 2022 and Shri Pavithran said that climate literacy programme in coordination with Gram Sabha is being held in Meenangadi. The first author also interviewed the newly elected Gram Panchayat President K E Vinayan on October 18, 2022. Shri Vinayan said that due to the Covid 19 outbreak, the activities under carbon-neutrality took a backseat but added that 90 per cent of the households in Meenangadi now uses LED bulbs. There are also discussions to scale up the carbon-mitigation strategies implemented at the Panachayat level to.

failure of LDF in the recent polls conducted to local governments. Leadership and regime change will have a huge impact on the future of carbon-neutrality in Meenangadi. It is to be noted that the carbon–neutral project was launched in 2016 then the LDF led by Ms. Vijayan was in power. But before the completion, the local government elections came in and the power equations changed. Ms. Vijayan who was a popular face among the residents of the panchayat was fielded as a candidate in the Block/ Intermediate Panchayat this time instead of Gram Panchayat. Though LDF had a dominance over Gram Panchayats, the third tier of the local governance, its representation and presence in the Intermediate/Block Panchayat and District Panchayat had been relatively poor. So, in the latest local government elections, the CPI(M) laid emphasis and attention on fielding candidates that have good prospects to get elected to Block and District Panchayats. Since Ms. Vijayan was a popular president adored by the people, the party realised that the popularity of the candidate at grassroot level can help her to win even if she is fielded at block level. Of course, the strategy worked well and Ms. Vijayan got elected to the Block Panchayat.[8] Had Ms. Vijayan been fielded in Gram Panchayat, perhaps LDF would have swept the local polls and returned to power and it would have in a way speeded the process to make Meenangadi carbon–neutral to an extent. Though UDF and LDF government have incorporated carbon-neutrality in their election manifestos, one needs to see whether the UDF government will be interested in taking up the carbon–neutral project way forward. Following the interaction with newly elected Panchayat president K E Vinayan,[9] it was revealed that they are clueless as there is a continuity deficit along with weak institutional memory. Weak institutional memory is a great deficit of the local governments in Kerala as a result there is no culture of continuity. The newly formed UDF government and elected members didn't get much time to study deeply about the carbon–neutral project and hence they are in a situation in which have to continue from the scratch.

17.3 Relevance of Carbon-Neutrality in Meenangadi and its Role in Saving Western Ghats

In 1957 Kerala had nearly 36 per cent of the land area was covered by forest but by 1990 it was drastically reduced to 12 per cent (Pillai, 2018). As per a study conducted by Indian Institute of Science, Kerala lost 906,400 hectares

---

[8] The following inputs were furnished by Dr. Jose George, who is a CPI (M) activist based in Manathavady, Wayanad and a former Professor of Civics and Politics with the University of Bombay. The author of this paper interacted with Dr. George on February 15, 2021.

[9] The first author interviewed the newly elected Gram Panchayat President K E Vinayan on October 18, 2022. Shri Vinayan said that due to the Covid 19 outbreak, the activities under carbon-neutrality took a backseat but added that 90 per cent of the households in Meenangadi now uses LED bulbs. There are also discussions to scale up the carbon-mitigation strategies implemented at the Panachayat level to block level.

(9064.4 sq.km) of the forest land between 1973 and 2016 (Ramakrishnan and Ramachandra 2016). States along Western Ghats including Kerala conducted large-scale mining, quarrying, construction of buildings and high rises along the hills and unsustainable farming and it made one of India's oldest hill ranges vulnerable to natural disasters. The massive floods of 2018 and 2019 in Kerala and the recurrent landslides in environmentally fragile areas of Wayanad and Meenangadi are due to the vast exploitation of Western Ghats. Large-scale urbanization led to a loss of forest cover and it led to scanty rainfall in places including Wayanad and Meenangadi. Bouts of drought has been plaguing Meenangadi Panchayat for years and it was a huge blow to the farmers and rural households in the Panchayat. Wayanad is in the news for agrarian crisis, farmer suicides (George and Krishnadas 2006; Münster 2012) floods and recurrent landslides. Meenangadi, which is in Wayanad is also going through the same crisis.

Though there have been legislations like Forest Act 2006, Environmental Impact Assessment Regulations 2006 and National Green Tribunal Act to scrutinize the impact of projects on environment before granting clearance and a tribunal to expedite and dispose of cases in connection with environmental issues, they have been effective only to some extent. Then came the Western Ghats Ecology Panel (Gadgil Committee led by Madhav Gadgil) in 2010 and many environmentalists and activists lauded it and hoped it would bring to an end to the exploitation of the fragile Western Ghats. Finally Gadgil report was rejected by the Union Ministry of Environment in 2011.

In its report submitted to the Union government on August 31, 2011, the Committee defined "*Western Ghats as the mountainous region encompassing 1.29 lakh square km, stretching 1490 kms from Tapi Valley in the north to Kanyakumari in south, with a maximum width of 210 km in Tamil Nadu and minimum of 48 km in Maharashtra*" (Gadgil 2011). The Committee proposed designating the entire zone as an ecologically sensitive area (ESA) and within ESA smaller grids were marked as ecologically sensitive zones (ESZ) I II and III. The Committee proposed regulations for these in the light of existing conditions and the threats that are likely to cause if exploitation continues unbated. The Committee also recommended stringent measures to prohibit the degradation of land and proposed ways to reclaim the Western Ghats (Nair and Moolakattu 2017).

However, political lobbying and strong opposition from all six states forced the Union government to reject it. Even more shameful was the resistance from Kerala as the ecclesiastical community[10] and politicians jointly came together to oppose Gadgil report on Western Ghats. The UDF and the LDF were hell bent on overthrowing the report. The unholy nexus between the

---

[10] The Syro-Malabar Catholic Church, the state's largest church which accounts for the maximum number of farmers living in villages located in Western Ghats strongly opposed the report. The Syro-Malabar church even suggested an international conspiracy behind the Gadgil report.

ecclesiastical community, political leadership as well as poachers and plunderers, mining mafia and quarry operators protested against the declaration that 37 per cent of the Western Ghats as "Ecologically Fragile" as it would put an end to all commercial activities in the said region. The Western Ghats region of Kerala especially Wayanad and Meenangadi has been in the news for recurring landslides and floods and human-animal conflicts (Viswanathan 2019). To save Western Ghats and its rich biodiversity carbon-neutrality is the way and to a climate-resilient approach is the means to achieve it. A radical shift from growth-centred paradigm to a Gandhi-Kumarappa sustainable framework is the perfect solution to this.

17.4 Mitigation Strategies to Achieve Carbon-Neutrality in Meenangadi

The carbon–neutral Meenangadi project envisions the reduction of human-induced carbon emission through people's lifestyle and sustainable development in this region. The plan is to achieve it by reducing carbon emissions in Meenangadi. The Biodiversity Management Committee (BMC) in Meenangadi has played a significant role in expanding the green cover in the village.

Tree Banking Scheme

Trees play a pivotal role in controlling carbon emissions and helps to sequestrate excess carbon. As expanding the green cover is one of the targets to be accomplished to become carbon–neutral, Meenangadi launched a Tree Banking Scheme. Under Tree Banking Scheme, anyone who owns a piece of land in Meenangadi are eligible beneficiaries for a loan of Rs. 50 (US $0.69) per year annually on the basis of its future market value. The loan is provided by Meenangadi Service Cooperative Bank on the guarantee of the Panchayat. As part of the carbon–neutral Meenangadi project, the state government has deposited a corpus fund of Rs. 10 crore (US $1.38 million) for the Tree Banking Scheme at the Meenangadi Service Cooperative Bank and this amount is used to disburse the loans. Under this scheme, each tree sapling can be pledged for Rs. 50 (US $0.69) per year for a period of 10 years. So, if a person pledges 100 trees in his or her land, the bank will pay them Rs. 5000 (US $68.90) per year for 10 years in the form of a loan.

The sapling between 3 and 5 years of age is distributed to beneficiaries under this scheme. The Panchayat has listed 34 species of trees that people can plant on their land and it includes mango (Mangifera indica L), jackfruit (*Artocarpus heterophyllus*), neem (*Azadirachta Indica*) lemon (*Citrus limon*) cinnamon tree (*Cinnamomum Verum*) arecanut (*Areca Catechu*) cashew (Anacardium occidentale) and banyan tree (*Ficus Benghalensis*) are some among them. As part of the MGNREGS, the Panchayat has distributed nearly three lakh saplings at free of cost through its MGNREGS nursery. The geo-tagging feature helps the Panchayat officials to monitor the growth of the sapling at six-month and one-year intervals. The saplings grown by the beneficiaries on their own are also taken into account under the scheme. The primary aim behind this scheme is to financially incentivize people, including farmers to grow and preserve trees. This way, farmers or even residents will be discouraged from axing trees, a practice that has been widespread in recent years and considered fatal to the biodiversity of Wayanad. The beneficiary has to start an account in the cooperative bank. Each beneficiary is given a unique registration number and each tree is geotagged and details digitally recorded.

Trees in Homesteads

Plants and trees serve as a natural source of removing carbon from atmosphere. The Panchayat carried a tree survey to estimate total carbon sequestration in Meenangadi. As part of the tree survey, the total number of trees, including their size, age, and species in each household in the Panchayat. Following the completion of the survey, a database of trees in each ward were constructed. As part of this exercise, average girth, and height of each species of trees and wood density of each tree species were collected and recorded. So far, 7526 trees in homesteads have been geotagged in 157 households in Choothuppara and Appad wards with the help of trained *Kudumbashree* workers and volunteers. As part of the survey, carbon sequestration levels were estimated for each tree species based on their age. Average carbon sequestration potential per homestead was calculated based on data collected from 3746 tree samples. It was extrapolated to the total number of households to find total carbon sequestrated in homestead trees in the Meenangadi. It was found that it accounts for 34 per cent of the total carbon stock of Meenangadi. Though the Panchayat has planted more than 3000 tree saplings in and around Meenangadi so far, it would take three to four years for the tree to start the carbon sequestration process. Therefore, the conservation of existing trees in homesteads is very important to attain carbon-neutrality. The trees that existed even before the carbon–neutral was launched are covered under the Tree Mortgage Scheme/Project. Under the Tree Mortgage Scheme, the trees that existed in homestead even before the launch of carbon–neutral project would be provided with interest-free loans considering trees as a security.

Organic Farming

Vegetable farming has also received a big push with around 70 acres of land now being used for cultivation. Meenangadi has today become self-sufficient in vegetable cultivation. One of the "Attakolli" Jaiva Park (organic farming land). Around Rs. 47 lakh (US $64,765) was spent on organic farming initiatives attempting to reverse the existing trends of using harmful pesticides and imports of vegetables and fruits. Organic farming was promoted not only to cultivate clean food but to also make the region self-sufficient in food produce.

Apart from planting trees, managing waste, preserving water resources, the carbon–neutral project has adopted various innovative methods to bring down the carbon levels. Around Rs. 80 lakh (US $110,238) was spent to set up an electric crematorium in the area. This has drastically brought down the use of wood for cremations. Plastic bags in the Panchayat markets have been replaced with handloom bags. Free distribution of bicycles to school students and encouraging people to use non-polluting vehicles are part of the project.

17.5 MGNREGS A Climate Resilient Approach Helping the Cause of Carbon-Neutrality in Meenangadi

The MGNREGS launched in 2006 in India is one of the world's largest social security programme. MGNREGS aims at enhancing the livelihood security of people in rural areas of India by guaranteeing a 100 days of wage-employment in a financial year to a rural household whose adult members, both men and women volunteer to do unskilled work. MGNREGS works are focused on the creation of durable assets to augment land and water resources, improve rural connectivity, and strengthen livelihood resources of the poor. It largely focuses on land and land development, drought proofing, renovation of traditional water bodies including water harvesting, groundwater recharge, and conservation, soil conservation and protection and so on. Thus, MGNREGS works have the potential to generate environmental benefits. Therefore, it is important to assess the "carbon-sequestration" potential, as a co-benefit, from MGNREGS (Ravindranath and Murthy 2018). In the case of Meenangadi, large-scale afforestation works are being carried out under drought proofing and afforestation under MGNREGS.

For instance, under Tree Banking Scheme, as part of MGNREGS, the Panchayat has distributed nearly three lakh saplings at free of cost to residents especially farmers through its MGNREGS nursery. As trees and plants have high carbon sequestration, this programme is proving beneficial to the environment and ecology of Meenangadi. Similarly, under MGNREGS, bamboo saplings are planted on the river sides, road sides and in schools as bamboos can capture and sequester significant amount of atmospheric carbon and thus help in mitigating climate change. Researches have shown that bamboo serves as a carbon sink (Giri et al. 2015). So sustainable utilization, conservation, and proper management of bamboo trees under MGNREGS in Meenangadi can make it an effective carbon sink besides fulfilling the diverse needs of rural households.

Social protection programmes like MGNREGS can effectively integrate climate risk management and thus help rural households to generate and invest in climate-resilient livelihood strategies. In the case of Meenangadi, as part of afforestation works, bamboo cultivation has got a boost. On the one hand bamboos have the potential to improve socio-economic status. The people in Meenangadi and Wayanad in general use bamboos to make handicrafts, furniture and other utility products and it has become a major livelihood for women in Wayanad. It has given the villagers financial and economical empowerment as in the case of MGNREGS. Apart from socio-economic potential offered, bamboo has a lot of potential when it comes to going carbon–neutral as bamboo has high carbon sequestration capacity. This potential of bamboo is being utilized by MGNREGS in Meenangadi by promoting bamboo cultivation. So, programmes like MGNREGS not only help poor households with poverty and marginalization but also helps to adopt absorptive, adaptive and transformative resilience to absorb the effects of climate risks, adapt to climate change impacts and transform their capacities and strategies to address growing climate changes (Kaur et al. 2019).

In addition, soil conservation, fodder development, afforestation, and drought proofing works have sequestered carbon, thus mitigating climate change (Esteves et al. 2013). Similarly, activities under MGNREGS in Meenangadi have been able to achieve this potential. The much-lauded tree banking scheme to increase green cover in Meenangadi, recreating forest and preserving groves, organic farming, bamboo park in schools and bamboo planting on the banks of rivers and all these activities are implemented under MGNREGS. In Meenangadi under MGNREGS, a total of 862 ponds were dug by MGNREGS workers and out of it 462 ponds are used for pisciculture. In drought proofing alone, 824 works have been carried out, 335 water bodies in the Meenangadi has been revived in the last six years and 708 water bodies have been preserved and conserved so far and all these activities were carried out MGNREGS. So, Meenangadi has too proven that MGNREGS has the huge potential to reduce vulnerability to climate risks. In the case of Meenangadi a huge investment is made for rural projects under MGNREGS. Wayanad where Meenangadi is located has showcased good performance when it comes to the implementation of MGNREGS since the beginning. The scheme was launched in Kerala on a pilot basis in two districts, Palakkad and Wayanad, in 2006 and at that time Wayanad score six points more than Palakkad (Chathukulam and Gireesan 2008). The average expenditure incurred per year for the last six years under MGNREGS in Meenangadi amounts to Rs 808.38 lakhs (US $1.11 million). The average person-days generated per year is 2,23,839. Details of activities undertaken under MGNREGS in Meenangadi (2015–2021) given in Table 17.1.

Under MGNREGS, drought proofing works such as afforestation and reforestation, and horticulture development are the activities carried out for improving the vegetation cover and biomass availability in the villages. Meenangadi has been actively involved afforestation and other drought proofing works. Programmes like bamboo park in schools and "*punyavanam programme*" to protect sacred groves in Meenangadi are a few among them. Meenangadi has won the 2018 Mahatma Award instituted by the Kerala government for the efficiency in implementing MGNREGS.

Table 17.1 Details of activities undertaken under MGNREGS in Meenangadi (2015–2021)

Sl. No	Name of the Activity/Work	2015–16	2016–17	2017–18	2018–19	2019–20	2020–21	Total
1	Rural Infrastructure	0	0	0	4	4	5	13
2	Drought Proofing	99	81	93	152	181	218	824
3	Flood Control	20	0	0	.0	0	0	20
4	Land Development	29	47	8	12	25	57	178
5	Playground	2	2	1	1	1	0	7
6	Renovating Water Bodies	50	68	42	30	68	77	335
7	Rural Connectivity	12	36	57	137	207	211	660
8	Rural Sanitation	0	2	37	66	46	45	196
9	Water Conservation	0	85	77	155	180	211	708
10	Works on Individual Land	547	447	337	485	427	295	2538
	Total	759	768	652	1042	1139	1119	5479

Source Complied and Computed from the MIS, MGNREGS, Ministry of Rural Development (MoRD), Government of India

17.6 Activities Undertaken in Meenangadi to Expand Green Cover

Bamboo Garden and Park

Schools in Meenangadi have also embraced carbon–neutral project. A notable one among them is a school named Navodaya Adivasi Aided Upper Primary School in Meenangadi which houses 50 rare species of bamboos, eight of them endemic and six of them under the endangered category. The school has a four-acre bamboo garden at present. This school has bagged the prestigious *Vanamitra Award* instituted by Forest Department 2019–2020. Meanwhile, it is interesting to note that the bamboo garden was started way back in 2002 by planting nearly 29 species. In 2016, when Meenangadi decided to go carbon–neutral, the Panchayat started to provide all support to the project through programmes such as Green India Mission under MGNREGS. The Panchayat has spent Rs. 15 lakh (US $20,669) for this bamboo garden. The school also possesses a medicinal plant garden with 30 species of plants, an arboretum with 84 tree species, and a collection of 26 wild edible fruit conservatory.

Preserving Sacred Groves in Meenangadi

Meenangadi has a total of 54 sacred groves[11] and the locals have realized the potential of these groves and have started preserving them. The transformation of nearly 34 acres out of the total 38 acres of the Manikavu Temple including a sacred grove attached to the temple into a green paradise. Manikavu temple is situated in Meenangadi, Sulthan Bathery Taluk of Wayanad district.

Back in 2012, much before the launch of Meenangadi carbon–neutral project, nearly 34 acres of land were lying abandoned for a long period of time and as years went by nearly it was filled with invasive plants. The thoughtful intervention of Meenangadi Panchayat and MSSRF helped to transform nearly 44 acres of abandoned area in the temple premises into a beautiful garden with endemic and medicinal plants. It was carried out under the "Punyavanam Programme" (and in the first phase the programme even got the funding support from Sir Dorabji Tata Trust, SDTT, Mumbai). The governing body of the temple, the Panchayat and MSSRF actively participated in the programme. With the involvement of the Panchayat, MGNREGS workers since 2016 have played a significant role in turning the Manikavu temple and the adjacent sacred grove with medicinal plants and 115 trees. Today it has turned into an ecosystem with deep flora and fauna and the members of the BMC have informed us that two elephants occasionally reside in this recreated forest.[12]

Setting up of Biodiversity Management Committee

As per Kerala Biological Diversity Rules 2008, Section 22 Sub Section (4) the Chairperson of the BMC shall be the chairperson of the local self-government and the secretary of the local self-government shall be the member secretary of the BMC, who shall maintain the records. Besides these six people are nominated as members of the local self-government to the committee of which two should be women (33 per cent) and one member (18 per cent) should belong to SC/ST categories (Kerala Biological Diversity Rules, 2008). All the six members nominated should be permanent residents of the panchayat jurisdiction and their names should be in the voters list.

As per the requirements, a BMC has been formed in the Panchayat since 2008. They have been actively involved in the making of Biodiversity Register for the Panchayat. As per the Biodiversity Register prepared by the Panchayat,

---

[11] Sacred groves, also called Kavu in Malayalam language, are rich abodes of biodiversity. Establishment of sacred groves was also seen as traditional efforts by the villages to conserve biodiversity and water resources. These groves had perennial water supply and thus supported human habitation. They also served as places for worshipping nature.

[12] The first author of this paper conducted an interview with Shri.O V Pavithran Master, Convenor of Meenangadi Biodiversity Management Committee and Shri. V Suresh, Former Chairman, Welfare Standing Committee, Meenangadi Gram Panchayat on March 5 and March 6, 2021 and October 16, 2021. The above mentioned details were provided by Pavithran and Suresh to the author.

the major ecosystem habitats prevalent in the area are wetlands, forests, plantations and homesteads. A total of 240 species of plants and trees have been identified from the forests and plantations of the Panchayat out of which 140 species were herbs, 55 species were trees. A total of 12 species of wild plants, 14 species of water plants, five species of wild relatives of crops and 14 species of wild ornamental plants were also identified (Jayakumar et al. 2018). The faunal diversity includes 88 species of birds, 20 species of mammals, 16 species of fishes, 15 species of reptiles, 12 species of aquatic animals and five species of frogs. Around 40 species of butterflies, five species of dragonflies and damselflies and 38 other species of insects were also documented in the biodiversity register (Jayakumar et al. 2018). In 2017–2018, the BMC, Meenangadi won the state biodiversity award for tree banking, conservation of ponds, promotion of organic farming implemented as part of carbon–neutral Meenangadi. The BMC also won the special jury mention in 2018 India Biodiversity Awards.

17.7 Conclusion

The concept of carbon-neutrality is not new to India. Attempts are underway in various places in India including, the Union Territory of Ladakh (Districts of Leh & Kargil) and Manjuli district in Assam to convert them into carbon–neutral regions. While it is interesting to notice that India is taking baby steps towards achieving carbon-neutrality, it is to be noted that the carbon-neutrality tag is being earned by placing it under a growth-centred trajectory. The approach is less like doing a carbon-trade within the four walls of the country. While India has not yet established a carbon market or carbon pricing policy, it is directly and indirectly engaged in carbon trade within their territories as in the case of Meenangadi. For instance, transport amount to more than 50 per cent of carbon emissions in Meenangadi. Instead of replacing it with eco-friendly transport systems, Meenangadi is planting trees on roadsides. In such cases, carbon-neutrality is achieved by balancing it within the growth-centred development. The Gandhi-Kumarappa framework which lays stress on *economy of permanence*, the foundation of which is total non-violence in production and consumption is missing not only in Meenangadi but also in many parts of the world. During the field visit to Meenangadi, it was evident that the neither the local community nor the politicians, panchayat functionaries, civil society organizations were aware of the Gandhi-Kumarappa perspectives on preserving the environment. They are not ready to move away from the growth-centric development and what they prefer is to place carbon-neutrality within growth-centric or development-centric paradigm.

The concept of carbon-neutrality is not in favour of growth-centred development paradigm. Since Meenangadi is trying to balance between growth-centred development and carbon-neutrality, it is highly doubtful that even if they earn the tag of carbon-neutrality by 2025 whether it will remain sustainable in a long run. Though neither the local community nor the politicians, Panchayat functionaries, civil society organizations were aware of the Gandhi-Kumarappa perspectives on preserving the environment and solidarity economy, they are familiar with environment and climate-related terminologies like carbon sequestration, carbon-positive, carbon-negative and greenhouse gas emissions and these words became the local political phraseology. Carbon-neutrality has become a topic of political agenda in Meenangadi and Kerala. The election manifestos of LDF and UDF mention about carbon–neutral Meenangadi. This itself denotes that carbon-neutrality is no longer a subject limited to environmentalists but also to the political machinery and thereby to the rest of the society (LDF and UDF Manifesto 2020). Meanwhile, attempts are being made to scale up and customize carbon-neutrality to Wayanad and rest of Kerala. The Kerala Institute of Local Administration (KILA) has developed a training module to introduce and familiarize people and functionaries of local governments on climate change resilient and carbon–neutral regime. The adversarial politics is also hampering the efforts of Meenangadi Gram Panchayat to earn the carbon–neutral tag. A consensus-based democratic approach along with an economy of permanence as advocated by Gandhi-Kumarappa is what Meenangadi and the rest of the world need if they want to go carbon–neutral in its true spirit.

Acknowledgements We are extremely grateful to the former and present elected functionaries of Meenangadi Gram Panchayat. We are also thankful to all those who cooperated with us for furnishing valuable data. We acknowledge the support of faculty and staff, Centre for Rural Management (CRM), Kottayam, Kerala, India for providing support and logistics.

Declaration There is no conflict of interest.

References

Alexander, Roy., Hope, Max., and Degg, Martin. (2007). 'Mainstreaming Sustainable Development—A Case Study: Ashton Hayes is going Carbon Neutral', *Local Economy*, 22(1): 62–74.

Ameerudheen, T. A. (2016, April 6). 'Issues of Land and Landlessness Echo in Adivasi Hamlets in Kerala as Elections Draw Near', *Scroll.in*. Available at: https://scroll.in/article/806211/issues-of-land-and-landlessness-echo-in-adivasi-hamlets-in-kerala-as-elections-draw-near Accessed at September 17, 2021.

Anand, Annu. (2017, July 5). 'Carbon Neutral Village', *The Statesman*,. Available at: https://www.thestatesman.com/features/carbon-neutral-village-1499295470.html. Accessed at September 26, 2021.

Bachan, Amitha H. K. (2019). Muthanga Tribal Rehabilitation Conflict, *Land Conflict Watch*

Bichard, Erik. (2013). *The Coming of Age of the Green Community: My Neighbourhood, My Planet*. Routledge.

Bohm, Steffen., Misoczky Ceci Maria., and Moog, Sandra. (2012). Greening Capitalism? A Marxist Critique of Carbon Markets. *Organization Studies*, 1–22.
Brown, Cameron. (2018, October 4). What's the Difference Between Carbon Neutral, Zero Carbon, and Negative Emissions? *Clentech Rising*.
Cancun agreement. (2010). *The Cancun Agreements: Outcome of the work of the Ad Hoc Working Group on Long-term Cooperative Action under the Convention*, United Nations.
Census of India. (2011). Office of the Registrar General and Census Commissioner India.
Chathukulam Jos., and John M.S. (2006). Issues in Tribal Development the Recent Experiences of Kerala, In Chandra Rath Govinda (Eds.), *Tribal Development in India*. SAGE Publications, New Delhi.
Chathukulam, Jos., and K. Gireesan. (2008). NREGS in Kerala: Evaluation of Systems and Processes. Technical Report, Centre for Rural Management (CRM), Kottayam, Kerala, India.
Chathukulam, Jos., Reddy Gopinath M., and Trinadharao, Palla. (2013). Formulation and Implementation of Tribal Sub-Plan (TSP) in Kerala, Centre for Economic and Social Studies, Hyderabad, India.
Clark, B., and Foster, J. (2006). The Environmental Conditions of the Working Class: An Introduction to Selections from Frederick Engels's "The Condition of the Working Class in England in 1844", *Organization & Environment*, 19(3): 375–388.
Department of Environment and Climate Change. (2014). State Action Plan on Climate Change. Kerala: Department of Environment and Climate Change, Government of Kerala
Department of Town and Country Planning. (2014). Integrated District Development Plan—Wayanad. Government of Kerala.
Energy and Climate Intelligence Unit. (2019). Countdown to Zero: Plotting Progress Towards Delivering Net Zero Emissions by 2050. ECIU, United Kingdom.
Engels, Friedrich. (1844). The Condition of the Working Class in England. Germany: Otto Wigand, Leipzig.
Esteves, T., Rao, K. V., Sinha, B., Roy, S. S., Rao, B., Jha, S., Singh, A. B., Vishal, P., Nitasha, S., Rao, S., and I. K, M. (2013). Agricultural and Livelihood Vulnerability Reduction through the MGNREGA. *Review of Rural Affairs, Economic and Political Weekly*, 48(52).
Gadgil, M. (2011). Report of the Western Ghats Ecology Expert Panel, Part I. New Delhi: Ministry of Environment and Forests, Government of India.
Gandhi, M. K. (1909). Hind Swaraj or Indian Home Rule. Ahmedabad: Navajivan Publishing House, 1996.
George, Jose, and Krishnaprasad, P. (2006). Agrarian Distress and Farmers' Suicides in the Tribal District of Wayanad. *Social Scientist*. 34(7–8): 70–85.
Giri, K., Mishra, G., Kumar, R., and Pandey, S. (2015). Role of Bamboo Forests in Carbon Sequestration and Climate Change Mitigation. *ResearchGate*.
Human Development Report. (2020). *The Next Frontier: Human Development and Anthropocene*, UNDP: New York.
Human Development Report. (2021). *Uncertain Times, Uncertain Lives: Shaping our Future in a Transforming World*, UNDP: New York.
IPCC. (2013). *Climate Change 2013: The Physical Science Basis. Contribution of Working Group I to the Fifth Assessment Report of the Intergovernmental Panel on*

Climate Change, Cambridge University Press, United Kingdom. https://doi.org/10.1017/CBO9781107415324

IPCC. (2021). Intergovernmental Panel on Climate Change (IPCC).

Jayakumar, C., Ushakumari S., Nair, S., Sridhar, R., Raju, S., Kumar, Dileep., and Paliath Nikhilesh. (2018). Carbon Neutral Meenangadi—Assessment and Recommendations, Thanal: Kerala.

Kaur, N., Agrawal, A., Steinbach, D., Panjiyar, A., Saigal, S,. Manuel,C., Barnwal, A., Shakya, C., Norton, A., Kumar, N., Soanes, M., and Venkataramani, V. (2019). Building Resilience to Climate Change through Social Protection: Lessons from MGNREGS, India. IIED Working Paper, IIED, London.

Kumar, G. P., Ashutosh A. M., Gupta, S., and Singh, B. S. (2009). Carbon Sequestration with Special Reference to Agroforestry in Cold Deserts of Ladakh. *Current Science*, 97(7): 1063–1068.

Kumarappa, J. C. (1946). Economy of Permanence. Wardha, C P; All India Village Industries Association.

Kyoto. (1997). *Kyoto Protocol to the United Nations Framework Convention on Climate Change*, United Nations.

Left Democratic Front (LDF)—Communist Party of India, (Marxist). (2020). *LDF Election Manifesto*. Thiruvananthapuram, Kerala.

Lindley, Mark. (2007). J C Kumarappa: Mahatma Gandhi's Economist, Mumbai: Popular Prakashan.

Morup, Tashi and Joshi, Sopan (2003, November 30). Ladakh on the Move. *Down To Earth*.

Münster, D. (2012). Farmers' Suicides and the State in India: Conceptual and Ethnographic Notes from Wayanad, Kerala. *Contributions to Indian Sociology*, 46(1–2): 181–208.

Nair, V. N. (2020). Solidarity Economics and Gandhian Economics: Can they Supplement Each Other? *Gandhi Marg*, 42(1&2): 83–106.

Nair, V. N., and Moolakattu, S. J. (2017). The Western Ghats Imbroglio in Kerala—A Political Economy Perspective. *Economic and Political Weekly*, 52(34).

Nair, V. N., and Moolakattu S. J. (2018). Revisiting the Discourse on Protection of Western Ghats from a Gandhi-Kumarappa Perspective. *Gandhi Marg*, 39(4): 311–330.

Nandi, Jayshree. (2019, April 1). A Village with Carbon-positive Tag. *Hindustan Times*

Nandi, Jayshree. (2020, August 16). Ladakh, Leh, Kargil to be India's first carbon neutral region: PM Modi. *Hindustan Times*

Newell, P. (2012). The Political Economy of Carbon Markets: The CDM and other stories. *Climate Policy*, 12: 135–139.

Newell, P., and Paterson, M. (2010). Climate Capitalism: Global Warming and the Transformation of the Global Economy. Cambridge, UK: Cambridge University Press.

Paris Agreement. (2015, Dec 13), in United Nations Framework Convention on Climate Change (UNFCCC), COP Report No. 21, Addenum, at 21, U.N. Doc. FCCC/CP/2015/10/Add, 1 (Jan. 29, 2016) [hereinafter Paris Agreement].

Raghunath, Arjun. (2021, Oct 26). 'Kerala Panchayat Offers Incentives for Planting Trees to become Carbon Neutral', Deccan Herald. Available at https://www.deccanherald.com/national/south/kerala-panchayat-offersincentives-for-planting-trees-to-become-carbon-neutral-1044421.html. Accessed on December 6, 2022.

Ramakrishnan, R., and Ramachandra, T. V (2016). Four Decades of Forest Loss: Drought in Kerala, Indian Institute of Science (IISc), Bengaluru, India.

Ravindranath, H. N., and Murthy K, Indhu. (2018). Estimation of Carbon Sequestration under MGNREGA: Achievement and Potential in India. Indian Institute of Science, Bengaluru, India.

Ray, Sunil. (2020). Cohesive Development as an Alternative Development Paradigm, In Ray S., Choudhary, N., & Kumar, R. K. (Eds.), *Theorizing Cohesive Development: An Alternative Paradigm* (1st ed.). Routledge.

Staff Reporter. (2021, February 11). ₹7,000-cr. special package for Wayanad. *The Hindu*.

Stockholm. (1972). *Report of the United Nations Conference on the Human Environment*, United Nations, New York.

Tiwari R. R. (2019). Gandhi as an Environmentalist. *The Indian Journal of Medical Research*, 149(Suppl), S141–S143.

United Democratic Front (UDF)—Indian National Congress (INC). (2020). *UDF Election Manifesto*, Thiruvananthapuram, Kerala.

Varma, Vishnu. (2020, October 22). 'In Kerala village, Farmers can Mortgage Their Trees for Interest-Free Bank Loans'. *The Indian Express*, p. 3.

Vishwanathan, Manoj. (2019, April 18). 'Save Western Ghats now, or it'll be too late'. *The New Indian Express*, p. 5.

CHAPTER 18

Nanominerals: A Climate Smart Tool in Livestock Feeding—A Review

Partha Sarathi Swain, S. B. N. Rao, D. Rajendran, Sonali Prusty, and George Dominic

18.1 INTRODUCTION

Livestock sector includes all the animals reared for human benefits and consumption. Livestock husbandry is an integral part of the human civilisation from ancient times (FAO 2007). Humans domesticated animals for more than 10,000 years (Dib 2010). Essence of livestock for quality human life is well realised and hence human race is more focussed to explore more and more profits from them. Animals provide diverse number of products utilised for human life, viz. wool, skin, meat, milk, eggs, medicines and many more. It plays a significant role in national and international economy and in socio-economic development. Particularly, in developing countries like India, animal husbandry has a pivotal role in rural economy which provides family incomes and employment, which is of utmost important particularly to the landless labourers, small and marginal farmers and women (FAO 2009).

Even though livestock aid to a quality life, the huge population is also a matter of concern for the human race, for instance, competition for grains and land, production of greenhouse gases leading to global warming. Cows and buffaloes are branded as one of the primary contributors of greenhouse gases

P. S. Swain (✉) · S. B. N. Rao · D. Rajendran · G. Dominic
Animal Nutrition Division, ICAR- National Institute of Animal Nutrition and Physiology, Bangalore, India
e-mail: parthavet@yahoo.com

P. S. Swain · S. Prusty · G. Dominic
Animal Nutrition Division, ICAR- National Dairy Research Institute, Karnal, India

© The Author(s), under exclusive license to Springer Nature Singapore Pte Ltd. 2023
S. Nautiyal et al. (eds.), *Palgrave Handbook of Socio-ecological Resilience in the Face of Climate Change*,
https://doi.org/10.1007/978-981-99-2206-2_18

in India. These animals play a major role in the emission of methane, a major offender towards global warming than the usual suspect carbon dioxide. Livestock releases a huge amount of methane during fibre fermentation through belching and flatulence. India's greenhouse gas emissions from agriculture sector, accounts for about 18% of the total between 1994 and 2007 (Swain et al. 2016a). Trough reports are scanty, in this chapter we have tried to put light on the possible effects of nanominerals in preventing environment pollution when used as a feed supplement.

18.2 Mineral Nutrition in Livestock Production

Minerals are the inorganic component of the diet which are soluble in acid, and required in the small quantities, but are indispensible for normal health production and reproduction. Mineral nutrition, at optimum dose, is critical for normal growth, reproduction, DNA synthesis, cell division, gene expression, photo-chemical processes of vision, wound healing, and ossification, augmenting the immune system of the body through energy production, protein synthesis and protection of membranes from bacterial endotoxins and lymphocyte replication and antibody production (Swain et al. 2016b). Trace minerals are supplemented in the poultry ration to obtain better responses (Aksu et al. 2012) which is also responsible for environmental issues in intensive poultry farming areas (Świątkiewicz et al. 2014). By using poultry manure, Mohanna and Nys (1998) found that the Zn content in soil was found in excess by 660% of plant Zn requirements, predisposing to phytotoxicity. Świątkiewicz et al. (2014) reported that Fe, Cu and Mn are always found in excess in plants as compared to their requirements. The bioavailability of trace elements present in low from the natural feed resources present in agro-ecosystems which fails to meet the requirement of animals and birds (Świątkiewicz et al. 2014). Phosphorus in plant origin feed ingredients are mostly not utilised by animals and birds due to lack of phytase, which accounts about 60–80 per cent of total Phosphorus (Abd El-Hack et al. 2018). So, dietary supplementations are practised to maintain the optimum health and production. Absorption of minerals is affected by many factors, for instance, net Zn absorption from daily oral administration is 12% in mature cows, 20% in 5 to 12 months calves and 55% in week-old calves (Miller and Cragle 1965). Thus, the remaining unabsorbed portion is excreted to the environment, which may lead to environment pollution (He et al. 2005; Sobhi et al. 2020; Yenice et al. 2015). Being low in bioavailability, inorganic salts are usually supplemented at higher doses to maintain the optimum productivity in animals and birds, which again causes excretion of unabsorbed fractions to the environment (Patra and Lalhriatpuii 2020; Swain et al. 2021a). Considering the animal population, excreted mineral quantity may be a matter of concern. Again this may be having severe consequences in the areas of intensive livestock and poultry production areas (Świątkiewicz et al. 2014). Hence, it is very apposite to develop better bioavailable mineral sources which should be preferably economical as the same would cause less excretion to the environment. Minerals are supplemented in its as inorganic salts and also as organic chelates. Study suggests better bioavailability of organic mineral salts ac compared to their inorganic sources, but higher cost limits their use in practical animal feeding (Zhao et al. 2014). Thus there is a scope for newer alternatives and nanominerals are one of them.

18.3 Nanominerals

Several attempts have been adopted to increase the bioavailability of the minerals, and use of nanominerals is one among them. Nanominerals are specially synthesised mineral particles whose size ranges from 1 to 100 nanometer (Feng et al. 2009), and owing to their smaller size, nanominerals exhibit novel physical, chemical and biological activities (Swain et al. 2015; Wang 2000), which may aid in improved efficiency of absorption of these nanominerals in animal models (Hassan et al. 2020). The nanoparticles (NP) are stable under high temperature and pressure (Stoimenov et al. 2002). Results advocate that use of nanominerals has shown to produce better results in terms of mineral retention as well as animal health and production when compared with conventional Zn sources (Elkloub et al. 2015; Hu et al. 2012; Mishra et al. 2014; Mohapatra et al. 2014; Sahoo et al. 2014; Swain et al. 2019a, 2021b, 2022; Wang et al. 2006).

Properties

These nanomineral particles are having higher potential than their conventional sources and hence, reduce the quantity required for optimum health and production as compared to its conventional inorganic salts (Sri Sindhura et al. 2014). By virtue of their small size, it is easier to be taken up by the gastrointestinal tract, so are more effective than the larger size ZnO at lower doses (Feng et al. 2009) and less toxic as compared to the conventional counterparts (Reddy et al. 2007). Nanominerals can cross the small intestine and thus enters into different systems like blood, brain, lung, heart, kidney, spleen, liver, intestine and stomach (Hillyer and Albrecht 2001). However, Translocation of NP through the epithelium is affected by many factors viz. size, shape, zeta potential, other ligand, surface chemistry, age and species of animal, intestinal health, dose of use etc. (des Rieux et al. 2006; O'Hagan 1996). Reports suggest that translocation of 100 nm NP is 15–250 times greater than that of micro molecules (Desai et al. 1996). Reports suggest that, nanominerals interact more effectively with organic and inorganic substances in animal body which may be due to their enhanced surface area (Zaboli et al. 2013). The chemical, catalytic or biological effects of NP are heavily governed by the particle size of the mineral NP (Rosi and Mirkin 2005). CuO NPs are transported quickly into cells compared with CuSO4 and CuO microparticles, (Gao et al. 2014).

Nanominerals as Climate Smart Tool in Animal Production System

As discussed earlier, by virtue of its small size and increased surface area, NP performs better than micro particles when used as animal feeds. This NP has been effective in producing several desirable effects as far as animal production is concerned. These particles tend to be retained more in the animals than its counterpart inorganic minerals (Patra and Lalhriatpuii 2020; Swain et al. 2016b, 2019a). These NP tend to alter the rumen fermentation in the fermentation vat of ruminants, the rumen, thus affecting the rumen fermentation as well (Swain et al. 2018). However, very few studies have been done in this regard and nanotechnology is still at its infancy and more studies on these particles needed to validate their responses further.

Improved Bioavailability and Reduced Faecal Excretion

A portion of the mineral intake is retained in the body and utilised for essential biological effects, called as mineral retention. Measuring the level of minerals in different organs like bone, plasma, liver, spleen, kidney, muscle etc. are reliable indicators of its mineral status and are assessed in trace mineral relative bioavailability (RBV) experiments (Underwood and Suttle 1999). Reports suggest that mineral retention is affected by the source of mineral. For instance, nano zinc (nano-Zn) is better retained as compared to inorganic Zn in rats (Swain et al. 2019a) and goats (Swain 2017). Reports suggest that mineral NP are better absorbed from the intestine and distributed in the tissues (Anwar et al. 2019; Patra and Lalhriatpuii, 2020; Swain et al. 2019a, 2021a). Ca and P on supplemented as NP i.e. nano dicalcium phosphate economised the ration and also prevented possible environment pollution with these metals (Patra and Lalhriatpuii, 2020). Hu et al. (2012) found higher retention of Se in the whole body of broiler chickens in nano-Zn supplemented birds as compared to sodium selenite groups. Se deposition efficiency is higher in Nano Se supplementation poultry birds as compared to that of Sodium selenite, indicating efficient retention of Nano Se (Meng et al. 2019; Mohapatra et al. 2014; Zhang et al. 2001). Swain et al. (2019a) observed better retention of Zn from nano-Zn sources as compared to that of Zn from Inorganic source, thus reducing the content excreted to the environment. The extra retained Zn from nano-Zn supplemented was also reflected by better Zn content in liver and serum. This NP not only reduces excretion of minerals to the environment but also improves the animal productivity which is well reflected from improved immunity, growth, blood biochemistry (Mishra et al. 2014; Mohapatra et al. 2014; Sahoo et al. 2014; Swain 2017; Swain et al. 2019b; Swain et al. 2022) of the animal. Due to better bioavailability the quantity of Zn retained from inorganic zinc supplemented at 25 ppm was similar to that of nano-Zn supplemented at 12.5 ppm i.e. nanominerals at half the dose is equivalent to fulldose of inorganic salts (Swain et al. 2019a), which was also reported by Patra and Lalhriatpuii (2020) with calcium phosphate NP. Lotfi et al. (2014) observed a maximum increase in bone quality attributes like tibia length, tibia volume, tibia breaking strength tibia diameter and bone weight in $MnSO_4$ NP supplemented birds.

Antibiotic Properties of Nanominerals

In the current scenario world is more focussed towards a green produce rather than products burdened with drug residues. In such cases, nanominerals can be proved promising to reduce antimicrobial drug usage. Antibacterial activity means the reagent that locally kills bacteria or slows down their growth, without being toxic to surrounding tissues (Swain et al. 2016b). Mineral NP of Zn, Ag and gold can be used as antibiotics, which can prevent products free from drug residues. Silver NP is been reported to have antimicrobial effect in the intestine of animals and also acts immune-modulator (Patra and Lalhriatpuii 2020; Swain et al. 2021a), thus limits antimicrobial use. Elkloub et al. (2015) observed improved serum antioxidant status due to supplementation of Silver NP at a dose of 2–10 mg/kg in poultry which is an indicative better health condition; never the less a decline in *Escherichia coli* count was reported by them due to silver NP supplementation in birds. Kulak et al. (2018) supplemented silver NP through drinking water and reported improved immunity in birds. Similarly, nano-Zn supplementation reduced the somatic cell count of mastitis milk in cows, which is due to the antibiotic effect of nano-Zn (Rajendran 2013). Zn NP is found to be effective not only against Gram-positive and Gram-negative bacteria (Arabi et al. 2012) but also against spores which are resistant to high temperature and high pressure (Rosi and Mirkin, 2005). The antimicrobial effect of zinc oxide (ZnO) is related to their electromagnetic effects as microorganisms are negatively charged and thus get attracted towards positively charged metal oxides, resulting in oxidisation and subsequently death of microbes (Arabi et al. 2012). Hence, nano-Zn can be used as an alternative to the antimicrobial drugs to prevent ailment in animals without having any drug-related resistant issues. Enhanced surface area of metal NP can be the reason behind the antimicrobial effect as compared to large-sized particles (Adams et al. 2006). Minerals in nano form also retard the bacterial adhesion and biofilm formation (Padmavathy and Vijayaraghavan 2008) which may also be a reason behind antimicrobial effect. Padmavathy and Vijayaraghavan (2008) studied the effect of nano-Zn of particle size 20 to 40 nm, 12 nm, 45 nm against *E.coli* and observed that nano-Zn has Better bactericidal activity than bigger ZnO particles, which might be due to the abrasiveness and the surface oxygen species of ZnO NP promoting its bactericidal effects. According to Rajendran et al. (2010), ZnO NP inactivates the proteins, responsible for nutrient transport; hence decreasing the membrane permeability, leading to cellular death. Gold NP may also be helpful in this attribute. For instance, Gold NP promotes innate immunity at <1 mg/kg dose in chicken (Sembratowicz and Ognik 2018). In this context further studies should be designed as per standardised protocols to validate the effects of metal NP as an alternative to antibiotics in animals either as therapeutics or as preventive as well.

Impact of Nanominerals on Rumen Fermentation

Of the total GHGs production, 18% is contributed by N_2O and CH_4. Annual methane emission globally is 535 MT (Houghton et al. 1996). It is the second most important greenhouse gas with global warming potential twenty-five times more than CO2 (Broucek 2014). Livestock accounts for 18% of global GHG emissions (FAO 2008) that comes from enteric fermentation and manure fermentation. Methane is produced from the fermentation of carbohydrates by microorganisms in the GI tract of herbivores. Ruminants are the major contributor of methane among the livestock which is generated in their rumen by methanogenic bacteria by reducing CO_2. There is about 50 kg of methane production per year from a buffalo (NATCOM 2004). In the form of methane a large part of feed energy (7–12%) is lost. Thus, any strategy to reduce methane would be beneficial for environment as well as reduce the loss of feed energy from ruminants.

Besides improved bioavailability nanominerals have been noticed to affect rumen fermentation pattern positively and reducing methane emissions as well. In vitro study (Abd El-Galil et al. 2018) revealed that nano-cobalt supplementation at 50% and 75% level of Cobalt in the control group increased DM, hemicellulose and cellulose digestion. The effect was related to the increased population of cellulolytic bacteria and cellulolytic enzyme activity in nano-cobalt supplemented groups. Swain et al. (2018) in goats observed that nano-Zn at 50 mg/kg DM altered the rumen pH, total protozoa count, fibre degradation and soluble zinc content without affecting the individual and total VFA profile. It has been documented that, Nano Zn can be supplemented at half the dose of inorganic Zn without affecting rumen VFA profile, rumen soluble Zn content in goats (Swain 2017; Swain et al. 2018). Increased microbial population also resulted in increased ammonia and total VFA concentration. Though no direct effect on methane production was there but altered VFA proportion might affect the CH_4 production. Nano Se at 0.3 mg/kg in sheep diet reduced ruminal pH (range of 6.68–6.80) and ammonia N concentration (9.95–12.49 mg/100 mL) but increased total VFA concentration (73.63–77.72 mM). The ratio of acetate to propionate was decreased due to the increasing of propionate concentration (Shi et al. 2011). Increased propionate slows down H_2 generation that may aid in CH_4 reduction. Direct potential of ZnO NP in reducing enteric methane production have also been revealed by in vitro gas production technique. Compared to the control, different levels of ZnO NP (100, 200, 500, and 1000 μg/g of feed) reduced CH_4, CO_2 and H_2S concentration by 9.14 to 46.85%, 4.89 to 42.79% and 9.33 to 53.65%, respectively in the total gas produced and no significant change in pH, redox potential and propionate to acetate ratio were there (Sarker et al. 2018). ZnO NP at high concentration (1000 μg/g) did not affect total gas production but reduced the enteric methane concentration likely due to the reduced action of methanogenic microbes or due to the adsorption of produced methane on the surface of ZnO NP. In vitro studies Study by (Riazi et al. 2019) revealed that nano-ZnO supplementation at 20 mg of Zn/kg feed reduced methane production and protozoa number but improved ($P < 0.05$) antioxidant capacity, microbial biomass production, digestibility and truly degraded substrate compared to non-supplemented feed. There was also reduced acetate to propionate ratio. Few studies with the objective to mitigate greenhouse gases from manure, reported nano-ZnO to reduce CH_4, CO_2 and H_2S from anaerobic storage of manure (Luna-delRisco et al. 2011; Mu et al. 2011). Release Zn^{2+} from nano-ZnO was the proposed reason for its inhibitory effect on methane generation from manure.

18.4 Conclusions

Nanominerals are having many beneficial aspects in animal husbandry. This ensures better retention of minerals and thus reduces the excretion to the environment, thus aid in preventing environmental pollution. Again this also tends to alter the rumen fermentation and reduce the enteric methane emission to the environment. Though, the studies on rumen fermentation are scare, but the use of nanominerals can be proved instrumental in reducing the methane production in rumen as well as preventing the discharge of unused minerals from the animals. But, further studies can validate these effects and nay also bring new aspects of these nanoparticles with a course of time (Table 18.1).

Table 18.1 Instances of Nano minerals supplementation in animal and poultry

Nanominerals	Animal/Bird	Salient findings	References
Nano Zn	Rat	Nano Zn supplementation reduced faecal excretion of zinc to the environment. higher Zn content in liver, bone, kidney, zinc level in serum, liver, bone and kidney suggesting its better bioavailability	Swain et al. (2019a)
Nano Zn	Rat	25 mg/kg of Zn from ZnI and 12.5 mg/kg of nano-Zn showed similar biological responses in terms of immunity, SOD-1 expression, hormonal profiles in rats	Swain et al. (2022)
Nano Cu	Poultry	Decreased Cu excretion in Nano Cu birds, body weight increased at 25 and 100% level on day 42, Reduction in lipid peroxidation and DNA oxidative damage in liver at 25 and 50% nano Cu level	Sawosz et al. (2018)
Nano Fe	Broiler birds	Increased Fe and ferritin content in blood and body at 4 mg/kg as Fe NP (80 nm)	Miroshnikov et al. (2017)
Nano Fe	Broiler birds	Fe NP @ of 2 mg/kg of body weight as I/M injection Improved the growth, blood parameters and tissue retention	Yausheva et al. (2016)
Nano Mn	Broiler birds	Mn NP increased the tibial bone weight, Mn-sulphate NP increased tibial length, volume, breaking strength and diameter	Lotfi et al. (2014)
Nano Cr	Layer chickens	Nano Cr improved egg quality as well as Cr and Zn retention, increased minerals level of Cr (liver, yolk, and eggshell), Ca and P (liver)	Sirirat et al. (2013)
Nano Se	Layer birds	Nano Se resulted in higher Se concentration in liver, breast muscle, pancreas and feathers as well as Improved GPx, erythrocyte catalase, SOD activities, and immunity	Mohapatra et al. (2014)
Nano Zn	Layer birds	Zn content in tibia, liver, and eggs increased, Egg mass increased in ZnO NP compared with the usual ZnO, SOD activity in liver increased, MDA content in eggs decreased	Abedini et al. (2017)
Nano Zn	Cow (Holstein Fresien)	Decline in somatic cell count in subclinical mastitis and resulted in improved milk production	Rajendran et al. (2013)

(continued)

Table 18.1 (continued)

Nanominerals	Animal/Bird	Salient findings	References
Nano Zn	Goats	Nano Zn at 50 mg/kg DM altered the rumen pH, total protozoa count, fibre degradation and soluble zinc content in goats	Swain et al. (2018)
Nano Se	Poultry	Nano Se supplementation improves organ Se level, growth, immunity and meat quality in poultry birds	Prusty et al. (2021)
Nano Zn	Cow	Nano Zn supplementation decreased somatic cell count as compared to ZnO and un-supplemented cows. The superoxide dismutase and plasma Zn concentrations in the cows provided ZnN were greater than those in the ZnO and control groups	Bakhshizadeh et al. (2019)
Nano Zn	Rabbit	Nano Zn significantly improved body weight, daily weight gain (DWG), feed conversion ratio (FCR) and nutrient digestibility, as well as decreased mortality, compared to ZnO and control groups. Nano-ZnO supplementation had increased hepatic Zn and Cu content and decreased faecal Zn content. Also nano-ZnO group recorded higher expression levels of genes encoding for metallothionein I and metallothionein II, interleukin-2 and interferon-γ in the liver of rabbits	Hassan et al. (2021)

Declaration The authors declare no conflict of interest.

REFERENCES

Abd El-Hack, M.E., Alagawany, M., Arif, M., Emam, M., Saeed, M., Arain, M.A., Siyal, F.A., Patra, A., Elnesr, S.S., and Khan, R.U. 2018. The uses of microbial phytase as a feed additive in poultry nutrition–A review. *Annals of Animal Science*, 18(3), pp.639–658.

Abedini, M., Shariatmadari, F., Torshizi, M.K., and Ahmadi, H. 2017. Effects of a dietary supplementation with zinc oxide nanoparticles, compared to zinc oxide and zinc methionine, on performance, egg quality, and zinc status of laying hens. *Livestock Science*, 203, pp.30–36.

Adams, L.K., Lyon, D.Y., and Alvarez, P.J. 2006. Comparative eco-toxicity of nanoscale TiO2, SiO2, and ZnO water suspensions. *Water Research*, 40(19), pp.3527–3532.

Aksu, D.S., Aksu, T., and Önel, S.E. 2012. Does inclusion at low levels of organically complexed minerals versus inorganic forms create a weakness in performance or

antioxidant defense system in broiler diets?. International Journal of Poultry Science, 11(10), pp.666.
Anwar, M.I., Awais, M.M., Akhtar, M., Navid, M.T., and Muhammad, F. 2019. Nutritional and immunological effects of nano-particles in commercial poultry birds. World's Poultry Science Journal, 75(2), pp.261–272.
Arabi, F., Imandar, M., Negahdary, M., Imandar, M., Noughabi, M.T., Akbaridastjerdi, H., and Fazilati, M. 2012. Investigation anti-bacterial effect of zinc oxide nanoparticles upon life of Listeria monocytogenes. Annals of Biological Research, 7, pp.3679–3685.
Bakhshizadeh, S., Aghjehgheshlagh, F.M., Taghizadeh, A., Seifdavati, J., and Navidshad, B. 2019. Effect of zinc sources on milk yield, milk composition and plasma concentration of metabolites in dairy cows. South African Journal of Animal Science, 49(5), pp.884–891.
Broucek, J. 2014. Production of methane emissions from ruminant husbandry: a review. Journal of Environmental Protection, 5(15), p.1482.
des Rieux, A., Fievez, V., Garinot, M., Schneider, Y.J., and Préat, V. 2006. Nanoparticles as potential oral delivery systems of proteins and vaccines: a mechanistic approach. Journal of controlled release, 116(1), pp.1–27.
Desai, M.P., Labhasetwar, V., Amidon, G.L., and Levy, R.J. 1996. Gastrointestinal uptake of biodegradable microparticles: effect of particle size. Pharmaceutical research, 13(12), pp.1838–1845.
Dib, M.G. 2010. Strategies for beef cattle adaptation to finishing diets, ractopamine hydrochloride utilization, and mature size genetic selection. Master's Thesis in Animal Science, University of Nebraska at Lincoln, USA, pp.1–105
El-Galil, A., Etab, R.I., and El-Bordeny, N.E.Y. 2018. Evaluation of nanocobalt particles addition in ruminant rations by in vitro gas production. Egyptian Journal of Nutrition and Feeds, 21(1), pp.91–102.
Elkloub, K., El Moustafa, M., Ghazalah, A.A., and Rehan, A.A.A. 2015. Effect of dietary nanosilver on broiler performance. International Journal of Poultry Science, 14(3), p.177.
FAO. 2007. The state of the world's animal genetic resources for food and agriculture. Food & Agriculture Org. Barbara Rischkowsky & Dafydd Pilling (Edited), Rome.
FAO. 2008. Livestock's Long Shadow (Rome: FAO).
FAO. 2009. The state of food and agriculture. Rome, FAO. www.fao.org/docrep/012/i0680e/i0680e.pdf.
Feng, M., Wang, Z.S., Zhou, A.G., and Ai, D.W. 2009. The effects of different sizes of nanometer zinc oxide on the proliferation and cell integrity of mice duodenum-epithelial cells in primary culture. Pakistan Journal of Nutrition, 8(8), pp.1164–1166.
Gao, C., Zhu, L., Zhu, F., Sun, J., and Zhu, Z. 2014. Effects of different sources of copper on Ctr1, ATP7A, ATP7B, MT and DMT1 protein and gene expression in Caco-2 cells. Journal of Trace Elements in Medicine and Biology, 28(3), pp.344–350.
Hassan, S., Hassan, F.U., and Rehman, M.S.U. 2020. Nano-particles of trace minerals in poultry nutrition: Potential applications and future prospects. Biological Trace Element Research, 195(2), pp.591–612.
Hassan, F.A., Kishawy, A.T., Moustafa, A., and Roushdy, E.M. 2021. Growth performance, tissue precipitation, metallothionein and cytokine transcript expression and

economics in response to different dietary zinc sources in growing rabbits. Journal of Animal Physiology and Animal Nutrition, 105(5), pp.965–974.
He, Z.L., Yang, X.E., and Stoffella, P.J. 2005. Trace elements in agroecosystems and impacts on the environment. Journal of Trace elements in Medicine and Biology, 19(2–3), pp.125–140.
Hillyer, J.F., and Albrecht, R.M. 2001. Gastrointestinal persorption and tissue distribution of differently sized colloidal gold nanoparticles. Journal of pharmaceutical sciences, 90(12), pp.1927–1936.
Houghton, J.T., Meira Filho, L.G., CAllander, B.A., Harris, N., Kattenberg, A., and Maskell, K., (eds). 1996. Climate change 1995: The science of climate change: contribution of working group I to the second assessment report of the International Governmental Panel on Climate Change (Vol. 2). Cambridge University Press. Cambridge, UK.
Hu, C.H., Li, Y.L., Xiong, L., Zhang, H.M., Song, J., and Xia, M.S. 2012. Comparative effects of nano elemental selenium and sodium selenite on selenium retention in broiler chickens. Animal Feed Science and Technology, 177(3–4), pp.204–210.
Kulak, E., Ognik, K., Stępniowska, A., and Sembratowicz, I. 2018. The effect of administration of silver nanoparticles on silver accumulation in tissues and the immune and antioxidant status of chickens. Journal of Animal and Feed Sciences, 27(1), pp.44–54.
Lotfi, L., Zaghari, M., Zeinoddini, S., Shivazad, M., and Davoodi, D. 2014. August. Comparison dietary nano and micro manganese on broilers performance. In Proceedings of the 5th International Conference on Nanotechnology: Fundamentals and Applications (Vol. 293).
Luna-delRisco, M., Orupõld, K., and Dubourguier, H.C. 2011. Particle-size effect of CuO and ZnO on biogas and methane production during anaerobic digestion. Journal of Hazardous Materials, 189(1–2), pp.603–608.
Meng, T., Liu, Y.L., Xie, C.Y., Zhang, B., Huang, Y.Q., Zhang, Y.W., Yao, Y., Huang, R., and Wu, X. 2019. Effects of different selenium sources on laying performance, egg selenium concentration, and antioxidant capacity in laying hens. Biological Trace Element Research, 189(2), pp.548–555.
Miller, J.K., and Cragle, R.G. 1965. Gastrointestinal sites of absorption and endogenous secretion of zinc in dairy cattle. Journal of Dairy Science, 48(3), pp.370–373.
Miroshnikov, S.A., Donnik, I.M., Yausheva, E.V., Kosyan, D.B., and Sizova, E.A. 2017. Research of opportunities for using iron nanoparticles and amino acids in poultry nutrition. GEOMATE Journal, 13(40), pp.124–131.
Mishra, A., Swain, R.K., Mishra, S.K., Panda, N. and Sethy, K. 2014. Growth performance and serum biochemical parameters as affected by nano zinc supplementation in layer chicks. Indian Journal of Animal Nutrition, 31(4), pp.384–388.
Mohanna, C., and Nys, Y. 1998. Influence of age, sex and cross on body concentrations of trace elements (zinc, iron, copper and manganese) in chickens. British Poultry Science, 39(4), pp.536–543.
Mohapatra, P., Swain, R.K., Mishra, S.K., Behera, T., Swain, P., Behura, N.C., Sahoo, G., Sethy, K., Bhol, B.P., and Dhama, K. 2014. Effects of dietary nano-selenium supplementation on the performance of layer grower birds. Asian Journal of Animal and Veterinary Advances, 9(10), pp.641–652.
Mu, H., Chen, Y., and Xiao, N. 2011. Effects of metal oxide nanoparticles (TiO2, Al2O3, SiO2 and ZnO) on waste activated sludge anaerobic digestion. Bioresource Technology, 102(22), pp.10305–10311.

NATCOM. 2004. India's National Communication of the UNFCCC. Ministry of Environment and Forests, Government of India, Delhi.

O'hagan, D.T. 1996. The intestinal uptake of particles and the implications for drug and antigen delivery. Journal of Anatomy, 189(Pt 3), p.477.

Padmavathy, N., and Vijayaraghavan, R. 2008. Enhanced bioactivity of ZnO nanoparticles—An antimicrobial study. Science and Technology of Advanced Materials, 9(1).

Patra, A., and Lalhriatpuii, M. 2020. Progress and prospect of essential mineral nanoparticles in poultry nutrition and feeding—A review. Biological Trace Element Research, 197(1), pp.233–253.

Prusty, S., Swain, P.S., Mishra, S.K., Das, J., Dubey, M., and Gendley, M.K. 2021. Effect of Nano Selenium Supplementation on Selenium Retention, Growth Performance, Immunity and Product Quality in Poultry: A Mini Review. Animal Nutrition and Feed Technology, 21(3), pp.605–620.

Rajendra, R., Balakumar, C., Ahammed, H.A.M., Jayakumar, S., Vaideki, K., and Rajesh, E. 2010. Use of zinc oxide nano particles for production of antimicrobial textiles. International Journal of Engineering, Science and Technology, 2(1), pp.202–208.

Rajendran, D. 2013. Application of nano minerals in animal production system. Research Journal of Biotechnology, 8(3), pp.1–3.

Rajendran, D., Kumar, G., Ramakrishnan, S., and Shibi, T.K. 2013. Enhancing the milk production and immunity in Holstein Friesian crossbred cow by supplementing novel nano zinc oxide. Research Journal of Biotechnology, 8(5), pp.11–17.

Reddy, S.T., Van Der Vlies, A.J., Simeoni, E., Angeli, V., Randolph, G.J., O'Neil, C.P., Lee, L.K., Swartz, M.A., and Hubbell, J.A. 2007. Exploiting lymphatic transport and complement activation in nanoparticle vaccines. Nature Biotechnology, 25(10), pp.1159–1164.

Riazi, H., Rezaei, J., and Rouzbehan, Y. 2019. Effects of supplementary nano-ZnO on in vitro ruminal fermentation, methane release, antioxidants, and microbial biomass. Turkish Journal of Veterinary & Animal Sciences, 43(6), pp.737–746.

Rosi, N.L., and Mirkin, C.A. 2005. Nanostructures in biodiagnostics. Chemical Reviews, 105(4), pp.1547–1562.

Sahoo, A., Swain, R.K., and Mishra, S.K. 2014. Effect of inorganic, organic and nano zinc supplemented diets on bioavailability and immunity status of broilers. International Journal of Advanced Research, 2(11), pp.828–837.

Sarker, N.C., Keomanivong, F., Borhan, M.D., Rahman, S., and Swanson, K. 2018. In vitro evaluation of nano zinc oxide (nZnO) on mitigation of gaseous emissions. Journal of Animal Science and Technology, 60(1), p.27.

Sawosz, E., Łukasiewicz, M., Łozicki, A., Sosnowska, M., Jaworski, S., Niemiec, J., Scott, A., Jankowski, J., Józefiak, D., and Chwalibog, A. 2018. Effect of copper nanoparticles on the mineral content of tissues and droppings, and growth of chickens. Archives of animal nutrition, 72(5), pp.396–406.

Sembratowicz, I., and Ognik, K. 2018. Evaluation of immunotropic activity of gold nanocolloid in chickens. Journal of Trace Elements in Medicine and Biology, 47, pp.98–103.

Shi, L., Xun, W., Yue, W., Zhang, C., Ren, Y., Liu, Q., Wang, Q., and Shi, L. 2011. Effect of elemental nano-selenium on feed digestibility, rumen fermentation, and purine derivatives in sheep. Animal Feed Science and Technology, 163(2–4), pp.136–142.

Sirirat, N., Lu, J.J., Hung, A.T.Y., and Lien, T.F. 2013. Effect of different levels of nanoparticles chromium picolinate supplementation on performance, egg quality, mineral retention, and tissues minerals accumulation in layer chickens. Journal of Agricultural Science, 5(2), p.150.

Sobhi, B.M., Ismael, E.Y., Elleithy, E., Elsabagh, M., and Fahmy, K.N.E.D. 2020. Influence of combined yeast-derived zinc, selenium and chromium on performance, carcass traits, immune response and histomorphological changes in broiler chickens. Journal of Advanced Veterinary Research, 10(4), pp.233–240.

Sri Sindhura, K., Prasad, T.N.V.K.V., Panner Selvam, P., and Hussain, O.M. 2014. Synthesis, characterisation and evaluation of effect of phytogenic zinc nanoparticles on soil exo-enzymes. Applied Nanoscience, 4(7), pp.819–827.

Stoimenov, P.K., Klinger, R.L., Marchin, G.L., and Klabunde, K.J. 2002. Metal oxide nanoparticles as bactericidal agents. Langmuir, 18(17), pp.6679–6686.

Swain P.S. 2017. Evaluation of nano zinc supplementation on growth, nutrient utilization and immunity in goats (*Capra hircus*). Ph.D. Thesis. ICAR-National Dairy Research Institute, Karnal, Haryana, India.

Swain, P.S., Dominic, G., Bhakthavatsalam, K.V.S., and Terhuja, M. 2016a. Impact of ruminants on global warming: Indian and global context. In Nautiyal S, Schaldach R, Raju K, Kaechele H, Pritchard B, Rao K (Eds) Climate Change Challenge (3C) and Social-Economic-Ecological Interface-Building (pp. 83–97). Springer, Cham.

Swain, P.S., Rao, S.B., Rajendran, D., Dominic, G., and Selvaraju, S. 2016b. Nano zinc, an alternative to conventional zinc as animal feed supplement: A review. Animal Nutrition, 2(3), pp.134–141.

Swain, P.S., Prusty, S., Rao, S.B.N., Rajendran, D., and Patra, A.K. 2021a. Essential nanominerals and other nanomaterials in poultry nutrition and production. Advances in Poultry Nutrition Research. IntechOpen. https://doi.org/10.5772/intechopen.96013.

Swain, P.S., Rao, S.B.N., Rajendran, D., Krishnamoorthy, P., Mondal, S., Pal, D., and Selvaraju, S. 2021b. Nano zinc supplementation in goat (Capra hircus) ration improves immunity, serum zinc profile and IGF-1 hormones without affecting thyroid hormones. Journal of Animal Physiology and Animal Nutrition, 105(4), pp.621–629.

Swain, P.S., Rajendran, D., Rao, S.B.N., and Dominic, G. 2015. Preparation and effects of nano mineral particle feeding in livestock: A review. Veterinary World, 8(7), pp.888.

Swain, P.S., Rajendran, D., Rao, S.B.N., Gowda, N.K.S., Krishnamoorthy, P., Mondal, S., Mor, A., and Selvaraju, S. 2022. Nano zinc supplementation affects immunity, hormonal profile, hepatic superoxide dismutase 1 (SOD1) Gene Expression and Vital Organ Histology in Wister Albino Rats. Biological Trace Element Research, pp.1–11. https://doi.org/10.1007/s12011-022-03355-8.

Swain, P.S., Rao, S.B.N., Rajendran, D., Pal, D., Mondal, S., and Selvaraju, S. 2019a. Effect of supplementation of nano zinc oxide on nutrient retention, organ and serum minerals profile, and hepatic metallothionein gene expression in wister albino rats. Biological Trace Element Research, 190(1), pp.76–86.

Swain, P.S., Rao, S.B.N., Rajendran, D., Poornachandra, K.T., Lokesha, E., and Kumar, R.D. 2019b. Effect of Nanozinc Supplementation on Haematological and Blood Biochemical Profiles in Goats. International Journal of Current Microbiology and Applied Sciences, 8(09), pp.2688–2694.

Swain, P.S., Rao, S.B.N., Rajendran, D., Soren, N.M., Pal, D.T., and Bhat, S.K. 2018. Effect of supplementation of nano zinc on rumen fermentation and fiber degradability in goats. Animal Nutrition and Feed Technology, 18(3), pp.297–309.
Świątkiewicz, S., Arczewska-Włosek, A., and Jozefiak, D. 2014. The efficacy of organic minerals in poultry nutrition: review and implications of recent studies. World's Poultry Science Journal, 70(3), pp.475–486.
Underwood, E.J., and Suttle, N.F. 1999. The mineral nutrition of livestock 3rd edition. CABI Publishing, New York.
Wang, B., Feng, W.Y., Wang, T.C., Jia, G., Wang, M., Shi, J.W., Zhang, F., Zhao, Y.L., and Chai, Z.F. 2006. Acute toxicity of nano-and micro-scale zinc powder in healthy adult mice. Toxicology letters, 161(2), pp.115–123.
Wang, Z.L. 2000. Characterisation of nanophase materials. Wiley- VCH Verlag GmbH, Weinheim, pp.13–14.
Yausheva, E.V., Miroshnikov, S.A., Kosyan, D.B., and Sizova, E.A. 2016. Nanoparticles in combination with amino acids change productive and immunological indicators of broiler chicken. Agricultural Biology, 51(6), pp.912–920.
Yenice, E., Mızrak, C., Gültekin, M., Atik, Z., and Tunca, M. 2015. Effects of organic and inorganic forms of manganese, zinc, copper, and chromium on bioavailability of these minerals and calcium in late-phase laying hens. Biological Trace Element Research, 167(2), pp.300–307.
Zaboli, K., Aliarabi, H., Bahari, A.A., and ABBAS, A.K.R. 2013. Role of dietary nano-zinc oxide on growth performance and blood levels of mineral: A study on in Iranian Angora (Markhoz) goat kids. International Advisory Board, 19.
Zhang, J.S., Gao, X.Y., Zhang, L.D., and Bao, Y.P. 2001. Biological effects of a nano red elemental selenium. Biofactors, 15(1), pp.27–38.
Zhao, C.Y., Tan, S.X., Xiao, X.Y., Qiu, X.S., Pan, J.Q., and Tang, Z.X. 2014. Effects of dietary zinc oxide nanoparticles on growth performance and antioxidative status in broilers. Biological trace element research, 160(3), pp.361–367.

CHAPTER 19

Climate Change Impacts on the Higher Altitude Forests of Indian Himalayan Regions: Nature-Based Solutions for Climate Resilience and Disaster Risk Reduction

Anwesha Chakraborty and Purabi Saikia

19.1 INTRODUCTION

Forests are the most complex terrestrial ecosystem and home to more than 80% of the terrestrial biodiversity. India is one of the 17 mega-biodiversity nations (Mittermeier and Mittermeier 2005) with rich and unique biodiversity with distinct climatic, altitudinal, and latitudinal gradients (Arisdason and Lakshminarasimhan 2016). The country holds 7–8% of all recorded species of the globe, including ~47,513 plant species belonging to all the important phyla, viz., angiosperms, gymnosperms, pteridophytes, and bryophytes and 91,000 species of animals encompassing all the major ecosystems including forests, grasslands, wetlands, coastal, and deserts (Arisdason and Lakshminarasimhan 2016; Gokhale 2015). India has a total of 21.67% of its total geographical area under forest cover, of which 3.02% are very dense, 9.39% moderately dense, and 9.26% open forests (ISFR 2019). Indian forests can be predominantly categorized into five major categories, viz., tropical, montane subtropical, montane temperate, subalpine, and alpine, primarily based on temperature (Champion and Seth 1968). They are further divided into 16 major types viz., tropical wet evergreen, tropical semi-evergreen, tropical moist

A. Chakraborty
Department of Life Sciences, Presidency University, Kolkata, India

P. Saikia (✉)
Department of Environmental Sciences, Central University of Jharkhand, Ranchi, India
e-mail: purabi.saikia@cuj.ac.in

© The Author(s), under exclusive license to Springer Nature Singapore Pte Ltd. 2023
S. Nautiyal et al. (eds.), *Palgrave Handbook of Socio-ecological Resilience in the Face of Climate Change*,
https://doi.org/10.1007/978-981-99-2206-2_19

deciduous, littoral and swamp, tropical dry deciduous, tropical thorn, tropical dry evergreen, subtropical broad-leaved hill, subtropical pine, subtropical dry evergreen, montane wet temperate, Himalayan moist temperate, Himalayan dry temperate, sub alpine, moist alpine scrub, and dry alpine scrub based on the differences in annual temperature, rainfall, and dry periods among the forests (Champion and Seth 1968).

The Indian Himalayan Regions (IHRs) offer a heterogeneous landscape along with one of the largest elevational and climatic gradients on Earth (Shooner et al. 2018). It has been categorized longitudinally into three major divisions, the Western, Central, and Eastern Himalayas. It extended from the extreme west to east covering a total of 12 Indian states, viz., Jammu and Kashmir, Ladakh, Himachal Pradesh, Uttarakhand, West Bengal, Sikkim, Arunachal Pradesh, Assam, Meghalaya, Manipur, Mizoram, Nagaland, and Tripura, which collectively encompass a total area of 23,034 (in 000' ha). IHRs have a highly diverse ecosystem with an altitudinal range of 500 to more than 5000 m with varying climatic regimes comprising different flora and fauna of great socio-economic importance (Singh and Rawat 2000). There is a peak in species richness in the intermediate elevation *i.e.,* the region between the actual lowlands and the lower limits along with the Himalayan altitude, that has resulted from a combined effect of different environmental conditions (Lomolino 2001). The Himalayan forests encompass all the major tropical, subtropical, temperate, and alpine bioclimatic zones (Saikia et al. 2016). A decrease in species richness with increasing altitude occurs due to the combined effects of prevailing climatic conditions and anthropogenic interventions (Carpenter 2005). Climatic conditions and topographic factors like altitude, latitude, and slope are equally responsible for species richness and their pattern of distribution in the North-Western Himalayas (Sharma and Raina 2013). Temperature and altitude play an essential role in the distribution and composition of plant species, and plant diversity usually decreases with increasing altitude and decreasing temperature (Champion and Seth 1968). There is a 1 °C drop in the mean temperature in the mountains with an altitude of every 270 m up to about 1500 m, which then faces a sharper decrease in mean temperature with a further rise in altitude (Singh and Singh 1987). Climate and soil are the major influencing factors of the species distribution and abundance in the Himalayas (Sharma et al. 2010). Solar radiation and snow cover stretch are important variables for determining the floristic diversity in alpine forests (Fischer 1990). Besides, radiation index and slope orientation play a pertinent role in the distribution of plants in mountainous terrains (Paudel and Vetaas 2014). Despite the abundance of natural resources and massive biodiversity in the IHRs, most people living in these mountain ecosystems are economically backwards and still live on a subsistence level (Singh 2006). The unscientific harvesting of forest resources, medicinal herbs, wild edible plants, and other non-timber forest produces (NTFPs) along with fuelwood collection, illegal timber felling, overgrazing by the livestock, and prevalence of natural disasters as an impact of ongoing global climate change leads to intense environmental damage that provoke the impacts of natural calamities in the region (Saikia et al. 2017; Singh 2006). Therefore, the present chapter attempts to give a detailed overview of varied vegetation types of the ecologically fragile and highly diverse higher-altitude forests of the IHRs and the climate change adaptation strategies and nature-based solutions of disaster risk reduction (DRR) in IHRs.

19.2 Forests Types and Forest Resources of IHRs

The Indian Himalayan region is geo-dynamically young mountains, significant from the climatic viewpoint and as a contributor of life, supplier of water to a major part of the entire Indian subcontinent, and also harbours a rich heritage of floral, faunal, human, and cultural diversity (Singh 2006). The IHRs accommodate almost one-tenth of the global familiar higher-altitude flora and fauna and 50% of India's native plant species (Padma 2014). Each of the Indian states of IHRs includes many groups of forest types within its territories of which the majority of the IHRs forest stretch is localized in the Indian Eastern Himalayas. Jammu & Kashmir, Ladakh, and Himachal Pradesh combined have nine forest-type groups, Uttarakhand alone has eight forest-type groups, and the Eastern and Northeastern Indian states combined have twelve forest-type groups, and the Sikkim Himalayas falls under the Himalayan (2) Biogeographic Zones and Central Himalaya (2C) Biotic Province having ~9 different forests types (Champion and Seth 1968). The Indian States that cover the IHRs, their respective area of forest cover, and the forest-type groups have been presented in Table 19.1 (Champion and Seth 1968; ISFR 2019). In terms of percentage of forest cover, the eastern Himalayan states have higher forest cover, e.g., Mizoram has the largest share in terms of forest cover (85.41%), followed by Arunachal Pradesh (79.63%), Meghalaya (76.33%), Manipur (75.46%), Nagaland (75.31%), and Tripura (73.68%). While, the Western Himalayan states, Uttarakhand has the largest forest cover (45.44%) followed by UT of J&K (36.66%), and Himachal Pradesh (27.72%) (Table 19.2).

Changes in altitude in India's Western Himalayas, are creating a very specific vegetation type including alluvial grasslands, subtropical forests, conifer mountain forests, and alpine meadows (Tewari et al. 2017). The forests of the IHRs are home to many rare, endangered, and threatened plant species, and the Indian states of Sikkim, Himachal Pradesh, and UT of Jammu & Kashmir are the richest in the number of threatened species (Mehta et al. 2020). The sub alpine to alpine forests of the northwestern part of the IHRs are considered home to a critically endangered orchid, *Dactylorhiza hatagirea* (D. Don) Soo, which has wide application in curing ailments in Ayurveda, Siddha, Unani, and folk medicine systems (Wani et al. 2020, 2021). The critically endangered plant species viz., *Saussurea costus* (Falc.) Lipsch, *Gentiana kurroo* Royle *Lilium polyphyllum* D. Don, and *Aconitum chasmanthum* Stapf, are endemic to the Kashmir Himalayas (Mir et al. 2020).

Table 19.1 Details of the area statistics and forest types in the forest stretch in the states of IHRs

Part of the IHRs	Indian states	Area of forest cover (in 000' ha)	Forest type groups
Indian Western Himalayas	Jammu & Kashmir and Ladakh	2299	Group 3: Tropical Moist Deciduous Forest Group 5: Tropical Dry Deciduous Forests Group 9: Subtropical Pine Forests Group 10: Subtropical Dry Evergreen Forests Group 12: Himalayan Moist Temperate Forests Group 13: Himalayan Dry Temperate Forests Group 14: Sub Alpine Forests Group 15: Moist Alpine Scrub Group 16: Dry Alpine Scrub

Part of the IHRs	Indian states	Area of forest cover (in 000' ha)	Forest type groups
Indian Central Himalayas	Himachal Pradesh	1126	Group 3: Tropical Moist Deciduous Forests
	Uttarakhand	3800	Group 5: Tropical Dry Deciduous Forests
			Group 9: Subtropical Pine Forests
			Group 12: Himalayan Moist Temperate Forests
			Group 13: Himalayan Dry Temperate Forests
			Group 14: Sub Alpine Forests
			Group 15: Moist Alpine Scrub
			Group 16: Dry Alpine Scrub
Indian Eastern Himalayas	West Bengal	1173	Group 1: Tropical Wet Evergreen Forests
			Group 2: Tropical Semi-Evergreen Forests
			Group 3: Tropical Moist Deciduous Forests
			Group 4: Littoral and Swamp Forests
			Group 5: Tropical Dry Deciduous Forests
			Group 8: Subtropical Broadleaved Hill Forests
			Group 9: Subtropical Pine Forests
			Group 11: Montane Wet Temperate Forests
			Group 12: Himalayan Moist Temperate Forests
			Group 13: Himalayan Dry Temperate Forests
			Group 14: Sub Alpine Forests
			Group 15: Moist Alpine Scrub

(continued)

Table 19.1 (continued)

Part of the IHRs	Indian states	Area of forest cover (in 000' ha)	Forest type groups
	Sikkim	336	
	Arunachal Pradesh	6725	
	Assam	1853	
	Meghalaya	946	
	Manipur	1699	
	Mizoram	1585	
	Nagaland	863	
	Tripura	629	

Total Area: 23,034 (in 000' ha)

Table 19.2 Total geographic area (sq. km) and area under very dense forest (VDF), moderately dense forest (MDF), and open forest (OF) of Himalaya states of India as per ISFR 2019

State/UT	Area (sq. km)					% of TGA	Change in forest cover w.r.t. ISFR 2017
	Total geographic area	VDF	MDF	OF	Total forest cover		
Arunachal Pradesh	83,743	21,095	30,557	15,036	66,688	79.63	−276
Assam	78,438	2795	10,279	15,253	28,327	36.11	222
Himachal Pradesh	55,673	3113	7126	5195	15,434	27.72	334
UT of J&K	53,758	4203	7952	8967	21,122	36.66	348
UT of Ladakh	169,421	78	660	1752	2490	1.47	23
Manipur	22,327	905	6386	9556	16,847	75.46	−499
Meghalaya	22,429	489	9267	7363	17,119	76.33	−27
Mizoram	21,081	157	5801	12,048	18,006	85.41	−180
Nagaland	16,579	1273	4534	6679	12,486	75.31	−3
Sikkim	7095	1102	1552	688	3342	47.10	−2
Tripura	10,486	654	5236	1836	7726	73.68	0
Uttarakhand	53,483	5047	12,805	6451	24,303	45.44	8

Similarly, a critically endangered deciduous tree species, *Lagerstroemia minuticarpa* Debberm. ex P.C. Kanjilal, prevalent in the Eastern Himalayan forests of Arunachal Pradesh and Sikkim (Adhikari et al. 2019). Around 44 threatened plant species including *Swertia chirayita* Roxb. ex Fleming Karsten, *Taxus wallichiana* Zucc., *Neopicrorhiza scrophulariiflora* Hook. F. (Prain), and *Sinopodophyllum hexandrum* (Royle) T.S. Ying have been documented from the forests stretch of Sikkim and North Bengal including Darjeeling, Jalpaiguri, Alipurduar, and Kalimpong districts (Kandel et al. 2019). The northeastern region has always been known to hold one of the richest reservoirs of genetic variability in terms of diverse flora (Rai et al. 2004). There are ~1549 endemic flowering plant species found in Himalayan forests (Pandit et al. 2007) excluding areas above the timberline, and on the other hand, Eastern Himalaya is richer in the number of endemic species as compared to the Western Himalaya where more endemism is found beyond the timberline (Dhar 2002).

19.3 Evidence of Climate Change Experienced by the Higher-Altitude Forests of IHRs

Climatic factors are considered an important determinant of vegetation distribution patterns, forest structure, and composition globally (Kirschbaum et al. 1996). Climate changes used to alter the forest configuration significantly which is considered important in the context of climate change because (i) forest destruction and land degradation contribute to *ca.*, 20% of global CO_2 emissions, (ii) it imparts a substantial opportunity to mitigate climate change, particularly through the REDD (Reducing Emissions from Deforestation and Degradation) and REDD + (reducing emissions from deforestation, forest degradation, and other forest management activities) mechanisms, and (iii) it will be affected in terms of biodiversity loss, limited biomass growth, forest productivity, and regeneration (Gopalakrishnan et al. 2011). Global warming is responsible for an overall rise in global average temperature, varying precipitation, and greenhouse gas (GHG) effects that have cumulatively brought plant species to suffer immensely (Agrawal and Gopal 2013). Sudden and unpredictable floods, drought, and landslides are the potential impacts of ongoing climate change in the fragile landscapes of the IHRs (Barnett et al. 2005; Saikia et al. 2017) that make the region more prone to the risk of food shortage and biodiversity loss (Xu et al. 2009). The impacts of climate change on IHRs are additionally exacerbated by unsustainable agricultural practices including Jhum cultivation, habitat destruction, fragmentation, loss, deforestation, overexploitation of forest resources, and extraction of non-timber forest products (NTFPs) (Saikia et al. 2016). The Himalaya Hindu Kush region would soon face a long-term surge in the atmospheric temperature, particularly above the ground with the glacier, due to changing climatic conditions (Barnett et al. 2005).

About 45% of Indian forests would be negatively affected in the coming future, and this impact will be maximum in the forests of IHRs (Gopalakrishnan et al. 2011). Forests of IHRs are vulnerable to climate change due to the decreasing trend of species diversity and changes in species composition although they have a high potential for biomass accumulation and carbon sequestration (Kumari et al. 2022). A high degree of vulnerability in the form of a lower tree density in the forest stretch, a considerably low biodiversity status, and increased forest fragmentation has been predicted for the mountainous, alpine, sub alpine, dry temperate, and moist temperate forests of IHRs as a repercussion of climate change (Gopalakrishnan et al. 2011). Climate change not only brings a negative impact on the biotic components of the high-altitude regions but also causes hardship to the social lives of the local inhabitants, by making vegetation and agriculture difficult due to the extreme weather events (Gupta et al. 2019). Floristic diversity, as well as their respective habitat, is rapidly declining, with serious consequences for the environment, biodiversity, and the socio-economic framework of local communities (Saikia et al. 2017). There is a greater vulnerability of the communities from middle and high-altitude zones, compared to those hailing from lower and very high altitudes of the IHRs (Gupta et al. 2019). Changes in the prevailing climatic conditions in the IHRs have proven consequences on the ecophysiological trait and the ultimate distribution pattern of the plants, which have further deleterious effects on the floral diversity of the region (Agnihotri et al. 2017). The plant species of the mountain ecosystems of the Himalayas have already started to shift towards higher altitudes due to the rise in global temperature (Padma 2014). The sharp elevational gradients in IHRs greatly enhance climatically-driven altitudinal shifts and contractions in the distribution ranges of many Himalayan plant species (Grabherr et al. 1994).

Global climate change has affected the productivity and carbon-regulating services of the forests, prolonged droughts, more pest invasions, and also a shift in the forest-type boundaries along latitudinal, altitudinal, temperature, and rainfall gradients (Kumar et al. 2020; Kaushik and Khalid 2011). The migration is towards the moist evergreen forest types in the northeastern Himalayas, while it is towards the dry deciduous forest types in the north-western Himalayas in the absence of anthropogenic influences (Ravindranath et al. 2005). The more vulnerable regions to climate change include the upper Himalayas, northern and central parts of Western Ghats, and parts of central India, while the northeastern Himalayan forests have been identified as the more resilient to climate change (Chaturvedi et al. 2010). The ongoing anthropogenic forest conversion in the IHRs due to the increasing population density and forest dependency may result in the catastrophic extinction of important biodiversity, which could be aggravated by global climate change (Pandit et al. 2007). The dense forest cover (>40% canopy cover) in the Indian Himalayas will be confined to only 10% of the land area by 2100 which may lead to a remarkable loss of endemic plant species (366 spp.) and higher animal species (35 vertebrates) of the region if the present rate of deforestation continues (Pandit et al. 2007). The increase in temperature has remarkable impacts on the phenology of various plant species and the distribution of species of *Abies* and *Rhododendrons* in Arunachal Pradesh (Bharali and Khan 2011). It also adversely affects the quantity of seed set in some plant species of *Primula* (Shimono and Washitan 2007).

Apart from unfavourable topographic conditions and natural calamities, one very essential concern is the constant threat from invasive alien species, which often cause a great loss to the native flora of the IHRs. The rich floristic composition of the Darjeeling Himalayas includes several exotic and endemic species apart from the native flora (Das 2004). This rich biodiversity of the Darjeeling Himalayas is invaded by various alien species including *Yushania maling* having definite negative impacts on the growth of the native plant species (Srivastava et al. 2018). A total of 66 invasive alien species belonging to 26 families have been recorded in the forests of IHRs, most of which, have invaded the American subcontinent (Moktan and Das 2013). An increased population of invasive weeds like *Lantana camara* L., *Chromolaena odorata* (L.) R.M. King & H.Rob., *Eupatorium* spp., *Parthenium hysterophorus* L., and native invaders like *Koenigia polystachya* (Wall. ex Meisn.) T.M. Schust. & Reveal has been reported in the forests of Uttarakhand as a repercussion of climate change (Negi et al. 2017).

19.4 Climate Change Adaptations in Higher-Altitude Forests of IHRs

Climate change has negative impacts not only on biodiversity at a global level but has equally lowered the quality of life of the native people inhabiting the hilly areas of the IHRs (Saikia et al. 2020). The phenomena of global warming coupled with climate change and their impact on any forest stretch is a very gradual process, thus necessary adaptations against the negative effects would need a long-term observation and a clear understanding of the cause of such dynamism (Tewari et al. 2017). It has been seen that as a part of adaptation to changing climatic conditions, phenology has a pronounced dependency on phylogenesis, and this is based on two factors *i.e.*, thermal memory and phenotypic plasticity (Kijowska-Oberc et al. 2020). The gradually increasing biodiversity loss can be minimized by integrating innovative and improved research methods, and strong management strategies with the implementation of suitable policies. Changing patterns of climate critically affect the indigenous people inhabiting the slopes of the Himalayas, who obtain food and cash from the forests of the IHRs. Various adaptive strategies have already been implemented by the local people utilizing their traditional knowledge and day-to-day experiences, like agroforestry, crop rotation, mixed-farming, crop diversification, traditional mixed-cropping, and others to combat extreme climatic events (Meena et al. 2019). Several adaptive measures have been undertaken by the local people of the Western Himalayas, like growing peas, cabbage, and cauliflower at a higher altitude, fruits like papaya, litchi, banana, and mango at middle altitudes, provide an optimum growing environment for the plants (Negi et al. 2017). Crop switching is an adaptation response to climate change, and it is also influenced by non-climatic factors such as market dynamics, pest occurrence, and land degradation (Tessema et al. 2019). Moreover, replacing the cultivation of *Phaseolus vulgaris* L. with *Macrotyloma uniflorum* (Lam.) Verdc. or *Glycine max* (L.) Merr. and *Vigna unguiculata* (L.) Walp. with that of *Cajanus cajan* (L.) Huth under the changing-of-cropping pattern strategy has proven to be effective in the mid-altitude of the Western Himalayas (Negi et al. 2017). Effective crop switching can prevent two-thirds of the potential damage from climate change in the agricultural sector (Costinot et al. 2016). Significant reversing effects of crop failure have been achieved by the locals of Western Himalaya by making use of improved or superior varieties of crops, organic manuring using weeds, and creating protective agricultural techniques for off-seasonal plants (Negi et al. 2017). Without any scientific exposure, and almost lack of knowledge of any modern agricultural tool/technique like zero tillage, snow water harvesting, glacial run-off water harvesting for agriculture, or any notable agro-advisory service, the natives have already adopted brilliant crop loss preventive measures (Meena et al. 2019). Thus, the traditional knowledge of our farmers and the native people inhabiting the forested areas of the IHRs must be safeguarded. Moreover, apart from technologies and tools, efficient policies and programs are also desirable for the benefit of people and also for protecting the degrading biodiversity due to harsh climatic conditions and achieving the goal of sustainable economic development in India (Pant et al. 2018). After commercial logging, natural forests of the Darjeeling district have been converted into 'taungya' plantations under which chosen species of high-value timber trees were planted after slash-burning forest lands (Shankar et al. 1998).

19.5 Nature-Based Solutions for Climate Resilience and Disaster Risk Reduction (DRR) in IHRs

Biodiversity conservation and improvement in ecosystem goods and services are considered the basis for perceiving solutions to global climate change, disaster risk reduction, and poverty alleviation to encourage a green economy (Pauleit et al. 2017). Proper solutions for climate resilience are required because the continuous rise in the global annual average temperature widely disturbs plants' metabolism leading to an imbalance in C:N ratio, this cumulatively leads to ill-developed plants with reduced nutrients and organic compounds essential for growth and development (Jha et al. 2020). Severe damage of *Pinus wallichiana* A.B.Jacks. in the forests of Jammu & Kashmir and Himachal Pradesh by a beetle (*Pityogenes scitus* Blandford, 1893) during the early 2000s, is an excellent example of increased activity of bugs and beetles due to changing climatic conditions in the Himalayan forests (Tewari et al. 2017). Climate change has also favoured the growth of invasives like *P. hysterophorus*, or *L. camara* which has deleterious effects on native plants (Negi et al. 2017).

The use of plant variants with suitably modified phenology and stress-resistant plants for reforestation programs can be a possible solution for climate resilience in the higher-altitude forest ecosystems of the IHRs. Another proposed solution to combat the rising temperature and naturally available water shortage is a replacement of rain-water-grown plants with irrigated plants (Minoli et al. 2019). Tribal people of Arunachal Pradesh have been seen to make their cattle graze on the desolated lands for dry fodder during times of drought, this is considered as a nature-based survival and resilience strategy (Saikia et al. 2020). Landslides in the hills as an outcome of a declining number of trees in hilly mountainous forests, while floods in the plains due to decreasing carrying capacity of water by our water bodies are very common disasters faced by the forests of IHRs (Saikia et al. 2017). The development of agroforestry-based natural resilience for DRR is very essential to minimize the impacts of these disasters. Afforestation and forest protection are desirable to reduce the lifetime risk faced by the locals since plant roots are highly potential to hold the soil particles together, preventing landslides (de Jesús Arce-Mojica et al. 2019). Locals of Arunachal Pradesh use stone walls during a flood, earthen/made-from-bamboo containers, and storage products, edibles, and other essential items from their reared livestock at times of disaster, which can also be considered effective nature-based DRR and disaster risk management (Saikia et al. 2020). Lastly, it should be mentioned that the development of policies and programs against natural disasters in the mountainous slopes of the IHRs is crucial. It is a vital step to include substantial adaptive strategies of disaster management into all policy-making and its practice under adaptation mainstreaming, considering the ever-increasing climate-based risks (Wamsler et al. 2017). However, due to the varying degree of vulnerability faced by plants growing at different altitude gradients, policy formulation should consider all the relevant subjects like altitude, livelihood options, available resources, and the varied intensity of pressure faced by the various plant species (Gupta et al. 2019).

19.6 Future Research Prospects and Recommendations

India is bestowed with its vast natural treasure that is still beyond our knowledge. A significant percentage of the total geographic area of India, including the majority of the Himalayan forests, is still unexplored. Therefore,

- A more focused study should be conducted to identify species that can be used for more economic purposes.
- Moreover, remote sensing-based studies in the hilly terrains of the IHRs may be done to get an overview of the unexplored forest resources.
- A complete understanding of the rich floral diversity of the IHRs shall lead to a better understanding of these unique landscapes and can also be used for the utmost benefit of the forest-dependent population.
- Only thorough knowledge and proper skills will lead to a better future by building up conservation strategies for the species endemic to the IHRs and also would enable scientists to develop elite approaches to control the growth of alien species that are potent threats to natural diversity.
- A better understanding of all the important factors contributing to the diversity of the species and its richness is desirable and will be a promising strategy to increase the land area under a forest stretch.
- Moreover, the implementation of stricter policies against anthropogenic activities should be implemented and maintained.
- Detailed climate change-related impacts and adaptations over the forest resources of IHRs help in devising climate-resilient strategies for this unique ecosystem of the world.

19.7 Conclusions

The forest ecosystems of the IHRs are home to extremely diverse, unique, and endemic biodiversity attributed to the prevailing climatic and edaphic conditions, the topography of this region, and other environmental factors. The species richness shows a peak in the intermediate zone in the Himalayan altitude and it used to be decreased with an increase in altitude and decrease in temperature. The forests of IHRs are known for their rich diversity of economically and traditionally important plants and the local tribes and the farmers who own lands of agroforestry have exceptional knowledge regarding the utilities of the different plant species. Plants can be a source of timber, fuelwood, and edibles which imparted significantly to the economy of the local people inhabiting the forest fringe areas. Forests of the Western Himalayas are highly disturbed as compared to the Eastern Himalayan forests due to the higher population density in the Western Himalayas. Similarly, forests of the Eastern Himalayas are decreasing at a higher rate due to the traditional Jhum cultivation, higher impacts of ongoing global climate change, and higher livelihood dependency of the forest-dwelling people. Large-scale forest protection, conservation, and restoration efforts are urgently required in the IHRs to avoid the upcoming disturbance-driven biodiversity losses in IHRs. It is also necessary to devise a national strategy considering the multiple vulnerabilities faced by the forests of IHRs to adapt to climate change impacts and also to further amplify ecological integrity, functionality, and sustainability.

References

Adhikari D, Tiwary R, Singh PP et al (2019) Ecological niche modeling as a cumulative environmental impact assessment tool for biodiversity assessment and conservation planning: a case study of critically endangered plant *Lagerstroemia minuticarpa* in the Indian Eastern Himalaya. J Environ Manage 243:299–307

Agnihotri P, Husain T, Shirke PA et al (2017) Climate change-driven shifts in elevation and ecophysiological traits of Himalayan plants during the past century. Curr Sci 112(3):595–601

Agrawal A, Gopal K (2013) Concept of rare and endangered species and its impact as biodiversity. Biomonitoring of water and waste water. Springer, India, pp 71–83. https://doi.org/10.1007/978-81-322-0864-8_7

Arisdason W, Lakshminarasimhan P (2016) Status of plant diversity in India: an overview. Central National Herbarium, Botanical Survey of India, Govt. of India, Howrah

Barnett TP, Adam JC, Lettenmaier DP (2005) Potential impacts of a warming climate on water availability in snow-dominated regions. Nature 438(7066):303–309

Bharali S, Khan ML (2011) Climate Change and its impact on biodiversity; some management options for mitigation in Arunachal Pradesh. Curr Sci 101(7):855–860

Carpenter C (2005) The environmental control of plant species density on a Himalayan elevation gradient. J Biogeogr 32(6):999–1018

Champion HG, Seth SK (1968) A revised survey of the forest types of India. Manager of publications, Govt. of India, Delhi

Chaturvedi RK, Gopalkrishnan R, Jayaraman M et al (2010) Impact of climate change on Indian Forests: a dynamic vegetation modelling approach. Mitig Adapt Strat Glob Change 16(2):119–142

Costinot A, Donaldson D, Smith C (2016) Evolving comparative advantage and the impact of climate change in agricultural markets: evidence from 1.7 million fields around the world. J. Political Econ. 124(1):205–248

Das AP (2004) Floristic studies in Darjeeling hills. Nelumbo 46(1–4):1–18

de Jesús Arce-Mojica T, Nehren U, Sudmeier-Rieux K, Miranda PJ, Anhuf D (2019) Nature-based solutions (NbS) for reducing the risk of shallow landslides: where do we stand? Int J Disaster Risk Reduct 41:101293

Dhar U (2002) Conservation implications of plant endemism in high-altitude Himalaya. Curr Sci 82:141–148

Fischer HS (1990) Simulating the distribution of plant communities in an alpine landscape. Coenoses 5:37–43

Forest Survey of India (2019) Indian State of Forest Report. Ministry of Environment, Forests, and Climate Change, Govt. of India, Dehradun

Gokhle Y (2015) Green Growth and Biodiversity in India. New Delhi: the Energy and Resources Institute, pp 1–13

Gopalakrishnan R, Jayaraman M, Bala G et al (2011) Climate change and Indian forests. Curr Sci 101(3):348–355

Grabherr G, Gottfried M, Pauli H (1994) Climate effects on mountain plants. Nature 369:448

Gupta AK, Negi M, Nandy S et al (2019) Assessing the vulnerability of socio-environmental systems to climate change along an altitude gradient in the Indian Himalayas. Ecol Indic 106:105512

Jha SK, Negi AK, Alatalo JM et al (2020) Assessment of climate change patterns in the Pauri Garhwal of the Western Himalayan Region: based on climate parameters and perceptions of forest-dependent communities. Environ Monit Assess 192(10):1–15

Kandel P, Chettri N, Chaudhary RP et al (2019) Plant diversity of the Kangchenjunga Landscape, Eastern Himalayas. Plant Divers 41(3):153–165

Kaushik G, Khalid MA (2011) Climate change impact on forestry in India. In: Lichtfouse E (ed) Alternative farming systems, biotechnology, drought stress, and ecological fertilization, Vol 6. Sustainable agriculture reviews, Springer Dordrecht Heidelberg London New York, pp 319–344

Kijowska-Oberc J, Staszak AM, Kamiński J et al (2020) Adaptation of forest trees to rapidly changing climate. Forests 11(2):123

Kirschbaum MUF, Cannell MGR, Cruz RVO et al (1996) Climate change impacts on forests. In: Watson RT, Zinyowera MC, Moss RH (eds) Climate Change 1995: The IPCC Second Assessment Report Scientific-Technical Analyses of Impacts, Adaptations, and Mitigation of Climate Change. Cambridge University Press, Cambridge

Kumar G, Kumari R, Kishore BSPC et al (2020) Climate Change impacts and implications: an Indian Perspective. In: Roy N, Roychoudhury S, Nautiyal S et al (eds) Socio-economic and Eco-Biological Dimensions in Resource use and Conservation—Strategies for Sustainability. Chapter 2, Springer International Publishing, Switzerland, pp 11–30

Kumari R, Kumar A, Saikia P et al (2022) Vulnerability assessment of Indian Himalayan forests in terms of biomass production and carbon sequestration potential in changing climatic conditions. In: Lackner M, Sajjadi B, Chen WY (eds)

Handbook of Climate Change Mitigation and Adaptation. Chapter 4, 3rd edition, Springer Nature Switzerland AG, pp 147–161

Lomolino MV (2001) Elevation gradients of species-density: historical and prospective views. Glob Ecol Biogeogr 10(1):3–13

Meena RK, Verma TP, Yadav RP et al (2019) Local perceptions and adaptation of indigenous communities to climate change: evidence from High Mountain Pangi valley of Indian Himalayas. Indian J Tradit Knowl 18(1):58–67

Mehta P, Sekar KC, Bhatt D et al (2020) Conservation and prioritization of threatened plants in the Indian Himalayan Region. Biodivers Conserv 29(6):1723–1745

Minoli S, Müller C, Elliott J et al (2019) Global response patterns of major rainfed crops to adaptation by maintaining current growing periods and irrigation. Earth's Future 7(12):1464–1480

Mir AH, Tyub S, Kamili AN (2020) Ecology, distribution mapping, and conservation implications of four critically endangered endemic plants of Kashmir Himalaya. Saudi J. Biol. Sci. 27(9):2380–2389.

Mittermeier RA, Mittermeier CG (2005) Megadiversity: earth's Biologically Wealthiest Nations. Cemex, Mexico

Moktan S, Das AP (2013) Diversity and distribution of invasive alien plants along the altitudinal gradient in Darjeeling Himalaya, India. Pleione 7(2):305–313

Negi VS, Maikhuri RK, Pharswan D et al (2017) Climate change impact in the Western Himalaya: people's perception and adaptive strategies. J Mt Sci 14(2):403–416

Padma TV (2014) Himalayan plants seek cooler climes. Nature 512:359

Pandit MK, Sodhi NS, Koh LP et al (2007) Unreported yet massive deforestation driving loss of endemic biodiversity in Indian Himalayas. Biodivers Conserv 16:153–163

Pant S, Rinchen T, Butola JS (2018) Indigenous knowledge on bio-resources management for a sustainable livelihood by the cold desert people, trans-Himalaya, Ladakh, India. Indian J Nat Prod Resour 9(2):168–173

Paudel S, Vetaas OR (2014) Effects of topography and land use on woody plant species composition and beta diversity in an arid Trans-Himalayan landscape, Nepal. J Mt Sci 11(5):1112–1122

Pauleit S, Zölch T, Hansen R et al (2017) Nature-based solutions and climate change–four shades of green. In: Kabisch N, Korn H, Stadler J et al (eds) Nature-Based solutions to climate change adaptation in urban areas, Linkages between Science, Policy, and Practice. Springer International Publishing, Switzerland, pp 29–49

Rai N, Asati BS, Yadav DS (2004) Conservation and genetic enhancement of underutilized vegetable crop species in the North-Eastern region of India. Leisa India 6:11–12

Ravindranath NH, Joshi NV, Sukumar R et al (2005) Impact of climate change on forests in India. Curr Sci 90(3):354–361

Saikia P, Deka J, Bharali S et al (2017) Plant diversity patterns and conservation status of Eastern Himalayan forests in Arunachal Pradesh, Northeast India. For Ecosyst 4:28

Saikia P, Kumar A, Khan ML (2016) Biodiversity status and climate change scenario in Northeast India. In: Nautiyal S, Schaldach R, Raju KV et al (eds) Climate Change Challenge (3C) and Social-Economic-Ecological Interface-Building. Chapter 8, Springer International Publishing, Switzerland, pp 107–120

Saikia P, Kumar A, Lal P et al (2020) Ecosystem-Based Adaptation to Climate Change and Disaster Risk Reduction in Eastern Himalayan Forests of Arunachal Pradesh, Northeast India. In: Dhyani S, Gupta AK, Karki M (eds) Nature-based Solutions for Resilient Ecosystems and Societies. Springer Nature, Singapore, pp 391–408

Shankar U, Lama SD, Bawa KS (1998) Ecosystem reconstruction through taungya plantation following commercial logging of a dry, mixed deciduous forest in Darjeeling Himalaya. For Ecol Manag 102(2–3):131–142

Sharma CM, Baduni NP, Gairola S et al (2010) Effects of slope aspects on forest compositions, community structures, and soil properties in natural temperate forests of Garhwal Himalaya. J For Res 21(3):331–337

Sharma N, Raina AK (2013) Composition, structure, and diversity of tree species along an elevational gradient in Jammu province of the north-western Himalayas, Jammu & Kashmir, India. J Biodivers Environ Sci 3(10):12–23

Shimono A, Washitan I (2007) Factors affecting variation in seed production in the heterostylous herb *Primula modesta*. Plant Species Biol 22(2):65–76

Shooner S, Davies J, Saikia P et al (2018) Phylogenetic diversity patterns in Himalayan forests reveal evidence for environmental filtering of distinct lineages. Ecosphere 9(5):e02157

Singh JS (2006) Sustainable development of the Indian Himalayan region: linking ecological and economic concerns. Curr Sci 90(6):784–788

Singh JS, Singh SP (1987) Forest vegetation of the Himalayas. Bot Rev The 53(1):80–192

Singh SK, Rawat GS (2000) Flora of Great Himalayan National Park, Himachal Pradesh. M/s Bishen Singh Mahendra Pal Singh, Dehradun, India

Srivastava V, Griess VC, Padalia H (2018) Mapping invasion potential using ensemble modelling. A case study on *Yushania maling* in the Darjeeling Himalayas. Ecol Model 385:35–44

Tessema YA, Joerin J, Patt A (2019) Crop switching as an adaptation strategy to climate change: the case of Semien Shewa Zone of Ethiopia. Int J Clim Chang Strateg Manag 11(3):358–371

Tewari VP, Verma RK, von Gadow K (2017) Climate change effects in the Western Himalayan ecosystems of India: evidence and strategies. For Ecosyst 4(1):1–9

Wamsler, C., Pauleit, S., Zölch, T et al (2017) Mainstreaming nature-based solutions for climate change adaptation in urban governance and planning. In: Kabisch N, Korn H, Stadler J et al (eds) Nature-Based solutions to climate change adaptation in urban areas, Linkages between Science, Policy, and Practice. Springer International Publishing, Switzerland, pp 257–273

Wani IA, Kumar V, Verma S et al (2020) *Dactylorhiza hatagirea* (D. Don) Soo: a critically endangered perennial orchid from the North-West Himalayas. Plants 9(12):1644

Wani IA, Verma S, Mushtaq, S et al (2021) Ecological analysis and environmental niche modelling of *Dactylorhiza hatagirea* (D. Don) Soo: a conservation approach for the critically endangered medicinal orchid. Saudi J. Biol. Sci. 28(4):2109–2122

Xu J, Grumbine RE, Shrestha A, Eriksson M et al (2009) The melting Himalayas: cascading effects of climate change on water, biodiversity, and livelihoods. Conserv Biol 23:520–530

PART III

Traditional Knowledge in Climate Resilience

CHAPTER 20

Traditional Ecological Knowledge *Versus* Climate Change Adaptation: A Case Study from the Indian Sundarbans

Sneha Biswas and Sunil Nautiyal

20.1 Introduction

Traditional Ecological Knowledge (TEK), sometimes termed as Local Ecological Knowledge (LEK), refers to the knowledge base developed over generations through direct interactions with the natural environment. TEK is considered a cumulative body of knowledge passed on to the next generation (Berkes et al. 2000). It is also referred to as one type of belief system or practice. Interaction with the natural environment is the key source for this knowledge (Berkes et al. 2000). As a result, it can be viewed as applied knowledge. The knowledge is considered to be highly localised and it is considered to be the knowledge's strength because of its local origin and the potential for providing a specific solution for the local context (Menzies 2006). Apart from the flow of knowledge, it also consists of world views, taboos, belief systems, etc. The more significant debate around TEK is whether it supports Western scientific knowledge or greatly deviates from it (Inglis 1993; UNFCCC 2013). Those in support of positivist approach of Western scientific knowledge generation questions its qualification as a source of knowledge. The larger consensus is on the fact that it is largely different from the Western scientific knowledge generation. In the process of generalisation, Western scientific knowledge often discards the specificity of a particular phenomenon. The right blend of

---

S. Biswas · S. Nautiyal (✉)
Centre for Ecological Economics and Natural Resources, Institute for Social and Economic Change, Nagarabhavi, Bengaluru, India
e-mail: nautiyal_sunil@rediffmail.com

© The Author(s), under exclusive license to Springer Nature Singapore Pte Ltd. 2023
S. Nautiyal et al. (eds.), *Palgrave Handbook of Socio-ecological Resilience in the Face of Climate Change*,
https://doi.org/10.1007/978-981-99-2206-2_20

both types of knowledge is often referred to as the solution to overcome the shortcomings of each stream of knowledge. The major areas where TEK is highly utilised are the areas of resource management, medicine, livelihood strategies and climate change and disaster risk reduction in the recent past.

TEK is a practical knowledge used by different indigenous communities against climate change. For example, indigenous communities in the semi-arid Sahel region of Africa use TEK for forecasting and monitoring resources against drought triggered by climate change (Nyong et al. 2007). They also recognise that TEK is generally participatory in nature and considers equity, economy and the environment simultaneously. A review paper identifies TEK as an important tool to maintain social-ecological resilience (Mallén and Corbera 2013). According to another paper, the governing agency of Vanuatu has already started implementing a combination of TEK and Western scientific knowledge in the seasonal forecast and other adaptation strategies. A similar approach can be seen in the case of Chad (UNFCCC 2013). UNFCCC (2013) proposes using phenological data, video recording of TEK, guided Participatory Rural Appraisal (PRA), and seasonal calendars are some ways we can incorporate TEK into the mainstream solutions for climate change. With the help of semi-structured interviews and PRA, it was found that agri-pastoralist group Dokpas of North Sikkim have observed the signs of changing climate in the mountain system; such as drying up of pasture land and how they have adapted to their changing pattern of storage of food during the winter season (Ingty 2017). The paper also suggests incorporating perception data from TEK in the downscaled or regional-level climate models. Similarly, a study by Hosen et al. (2020) found that people observed changes in climate through their traditional knowledge and how they have adapted to this with the help of changes in land use and agricultural practices (crop-animal farming, shifting cultivation etc.) in Borneo island. Similar practices can be observed by the Inuit population who use TEK in climate change adaptation by changing their hunting pattern, methods of hazard avoidance and emergency preparedness (Pearce et al. 2015). However, it is equally essential that TEK should be dynamic, evolving and flexible. On a similar line, the present paper explores the usage of TEK in the use of different resource units (land, aquatic, livestock and forest) of Indian Sundarbans.

Empirical Studies from Sundarbans Exploring TEK

Studies exploring traditional ecological knowledge in the Sundarbans can be categorised into four main categories: its usage in the agriculture sector, its use among the fishing community, medicinal plants and other aspects such as cultural belief system. Mukhopadhyay and Roy (2015) found that cultivating salt-tolerant crops is one of the adaptive strategies practised in Indian Sundarbans. In another study, the main objective was to understand the use and practice of indigenous knowledge in agricultural practices against the impact of climate change (Sarkar and Padaria 2011). With an in-depth survey and participant observation, they have identified different practices in plant protection, soil management, and the cultivation process. Some of the practices identified by them were the application of neem water solution as an organic insecticide; the use of washed earthworm water to stop the shedding of the lemon tree flower; washing off fog to stop the curling of crops; land shaping; increasing the height of the sides of the cultivating land using the dugout pond by side of it; digging soil with plough or spade to increase soil water retention; the practice of dhibi cultivation (by creating 2.5 metre dhibi); pot irrigation; application of liquid fertiliser made of faeces of animals; vermicomposting; storage in mud house; sprayer in kerosene container and use of lime in the pond water to clean the water, etc.

A recent study conducted by Roy et al. (2020) tried to identify TEK in fishing techniques used by the fishermen of Sundarbans through an ethnographic study. They have identified seven such useful techniques with the help of experts. Documentation of different types of nets (cast net, seine net, bag net, etc.) used by the fishermen of the Sundarbans area is also noticed in the literature (Ray 2013). While Sen and Pattanaik (2017) have not explicitly focused on fishing occupation, they have explored the methods in traditional livelihoods in Sundarbans, i.e. fishing, crab collection and honey collection. It holds a descriptive analysis of the traditional practices found in this aspect. Overall, qualitative methods such as ethnography, participant observation, and in-depth surveys are followed to get information about using traditional ecological knowledge. Pramanik and Nandi (2014) have used ethnographic data from two decades and documented that, crab catchers and fisher folk have developed a wealth of traditional knowledge especially in the identification of fishing zone, use of lunar calendar, weather prediction, taboos related to fish-eating habit and belief system. People's knowledge about weather and tides was also identified as a crucial component by (Srivastava and Sahay 2006). 'In Sundarbans, the traditional knowledge, social-religious practice, belief system and ethical values of the fisherfolk have contributed to the sustainability of forest and fishery resources of this region' (Pramanik and Nandi 2014). Due to the geographical similarity of the two places, studies from the Bangladesh Sundarbans area are also taken into consideration. Studies on the other part of Sundarbans (Bangladesh) also looked at the importance of indigenous knowledge against disasters. It was found that the Munda Tribe who are indigenous community were largely affected by the cyclones but eventually, they adapted through diversification of livelihood (Roy 2018). Titumir et al. (2020) made an honest attempt in capturing TEK usage in different livelihood sectors of Sundarbans area. They looked for the practices followed and suggested some policy recommendations and an alternative framework involving high importance to TEK to maintain the sustainability of the region. Mondal et al. (2012) looked into ethnobotanical importance and identified 35 species of mangrove plants (e.g. sundari, henthal, golpata etc.) that form medicine to cure some common as well as chronic ailments like fever, malaria, cold and cough, bronchitis, asthma, skin diseases, ulcers, leprosy, smallpox, diarrhoea, dysentery, diabetes, antifertility, etc. Another study from Bangladesh has explored the socio-cultural belief system of the people taking into consideration the folklore (Uddin and Habib 2019). This particular paper argued for bottom-up approach as against the top-down conservation approach taken to help people save their commons.

Theoretical Underpinning of TEK as an Adaptation Strategy

The present study is built on the theoretical framework of environmental determinism which implies that human decisions and culture are highly determined by the natural environment. It views the natural environment as the basic factor controlling human achievement (Lewthwaite 2004). The idea is supported by the emergence of the concept of Social Darwinism which again supports better bio-physical conditions as the yardstick of survival and development. Major proponents of environmental determinism can be found in the early twentieth century. The theory is well explored in the works by Ratzel, Semple, Huntington, etc. Huntington expressed that Nature is the cause and culture is the effect not the other way around. The word determinism itself imposes a cause and effect relationship. The impact of climate is well explored in the works of these environmental determinism proponents. Hardin (2009) emphasised that the impact of nature would be more visible for people who are directly dependent on it i.e. farmers, fishermen, hunter-gatherers. Life in Sundarban, people's lives are influenced by environmental determinism (Mukhopadhyay 2016). However, in some cases physical causation is unavoidable. For example, in case of natural disasters, we can only take precautions but we are not able to completely avoid nature's dictate in this case. In a later phase, environmental determinism suffered from vast criticism for promoting racism. Due to its harsh critics, some prefer to limit the reference to nature as mere 'influence' rather than being the determinant.

20.2 Data and Methodology

To understand the role of TEK as climate change adaptation in different sectors of resource usage in Indian Sundarbans, the researchers have followed a qualitative approach based on data collected from primary sources. Additionally, some secondary sources of data have been used to supplement the primary data from various government portals. Out of 19 blocks spanning the Indian Sundarbans, two blocks namely Kultali and Gosaba were purposively selected based on their geographical locations (see Fig. 20.1). A household survey consisting of 200 samples was conducted following geographical clustering. Further, in-depth interviews were conducted with 12 individuals ($N = 12$) consisting of farmers, fishermen and honey collectors. However, to explore

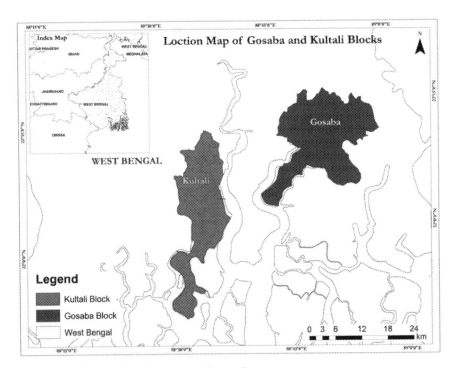

Fig. 20.1 Geographical location of the study area

the present research question, the authors have mostly depended on the qualitative data collected from these in-depth interviews, supported by data from the household surveys. The in-depth interviews have been analysed with the help of Atlas.ti software. Quotations were created from the data and then they were labelled to different codes. The codes have resulted into formation of different themes. The emerging themes have been analysed according to the different resource unit sectors of a Social-ecological system (SES).

20.3 The Social-Ecological System of Indian Sundarbans

Social-ecological system (SES) can be described as a whole system having two separate entities i.e. bio-physical system and the social system and interaction between them, or it can be considered as the interaction between two separate functioning systems. Folke et al. (2010) define SES as an 'integrated system of ecosystems and human society with reciprocal feedback and interdependence'. In order to understand the Indian Sundarbans as an SES, an overview of both of its bio-physical and social component is described below (see Fig. 20.2).

Indian Sundarbans is situated in the world's largest delta i.e. Ganges–Brahmaputra-Meghna delta covering an area of 4262 sq. km. It is part of the

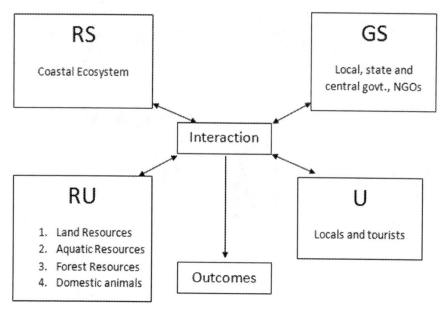

Fig. 20.2 Social-ecological system, adopted from Ostrom (2009)

larger SES known as Sundarbans, a major part of which lies in Bangladesh. Historically, Sundarbans was a place of dense mangrove forest with very sparse settlements. Reclamation of land started around 1830 with the motif of revenue collection which went on until 1875 (British Act VII). At present, Indian Sundarbans account for around 85 percentage of the mangroves found all over India (WWF- Vision). Also, 70 percentage of the mangrove species identified all around the globe can be found here. Among the faunal diversity estuarine and coastal fish, honeybees (*Apis dorsata*) and crabs are the resources that hold economic importance. The faunal diversity of Indian Sundarbans includes seven species of amphibians, 59 species of reptiles, 315 species of avian species and around 165 species of fish (Gopal and Chauhan 2006). It is also home to rare and globally threatened species such as estuarine crocodile, fishing cat, water monitor lizard, Gangetic dolphin, etc. Glimpses of Biodiversity in Indian Sundarbans are given in Table 20.1.

In 1930, an area falling below the imaginary Dampier-Hodges line was demarcated as Sundarbans. Currently, the Indian Sundarbans is situated in the eastern corner of the country, which covers 19 blocks spreading across two districts (around 9600 sq. km area) of the state of West Bengal. The Reserved Forest area, i.e. Sundarbans Biosphere Reserve (SBR), consists of one National Park and three wildlife sanctuaries covering an area of 4260 sq. km. The reserved forest comes under the jurisdiction of the state forest. Apart from this, there is a separate ministerial Department of Sundarbans Affairs under the state government of West Bengal to look after the developmental needs

Table 20.1 Biodiversity of Indian Sundarbans

Flora and Fauna	No. of species
True Mangrove species	26
Mangrove associates	29
Back Mangrove species	29
Total Flora species	**84**
Vertebrate	481
Hemichordate	1
Invertebrate	1104
Mammals	58
Avian species	248
Reptiles	55
Protozoan species	106

Source www.sundarbanbiosphere.org

of the people residing in Indian Sundarbans. According to the 2011 Census, Indian Sundarbans has a high population density, i.e. 1029 persons per sq. km and a decadal growth of population at 13.93%. Almost one-third of the population in the study area belongs to the disadvantaged section (Scheduled Caste and Scheduled Tribes). It was also observed that around 51% referred to agriculture as their primary occupation, 17.2% as fishing, while 10% as daily wage labourers and the remaining 21.5% stated some other job as their primary occupation.

20.4 Results

Family Structure

The family size of the respondents ranged from five to eight members in each family. Respondents mostly belonged to joint families with co-living arrangements (Parents living with married son, daughter-in-law and grandchildren). The age of the respondents ranged from 41 to 68. Elderly people were purposively chosen to get insights into their years of experience. All the respondents were permanent residents living in the same place from childhood. However, a few of the respondents' ancestors were migrants from the neighbouring country of Bangladesh. Land fragmentation became a problem for families who depend solely on agriculture. In most of the cases, it was observed that there is unwillingness among the present generation to take up the traditional or ancestral occupation. But most of the respondents are carriers of the same occupation over generations. One of the respondents replied, 'we have only one option that is agriculture, what else choice do we have! There is nothing like ancestral'. On the other hand, fishing and honey collection is avoided by the new generation as it needs patience and experience to learn the techniques. Many from the new generation are keen on learning some other skills which they refer to as 'haater kaj'.

Land Resources

Multiple themes emerged while analysing the land resources as agriculture is the main source of income for most of the people living in Indian Sundarbans. The household survey shows that 51% of households practice agriculture as their primary occupation. Although, a large proportion who depend upon cultivation are marginal farmers, small farmers and agricultural labourers. The amount of land cultivated ranged from 1 to 7 bighas in the study area. The lack of an irrigation system and high dependence on monsoon water makes the cultivation system highly mono-cropped. Hence farmers of the region in general practice paddy cultivation during the Kharif season. The major themes which have emerged in this area are general cropping patterns, tools, machinery and fertilisers, changes in the production or yield, weather-related issues and the cultivation process.

Changes in Crop Calendar

Majorly two types of paddy are cultivated in this widely mono-cropped area i.e. Aman (monsoon) and Boro (winter). The usual crop calendar starts with sowing Aman paddy in the monsoon season (in June-July); after its ripening in November, they'll sow Boro paddy, which will be ready to harvest by March–April. Many avoid Rabi cultivation due to the shortage of water. Among pulses, moong and grass peas are preferred by the farmers. Vegetables are grown either on a sustenance basis or some grow it for sale in the market. Some of the vegetables are grown along the paddy field, referred to as 'aal' in Bengali. Homestead gardening is practised to grow vegetables. Bitter gourd is one such vegetable grown along the 'aal' and cultivated twice a year like a paddy. Some other vegetables cultivated are different varieties of beans—flat beans, Indian beans or string beans, pumpkin, brinjal, watermelon, chilli, okra, onion, potato, bottle gourd, cauliflower, cabbage, etc. Some crops which are sown during winter are Indian beans and flat beans; crops that are planted during summer are yam, turmeric, bitter gourd and bottle gourd. The crop calendars for two periods have been given in the following tables (see Table 20.2 and 20.3).

Analysing the crop calendars for two points of time showed the changes in crop i.e. change in sowing, harvesting and duration of the crops. For example, the sowing of Boro rice (December) is observed to precede by one month of the sowing season of moong (January) in the 1990 table. On the other hand, at present, the sowing time of moong seems to be delayed for a month. Vegetables like flat beans, okra and brinjal take longer time to produce; The sowing period of Aman rice is also observed to be delayed for a month. On the other hand, onion and bitter gourd are observed to maintain more or less the same seasonal calendar. Although cultivation in Indian Sundarbans has been highly influenced by the monsoon, a more intense reliance

Table 20.2 Crop calendar of the study area as described by the farmer for the year 1990

1990	Jan	Feb	Mar	Apr	May	Jun	July	August	Sep	Oct	Nov	Dec
Aman Rice							▓	▓				
Boro Rice				▓								▓
Moong	▓				▓							
BG* Gourd*					▓	▓						
BG**		▓								▓	▓	
Beans^						▓	▓					
Okra						▓						
Brinjal										▓	▓	
Onion	▓											▓

*Bitter Gourd (Monsoon season), **Bitter Gourd (Winter season), ^Flat beans

Table 20.3 Crop calendar of the study area as described by the farmer for the year 2019

2019	Jan	Feb	Mar	Ap	Ma	Jun	Jul	Aug	Sep	Oct	Nov	Dec
Aman						▓	▓				▓	
Boro											▓	▓
Moong		▓							▓			
BG*					▓	▓						
BG**	▓									▓	▓	▓
Flat Beans						▓	▓	▓			▓	
Okra						▓	▓	▓				
Brinjal										▓	▓	
Onion	▓										▓	

*Bitter Gourd (Monsoon season), **Bitter Gourd (Winter season)

on the monsoon and post-monsoon season can be observed. It implies farmers' increased reliance on monsoon, as reported by them and testified by our empirical field study.

Tools, Machinery, Fertilisers

Changes in machinery, tools, and fertilisers have changed agricultural production. However, there were contentions among the farmers regarding using chemical fertilisers. One of the farmers informed us that they observe good yield from using chemical fertilisers. While on the other hand, farmers blamed the increased use of fertilisers and insecticides for soil degradation. Most farmers currently prefer to use chemical fertilisers or a combination of chemical and organic fertilisers in their fields. Cow dung is widely used as organic fertiliser. As one respondent quoted, 'We spread cow dung in the field. Apart from that, after cleaning the yard of our house, whatever debris we get, we dump it in the field'. Excreta of other animals is also reported to be in use (Sarkar and Padaria 2011). Other organic sources, such as dried rice straws, are kept to be rotten, and, when they form a lump, they are spread in the field as fertiliser. A shift in the usage of machinery has also taken place. Power tillers are used instead of ploughs and oxen. Water pumps are also utilised collectively by many farming families of Indian Sundarbans.

Weather or Climate-Related Issues in Agriculture

Climate or weather, in general, have caused much suffering to the farmers of Indian Sundarbans. The majority of the respondents were unanimous about the negative change in production. The majority agreed that the total yield has become around half in the past few decades due to changed weather conditions. Major weather-related problems are remarkable water shortage, increased temperature, decreased humidity, and increased soil salinity. The saline nature of the soil, which by recurring disaster situations, is most prominent during the winter because of less humidity. The drying weather and shortage of water are reported to cause dry leaves and dying plants. It is also held responsible for the reduction in yield. Production is primarily dependent upon rainwater. Irrigation is difficult as the rivers' water is also saline in nature. The only way of irrigation is the collection of rainwater in ponds, lakes, bills, small creeks, etc. Although ponds and bills are abundant in study areas, those alternatives also have limitations. Farmers are left to lack of enough water to sustain their water demand.

On the other hand, disaster risk is a well-known factor recognised by all farmers. The impact of disasters can be seen in property loss or loss of livelihood. The impact of disasters is undisputed among the residents of the Indian Sundarbans. The most recent devastating disaster was found to be Aila which came into discussion several times during the household surveys and in-depth interviews. However, the elderly remember another flood occurring in 1986 to be more overwhelming. However, it was found that the recent cyclones (2009, 2014, 2019) carried the most devastating effect on the agricultural fields. The presence of saline water in the area is causing the reduction of disappearance in the production of some traditional crops. One of the respondents recognised that natural disasters are very common to Indian Sundarbans, they have internalised them and adapted accordingly. Years of experience have taught them that agricultural land returns to its productivity (paddy production) after 3 or 4 years of floods or water stagnation. Although farmers employ people to dig out the 'aal' and pass the immediate water stagnation, the effect of it remains long-lasting. Aal is an elevated strip surrounding the main cultivable piece of land. Farmers build a little high (around 1–1.5 feet) barrier surrounding their cultivating land to protect against saline water intrusion. This barrier is further used to grow vegetables along with the paddy. Apart from the lack of water and dryness, the low land elevation prevents farmers from cultivating a few varieties of crops. However, changes in weather and other temporal alterations lead to adaptation in cultivation. People also take responsibility for reducing the yield of the increasing population and fragmentation of lands. Table 20.5

summarises climate change adaptation strategies found in different livelihood sections.

Observed Changes in Production

Kharif season produces 4 to 5 sacks of rice from one bigha of land. Some farmers said they used to have, on average, 10–14 bags of rice production in a year. The total yield of crops is reported to be half what it used to be two to three decades ago. A general consensus was found among the farmers regarding the drastic change in chilli production. After the 2009 cyclone, the area observed a sharp decline in chilli production, which used to be the major Rabi crop. Gosaba Block Land and Land Reforms department also confirmed the claim. Reductions in the yield of some lentils are also reported. The production of watermelon and Khesari daal is also reported to be reduced. Previously, people were satisfied with the production even with less fertilisers and insecticides. Some farmers contradicted this statement and said the yield was less in the past due to more usage of organic fertilisers. At the same time, most have opined that the previous weather condition was suitable for cultivation.

Livestock Resources

Livestock is a resource that is seen as the second source of income. It is also an instant source of cash in times of need. Cow, poultry, duck and sheep are the major types of livestock owned by the people of the study villages. Training facilities are available from the government and non-government organisations to train the family members, as livestock rearing is a viable source of income. Block office (in Gosaba) also provides medicine and treatment for diseases of domestic animals. However, some still use traditional medicines such as lime and turmeric are fed to chicken to increase their immunity. In most cases, people prefer to get medicines from the block office itself.

The changing weather scenario and increased land pressure have contributed to the reduction in the livestock population, especially cattle. Families with four to five cattle have now been reduced to only one or two. Fragmentation of land is an inevitable phenomenon, and with the increased pressure on land (extension of agricultural land and settlements), there are hardly any places left for cattle to graze. The reduced number of livestock is also due to the introduction of machinery (tractors and power tillers etc.) for agricultural purposes. Figure 20.3 shows the total number of livestock in the study area according to the latest census. Besides, livestock takes up a larger area in the house premises and involves a lot of time and attention. A dedicated family member is also required to look after the livestock; in most cases, they are the female members of the households.

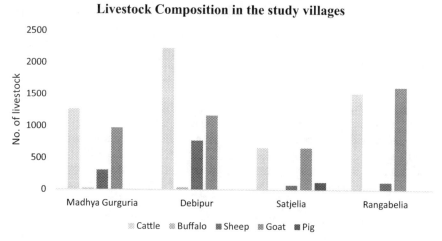

Fig. 20.3 Livestock composition of the study area (*Source 20th Livestock Census 2019, Ministry of Animal Husbandry and Dairying, Government of India*)

Aquatic Resources

Fishing in the dense Sundarbans forest has traditionally been a source of income for the Scheduled Caste community since the late nineteenth century when settlement began in the Indian Sundarbans. However, in recent years, fishing has adopted other communities as well, due to a lack of enough livelihood opportunities. The fishing season generally extends for Nine months *Ashwin* to *Falgun* (June to February). Generally, this fishing activity is done in groups. Usually, groups of four to five people twice a month sail to the rivers in a small boat for 78 days continuously while carrying minimum supplies such as food, net, etc. Some use the ponds to cultivate fish (putting small fishes in the month of *Ashwin* and catching the matured fishes during Poush); others create channels to cultivate shrimps. Another aquatic livelihood of the community in this region is collecting sea shells used to make jewellery.

Boats and nets are the common tools used for fishing. Fishing boats are usually smaller in size and made of wood (known as *dingi nouka* in local language) and contain only necessary things such as utensils, food items (rice, daal, oil), freezer, etc. Fishermen also use *Donga* i.e. a dugout canoe made from a palm tree trunk. They have to adjust on the boats for everything during their trip, which lasts about 8–9 days. Fishermen either own their boats or take them on rent. Building one's boat costs around Rs. 30,000–40,000. Boats are also bought from Kolkata. The profit made from the journey is divided among the boat owner, net owner, pump owner and fishermen.

Many types of nets are found that people use for fishing. These *include Berjaal, Khepla jaal, Manipul Net (Multifilament) or Nylon jaal, Pata jaal* etc. (known as seine net, stake net, cast net, bag net). Fishing nets are usually long enough to be spread around an ample area. For example, Ber jaal is 150 feet long. The fishing net, which is closed on one side, is called Pata jaal. Nets are both bought from Kolkata and stitched by hand, and the fishermen repair the nets themselves. Fishing nets need to be handled carefully and checked regularly. Fishermen check the nets after every six hours of ebb and tide. Fash jaal or Monofil net (Seine net) called monofil net as it is made of monofilament or nylon. Monofil nets are widespread to be used by fishermen. Pata jaal, or stake net used both for catching crabs and fish, in the river creeks. It is around 100 to 150 metres in length. It is positioned according to the width of the creeks. Fishermen put the pata jaal on the mouth of khari with the help of one long stick.

> Fish gets stuck during the tide and when we leave the net, the net gets closed, and all the fishes are stuck, after which we can collect them. It is estimated fishermen can catch one to two quintal fish with the help of pata jal.

Nets are owned mainly by the fishermen, while boats are sometimes rented by those who cannot afford to build one's own one. The usual fish varieties which can be found are prawns, bhetki, tengra, dadri and java (expensive 3000 per kg). Approximately 40–50 types of fish are caught in one trip.

The Fishing Process

Fishermen, crab collectors and honey collectors are people who deal with the forest and are referred to as the people who do jungle' jongol kore' (Mukhopadhyay 2016). At first, fishermen have to collect BLC (Boat Licence Certificate) from the forest department or the 'foresters'. The limited number of BLC (923) in the whole Sundarbans has resulted in competition over the years and business for the licence holders. The fishing season extends about Nine months (*Ashar to Ahswin*) from June-July to February–March. Fishermen (usually four to six people) arrange a boat, net, food, water and other necessary goods for a journey extending six to nine days. Fishing trips are generally arranged twice a month. Once the 'gon' starts, i.e. the tenth day after the full moon or no moon, the journey begins. The reason for selecting that particular day is because waves or river flow begin to increase on that specific day. The higher flow (due to the tidal impact caused by the position of the moon) helps the fishermen both at rowing boats and better catching of fishes. The river flow decreases from the second day (*Dwitia*) and becomes significantly less by the fifth or sixth day. Fishermen return by the time the flow subsides. The selection of their journey date according to the lunar calendar is important as it decides the amplitude of waves in the river, favouring their rowing of boats. Upon reaching a suitable place between the creeks, they anchor their boats in the river. Meal preparation, bathing and other daily necessities are done on the boat or by the side of it. They bathe in the saline river water and fail to maintain their timetable, and all of their daily activities depend upon the time of tide and ebb. A fisherman describes a particular first day at fishing,

On the first day, we cast our nets (Fash Jaal, Ber Jaal or Pata Jal) around evening, then we have our meals; and then we wait for the tidal event to end (Either Tide or Ebb).

After every six hours, they keep changing the position of the nets. First, they pull the net and check if to see whether fishes are captured or not, then they keep the trapped fishes inside the cold storage and then again spread their nets.

It doesn't matter what the tide and ebb are; even in the middle of the night, we have to spread out our net... We wait at each place for two days, then move ahead to the next creek.

Experienced fishermen informed that earlier, the amount of fish caught was higher than at present. The reason provided by them is the rampant growth of tourism in the Indian Sundarbans area. The tourist season starts in November and lasts till March when fishes are at their juvenile stage. According to them, the increased pollution level in the river and the congestion caused by the tourist boats (run on diesel) are the reasons for fish stock reduction. They observed an inverse relationship between the amount of fish stock and the number of tourists over the period of time. 'Juvenile fish are developed during the month of Chaitra (March–April). When these small fishes are moving across the river, boats are moving over them, killing the juvenile fish/ how will they (the fishes) survive?' The reason for such depletion may also be attributed to the dying river, mechanisation in the fishing process, increase in riverine transport, etc. Fishing and crab collection in the river channel can pose a high risk in terms of safety; however, this activity brings a high profit. The limited number of BLCs has restricted the number of fishermen entering the forest. Sometimes clashes take place between the forest department and the fishermen.

Forest Resources

The most excellent resource of the Sundarbans area is mangroves. The name of Sundarbans itself is derived from one of its mangrove species, i.e. *Sundari*. Different forest acts impacted the nature of the use of the forest by the locals. West Bengal State Forest Department (FD) is tasked with safeguarding the protected areas. The only forest product which is left for collection is honey. Earlier, people used to collect Golpata (leaves) and Gab (one type of resin). Now, people can only collect honey from the forest, which they must sell back fully or partially to the FD at a meagre amount (Rs. 90–130 per kilo). The honey collection season is restricted to only one to two months, i.e. March to April. Apart from it, people of Indian Sundarbans are not allowed to collect other NTFPs. Strict regulation is the result of Project Tiger since 1973 and the declaration of Indian Sundarbans National Park.

While fishermen are used to travelling by small *dingi* boats for their expeditions, honey collectors usually take bigger boats on their honey collection trips, sometimes ranging from 60 to 70 boats together. Honey collection is done (allowed) only once annually in the month of Chaitra, i.e. March–April. The whole month is reserved for honey collection. As one of the respondents narrated the process,

> Eight to nine people team up and collect one pass (BLC) by paying Rs. 50000. Once we get the pass, we set for one month's journey into the forest. After that, we have to go deep inside the forest to collect honey. Earlier we used to stay for one month, but now we stay for 22 days in the forest and come back. While returning, we must sell some amount to the forest department and keep the remaining honey.

While on the other hand, individual honey collector enters for a few days (four to five days) in the Chaitra month. Once they enter the forest, they will be busy with honey collection from morning to evening. '*After getting a pass from forester, we enter the jungle and keep ourselves busy in madhu kata from morning to afternoon, from morning to afternoon daily, and only come during lunchtime. One person waits and cooks for us in the boat*'. Honey collectors have to keep track of their way inside the forest constantly. They have to be sure which side their boat is (right or left), which side (N-S-E-W) they are heading etc. The navigation technique results from years of experience gathered through numerous trips. The only tool they carry with them inside the forest is a stick or a big hand knife or dagger called locally 'da'.

Honey collection is the riskiest affair among the three (fishing, crabs and honey) as they have to enter deeper forests. Still, they are often left with a significantly lower profit margin. Other than the significant resources used widely by the people of Indian Sundarbans, medicinal plants grown naturally are used to treat basic ailments such as the common cold, fever, etc. (Table 20.4).

The Legacy of Bonbibi

The legacy of Bonbibi in Sundarbans stayed for a long time. Initially, she was a goddess preached by Muslim Pirs. Later, she became the epitome or the symbol of the goddess of the forest for people from different religions. She is celebrated by people dependent on Sundarbans for their livelihoods. Her legacy is imprinted in the famous story of 'Bonbibi Johurinama' in which she saves a destitute boy named Dukhi from the treacherous Dakhin Roy (who also takes the form of a tiger). Dakhin Roy is a half-human and half-tiger, always worried about his territory in the forest. The symbolism is that every common man entering the forest relates with Dukhi. Dakhin Roy's rage can be related to people's greed. The tale of Bonbibi, her brother Shah Jongoli, Dakhin Roy and Dukhi provides guidelines for people on how to behave while entering into a forest.

Table 20.4 Major areas of usage of TEK in

Areas	Usage of TEK
Organic fertilisers	Farmers use their traditional knowledge to use household waste, cow dung, and rotten rice straws as organic fertilisers
Nets (Fash jaal, pata jaal, ber jaal)	Nets are placed according to the width of creeks, nets are managed according to the time of tide and ebb
Boats	Fishermen build their boats out of palm trees and other woods
Dagger	Used for protection as well as clearing the jungle path during the honey collection
Sailing	Dependence on the lunar calendar for getting favourable tidal conditions, support on daily tide and ebb for fishing
Direction	Use experience gathered over the years to determine the direction they're going into the forest and how far they're from their boats
Medicinal plants	Use different varieties of Tulsi as a remedy for common cold, flu, etc

People living near the forest or temple of Bonbibi participate in the celebration of worshipping her. Bonbibi pujo takes place towards the end of Falgun month. One of the respondents particularly referred that Bonbibi is worshipped on the 28th day of the Bengali month Falgun (February–March). Both Bonbibi and Dakhin Roy are worshipped. There are several temples built near the forest in each village. Brahmin pundits are invited to do the puja rituals for Dakhin Roy and 'puthi' referred to Bonbibi Johurinama has to be read. Donations for the festival is collected from all the nearby neighbours and villagers and celebrated like other festivals. Songs and acts are performed on 'Bonbibi Kahini' meaning the story of Bonbibi. Everyone worships Bonbibi as she has become a cultural symbol of Sundarbans area. Furthermore, whatever the occupation people know that their lives are deeply linked to the forest. They cannot avoid forests while living in this archipelago. People express their dependence and faith on the goddess saying.

> …means she is actually the goddess of the forest, and we have to totally depend on the forest. Majority of the people here depend upon either on cultivation or on the forest. This is the reason why we worship her.

Another fisherman confirmed his faith by narrating '*We go to the forest keeping faith on Bonbibi only. Those who have faith from their heart they survive, those who don't have faith face danger*'. Forest is nearby to the villages surveyed. Many opt for fishing, crab collection or prawn seed collection as their source of income. It is heard many times, crab collecters are attacked by tigers. Another farmer explained,

> I have some friends who are fishermen. Sometimes I used to go to fishing along with them. Sometimes I didn't go along with them as well. Sometimes I

got to know that tiger has taken someone during crab collection. Poor people, how they will survive (have food)! Within one and half km from this village, I had a friend named Sanjay Sardar (belonging to Munda community) who went to catch crabs in the jungle. Tiger took him. After that, his family is in dire condition.

People believe that nothing happens to those who have deep belief in her. Those who do not have true faith on her, fall into danger. How faith on the forest goddess is deeply embedded in the mind of the people, it can be expressed through the narration of one of the fishermen who believes that several times when his 'dingi' boat got drowned was due to his greater greed. The belief that forest goddess will only allow those with pure heart and who are content with the just amount of forest resources. The forest goddess (mother nature) will remind whenever one becomes greedy. It is confirmed from the locals' perception that Bonbibi has enough for the needy, not for the greedy.

20.5 Adaptations to Climate Change and Disaster-Related Issues

The main objective of the present paper was to identify the usage of TEK as adaptation strategy against the issues posed by weather and disaster-related issues and in resource usage in the Indian Sundarbans area. Certain changes in the production pattern have been observed due to environmental hazards and weather variations. Similarly, certain adaptation strategies have been identified in agriculture, fishing, honey collection, disaster management and community-based initiatives. Table 20.5 summarises climate change adaptation strategies found in different sections.

In agriculture, the adaptation is related to changes in cropping pattern, mono-cropping, elevating the side of the land in order to prevent saline water

Table 20.5 Climate induced changes and adaptation strategies among different livelihood options

Sector	Climate change adaptation strategy
Agriculture	Cultivating mostly Kharif crops, cultivating varieties that require less water, sowing vegetables along the 'aal' of the agricultural field
Fishing	Selective season fishing limited livelihood opportunity have forced people from different castes to accept fishing as an occupation
Honey	Reduced days of honey collection
Disaster	Wait for natural regeneration of land, social capital, stocking of food and other necessary goods
Community-based initiatives	Repairing embankment, consciousness about protection of forest resources, festivals

intrusion. Fishing although remained more traditional in technique, has to adapt to a particular time of the year due to the existing government policies. Their total fish catch is also influenced by the change in fish stock in the river. Besides, limited income opportunities have pushed people from different castes to opt for fishing as a means of livelihood. Similarly, like fishing, honey collection saw a reduction in the number of days of honey extraction to adjust to the changed weather condition and wait for the regeneration of the land. In the case of adaptation to disaster, people are mostly dependent upon social capital. In the end, community-based initiatives mostly revolved around protection of the environment.

20.6 Discussion and Policy Implication:

Lives of the people living in Sundarbans are guided by environmental determinism (Jalais 2014; Mukhopadhyay 2016). The qualitative assessment of this paper enforces this concept demonstrating the dependence on nature in different livelihoods. The major themes which emerged from the qualitative assessment revolved around weather-related problems, season or time, tools and processes. The major impediment in cultivation was found to be natural hazards and changes in weather conditions. Farmers have adapted to the saline water intrusion by elevating the surrounding area of a cultivable land known as 'aal' and using that strip for vegetable cultivation. Further, farmers manage water shortage for cultivation by collecting water into small ponds, however further state intervention is required to improve the situation. According to Hazra et al. (2015) 80.53% of the gross cropped area in Indian Sundarbans remains unirrigated. Additionally, given the bio-physical vulnerability of the region climate resilient farming should be introduced and promoted. Although fishing and crab collection cover a smaller percentage of the total occupational structure, it is an integral part of the larger social-ecological system which is disturbed at present mostly due to human intervention. It was found that fishing, crab collection and honey collection remains dependent upon traditional tools and methods. Existing literature claims that TEK adapts with change in generation but in the case of fishing and honey collection it was observed that people are dependent upon years of experience and the traditional tools and methods. Along with the usage of traditional tools and methods, their livelihood is also influenced and shaped by their belief in the goddess of the forest, Bonbibi. Reverence towards nature results from their belief system. The hardships and dependence upon traditional tools in the case of fishing and honey collection are partly guided by state directives. Conflicting actions are observed as the result of stringent state directives and the dire economic condition of the people. The double-dealing of the state is exposed when a large number of engine boats and lodges are encouraged in the name of tourism, but on the other hand, fishermen are still required to use hand rowing boats to reduce river pollution. Improvement in irrigation facilities, drainage system, marketing facilities for the produced crops, smooth process of compensation and increase in boat license certificates are some of the aspects which need to be looked at. One needs to take into consideration the development and continuity of use of TEK in the changing socio-political set-up. As TEK is also shaped by the political set-up, power relations and historical change (Butler, 2006).

References

Berkes F, Colding J, & Folke C (2000) Rediscovery of traditional ecological knowledge as adaptive management. Ecol Appl 10(5): 1251–1262.

Butler C (2006) Historicizing indigenous knowledge: practical and political issues. In: Menzies C R (Ed) Traditional ecological knowledge and natural resource management. Lincoln and London, University of Nebraska Press, p 107–126.

Folke C, Carpenter S R, Walker B, Scheffer M, Chapin T & Rockström J (2010) Resilience thinking: integrating resilience, adaptability and transformability. Ecol Soc 15(4): 20. http://www.ecologyandsociety.org/vol15/iss4/art20/.

Gopal B, & Chauhan M (2006) Biodiversity and its conservation in the Sundarban Mangrove Ecosystem. Aquat Sci 68(3): 338–354.

Hardin G L (2009) Environmental Determinism: Broken Paradigm or Viable Perspective? PhD Thesis. East Tennessee State University, Tennesse, USA. Paper 1839. https://dc.etsu.edu/etd/1839.

Hazra S, Bhadra T, & Roy S P (2015) Sustainable water resource management in the Indian Sundarban Delta. In: Proceedings of the international seminar on challenges to ground water management: vision, 2050, pp 324–332.

Hosen N, Nakamura H, & Hamzah A (2020) Adaptation to climate change: Does traditional ecological knowledge hold the key?. Sustainability 12(2): 676.

Huntington H P (2000) Using traditional ecological knowledge in science: methods and applications. Ecological applications 10(5): 1270–1274.

Inglis J (Ed) (1993) Traditional ecological knowledge: Concepts and cases. IDRC.

Ingty T (2017) High mountain communities and climate change: adaptation, traditional ecological knowledge, and institutions. Clim Change 145(1-2): 41–55.

Jalais A (2014) Forest of tigers: people, politics and environment in the Sundarbans. New Delhi, Routledge.

Lewthwaite Gordon R (2004) Environmental Determinism. In: Smelser Neil J and Baltes Paul B (Hrsg): International Encyclopedia of the Social & Behavioral Sciences. (Originalausgabe: 2001) Elsevier, Oxford, p 4607–4611.

Menzies C R (Ed) (2006) Traditional ecological knowledge and natural resource management. Lincoln and London, University of Nebraska Press.

Mondal B, Sarkar N C, Mondal C K, Maiti R K, & Rodriguez, H G (2012) Mangrove plants and traditional Ayurvedic practitioners in Sundarbans region of West Bengal, India. Crop Res 13(2): 669–674.

Mukhopadhyay A (2016) Living with Disasters: Communities and Development in the Indian Sundarbans. Cambridge, Cambridge University Press. https://doi.org/10.1017/CBO9781316227572.

Mukhopadhyay R, & Roy S B (2015) Traditional knowledge for biodiversity conservation, maintain ecosystem services and livelihood security in the context of climate change: Case studies from West Bengal, India. Biodiversity J 6(1–2): 22–29.

Nyong A, Adesina F, & Elasha B O (2007) The value of indigenous knowledge in climate change mitigation and adaptation strategies in the African Sahel. Mitig Adapt Strateg Glob Chang 12(5): 787–797.

Ostrom E (2009) A general framework for analyzing sustainability of social-ecological systems. Sci 325(5939):419–422. https://doi.org/10.1126/science.1172133

Pearce T, Ford J, Willox A C, & Smit, B (2015) Inuit traditional ecological knowledge (TEK), subsistence hunting and adaptation to climate change in the Canadian Arctic. Arctic 233–245.

Pramanik S K, & Nandi N C (2014) Traditional knowledge of fishermen communities of Sundarban, West Bengal, Traditional Knowledge 21.

Ray T (2013) Indigenous fishing knowledge of Sundarban. Lokaratna 5&6.

Roy A, Sinha A, Manna R K, Aftabuddin M D, & Das S K (2020) Traditional knowledge of the fishermen community of Indian Sundarbans: An assessment of rationality and effectiveness. Indian J Fish 67(2): 94–101.

Roy S (2018) Livelihood resilience of the indigenous Munda community in the Bangladesh Sundarbans forest. Handbook of climate change resilience. Springer Nature, Germany, 1-22.

Ruiz-Mallén I, & Corbera E (2013) Community-based conservation and traditional ecological knowledge: implications for social-ecological resilience. Ecol Soc 18(4).

Sarkar S, & Padaria R N (2011) Understanding indigenous knowledge system in coastal ecosystem of West Bengal. J Community Mobilization Sustain Dev 6(1): 019–024.

Sen A, & Pattanaik S (2017) How can traditional livelihoods find a place in contemporary conservation politics debates in India? Understanding community perspectives in Sundarban, West Bengal. J Political Ecol 24(1): 861–880.

Srivastava P, & Sahay V S (2006). Traditional Ecological Knowledge Vis-à-vis Sustainable Development in Suderbans Development in Suderbans. Orient Anthropol 6(1): 192–199.

Titumir R A M, Afrin T, & Islam M S (2020) Traditional Knowledge, Institutions and Human Sociality in Sustainable Use and Conservation of Biodiversity of the Sundarbans of Bangladesh. In: Managing Socio-ecological Production Landscapes and Seascapes for Sustainable Communities in Asia. Springer, Singapore. pp 67–92.

Uddin M, & Habib K A (2019) Traditional Knowledge, Culture, and Biodiversity of Sundarban UNESCO World Heritage Site. J Ocean Cul 2: 74–86.

UNFCCC (2013) Best practices and available tools for the use of indigenous and traditional knowledge and practices for adaptation, and the application of gender-sensitive approaches and tools for understanding and assessing impacts, vulnerability and adaptation to climate change. Technical Paper, (FCCC/TP/2013/11) 1–62.

CHAPTER 21

From Resilience to Vulnerability: Indigenous Agri-Food Systems of Wayanad District

V. Swaran, C. S. Chandrika, and Chubamenla Jamir

21.1 Introduction

More than 54% of the world's population currently lives in urban areas. However, Indigenous Knowledge Systems (IKS) in their largely rural and traditional settings remain as islands of resilience against Global Environmental Change (GEC), in this case, considered as the human-induced and natural changes affecting life on earth (UNDESA 2018). Indigenous knowledge can be broadly defined as the 'local knowledge held by indigenous peoples or local knowledge unique to a given culture or society' (Warren et al. 1995 cited in Berkes 2008, p. 9). This knowledge stems from the localised contexts of

---

V. Swaran (✉)
Centre for Technology Alternatives for Rural Areas (CTARA), IIT Bombay, Mumbai, India
e-mail: swaran@iitb.ac.in

Watershed Support Services and Activities Network (WASSAN), Hyderabad, India

V. Swaran · C. Jamir
Department of Natural and Applied Sciences, TERI School of Advanced Studies, New Delhi, India
e-mail: cj@chubamenlajamir.com

V. Swaran · C. S. Chandrika
M. S. Swaminathan Research Foundation Community Agrobiodiversity Centre, Wayanad, Kerala, India
e-mail: cschandrika.writer@gmail.com

© The Author(s), under exclusive license to Springer Nature Singapore Pte Ltd. 2023
S. Nautiyal et al. (eds.), *Palgrave Handbook of Socio-ecological Resilience in the Face of Climate Change*,
https://doi.org/10.1007/978-981-99-2206-2_21

communities and their long-term interactions with their local natural resources (Gadgil et al. 1993, p. 151; UNESCO, n.d.). Agrobiodiversity is an important domain where the application of IKS is prevalent across the globe (Diaz et al. 2015; Kahane et al. 2013).

Within agrobiodiversity, the Neglected and Underutilised Species (NUS) are increasingly relevant in the context of resilience (Chivenge et al. 2015). The terms neglected and underutilised have commercial subtexts suggesting these crops are largely outside the mainstream agri-food system. Certain attributes of these crops, like their medicinal and nutritional properties, remained emic knowledge of communities who preserved them, even as mainstream agriculture limited its focus on yield and supply chain amenability of crops and livestock (Pieroni and Price 2006). This knowledge set, widely acknowledged as a source of resilience for the community, is in a vulnerable position due to the pressures from the GEC, like fragmentation of forests and urbanisation apart from the climate crisis (ISSC and UNESCO 2013; Johns 2007). Any changes to the local socio-ecological contexts and community interaction with natural resources might limit the scope of IKS' applications, especially in terms of resilience. Resilience in this context could be taken as "a measure of the persistence of systems and of their ability to absorb change and disturbance and maintain the same relationships between populations" (Holling 1973, p. 14).

One way to counter the vulnerability is to combine informal but contextualised IKS with tools and techniques of formal scientific knowledge. In the context of the climate crisis, the IKS plays an important role in identifying and interpreting the bioindicators by providing information on "what to look for and how to look for what is important" (Berkes 2008, p. 162). Such observations are local concerns and livelihood priorities, making them helpful in assessing community-level impacts and options for adaptation (Nakashima 2010; Orlove et al. 2010).

This study, carried out in 2015, delves into the traditionally accumulated knowledge systems of Adivasi communities in the Wayanad district of Kerala, India. It documents the current state of traditional cropping systems, food habits among three Adivasi communities, and how it is sustained, transformed and influenced by local weather anomalies and societal changes. This study's results may help synergise conservation, economic empowerment of rural areas and attainment of food security at large.

Study Area—Geography and Demography

Wayanad district is situated in the northeast of Kerala state, Southern India (see Fig. 21.1). The entire 2130 Km^2 area of the district is located at an elevated table that forms the southern end of the Deccan Plateau. Its elevation varies from 700 to 2100 metres above the Mean Sea Level. The elevation drops west to east, with flat and open terrain (see Fig. 21.2). This topography makes the western part of the district wetter than the east.

21 FROM RESILIENCE TO VULNERABILITY: INDIGENOUS ... 353

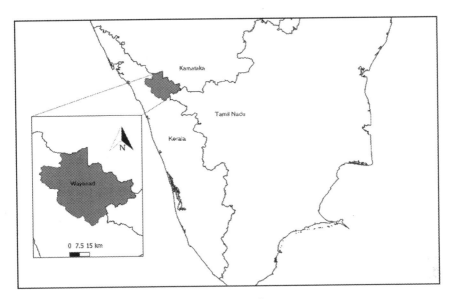

Fig. 21.1 Location of Wayanad district in Kerala

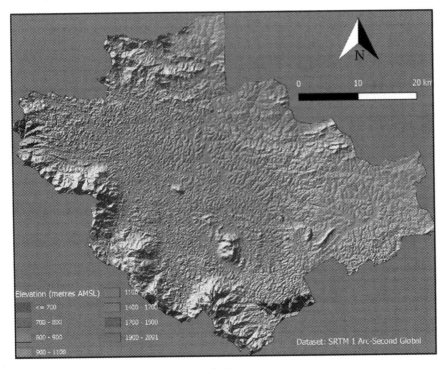

Fig. 21.2 Elevation profile of Wayanad district

Forests, some of them a part of the Nilgiri Biosphere Reserve, constitute about 40% of the district's area. The district is home to about 300 species of an estimated 2000 species of flowering plants endemic to the Western Ghats. All three talukas in the district are assigned to Ecologically Sensitive Zone 1, warranting the highest degree of protection, by the Western Ghats Ecological Expert Panel underpinning its social and ecological significance of the district (Western Ghats Ecological Expert Panel Report 1 2011). Besides, the district is one of the four climate change hotspots identified in the Kerala State Action Plan on Climate Change (Department of Environment and Climate Change 2014, p. 39).

Demography

As per the 2011 census, about 96% of the district's population lives in rural areas with agriculture as their primary occupation. About 48% of the total cropped area of the district is under plantation crops of which coffee constitute 80% of the area. Rice is the major cereal grain cultivated in Wayanad and nearly 26% of them are local varieties (Department of Economics and Statistics 2015). The ten different Adivasi communities of Wayanad make 18.53% of its population, which is the highest in any district of Kerala (District Census Handbook Wayanad 2014). This study is carried out among three groups, Paniya, Kuruma and Kattunayakan. Paniya is the largest single Adivasi community in Kerala and most of them live in the Wayanad District. They have a distinct language of their own, closely related to Malayalam. Originally, they were wage labourers to landlords in the district (Thurston and Rangachari 1909).

Kuruma community traditionally practised slash and burn agriculture, wood cutting and collection of minor forest produce. Their population is distributed within a radius of about 30 km, including the eastern part of Wayanad and the western part of the Gudalur taluk of Tamil Nadu. Their language is Malayalam with a spattering of Kannada and Tamil words. Agriculture is the main occupation of this settled land-owning community. The main crop is paddy, which is cultivated in the fallows and flat lands as well as on moderate slopes. Hunting and fishing were as important to them as agriculture as a source of food (Narayanan et al. 2011).

Kattunayakan community is one of the 75 Adivasi groups of India designated as 'Particularly Vulnerable Tribal Groups' by the Ministry of Tribal Affairs, Govt. of India. The term Kattunayakan is a portmanteau of the terms 'Kadu'—meaning forest, and 'Nayakan'—meaning leader or headman (Thurston and Rangachari 1909). They speak the Kattunaikka dialect, which is close to Kannada. Rice and finger millet are their staple cereals supplemented by roots and tubers (Narayanan et al. 2011).

21.2 Methods

Through questionnaire-based interviews, contextual inquiry and literature review, the present state of the indigenous knowledge in food habits and livelihood practices, perceptions of rainfall and temperature among the selected three Adivasi groups was carried out. The reviewed literature also includes the data collected by MS Swaminathan Research Foundation-Community Agrobiodiversity Centre (MSSRF-CAbC) staff during various field surveys apart from published documents. Contextual inquiry was the primary mode of data collection. Contextual inquiry can be applied to a specific locality, case or social setting. This method sacrifices the breadth of population coverage and statistical generalisability to explore issues in depth (Booth et al. 1998). In this approach, semi-structured interviews are held where interviewees are interviewed in their context when doing their tasks, with as little interference from the interviewer. Apart from the questions in the questionnaire, impromptu questions were asked. Ten individuals from Paniya, Kuruma and Kattunayakan communities were interviewed in this study. Purposive sampling is used wherein those who can and are willing to provide the required information by virtue of knowledge or experience are interviewed (Tongcon 2007). The sample sizes are determined based on theoretical saturation, 'the point in data collection when new data no longer brings additional information to the research questions' (Holloway 1997). A semi-structured questionnaire in English was prepared which was later translated into Malayalam which is the lingua franca of the interviewees. In addition, a few closed-end questions were included to understand their socio-economic status. Open-ended questions can capture judgements and perceptions and allow complex analyses of often non-quantifiable cause-and-effect processes (Tongcon 2007). A narrative approach was adopted to collect interviewees experiences with farming and food habits. Interview data recorded in the questionnaire response sheets were coded and manually analysed. Additionally, existing literature, previous studies, and field notes from the researchers at the MSSRF-CAbC were compiled.

21.3 Results and Discussions

The knowledge and dependence of various Adivasi communities on local flora and fauna are well studied and documented (Narayanan et al. 2011). The green Leafy Vegetables (LV's) were once extensively used by the Adivasis and are inexpensive, easy to cook and rich sources of macro- and micronutrients. A decline in the usage of local food sources and subsequent erosion of knowledge and cultural values associated with wild food plant use in Wayanad has been reported due to factors like social stigma and discrimination against the Adivasi communities and cultural exchanges (Pradeepkumar et al. 2013). Collection and consumption of wild food plants, which form an important part of local diets and are used as famine foods and medicines, are increasingly stigmatised as symbols of poverty and 'tribalness'. This change in food habits and lifestyle of Adivasis has increased the incidence of nutrition deficiency-related diseases among them (Shrinivasa et al. 2014).

Besides vegetables, Wayanad has many traditional rice varieties with aromatic and medicinal values, drought/flood resilient and pest resistant properties. However, most of these are on the verge of extinction, with most of them being conserved by a few tribal communities (Keralabiodiversity.org, n.d.). Rice varieties are particularly protected by tribal communities practising family farming and by others holding land whereas those communities owning marginal area of land or no land at all collects the wild edibles either from open-landscape or forests (Chandrika and Nandakumar, n.d.; Narayanan et al. 2011; Suma 2014).

The aforementioned is the background with which this study was carried out. It reveals the different pathways through which traditional knowledge and food influence the three communities. Distinctions were observed in how each community harnessed its traditional knowledge with respect to the use of flora and fauna.

Food Habits and Influence of Traditional Knowledge

Paniya and Kattunaikka community predominantly applies the traditional knowledge and practices related to plants and small animals for sourcing food. In contrast, the Kuruma community utilises the knowledge for medicinal and agricultural purposes. Rice is a staple food of all three communities, akin to the majority in Kerala, though traditional and wild food was a major ingredient in their diet (Thurston and Rangachari 1909). Paniya consumed 72 types of LV's, 25 types of Mushrooms, 19 varieties of roots and tubers, 48 types of fruits, nuts and seeds, 36 varieties of fish and eight types of crustaceans. The Kattunaikka consumes 35 types of LVs, 25 types of tubers and roots and 37 types of fruits and seeds (Narayanan et al. 2011). Most of the roots and tubers they consume are *Dioscorea* and *Colocasia* varieties and locally available peas. The present study documents only a fainter reflection of their rich dietary basket described by previous studies (see Table 21.1). Reasons for this can range from a change in dietary patterns to the comparatively brief study period and the general reluctance observed among the respondents to disclose their food habits.

For treating minor illnesses Paniya and Kuruma communities use medicinal plants like Brahmi (*Bacopa monnieri*), Five-leaved chaste tree (*Vitex negundo*), Pepper leaf, Willow-leaved Justicia (*Gendarussa vulgaris*), Borage, Fringed Rue (*Ruta chalepensis*), Aatalotakam (*Adhatoda vasica*), Tulsi, Dried Ginger. Regardless of the service of government and private hospitals (or the monthly household visit of government-deployed doctors as in the case of the Kattunaikka people at Rattakolly), reliance on traditional knowledge is still prevalent.

In the Kuruma community, elders have comprehensive knowledge of medicinal plants and their treatment processes. Most community members are

Table 21.1 Some of the traditional food reported to be consumed in the present study

Category	Types reported
Leafy Vegetables	Leaves of Colocasia
	Wild Senna
	Indigenous spinaches
	Sessile joyweed
	Wood sorrel varieties
	Water hyssop
	Lilac tassel flower
	Vegetable fern
Traditional sources of meat	Crabs
	Snails ('*Noonji*' in local dialect)
	Small birds (Coucal, Wild hen, Crane, Dove and Myna)[1]

aware of herbs to treat conditions like fever and headache, but they approach the expert elders for other diseases. The experts guarded this knowledge as a community secret and were unwilling to reveal anything apart from the aforementioned medicinal plants. One interviewee explained about a concoction given for postnatal health care of women made of 'Thumpa' (*Leucas cephalotes*), 'Kattu Mangappuly' (Souring agent made out of Wild Mango), and Garlic (*Allium sativum*).

Factors Influencing Food Consumption and Conservation

Economic and Cultural Dynamics

Landholding size, annual average income and alternate livelihood options influence how IKS, food habits and conservation practices are linked, as shown in Table 21.2.

With their relatively larger land holdings, six of the ten Kuruma families engaged in the present study are doing paddy farming, and all use the harvest for self-consumption. However, farming is identified as the primary source of income by just one of the ten interviewed, with all but one (a state government employee) working as un/semi-skilled labourers. Traditionally landless, and as evident from the statistics given in Table 21.2, Paniya and Kattunaikka people are not engaged in farming. Almost all of them are employed as daily-wage labourers. Some of the Kattunaikka families in the Muthanga area grew pepper vines and ginger in their homesteads, whose produce they used to sell in local shops. Fair Price shops run by the government are the primary rice

---

[1] Though the Wildlife (Protection) Act 1972 has largely led to withdrawal from hunting it is still rarely practised. On most occasions, the catches are small birds found nearby their houses.

Table 21.2 Economic status and application of traditional knowledge

Community	Average landholding size (in cents²)	Major land holding type	Average annual income (in INR)	Prevalent application of TK
Paniya	13.65	Joint	3066	Food and Medicine
Kuruma	163	Single	4200	Farming and Medicine
Kattunaikka	7³	NA⁴	NA⁵	Food

source for these two groups. Their land holding size is meagre that sometimes they cannot keep vegetable gardens at home and have to depend on the open market or collect tubers, legumes and LVs. A major factor influencing their dietary choice is hence its seasonal availability. Therefore mushrooms, LVs and crabs are consumed mainly in the monsoon season, whereas tubers are during the summer.

Consumption of LVs and other indigenous edibles by the Kuruma community is much less these days as their dietary preference and livelihood options have changed. Many reported finding time for food collection amidst their day job is difficult. But they continue to utilise many of the local floras for medicinal purposes. A large body of IKS centred around the paddy farming practice of Kuruma families. Cheeyach, Veliyan, Valichoori, Gandhakshala, Adukkan, Mullan Kayama, Kalladiyan, Chettu Veliyan, Mannu Veliyan, Kutty Veliyan, Chomala., Thondi, Mullan Puncha and Koranda are the indigenous rice varieties reported to be in cultivation during this study.

More than a source of food, the Kuruma community consider these landraces a part of their cultural heritage. This is a key factor that keeps them in circulation despite the reported economic burdens of farming operations. Some indigenous rice varieties are unavoidable during certain rituals. 'Kayama'

---

[2] 1 cent = 40.467m².

[3] Average area of the land they told to have occupying.

[4] Half of the interviewees in Muthanga forest area didn't have any title deed though they've the privileges under The Scheduled Tribes and Other Traditional Forest Dwellers (Recognition of Forest Rights) Act, 2006.

Those in Rattakolly neither have title deed or nor such rights and one family over there lives in the *Poramboke* land (As per the Kerala Land Conservancy Act, 1957, *Poramboke* is the unassessed lands which are the property of Government and are reserved for public purposes or for the communal use of villagers).

[5] Most of them didn't know their annual income which suggested against calculating the average.

is used for 'para nirakkal'[6] ritual during birth and death, whereas the red-coloured rice (Chennellu[7]) is given to the newborn on the 15th day after the birth. Each of these varieties has an associated ecological knowledge with it. For example 'Veliyan' is considered a flood-tolerant while 'Adukkan' and 'Kalladiyan' are drought tolerant. Due to this peculiarity of 'Kalladiyan', it is used for 'Podivitha' (dry sowing during March–April). Also, using their emic knowledge, it is claimed that when they anticipate threats from pests and birds, 'Mullan Kayama' is sowed that season as its thorn-like structures help deter any bird attack.[8]

Physical Factors

To an extent, changes in local land cover, like increased built-up areas, have affected the availability of indigenous food sources. As per the interviewees, plantain cultivation has led to dwindling paddy cultivation. Once planted in a field, plantain reduces the water availability to the neighbouring paddy fields, forcing them to quit paddy and switch to plantain. Likewise, constructions upstream are believed to have broken the flow of the perennial stream channel used for irrigating the fields. Since the fields are rainfed, the channel serves as an alternate water source. Four interviewees have plantain cultivated in the fields where they used to have paddy earlier.

Additionally, farmers attribute the change in local weather patterns as reason for switching crop varieties. Unseasonal rain during Podivitha is hindering the process and has forced them to shift from sowing 'Veliyan' (six months crop duration) to 'Valichoori'. Damages to crops by wild animals and birds, pest infestations, heavy wind, and anomalies in rain and temperature are taking a toll on farmers' revenue. The farmers also face other issues like non-availability of agro machinery.

Apart from engaging in daily-wage labour, some Kattunayaka families in the Muthanga colony have subsistence livestock and poultry. Selling eggs and milk to nearby shops and cooperative societies is a supplementary source of income for them. Being located in the forest fringes, these possessions and the people themselves are vulnerable to the constant raids of Elephants, Porcupine, Wild Boars, Bison and Monkeys.

Perceptions on Weather Pattern

Data regarding the perception of people belonging to the three communities were collected using questionnaire-based interviews. Except for three interviewees from Paniya, two from Kuruma, and three from the Kattunaikka community, the rest perceived some recent changes in their local climatic conditions (see Table 21.3).

---

[6] 'Para' is a vessel used to measure paddy and 'Para Nirakkal' is the filling up of 'Para' with paddy.

[7] Leena Kumari (n.d.) notes that 'Chennellu' is used for the treatment of jaundice and diarrhoea.

[8] I owe a debt of gratitude to Mr. Girigan Gopi, Principal Scientist MSSRF-CAbC, to draw my attention to this point.

Table 21.3 Perceptions of changes in local climate

Community	Perceptions
Paniya	• The temperature increased in both summer and winter. However, peak winter shifted from December to January • Blossom showers[9] have decreased • Rainfall decreased; the intensity of rainfall increased

Kuruma	• Irregular rainfall
	    - Scanty/No rainfall after sowing and heavy rain close to the harvest period
	    - Heavy rain when rice puddle is transplanted
	• The temperature increased in summer and winter
	• The intensity of rainfall increased

Kattunaikka	• Winter temperature increased • Increase in rainfall and its intensity

Trend Analysis Results

To identify any trends existing in the rainfall pattern of the district provided by the India Meteorological Department, Mann–Kendall test has been carried out for monthly rainfall. This test is suitable for measuring short-duration trends and does not assume any distribution for the variables. The MAKESENS template of the Finnish Meteorological Institute was used for the analysis (Salmi et al. 2002).

Total Monthly Precipitation
   Region: Wayanad District.
   Maximum: 1256.1 mm (July 2009).
   Minimum: 0 mm (January (2008, 2009, 2011, 2013); February 2006; December (2004).

At a 95.5% confidence level, rainfall in December is increasing. Whereas in May (see Fig. 21.3) and January, it is decreasing and probably decreasing trends with 97.7% and 92.2% confidence levels, respectively. However, the correlation coefficient for January and December is too low at 0.10 and 0.04, respectively, whereas that of May is 0.47 (see Fig. 21.4).

Perceptions Compared to Recorded Data

Only with some parameters was a concurrence observed between the recorded rainfall and interviewees' perceptions. Unlike the rest of the state, which used to have three cropping seasons, Wayanad has two seasons wherein sowing is

---

[9] First shower of the year which usually starts during mid-February (in the Malayalam month of *Kumbham*) and is quite decisive for agricultural activities in Wayanad, including the flowering of coffee plants.

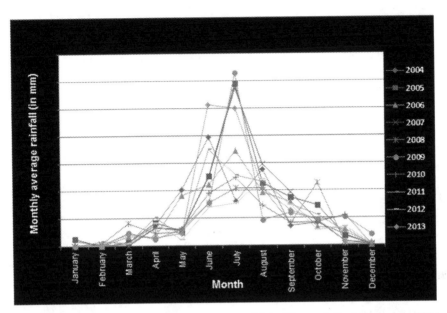

Fig. 21.3 Time series of monthly average rainfall in Wayanad district between 2004 and 2013

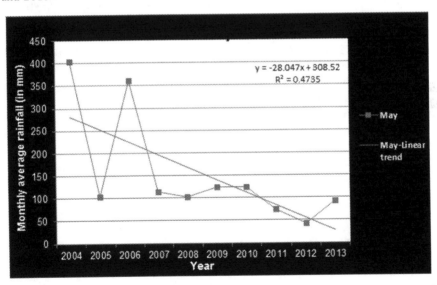

Fig. 21.4 Time series of rainfall trend in May between 2004 and 2013 in Wayanad district

done during May/June and January/February and harvesting is carried out in November/December and March/April. With regard to this, the perception is that rainfall has become scanty after sowing and heavy during harvesting. This observation agrees with decreasing and increasing trends of rainfall in May and December, respectively.

21.4 Conclusion

This study has tried to analyse and document the interlinkages between the indigenous knowledge systems, food security and rainfall in the three indigenous communities of the Wayanad district in Kerala. The analysis of rainfall trends carried out in 2015, could be done at a much finer scale now, which makes a more rational comparison with the perceptions.[10]

We showed that the indigenous knowledge systems developed around the communities' traditional sources of livelihood and lifestyle are still in vogue albeit have eroded over the years and why it is important to bolster them for building resilient communities. Now we suggest some possible avenues through which the IKS can be applied for the socio-ecological development of society at large.

Agrobiodiversity emerges as a key theme through which the IKS could be applied for socio-ecological development. Globally, there is a shift in the thinking of key stakeholders of agri-food systems towards agroecology. Indigenous knowledge and cropping principles are key components of agroecological principles (Wezel et al. 2020). This should be an avenue for leveraging the IKS. Given that in India agriculture is a state subject of the constitution, state governments have the mandate to innovate the agri-food systems of respective states. Millets, which used to be a key staple in Wayanad, is the key channel through which this transition is currently underway in many states, especially in the rainfed agriculture systems (Niyogi 2020). While these small grain cereals are getting back to public policy attention, they should be used as an entry point to appreciate the diversity of other crops and their associated IKS. If not, it risks the transition becoming lopsided and narrow (Swaran 2022).

Appreciating the IKS by explicit acknowledgement and awareness building of its ecosystem services could be one way of reversing the erosion. Some of the existing neglected or underutilised institutional and policy mechanisms could be deployed for this purpose. People's Biodiversity Register—the document on indigenous/traditional knowledge and biological resources to be prepared by local governments under the Biodiversity Act 2002—can be a platform for synergising applications of IKS (Gadgil 2000). It could be used

---

[10] The original study has estimated the changes in monthly average daytime and nighttime temperature using the AIRS dataset (AIRS Science Team/Joao Teixeira 2013). It has been abandoned in this study in view of the usefulness of making assertions based on a single datapoint for an approximately 111 Km^2 area.

to generate awareness among the community and local administration of their local resources and its relevance to food systems. Setting up of social enterprises for prudent utilisation of NUS while adhering to the access and benefit clauses of the Biodiversity Act and other relevant community norms can be explored. Likewise, an evaluation of ecosystem services of these resources could be looked into within the Payments for Ecosystem Services framework. The key to all these strategies would be to stay cautious about using these measurements as a pretext for the wanton exploitation of natural resources, which for the community would be beyond any monetary valuation (Ludwig 2000).

Acknowledgments The authors are grateful to the participants of this study from *Paniya, Kuruma* and *Kattuanayakan* communities for kindly sharing their traditional knowledge and practices. V.S. would like to thank Ms. Bindu Joseph, Mr. K.D. Pradeep, Mr. P.C. Babu, and all the other staff members of MSSRF-CAbC, Wayanad.

References

AIRS Science Team/Joao Teixeira. (2013). AIRS/Aqua L3 Daily Standard Physical Retrieval (AIRS-only) 1 Degree x 1 Degree V006, Greenbelt, MD, USA, Goddard Earth Sciences Data and Information Services Center (GES DISC), Accessed: https://doi.org/10.5067/Aqua/AIRS/DATA303.

Berkes, F. (2008). *Sacred Ecology*, 2nd ed. London and New York: Routledge, p. 9.

Booth, D., Holland, J., Hentschel, J., Lanjouw, P., and A. Herbert. (1998). *Participation and Combined Methods in African Poverty Evaluation: Renewing the Agenda*. Department for International Development, Issues, February, pp. 127, ISBN 86192 027 X.

Chandrika, C.S., and Nandakumar, P.M. (n.d.). Paniya Adivasi Women's Innovative Livelihood Development Endeavours in Farming. In: *Empowering Women in Agriculture*, 1st ed. ACCESS Development Services, pp. 50–67.

Chivenge, P., Mabhaudhi, T., Modi, A.T., and Mafongoya, P. (2015). The Potential Role of Neglected and Underutilised Crop Species as Future Crops Under Water Scarce Conditions in Sub-Saharan Africa. *International Journal of Environmental Research and Public Health*, 12(6), pp. 5685–5711.

Department of Economics and Statistics. (2015). *Agricultural Statistics 2013–2014*. Thiruvananthapuram: Government of Kerala.

Department of Environment and Climate Change. (2014). *Kerala State Action Plan on Climate Change*. Thiruvananthapuram: Government of Kerala, p. 39.

Díaz, S., Demissew, S., Carabias, J., Joly, C., Lonsdale, M., Ash, N., Larigauderie, A., Adhikari, J.R., Arico, S., Báldi, A., and Bartuska, A. (2015). The IPBES Conceptual Framework—Connecting Nature and People. *Current Opinion in Environmental Sustainability*, 14, pp. 1–16.

Directorate of Census Operations—Kerala. (2014). *District Census Handbook Wayanad*. Available at: https://censusindia.gov.in/2011census/dchb/3203_PART_B_WAYANAD.pdf [Accessed 14 July 2015].

Gadgil, M. (2000). People's Biodiversity Registers: Lessons Learnt. *Environment, Development and Sustainability*, 2(3), pp. 323–332.

Gadgil, M., Berkes, F., and Folke, C. (1993). Indigenous Knowledge for Biodiversity Conservation. *Ambio*, pp. 151–156.

Hiwasaki, L., Luna, E., and Shaw, R. (2014). Process for Integrating Local and Indigenous Knowledge with Science for Hydro-Meteorological Disaster Risk Reduction and Climate Change Adaptation in Coastal and Small Island Communities. *International Journal of Disaster Risk Reduction, 10*, pp. 15–27.

Holling, C.S. (1973). Resilience and Stability of Ecological Systems. *Annual Review of Ecology and Systematics, 4*, p. 14. Available at: http://www.jstor.org/stable/2096802.

Holloway, I. (1997). *Basic Concepts for Qualitative Research*. Oxford: Blackwell Science.

ISSC and UNESCO. (2013). *World Social Science Report 2013: Changing Global Environments*. Paris: OECD Publishing and UNESCO Publishing.

Johns, T. (2007). 15. Agrobiodiversity, Diet, and Human Health. In: *Managing Biodiversity in Agricultural Ecosystems*. Columbia University Press, pp. 382–406.

Kahane, R., Hodgkin, T., Jaenicke, H., Hoogendoorn, C., Hermann, M., d'Arros Hughes, J., Padulosi, S., and Looney, N. (2013). Agrobiodiversity for Food Security, Health and Income. *Agronomy for Sustainable Development, 33*(4), pp. 671–693.

Keralabiodiversity.org. (n.d.). *Conservation of Rice Varieties of Wayanad* [online]. Available at: http://keralabiodiversity.org/index.php?option=com_content&view=article&id=2&Itemid=243 [Accessed 5 January 2015].

Ludwig, D. (2000). Limitations of Economic Valuation of Ecosystems. *Ecosystems*, pp. 31–35.

Nakashima, D. (ed.). (2010). *Indigenous Knowledge in Global Policies and Practice for Education, Science and Culture*. Paris: UNESCO.

Narayanan, M., Anilkumar, N., and Balakrishnan, V. (2011). Uses of Wild Edibles Among the Paniya Tribe in Kerala, India. In: *Conservation and Sustainable Use of Agricultural Biodiversity: A Sourcebook*. CIP-UPWARD, pp. 100–108.

Niyogi, D.G. (2020). India's Millets Policy: Is It Headed in the Right Direction? *Mongabay*. Available at: https://india.mongabay.com/2020/07/indias-millets-policy-is-it-headed-in-the-right-direction/ [Accessed 6 October 2022].

Orlove, B., Roncoli, C., Kabugo, M., and Majugu, A. (2009). Indigenous Climate Knowledge in Southern Uganda: The Multiple Components of a Dynamic Regional System. *Climatic Change, 100*(2), pp. 243–265.

Pieroni, A., and Price, L. (eds.). (2006). *Eating and Healing: Traditional Food as Medicine*. CRC Press.

Pradeepkumar, T., Indira, V., and Sankar, M. (2013). Nutritional Evaluation of Wild Leafy Vegetables Consumed by Tribals in the Wayanad District of Kerala. *Proceedings of the National Academy of Sciences, India Section B, 85*(1), pp. 93–99.

Salmi, T., Määttä, A., Anttila, P., Ruoho-Airola, T., and Amnell, T. (2002). *Detecting Trends of Annual Values of Atmospheric Pollutants by the Mann-Kendall Test and Sen's Slope Estimates—The Excel Template Application MAKESENS*. Publications on Air Quality No. 31, Report Code FMI-AQ-31. Helsinki: Finnish Meteorological Institute.

Shrinivasa, B., Philip, R., Krishnapali, V., Suraj, A., and Sreelakshmi, P. (2014). Prevalence of Anemia Among Tribal Women of Reproductive Age-Group in Wayanad District of Kerala. *International Journal of Health & Allied Sciences, 3*(2), pp. 120–120.

Suma, R. (2014). Customary vs State Laws of Land Governance: Adivasi Joint Family Farmers Seek Policy Support The Case of Kurichya Joint Families in Wayanad, Southern India. In: *Family Farming and People-Centred Land Governance: Exploring Linkages, Sharing Experiences and Identifying Policy Gaps*. International Land Coalition.

Swaran, V. (2022). Rich Millet Poor Millet: The Irony in Our Consumerism. *The Bastion*. Available at: https://thebastion.co.in/politics-and/environment/resource-management/rich-millet-poor-millet-the-irony-in-our-consumerism/ [Accessed 17 October 2022].

Thurston, E., and Rangachari, K. (1909). *Castes and Tribes of Southern India*. Madras Government Press.

Tongcon, M. (2007). Purposive Sampling as a Tool for Informant Selection. *Ethnobotany Research & Applications*, .5, pp. 147–158.

UNDESA. (2018). *Revision of the World Urbanization Prospects*. United Nations Department of Economic and Social Affairs. Available at: https://www.un.org/sw/desa/68-world-population-projected-live-urban-areas-2050-says-un [Accessed 15 June 2022].

UNESCO. (n.d.). *What Is Local and Indigenous Knowledge | United Nations Educational, Scientific and Cultural Organization* [online]. Available at: http://www.unesco.org/new/en/natural-sciences/priority-areas/links/related-information/what-is-local-and-indigenous-knowledge [Accessed 13 November 2020].

Warren, D.M., Slikkerveer, L.J., and Brokensha, D. (eds.). (1995). *The Cultural Dimension of Development: Indigenous Knowledge Systems*. London: Intermediate Technology Publications.

Western Ghats Ecological Expert Panel Report 1. (2011). Ministry of Environment and Forests, Government of India [online]. Available at: http://www.moef.nic.in/downloads/public-information/wg-23052012.pdf [Accessed 9 February 2015].

Wezel, A., Herren, B.G., Kerr, R.B., Barrios, E., Gonçalves, A.L.R., and Sinclair, F. (2020). Agroecological Principles and Elements and Their Implications for Transitioning to Sustainable Food Systems: A Review. *Agronomy for Sustainable Development*, 40(6), pp. 1–13.

CHAPTER 22

Ethnobotanical Knowledge and Floristic Diversity of South East Indian Coastal Region

Nallakaruppan Nagaraj, Veluchamy Chandra, Sekaran Manoj, Nachiappan Kanagam, Sunil Nautiyal, Thiagarajan Kalaivani, and Chandrasekaran Rajasekaran

22.1 Introduction

World Health Organization (WHO) states that about 80% of the global population depends on traditional medicines for basic healthcare needs (Bannerman 1982). In India, over thousands of years, local healers/practitioners have used folk medicines, thereby contributing to human health care services in remote rural places by treating and preventing various ailments. Traditional ethnobotanical knowledge (TEK) refers to the knowledge, specific to a place or a particular population derived from their experience and traditions, which are unique in making it different from the scientific approach (Usher 2000). Traditional Ethnobotanical Knowledge (TEK) holds a community heritage gained solely by the expertise attained by the healer over time. TEK documentation

---

N. Nagaraj · V. Chandra · S. Manoj · N. Kanagam · T. Kalaivani ·
C. Rajasekaran (✉)
Department of Biotechnology, School of Bio Sciences and Technology, Vellore Institute of Technology, Vellore, Tamil Nadu, India
e-mail: drcrs70@gmail.com

N. Kanagam
Department of Biotechnology, Sri Venkateswara College of Engineering, Sriperumbudur, Tamil Nadu, India

S. Nautiyal
Centre for Ecological Economics and Natural Resources, Institute for Social and Economic Change, Bengaluru, India

© The Author(s), under exclusive license to Springer Nature Singapore Pte Ltd. 2023
S. Nautiyal et al. (eds.), *Palgrave Handbook of Socio-ecological Resilience in the Face of Climate Change*,
https://doi.org/10.1007/978-981-99-2206-2_22

367

is proved to be a time and cost-efficient approach that supports the conservation and utilization of floristic diversity. Though affordability, accessibility, and availability of traditional medicine are simple, scientific evidence is lacking in evaluation and efficacy (Sivaperumal et al. 2009). Hence, the collection, documentation, and conservation of ethnobotanical plants and seaweed have accelerated. Novel bioactive compounds isolated from medicinal plants and marine seaweeds act as 'leads'; which can be further developed into effective drugs with the combined efforts of botanists, ethno-pharmacologists, microbiologists, and ethnobotanists (Pushpangadan and Atal 1984; Srivastav et al. 2011).

Coastal areas are defined as transitional zone lying between land and sea. A habitat with a specific structure, complexity, and energy flow exists because of the transitory character of the land and ocean's varied points of contact. The main ecosystems in these areas include mangroves, wetlands, reefs, marshes, seagrass, estuaries, bays, and dunes. They are home to various plants and animals and provide crucial ecosystem services (FAO 1998). The local people use psammophytes for ethnomedical applications. Due to their proximity to the ocean, ideal tourist locations, resource endowments, and over-exploitation, tropical regions' coastal flora and fauna were severely hampered by anthropogenic activities (Michel and Pandya 2010).

The primary goal of this investigation is the ethnobotanical application of psammophytes and seaweeds by the local inhabitants of coastal regions of Tamil Nadu and Pondicherry in southeast India.

Similar TEK documentation studies for endangered species have been recorded throughout India (Rajan et al. 2002; Ganesan et al. 2004; Sandhya et al. 2006; Ignacimuthu et al. 2006). However, there is no adequate documentation of indigenous ethnobotanical knowledge in the study area. Very little work has reportedly been done in India and other countries thus far (Navaneethan et al. 2011). Our present study aims at baseline data collection about cross-cultural TEK to understand the human–ecosystem interaction to create better policy formulations and socio-ecological development in this region.

22.2 Problem Statement

Coastal ecosystems fall among the most vital ecological and socio-economical ecosystems on the planet. Marine habitats from the intertidal zone out to the continental shelf break are estimated to provide over US$ 14 trillion worth of ecosystem goods (e.g., food and raw materials) and services (e.g., disturbance regulation and nutrient cycling) per year (Costanza et al. 1997). Climate change is the major reason for causing humongous changes in the marine ecological community, which have been unusual over the past 1000 years. The causative factors of climatic changes are the emission of greenhouse gases such as CO_2, CH_4, O_3, and N_2O. Therefore, it is important to address the adverse effects of climatic variations like biodiversity threats, rising ocean levels, physicochemical characteristics, etc. (Neelmani et al. 2019). The physiology, phenology, and population dynamics of marine biodiversity change in response to changing oceanic conditions, including temperature (Pauly 2010). These responses to ocean-atmospheric changes have been projected to lead to altered patterns of species richness (Cheung et al. 2009), changes in community structure (MacNeil et al. 2010), ecosystem functions (Petchey et al. 1999), and consequential changes in marine goods and services (Cheung et al. 2010). Anthropogenic activity influences, as well as rising sea levels, rising sea temperatures, and other climate-related changes to the ocean, such as prevailing waves, storm waves, and surges, would drastically impair ecosystem resilience. Sea level rise and rising seawater temperatures are expected to hasten beach erosion and degrade natural coastal defences like mangroves and coral reefs, wreaking havoc on the socio-economic elements of coastal communities. To control the effects of climate change on the coastal region, integrated approaches at multiple levels are required (UNFCC 1992).

22.3 Study Area

South East Indian Coastal Region

The study area lies along the South East Indian coastal region ranging from Pulicate to Kanyakumari, covering around 1200 km, lying between 80.32839°E longitude 13.40789°N latitude and 77.17469°E longitude 8.23306°N latitude depicted in Table 22.1. The coastal districts included in the study include both Tamil Nadu and Pondicherry states. The South East Indian coastal region stands as the country's second-longest coastline, which accounts for about 15% of the total Indian coastal length. The coastal area of South East India appears to be a coast of emergence that uplifts the land to lower the sea level. This region covers low-lying lagoons, marshes, beaches, and deltas rich in mangrove forests (Coastline of India/Coastal Plains of India).

Location and Climate

There are 13 districts from Tamilnadu and two from Pondicherry (Thiruvallur, Chennai, Kancheepuram, Villupuram, Pondicherry, Cuddalore, Karaikkal, Nagapattinam, Thiruvarur, Thanjavur, Pudukkottai, Ramanathapuram, Thoothukudi, Tirunelveli, and Kanyakumari) share the coastline of South East India. This region experiences three climate seasons (i) Summer (March–June), (ii) rainy (July–October), and (iii) winter (November–February). Rainfall pattern over the coastal areas is always higher when compared to the inland regions due to the presence of rainfall-causing system starting from the Bay of Bengal with an average temperature around 28–40 °C (https://www.researchgate.net/figure/Distribution-of-rainfall-in-west-coast-south-to-north-and-east-coast-of-India).

Table 22.1 Geographical and floristic details of South East Indian Coastal line

DISTRICT	SITE	Latitude	Longitude	Angiosperm	Seaweed
Thiruvallur	Pulicat	13.40789°N	80.32839°E	✓	×
Chennai	Ennore Port	13.29044°N	80.34742°E	✓	✓
	Santhom Beach	13.03161°N	80.27989°E	×	×
	Copper Beach	12.86153°N	80.24956°E	✓	×
Kancheepuram	Muttukadu	12.79531°N	80.25022°E	✓	×
	Mahabalipuram	12.61567°N	80.19894°E	✓	×
Villupuram	Marakanam	12.20144°N	79.97042°E	✓	×
Pondicherry	Kalapet	12.02169°N	79.86461°E	✓	×
Cuddalore	Cuddalore OT Beach	11.68506°N	79.77219°E	✓	×
	Samiyarpettai Beach	11.54828°N	79.75361°E	✓	×
	Pichavaram	11.43197°N	79.73197°E	✓	×
Nagapattinam	Poompuhar	11.14261°N	79.85722°E	✓	✓
	Vedaranyam	10.27491°N	79.81694°E	✓	✓
	Kodiyakkarai Vedaranyam	10.30036°N	79.84917°E	✓	✓
Karaikal	Karaikal Beach	10.92017°N	79.85203°E	✓	×
	Karaikal port	10.91083°N	79.84931°E	✓	×
Thanjavur	Manora	10.26611°N	79.30389°E	✓	✓
	Manammel kudi	10.03872°N	79.26100°E	✓	✓
Pudukkottai	Mimisal	9.92244°N	79.15211°E	✓	✓
Ramanathapuram	Karankadu	9.64639°N	78.96522°E	✓	✓
	Ucchipuli	9.32990°N	80.01862°E	✓	✓
	Thonithurai	9.26656°N	79.03842°E	✓	✓
	Pamban Ramnad side	9.28175°N	79.18533°E	✓	✓
	Pamban Island Palk Bay side	9.28404°N	79.21417°E	✓	✓
	Pamban island Gulf of Mannar side	9.27444°N	79.20542°E	✓	✓
Ramanathapuram	Arichalmunai Palk Bay Side	9.26636°N	79.32175°E	✓	✓
	Arichalmunai Gulf of Mannar Side	9.23561°N	79.32325°E	✓	✓
	Vivekananda mandapam	9.25808°N	79.22133°E	✓	✓
	Villuvandi thirtham	9.29272°N	79.26244°E	✓	✓
	Olaikuda	9.29425°N	79.32608°E	✓	✓
	Olaikuda aginithirtham	9.17268°N	79.19215°E	✓	✓
	Dhanushkodi	9.17775°N	79.41731°E	✓	✓
	Kothandaramar koil	9.01428°N	76.96527°E	✓	✓

(continued)

Table 22.1 (continued)

DISTRICT	SITE	Latitude	Longitude	Angiosperm	Seaweed
	Vedalai	9.26333°N	79.10117°E	✓	✓
	Kilakarai	9.22722°N	78.78556°E	✓	✓
	Vembar	9.07653°N	78.36506°E	✓	✓
Thoothukudi	Vaippar	9.01717°N	78.26750°E	✓	✓
	Muyal Theevu	8.77356°N	78.18978°E	✓	✓
	Tuticorin Port beach	8.74428°N	78.17039°E	✓	✓
	Kayalpattinam	8.56006°N	78.13219°E	✓	✓
	Manapad	8.37250°N	78.06361°E	✓	✓
Tirunelveli	Uvari	8.27997°N	77.89483°E	✓	✓
	Idinthakarai	8.17722°N	77.74500°E	✓	✓
Kannyakumari	Kannyakumari sunrise Point	8.08308°N	77.55239°E	✓	✓
	Muttom	8.12431°N	77.31389°E	✓	✓
	Cholachal	8.17231°N	77.25639°E	✓	✓
	Inayam	8.24037°N	77.77305°E	✓	✓
	Thengapattinam	8.23306°N	77.17469°E	✓	✓

Socio Economic Profile

The coastal line, which comprises the beach and neighbouring sand dunes, is also essential to fisherfolk's social and daily activities. Furthermore, fishermen prefer to build their homes in elevated coastal areas (primarily on dunes), as this provides them with protection from storms and waves and a direct view of the sea, which is essential for determining wind direction and weather conditions for day-to-day fishing activity. Hence, communities have traditionally built their hamlets on sand dunes and high areas along the coast of Tamil Nadu. For instance, Manapad is one such experimental site in our study area. In total, the fishermen population along the South East Indian coastal region is around 1.05 million, i.e., 1:10 ratio, colonized in the vicinity of major and minor fishing harbours and fish landing centres. The primary livelihood activities of this community are fishing, shipping, construction of small fishing boats, fishing net making and tourism, etc. Besides, they are partially involved in agricultural/horticultural activities such as the plantation of cash crops like casuarina, coconut, and palmyra species. The majority of nuclear power stations are being built along the coast. The younger generation of the fishing community is abandoning their traditional fishing activities and opting their jobs in these nuclear power stations.

22.4 METHODOLOGY

Bioresources

A field study was conducted across three seasons. The methodology includes collecting and identifying the plants and their distribution; algae and microbial diversity; seawater (coastal regions); and soil samples to analyse the physicochemical parameters of it further.

Available bioresources in the field were further analysed for their traditional ethnobotanical knowledge by surveying the fisherman community to understand the bioresource contribution to the climatic variation in the coastal region.

Physicochemical Parameters

Water and soil samples were analysed to study the parameters like temperature, pH, and salinity. Water (100 ml) and soil samples (20 g) were collected from 0 to 10 cm depth at our study sites during the three distinct climate seasons—summer (March–June), rainy (July–October), and winter (November–February).

Soil samples (5 g) were added to 25 mL of distilled water, and shaken for 1 h. The soil solution was then filtered through a filter paper (Whatman No. 40, 110 mm), and the pH and salinity content of both water and soil solution were measured (Hwang et al. 2016) using a pH meter (Aquasol digital, Rakiro biotech Mumbai, India).

These collected data are helpful for the prediction of climate variability of the selected study region, as tabulated in Table 22.2 and Fig. 22.1. Climate changes are correlated with simultaneous shifts in oxygen content, temperature, circulation, pH, nutrient input, precipitation, and ocean acidification, with potentially wide-ranging biological effects. Moreover, climate changes may alter the energy and material outflows and biogeochemical cycles, affecting the overall ecosystem and services upon which people and communities depend (Doney et al. 2012).

Plants

In the angiosperm diversity study conducted along the sand dunes of the South East Indian coastal region, a total of 43 families of plant species were identified and collected. Among them, Fabaceae, Malvaceae, and Euphorbiaceae were the dominant families in the region. The local fisherman community was found to utilise the available plants in traditional medicinal practices for the treatment of diseases like pain, cold, cough, fever, headache, toothache, skin allergy, wounds and swelling, snake bites, gastrointestinal disorders, diabetes, jaundice, chickenpox respiratory disorders, tuberculosis, and malaria. The family distribution of the collected plant species for all three seasons is tabulated in Table 22.3 and Fig. 22.2.

Table 22.2 Physicochemical parameters of water and soil samples from South East Indian Coastal line

Site	Water										Soil					
	Temperature			pH			Salinity				pH			Salinity		
	S	R	W	S	R	W	S	R	W		S	R	W	S	R	W
Pulicate	32.2	24.4	28.2	7.98	7.98	8.41	30.9	30.3	25.1		8.52	7.98	7.99	27.2	2.44	24.6
Ennore Port	30.8	29.3	29.8	8.06	8.06	8.34	30.9	30.1	27.9		8.00	7.06	8.09	28.9	8.10	25.0
Santhom Beach	32.8	29.5	29.8	8.07	8.07	8.5	29.2	31.1	31.3		7.86	8.07	8.23	30.6	3.19	26.1
Copper Beach	30.7	30.3	26.7	8.01	8.01	8.36	30.1	29.6	24.2		9.70	8.01	8.80	32.9	5.40	24.8
Muttukadu	30.4	30.1	25.1	7.95	7.95	8.32	31.0	25.0	28.2		9.27	7.65	8.25	35.8	5.26	24.7
Mahabalipuram	31.4	31.1	24.1	7.95	7.95	8.29	32.1	27.2	36.4		7.42	7.55	8.25	34.6	9.84	25.6
Marakanam	34.1	29.6	28.1	8.01	8.01	6.13	31.9	25.2	34.2		8.05	8.11	8.14	23.2	7.68	28.8
Kalapet	33.2	25.0	26.7	8.03	8.03	8.77	31.2	26.4	31.2		7.74	7.87	8.44	27.0	6.90	29.9
Cuddalore OT Beach	32.5	27.2	27.2	7.95	7.95	8.31	30.4	29.3	30.0		7.56	8.43	8.58	26.8	7.45	26.4
Samiyarpettai Beach	31.9	25.2	28.5	8.01	8.01	8.34	30.9	29.7	35.9		8.05	7.95	7.84	28.1	4.34	25.3
Pichavaram	30.1	26.4	26.9	8.04	8.04	8.90	30.9	30.0	28.1		7.89	8.31	8.01	25.1	4.49	20.2
Poompuhar	33.3	29.3	29.1	10.5	10.5	8.22	28.5	26.2	29.2		7.45	8.84	8.17	27.9	8.15	23.3
Vedaranyam	31.4	29.7	27.3	7.69	7.69	8.17	32.5	25.8	27.4		7.72	9.50	8.61	31.3	8.87	22.5
Kodiyakkarai Vedaranyam	32.4	30.0	26.4	10.2	10.2	8.60	31.0	29.7	31.9		7.46	7.69	8.11	24.2	7.24	14.4
Karaikal Beach	32.7	26.2	29.4	10.5	10.5	8.34	31.4	30.3	36.0		8.67	8.15	8.12	28.2	5.56	21.3
Karaikal port	30.5	25.8	27.7	9.96	9.96	8.31	30.3	30.1	35.0		8.20	9.45	8.17	36.4	5.80	19.6
Manora	31.5	29.7	28.5	9.98	9.98	8.10	30.7	31.1	27.9		7.84	9.66	8.12	34.2	12.2	23.1
Manammel kudi	31.3	26.1	25.6	9.54	9.54	8.57	31.6	28.6	31.7		8.23	8.76	8.35	31.2	5.84	24.6
Mimisal	33.0	31.4	23.7	8.12	8.12	8.61	30.8	29.2	29.5		7.83	9.98	8.21	30.0	4.10	23.2
Karankadu	30.3	31.2	28.3	11.1	11.1	8.36	32.2	31.3	28.4		9.04	9.24	8.04	35.9	5.94	21.6
Ucchipuli	33.7	26.3	29.1	11.3	11.3	8.34	32.2	28.1	28.2		8.64	8.12	7.95	28.0	2.64	20.9
Thonithurai	26.2	23.5	26.2	8.67	8.67	8.34	32.2	28.3	26.7		8.56	9.07	8.82	29.0	7.32	20.8

(continued)

Table 22.2 (continued)

Site	Water Temperature S	R	W	pH S	R	W	Salinity S	R	W	Soil pH S	R	W	Salinity S	R	W
Pamban Ramnad side	28.5	28.8	27.7	9.05	9.05	5.02	30.8	21.2	25.1	8.01	7.28	8.27	27.4	3.56	20.8
Pamban Island PB side	32.5	28.5	27.5	11.0	11.1	8.11	32.8	29.2	24.1	8.66	9.95	8.22	31.9	3.25	23.1
Pamban island GM side	31.0	26.4	28.1	10.9	10.9	8.08	30.7	30.1	28.1	8.55	7.89	8.97	36.0	5.04	22.1
Arichalmunai PB Side	31.4	28.7	27.4	10.5	10.5	8.25	30.4	29.2	26.7	8.74	8.53	8.13	35.0	1.54	20.8
Arichalmunai GM Side	30.3	28.6	26.3	9.56	9.56	8.01	31.4	29.0	27.2	10.4	8.56	7.99	27.9	4.26	21.1
Vivekananda mandapam	30.7	29.2	29.7	10.8	10.8	8.31	34.1	26.8	28.5	9.72	9.84	8.16	31.7	3.27	18.6
Villuvandi thirtham	31.6	31.3	28.6	11.9	11.9	9.01	33.2	30.6	26.9	9.45	9.09	8.42	29.5	5.14	23.2
Olaikuda	30.8	28.1	26.8	11.5	11.0	9.22	32.5	28.6	29.1	10.3	10.0	8.17	28.4	5.05	21.8
Olaikuda aginithirtham	32.2	28.3	29.2	11.2	11.2	8.37	31.9	29.2	27.3	9.11	9.19	8.17	28.2	1.44	22.0
Dhanush kodi	32.2	21.2	27.2	11.2	11.2	8.41	30.1	31.3	26.4	8.54	10.2	8.13	35.2	3.94	23.3
Kothandaramar koil	30.7	29.2	25.7	10.1	10.1	7.98	33.3	28.1	29.4	8.78	9.66	8.15	31.0	2.55	9.14
Vedalai	30.1	30.1	29.1	10.1	10.1	8.2	31.4	28.3	27.7	9.74	9.05	8.14	28.1	3.17	9.31
Kilakarai	33.1	29.2	27.1	9.47	9.47	8.16	32.4	21.2	28.5	8.46	8.47	8.74	24.8	4.01	9.53
Vembar	32.0	29.0	26.0	11.5	11.8	8.25	32.7	32.2	25.6	9.48	10.12	8.07	29.1	5.44	10.5
Vaippar	31.1	26.8	29.1	9.95	9.95	8.44	30.5	32.2	23.7	9.29	8.95	8.78	35.8	5.95	9.51
Muyal Theevu	30.3	30.6	27.3	10.8	10.8	8.35	32.2	30.7	28.3	9.20	8.76	8.10	28.0	5.99	9.90
Tuticorin Port beach	32.6	30.9	23.6	7.98	7.98	8.40	30.8	30.1	29.1	10.8	7.98	8.70	36.3	5.14	10.2
Tiruchendur	30.9	29.2	28.9	8.01	8.01	8.7	32.8	33.0	26.2	9.45	7.61	8.03	27.6	4.14	9.40
Manapad	30.9	27.2	29.9	8.03	8.03	8.66	30.7	30.3	27.7	9.23	9.03	7.96	28.9	3.98	10.4
Uvari	29.2	27.2	23.2	10.4	10.4	8.23	30.4	33.7	28.1	9.75	9.41	8.01	29.2	3.10	9.59

Site	Water Temperature S	R	W	Water pH S	R	W	Water Salinity S	R	W	Soil Temperature S	R	W	Soil pH S	R	W	Soil Salinity S	R	W
Idinthakarai	30.1	28.1	30.1	7.84	7.84	8.34	31.4	26.2	27.4	8.42	7.94	7.85	36.6	4.82	7.61
Kannyakumari sunrise Point	31.0	30.6	29.0	9.13	9.13	8.23	33.1	28.5	26.3	9.36	8.13	8.21	29.4	4.06	4.72
Muttom	32.1	30.4	31.1	7.86	7.86	8.44	32.0	32.5	29.7	10.2	8.86	8.15	28.1	3.14	8.51
Cholachal	31.9	30.6	30.9	8.01	8.01	8.34	31.1	31.0	28.6	8.75	8.71	8.12	29.8	2.15	4.80
Inayam	31.2	30.6	29.2	7.89	7.89	8.29	30.3	31.4	26.8	8.23	8.89	8.06	37.5	3.90	14.9
Thengapattinam	30.4	29.9	30.2	8.24	8.02	8.12	32.6	30.3	29.2	9.08	8.07	8.54	33.4	4.01	12.7
Mean average	31.36	28.54	27.73	9.31	9.33	8.26	31.43	29.12	28.7	8.73	8.65	8.22	30.4	5.01	18.4

S—Summer Season; R—Rainy Season; W—Winter Season

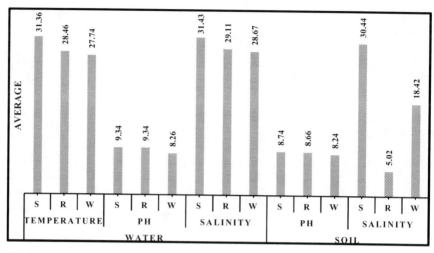

Fig. 22.1 Visualization of seasonal indices

Table 22.3 Seasonal family-wise distribution of the angiosperms

Class	Dicotyledon						Monocotyledon					
Season	Summer		Rainy		Winter		Summer		Rainy		Winter	
Family (No.)	30		32		33		6		9		10	
Species (No.)	76		89		90		14		16		18	
Life span	A	P	A	P	A	P	A	P	A	P	A	P
Species (No.)	18	58	31	58	32	58	2	12	4	12	6	12

A—Annuals, P—Perennials

Algae

In the seasonal field study of the South East Indian coastal region, we have identified and collected around 15 different algal species divided into Chlorophyceae, Phaeophyceae, and Rhodophyceae. The collected algal species are potent metabolites for medicinal and commercial uses. Drug manufacturing industries are interested in isolating bioactive compounds from algae as they show potential bio-activity, including anti-microbial, anti-oxidant, and anti-cancer agents. Some of the collected algae-like Sargassaceae, Gracillariaceae, and Ulvaceae species, are used commercially for the production of gels, food, animal feed, and fertilizers on an industrial scale, as well as on a small scale; it is also cultivated along the coastal region, providing daily wages work for the local inhabitants. The family-wise algal distribution over all the seasons are tabulated in Table 22.4 and Fig. 22.3.

Fig. 22.2 Distribution of plants in coastal sand dunes **a** Palm trees and sand dune pattern, **b** *Ipomea pescarprae*, **c** *Spinifex littoreus*

Table 22.4 Seasonal family-wise distribution of algal species

Class	Summer F	Summer S	Rainy F	Rainy S	Winter F	Winter S
Chlorophyceae	6	8	2	12	4	19
Phaeophyceae	4	5	2	7	5	12
Rhodophyceae	5	12	2	17	6	23

F—Family; S—Species

22.5 TEK OF BIORESOURCES BY FISHERMAN COMMUNITY

The local fisherman community living along the South East coastal region shared their traditional knowledge about the availability of plants and algal species in different seasons. Some species are not available in certain climatic conditions. This statement is also justified by our sample collection (water, soil, angiosperms, and seaweeds) and their physicochemical parameters analysed season-wise. In Table 22.5, the list of season-wise distribution of plants and algal species was recorded from the TEK of local inhabitants.

Fig. 22.3 Different seaweed collection a *Dictyota cervicornis*, b *Sargassum cineserum*, c *Padina boergerenii*, d *Gracillaria corticata*

This table includes data generated by analysing the physicochemical parameters of soil and water samples collected in the summer, rainy, and winter seasons. Comparing this data with the TEK of bioresource availability shared by local people, can correlate the link between physicochemical parameters and the distribution of bioresource. In summer, the recorded average temperature of the water is 31.36 °C, pH 9.34, salinity 31.43 ppt, and for soil, the average pH value is 8.73, and salinity is around 30.4 ppt. As reported by locals, the most common algal and plant species present in the coastal areas during the summer season are *Acanthophora dendroides, Sargassum wightii, S. licifolium, Gracilaria edulis, Caulerpa fastigata, Ulva reticulata* and *Abutilon indicum, Acacia nilotica, Albizia. saman, Azardicata indica, Borassus flabellifer, Calotropis gigantea, Cocos nucifera, Datura metal, Ficus benghalensis, Ipomea pescarprae, Nerium oleander, Ricinus communis, Sida cordifolia, Spinifex littoreus, Achyranthus aspera, Aristida setacea, Leucas aspera, Tephrosia purpurea, Tridax procumbens* respectively. But during the rainy season, when temperature, pH and salinity reduce, species like *Sargassum wightii, S. licifolium, Gracilaria edulis, Caulerpa fastigata* were reported the other species of algae disappeared. Other species found were *Acanthophora spicifera, Sargassum tenerrimum, Gracilaria verucosa, Halimeda lacrimosa*, and *Ulva compressa*. Similarly, in plants *Cassia auriculata, Casurina equestifolia, Cymodan dactylon, Cyperus esculantus, Opentia dilleni, Oldenlandia umbellata, Salicornia brachiata, Sida cordifolia, Boerhavia diffusa, Catharanthus rosus, Phyllanthus niruri* and *Solanum nigrum* were the new species reported by local inhabitants. In winter, when temperature, pH and salinity reduce further, some new algal strains like *Sargassum plagiophylum, Gracilaria corticata,* & *Caulerpa peltata* and plant species like *Aerva lanata, Aristida setacea, Cisscus quadrangularis, Cyperus esculentus,* and *Euphorbia hirta,* were reported to be found along the coast. So, the survey taken from coastal inhabitants connects with the physicochemical parameters like pH, temperature and salinity of water and soil. Based on the varying parameters few species disappear and few new species grow widespread in the study area.

In India, the local fisherman community residing along the South East coast shared their traditional knowledge about the application of flora and algae for curing various ailments. Survey results show that 28 angiosperms and 20 algal species were used for treating major diseases like cancer, asthma, diabetes, piles and hypertension. In addition, around 73 angiosperms and nine algae were used to treat minor ailments like snakebites, insect bites, lesions, cold, cough, fever, and wound healing. Extraction of solvent from algae and plants yields many bioactive compounds like carotenoids, flavonoids, alkaloids, saponins, fatty acids, amino acids, and carbohydrates that have potent anti-microbial properties against virulent strains of bacteria and fungi that are pathogenic to humans (see Table 22.6). So, this knowledge transfer of TEK of local bioresources by proper documentation and awareness can greatly support the pharmaceutical industries for novel drug discovery.

22.6 Effect of Physicochemical Properties on Floristic Diversity

The data of the one-year study (from June 2018 to May 2019) were pooled for three seasons and analysed for seasonal variations concerning rainy, winter, and summer. Physicochemical parameters are reported in Table 22.2 and Fig. 22.1. The results showed a significant difference in physicochemical properties in different seasons. Most of the parameters tested were slightly higher in the summer compared to the other two seasons.

In seaweeds, different physicochemical parameters play a vital role in their growth and development. Many climatically sensitive environmental variables, like temperature (Lüning and Neushul 1978), salinity (Steen 2004), and pH (Kuffner et al. 2008), are known to affect seaweed survival, growth, and reproduction. Our study indicates that the post-rainy/pre-summer season provides adaptable conditions for the growth of seaweeds because, in this season, the average water temperature is approx. 30 °C, with pH 8.5 and salinity 31 ppt. Comparatively, the seaweed species distribution pattern is very less in mid and late-summer seasons than in rainy and windy seasons.

In-plant diversity study, the soil pH averagely extended up to 8.7 in the summer season, and the soil's salinity was nearly 30 ppt. The increase of pH and salinity in the dune declines the CO_2 fixation and microbial action in the

Table 22.5 Season wise distribution of bioresources with their varying physiochemical parameters

Season	Physiochemical parameters							TEK knowledge from the survey
		Temperature		pH		Salinity		
		High	Low	High	Low	High	Low	
Summer (March–June)	Water	33.7	26.2	11.9	7.69	34.1	28.5	**Algae:** *Acanthophora dendroides, Sargassum wightii, S.licifolium, Gracilaria edulis, Caulerpa fastigata, Ulva reticulate*
	Average	31.36		9.31		31.43		
	Soil	High	Low	High	Low			**Plants:** *Abutilon indicum, Acacia nilotica, Albizia ssaman, Azardicata indica, Borassus flabellifer, Calotropis gigantea, Cocos nucifera, Datura metal, Ficus benghalensis, Ipomea pescarprae, Nerium oleander, Ricinus communis, Sida cordifolia, Spinifex littoreus, Achyranthus aspera, Aristida setacea, Leucas aspera, Tephrosia purpurea, Tridax procumbens*
		10.8	7.42	37.5		30.4	24.8	
	Average	8.73						
		Temperature		pH		Salinity		
		High	Low	High	Low	High	Low	
Rainy (July–October)	Water	31.4	21.2	11.9	7.69	33.7	21.2	**Algae:** *Acanthophora spicifera, Sargassum weightii, S. tenerrimum, S. licifolium, Gracilaria edulis, G. verrucosa, Caulerpa fastigata, Halimeda lacrimosa, Ulva compressa*
	Average	28.54		9.33		29.12		

Season	Physiochemical parameters						TEK knowledge from the survey
	Soil	pH		Salinity			**Plants:** *Abutilon indicum, Albizia saman, Azardicata indica, Borassus flabellifer, Calotropis gigantea, Cassia auriculata, Casurina equestifolia, Cocos nucifera, Cymodan dactylon, Cyperus esculantus, Datura metal, Ficus benghalensis, Ipomea pescarprae, Nerium oleander, Opuntia dilleni, Oldenlandia umbellata, Ricinus communis, Salicornia brachiata, Sida cordifolia, Spinifex littoreus, Acalypha indica, Achyranthus aspera, Boerhavia diffusa, Catharanthus rosus, Leucas aspera, Phyllanthus niruri, Solanum nigrum, Tephrosia purpurea, Tridax procumbens, Tribulus terrestris*
		High	Low	High	Low		
		10.12	7.06	12.2	1.4		
	Average	8.65		5.01			
Winter (November-February)	Water	Temperature		pH		Salinity	**Algae:** *Acanthophora dendroides, A. spicifera, Sargassum wightii, S. tenerrimum, S. licifolium, S. plagiophylum, Gracilaria edulis, G.corticata, G.verucosa, Caulerpa fastigata, C. peltata, Halimeda lacrimosa, Ulva compressa, U.reticulata*
		High	Low	High	Low	High Low	
		31.1	23.2	9.22	5.02	36.4 23.7	
	Average	27.73		8.26		28.7	

(continued)

Table 22.5 (continued)

Season	Physiochemical parameters					TEK knowledge from the survey
	Soil	pH		Salinity		
		High	Low	High	Low	
		8.97	7.84	29.9	4.80	**Plants:** *Abutilon indicum, Acacia nilotica, Albizia saman, Azardicata indica, Achyranthus aspera, Aerva lanata, Aristida setacea, Borassus flabellifer, Boerhavia diffusa, Calotropis gigantea, Cassia auriculata, Casurina equestifolia, Cisscus quadrangularis, Cocos nucifera, Cymodan dactylon, Cyperus esculentus, Catharanthus roseus, Datura metal, Euphorbia hirta, Ficus benghalensis, Ipomea pescarprae, Leucas aspera, Nerium oleander, Opuntia dilleni, Oldenlandia umbellata, Phyllanthus niruri, Ricinus communis, Salicornia brachiata, Sida cordifolia, Solanum nigrum, Spinifex littoreus, Tephrosia purpurea, Tridax procumbens, Tribulus terrestris*
	Average	8.22		18.4		

Table 22.6 Utilisation of bioresources by the fisherman community in treating major and minor diseases

Category	Diseases	Angiosperm	Algae
Minor	Insect bites	10	–
	Backache	5	–
	Lesions	8	–
	Cold	10	1
	Cough	12	3
	Fever	17	4
	Indigestion	5	1
	Wound	6	–
Major	Cancer	5	8
	Asthma	3	2
	Piles	4	–
	Snakebite	2	–
	Diabetes	8	10
	Hypertension	6	–

soil (Sivansan et al. 1993), which affects plant growth and development, with only a smaller number of plant species getting reported in the summer season compared to rainy and winter season.

22.7 Climatic Modulations of South East Indian Coastline

The growth of industrialization in the coastal region is rapidly increasing the benefits of low-cost marine transportation and direct usage of marine resources. On the other hand, much anthropogenic activity leads to unpredictable activity on climate in coastal areas like Coastal inundation, declining pH, coastline erosions, and rapid increases in salinity and temperature (Cherian et al. 2012).

Climatic variations, industrialization, and various anthropogenic activities are slowly hampering the coastal biodiversity, resulting in adverse effects that affect the livelihood of the fishing and allied-fishing community (https://www.pmfias.com/coastline-of-india-indian-coastline-coastal-plains-of-india/). Policy constraints, lack of TEK of bioresources, and socio-economic problems force the youths of the fisherman community to find other job opportunities in metro cities and towns. Therefore, migration of coastal communities to the urban area seems rapid, posing a serious threat to the loss of TEK within the older generation. The following recommendations may help restore and conserve TEK and bioresources present on this site.

22.8 Conclusion

The South East Indian coastal region comprises a huge number of plant and algal species with an ethnomedicinal value that can be used to treat various diseases. However, references or databases on traditional ethnobotanical knowledge (TEK) are almost nil among the local fisherman community. The outcome of this study becomes a document and conservation of TEK of the fishermen's community; it will be gaining importance in the current era. The results/data obtained from the study region help predict the climate variation and physiochemical changes in the prevalence of angiosperms and seaweeds. These results can be a potent sustainable management option for ensuring India's coastal sand dune habitats and marine seaweeds.

Acknowledgements The authors would like to extend our profound gratitude to our management, Vellore Institute of Technology, for their constant support and encouragement in all our endeavours. Also, we extend our sincere thanks to DST—SPLICE—Climate Change Programme (DST/CCP/NCC&CV/133/2017 (G), Govt. of India, for providing financial support in carrying over this project successfully till date.

Declaration There is no conflict of interest.

References

Bannerman RH (1982) Traditional medicine in modern health care. World Health Forum 3(1): 8–13.

Cherian A, Chandrasekar N, Gujar AR, Rajamanickam GV (2012) Coastal erosion assessment along the southern Tamilnadu coast, India. International Journal of Earth Sciences.

Cheung WW, Lam VW, Sarmiento JL, Kearney K, Watson R, Pauly D (2009) Projecting global marine biodiversity impacts under climate change scenarios. Fish and Fisheries 10(3): 235–251.

Cheung WW, Lam VW, Sarmiento JL, Kearney K, Watson REG, Zeller D, Pauly D (2010) Large-scale redistribution of maximum fisheries catch potential in the global ocean under climate change. Global Change Biology 16(1): 24–35.

Costanza R, d'Arge R, De Groot R, Farber S, Grasso M, Hannon B, Limburg K, Naeem, S, O'neill RV, Paruelo J, Raskin RG (1997) The value of the world's ecosystem services and natural capital nature. 387(6630): 253–260.

Doney SC, Ruckelshaus M, Emmett Duffy J, Barry JP, Chan F, English CA, Galindo HM, Grebmeier JM, Hollowed AB, Knowlton N, Polovina J (2012) Climate change impacts on marine ecosystems. Annual Review of Marine Science 4:11–37.

FAO (1998) Issues, perspectives, policy and planning processes for integrated coastal area management. Rural development through entrepreneurship. Food and Agricultural Organization of the United Nations, Rome, Italy http://www.fao.org/docrep/.

Ganesan S, Suresh N, Kesavan L (2004) Ethnomedicinal survey of lower Palani Hills of Tamil Nadu. Indian Journal of Traditional Knowledge 3(3): 299–304.

Hwang JS, Choi DG, Choi SC, Park HS, Park YM, Bae JJ, Choo YS (2016) Relationship between the spatial distribution of coastal sand dune plants and edaphic factors in a coastal sand dune system in Korea. Journal of Ecology and Environment 39(1):17–29.
https://www.pmfias.com/coastline-of-india-indian-coastline-coastal-plains-of-india/.
https://www.researchgate.net/figure/Distribution-of-rainfall-in-west-coast-south-to-north-and-east-coast-of-India.
Ignacimuthu S, Ayyanar M, Sankara Sivaraman K (2006) Ethnobotanical investigations among tribes in Madurai District of Tamil Nadu (India). Journal of Ethnobiology and Ethnomedicine 2: 25–30.
Kuffner IB, Andersson AJ, Jokiel PL, Rodgers KS, Mackenzie FT (2008) Decreased abundance of crustosecoralline algae due to ocean acidification. Nature Geosci. 1:114–117.
Lüning K, Neushul M (1978) Light and temperature demands for growth and reproduction of Laminarian gametophytes in southern and central California. Marine Biology 45:297–309.
MacNeil MA, Graham NA, Cinner JE, Dulvy NK, Loring PA, Jennings S, Polunin NV, Fisk AT, McClanahan TR (2010) Transitional states in marine fisheries: adapting to predicted global change. Philosophical Transactions of the Royal Society B: Biological Sciences 365(1558): 3753–3763.
Michel D, Pandya A (2010) Coastal zones and climate change. The Henry L. Stimson Center, Washington, DC, p 20036.
Navaneethan P, Nautiyal S, Kalaivani T, Rajasekaran C (2011) Cross-cultural ethnobotany and conservation of medicinal and aromatic plants in the Nilgiris, Western Ghats: A case study. Medicinal Plants-International Journal of Phytomedicines and Related Industries 3(1): 27–45.
Neelmani RC, Pal M, Sarman V, Vyas UD, Muniya TN (2019) Impacts of climate change on marine biodiversity.
Pauly D (2010) Gasping Fish and Panting Squids: Oxygen, Temperature and the Growth of Water-Breathing Animals. International Ecology Institute, Oldendorf/Luhe, Germany, p xxviii.
Petchey OL, McPhearson PT, Casey TM, Morin PJ (1999) Environmental warming alters food-web structure and ecosystem function. Nature 402(6757): 69–72.
Pushpangadan P, Atal CK (1984) Ethno-medico-botanical investigations in Kerala—Some primitive tribal of Western Ghats and their herbal medicine. Journal of Ethnopharmacology 11(1): 59–77.
Rajan S, Sethuraman M, Mukherjee PK (2002) Ethno biology of the Nilgiri Hills, India. Phytotherapy Research 16(2): 98–116.
Sandhya B, Thomas S, Isabel W, Shenbagarathai R (2006) Ethnomedicinal plants used by the valaiyan community of Piranmalai hills, Tamil Nadu, India—A pilot study. African Journal of Traditional, Complementary and Alternative Medicines 3(1): 101–114.
Sivaperumal R, Ramya S, Ravi AV, Rajasekaran C, Jayakumararaj R (2009) Herbal remedies practiced by Malayali's to treat skin diseases. Environment & We an International Journal of Science & Technology 4(1): 35–44.
Sivansan K, Mithyantha MS, Natesan S, Subharayappa CT (1993) Physiochemical properties and nutrient management of red and literate soils under plantation crops in southern India. NBSS Publication 37: 280.

Srivastav S, Singh P, Mishra G, Jha KK, Khosa RL (2011) Achyranthes aspera-An important medicinal plant: A review. Journal of Natural Product and Plant Resources 1: 1–14

Steen H (2004) Effects of reduced salinity on reproduction and germling development in Sargassum muticum (Phaeophyceae, Fucales). European Journal of Phycology 39: 293–299.

UNFCC (1992) United Nations framework convention on climate change, Bonn, FCCC/ INFORMAL/84, p 24.

Usher P (2000) Traditional ecological knowledge in environmental assessment and management. Arctic 53: 183–193.

PART IV

Urban Sustainability and Resilience

CHAPTER 23

Urban Green Spaces for Environmental Sustainability and Climate Resilience

Amit Kumar, Pawan Ekka, Manjari Upreti, Shilky, and Purabi Saikia

23.1 Introduction

The momentum and scale of larger rising cities can invariably accelerate tremendous stress on the immediate environment and propound significant challenges for sustainable development (Cohen 2006). Green and sustainable cities present favourable circumstances in terms of applications of new technologies such as public transit, green buildings, and design and also bring about crucial lifestyle changes, including the use of bicycles, walking, and energy-efficient technologies (Beatley 2000). Urban Green Spaces (UGSs) are predominantly vegetated, open green spaces in urban areas that provide potential natural recreational spaces that positively influence the urban environment (Levent and Nijkamp 2004). UGSs and Green Infrastructures (GIs) in combination with engineered infrastructure are considered cost-effective options to provide improved air quality, reduced soil erosion and noise pollution, offset the urban heat island (UHI) effects, help to manage water quality, and offer benefits to physical and mental health (Hernández and Wielgołaska 2021). The quantity and the quality of UGSs determine their efficiency in different activities, experiences, and perceived benefits to the users (Bailey et al. 2004).

A. Kumar (✉) · M. Upreti
Department of Environmental Sciences, Central University of Jharkhand, Ranchi, India
e-mail: amit.kumar@cuj.ac.in

P. Ekka · Shilky · P. Saikia
Department of Geoinformatics, Central University of Jharkhand, Ranchi, India

© The Author(s), under exclusive license to Springer Nature Singapore Pte Ltd. 2023
S. Nautiyal et al. (eds.), *Palgrave Handbook of Socio-ecological Resilience in the Face of Climate Change*,
https://doi.org/10.1007/978-981-99-2206-2_23

Moreover, location, distribution, accessibility, and proximity are some conditions that favour the use of UGSs more effectively (Herzele and Wiedemann 2003). The green regional density and connectivity of UGSs determined the cooling capacity and pollution abatement in an urban ecosystem (Hernández et al. 2018). UGSs offer six major ecosystem services, *i.e.* air purification, regulation of microclimate, noise pollution abatement, rainwater drainage and groundwater recharge, and recreational and cultural values (Bolund and Hunhammar 1999). Besides, temperature regulation, water supply, aesthetic beauty, habitats, and foods for wildlife, improved quality of life, sense of identity, and provision of land for housing, economic, and commercial activities are some additional benefits provided by the UGSs (Jansson 2014; Elmqvist et al. 2016). UGSs are resilient to stress instigated by both urbanisation and climate change and serve as a buffer against natural disasters and disturbances (Hernández and Wielgołaska 2021). It tends to greatly impact local climatic energy budgets through evapotranspiration and shading as it provides higher albedo values (Armson et al. 2012). Vegetation can result in a temperature drop of *ca.*, 2 to 8 °C compared to the surrounding built environments by the cooling effects of evapotranspiration (Taha 1997). The areas dominated by different tree species have less diffuse and direct solar insolation than open spaces with herbaceous plants, asphalt, or other artificial surfaces (Száraz 2014).

UGSs primarily consisted of reserved forests, protected forests, green belts, urban parks, green strips, playgrounds, and roadside plantations (Urban Greening Guidelines 2014). Under the provision of the Indian Forest Act 1927, reserved forests are the areas holding full protection with prohibition to all anthropogenic activities unless permitted, while protected forests are found in the urban and peri-urban areas secured by appropriate fencing, where construction activities are strictly prohibited (Government of India 1927). Protect forests are primarily under the government jurisdiction with defined restricted uses necessary for sustained conservation (Molnar et al. 2004). On the other hand, green belts are the invisible lines encircling a certain area (McMichael 2000) to protect natural or agricultural lands restricting direct metropolitan growth (Ozyavuz 2012). The form and content of green belts vary from place to place. Its planning incorporates bio-aesthetics, site topography, meteorological, and ecological circumstances that greatly aid in pollution abatement and air quality improvement (Abbasi et al. 2004). The effectiveness of a green belt for interception and pollution retention depends on the plant composition, shape, cluster size, wetness, surface texture, nature of the ambient pollutants, and the intercepting plant parts (Ingold 1971). Green strips are developed on vacant land and along arterial roads with a boundary of residential areas that a complex landscape could accompany to significantly improve the roadside pedestrian environment (Jim 2013). Urban parks are the specified open areas, generally reserved for public use, vegetation and water dominated, more extensive in size, that can also be modified into the shape of smaller pocket parks (Konijnendijk et al. 2013). Public UGSs, including gardens and playgrounds, are accessible to all citizens that aim to improve the quality of open spaces with community participation in spatial planning and management (WHO 2017). Trees planted along the roads within the right-of-way and on the central verge are considered roadside plantations that include avenue plantations, group plantations, mixed plantations, and informal plantations (Urban Greening Guidelines 2014). Avenue plantation is one of the essential practices of growing trees along the roadside. The canal side is crucial in increasing aesthetic value and maintaining the ecological balance in an urban area (Kumar et al. 2022). It is the greenery source in urban areas and holds effectiveness for climate stabilisation through tree shading and heat absorption (Mandal et al. 2019).

On the other hand, a group of 3–4 or more tree species planted at specified intervals instead of planting in avenues is referred to as a group plantation, where a selection of tree species is made according to the environmental conditions of the plantation site (Randhawa and Mukhopadhyay 2001). Mixed plantations are the plantation of multiple plant species in an area according to the local edaphic and climatic conditions (Pancel 2016). UGSs in the forms of parks, urban farms, and community gardens serve as a health-promoting setting for the urban population (Braubach et al. 2017) that positively impacts the ecosystem and human health, social coherence, and community support (Mckinney 2002). Therefore, it is crucial to ensure that the city's allocation of public green areas is equitable and that all population groups may easily use them (WHO 2017). Sustainable development of cities challenges the establishment of UGSs as ~50% of the global population lives in cities (Goi 2017), and the rate of migration from rural to urban and international migration to developed countries is quite high (Haq and Shah 2011). UGSs are one of the prime ways to bridge the gap between city dwellers and the environment and sustainable development (Saha 2017). The major concerns of a sustainable environment are incorporated into sustainable development from local to global scales through the management, conservation, planning, design, protection, policy formation, and implementation of UGSs (Levent and Nijkamp 2004). Therefore, the present chapter aims to summarise the role of UGSs in urban ecosystems and their contribution to environmental sustainability and climate resilience. It provides a framework for the need for UGSs in rapidly growing cities. It elucidates the integrated three-dimensional approach, including social, economic, and environmental aspects for sustainable utilisation and management of UGSs.

23.2 Consequences of Urbanisation

Urban areas accommodate ~54% of the global population; if the trend continues, more than 70% of the global population will live in cities by 2050 (Ritchie and Roser 2018). The urban amenities including better livelihood, education, employment, and other economic opportunities act as a centripetal force that led to rising urbanisation and population growth (Zope et al. 2015). It further leads to the shortage of natural resources and the emergence of environmental issues, including reducing green spaces, rising pollution, destruction of wildlife habitats, and disruption of the hydrological cycle (Cobbinah et al. 2015). Rapid urbanisation transforms non-urbanised areas, primarily in the peri-urban regions, for various urban developmental activities, including housing, social amenities, industrial, and other urban land uses (Kumar et al. 2011; Ritchie and Roser 2018). The concept of sustainability is often overlooked in urban areas to cater to the fundamental needs of a rapidly urbanising population, which leads to unplanned infrastructure development (Cobbinah et al. 2015). That led to the irreversible transformation of agricultural lands, green cover, floodplains, and wetlands into built-up land that often results in urban disasters, including flash floods, heatwaves, winter fog, and so on (Kumar et al. 2020). In addition, urban areas produce more than 70% of the world's fossil fuel-based CO_2 emissions, which is a significant contributor to climate change (Kumar and Kumar 2019). The phenomenon of higher temperatures in urban areas relative to neighbouring rural regions was caused by the growing loss of UGSs and rapid land change (Revi et al. 2014), which was exacerbated by the rise in the number of hot days (Choudary and Kumar 2020). UHI is primarily regulated by the size and quality of vegetated surfaces as it absorbs a high amount of incoming solar radiation (Zhang et al. 2014). The cooling effect of UGSs as a measure of evapotranspiration regulated UHI effects in an urban area (Qiu et al. 2017), while the inverse phenomenon exacerbates the impact of urban thermal comfort (Choudary and Kumar 2020) and heat-related mortality (Lee et al. 2014).

Extreme weather events become more intense and long-lasting due to climate change, and the vulnerability to climate-related hazards might vary based on the geomorphological and topological aspects of cities (Revi et al. 2014). Around 400 million global population live within a 20 m boundary of the sea or 20 km of the coast (Small and Nicholls 2003), and the major Indian cities, including Kolkata, Mumbai, and Chennai, located near the coastal regions, are susceptible to floods, cyclones, and storm surges (Revi 2008). Heavy rainfall and storm surges are mainly responsible for floods in urban areas, resulting in property and infrastructure damage, waterlogging, and loss of economic and livelihood opportunities that would eventually give rise to waterborne diseases due to contamination of drinking water (Sharma and Tomar 2010). There is an increase in the rate and volume of surface runoff of rainwater due to higher impervious surfaces and rapid loss of vegetation, wetlands, and floodplains in the urban proximity, which often leads to urban flooding (Zope et al. 2015; Lal et al. 2020). Most global cities face a severe challenge of air pollution, mainly due to unprecedented population growth, various economic activities, and land-use transformation in urban areas (Mayer 1999). Intensive anthropogenic activities such as the burning of fossil fuels and industrial activities lead to a higher concentration of O_3, NO_2, and VOCs in the atmosphere and act as a precursor for photochemical smog, which has significant human health consequences and harmful impacts on the plants (Száraz 2014; Mayer 1999). Urban air pollution in the developing world will considerably increase by 2050 under current conditions, causing significant devastating effects on the quality of life of the global urban population (Elmqvist et al. 2016). Uncontrolled discharge of industrial effluents exacerbates the water contamination issue, forcing a substantial portion of the urban population to drink contaminated water, resulting in various health issues and high treatment costs. As a result, in important Indian cities like Mumbai and Delhi, there is only 30% of the capacity for sewage treatment of the expected need (Ahluwalia et al. 2014). Global urbanisation would significantly impact the ecosystem services of the freshwater ecosystems, including maintenance of water quantity and quantity (Elmqvist et al. 2016). Noise pollution is another major problem in cities arising mainly due to high traffic congestion, construction, heavy machinery in industries, and other human activities, which are responsible for negative impacts on human health by causing mental stress and on wildlife (Kumar and Pandey 2013).

Urban biodiversity declines over time due to population expansion, changing land use, climate change, and biological invasion (Hernández and Wielgołaska 2021). The accelerated rate of urbanisation hinders the resistance and resilience of urban ecosystems and biodiversity at varying intensities (Smith et al. 2018). The direct impacts on biodiversity and ecosystem services are primarily attributed to habitat loss and destruction, altered disturbance regimes, land degradation, and other physicochemical transformations caused by the unplanned expansion of urban areas (Elmqvist et al. 2016). Urbanised environments face ongoing challenges to their rich biodiversity of unique and endangered species, which typically drives out native species, replaces them with weedy non-native species, and accelerates some of the highest local extinction rates (Mckinney 2002). Physical changes cause a steeper decline in natural habitats from rural to urban core locations. The number of urban lands around protected areas is expected to increase globally more than three times between 2000 and 2030 (Elmqvist et al. 2016). Pavements and buildings cover more than 80% of the major urban cities (Blair and Launer 1997), while the vegetated areas are progressively partitioned into smaller but more numerous residual patches (Collins et al. 2000).

23.3 Green Cities: A Nature-Based Solution (NBS) Towards Urban Sustainability

Urbanisation provides opportunities for economic, social, and environmental reforms (Ramachandra 1999), where UGSs possess a great potential to sustainably manage the issues of rapid and unplanned urbanisation (Kumar et al. 2022). Vegetation cover helps in the rainwater interception, storage, and infiltration, thereby maintaining the hydrological cycle (Gill et al. 2007). The incorporation of green and blue infrastructure improves the quality of life in urban areas, lessens their ecological footprints, and adapts to global climate change (Kabisch et al. 2017). Green cities have a high coping capacity to deal with the existing and expected negative consequences of climate change through absorbing incoming direct and diffuse solar radiation, tree shading, and evapotranspiration (Száraz 2014). Urban vegetation has a cooling effect of 1 °C during the day and it varies according to the size of parks and their floral composition (Kabisch et al. 2017). Incorporating blue infrastructure with UGSs can significantly mitigate UHI effects (Kabisch et al. 2017). Properly planned and managed UGSs may provide habitat to the species that have been affected by land-use change in cities (Elmqvist et al. 2016). Water conservation is another benefit associated with UGSs, as the vegetated cover provides opportunities for water retention by controlling the runoff and increasing the infiltration rate (Gill et al. 2007). The New York State watershed is one of the most significant natural resources, supplying *ca.* 1.3 billion gallons of potable water to ~9 million people every day (Elmqvist et al. 2016). UGSs can boost air quality by eliminating contaminants such as O_3, SO_2, NO_2, CO, and particulate matter smaller than 10 μm, and different plant species have varied abilities to reduce noise through the reflection and refraction of sound waves and disperse the sound energy in the tree rows (Baggethun et al. 2013). NbSs are actions that are supported by nature that can be predominantly defined as the use of solutions based on ecological principles to address climate change impacts, natural disasters, public health, and food security, through the delivery of multiple ecosystem services and adaptations towards climate change (Longato and Geneletti 2019). It facilitates the protection, long-term management, and restoration of natural or modified ecosystems in both urban and rural locations. It also delivers several benefits and co-benefits like stormwater mitigation, biodiversity enrichment, and social well-being (Watkin et al. 2019). NbSs are increasingly being deployed in metropolitan settings to optimise resilience, assist sustainable development, and protect biodiversity by incorporating grasses, shrubs, and trees to reduce UHI effects, environmental pollution, waterborne infections, respiratory diseases, and distress for people who live nearby (ICLEI 2017). It also incorporates ponds and wetlands which provide benefits of water storage, infiltration, and reuse, evapotranspiration, aquifer recharge, and habitat for flora and fauna (Roy et al. 2008). It can stimulate economic growth, make cities more appealing, and improve the existing environmental conditions and human well-being by restoring damaged ecosystems to enhance ecosystem resilience and better delivering pivotal ecosystem services (DGRI 2015).

Disaster risk reduction is vital for sustainable urban development (Dhyani et al. 2020). Depending on the geographical areas, various NbSs may be developed and applied to increase the effectiveness of urban risk resilience (Baggethun et al. 2013). These solutions may be alternatively or complementary effective on their own but, networking of conventional grey infrastructure with other blue-green spaces can be highly effective (Mukherjee and Takara 2018). For example, cities in hilly regions efficiently mitigate the risk of landslides and avalanches by implementing afforestation initiatives along slopes to stabilise soils (Stokes et al. 2014). Increasing UGSs facilitate climate change adaptation and enhance urban resilience against droughts, floods, and heat waves (DGRI 2015). Additionally, the restoration and management of salt marshes, wetlands, and blue infrastructures including rivers, lakes, and seashores create a natural barrier between water and land that protects against erosion, serves as a habitat for various species, and soaks up extra rainwater to reduce the risk of water logging and flooding in urban areas (ICLEI 2017). Green surfaces, green roofs, and riparian forests reduce flood risks by absorbing excess stormwater in the floodplains (Gupta et al. 2019), while healthy wetlands and mangroves increase the resilience of coastal cities against cyclones, tsunamis, and flooding (Marois and Mitsch 2015). Green walls and roofs also improve the landscape's aesthetic appeal and support urban biodiversity, promote residents' thermal comfort by reducing heat waves, and minimise building cooling costs in addition to small-scale climate mitigation through improved carbon storage (ICLEI 2017).

NbSs should be used to encourage urban redevelopment by constructing green belts, public green spaces, community gardens, and urban farms on vacant, deteriorated, and underutilised residential and industrial areas (DGRI 2015). Improving water quality, property values, and UGSs may be aided by restoring damaged and degraded urban ecosystems to near-natural conditions (ICLEI 2017). Green and blue infrastructure networks must be renewed and regenerated through the progressive ecological restoration of the river and its tributaries within a 300–500 m radius to protect the surrounding from any disaster (Rohilla et al. 2017; WHO 2017). The number of inner-city routes should be reduced to pave the way for additional greenways to improve air quality, promote healthy lifestyles, and encourage a shift from using cars to cycling and walking (ICLEI 2017). UGSs in and around metropolitan areas store carbon, maintain the microclimate, improve air quality, minimise noise, and provide an excellent platform for social cohesion, recreation, mental peace, quality of life, and calmness with its aesthetic appeal to city dwellers (Elmqvist et al. 2016). Furthermore, spending time in nature and near-natural elements may boost mental health and well-being (ICLEI 2017).

23.4 Sustainable Adaptation Strategies for the Development of UGSs

UGSs have become an essential component of urban environments; therefore, planning for them must consider both the socioeconomic and ecological facets of human well-being (Thompson 2002). Indian cities are now losing UGSs and are vulnerable to the effects of climate change due to the pressure of urbanisation (Govindarajulu 2014). UGSs and GIs help reduce energy demand and emission of greenhouse gasses, which benefits climate change mitigation and contributes towards sustainable urban development (Dhyani et al. 2020). It regulates rainfall, maintains temperature and water tables, and provides wildlife habitats and ecosystem functions (Elmqvist et al. 2016).

GIs is a strategically planned network of all kinds of green spaces, including natural and semi-natural areas with other environmental features designed and managed to deliver a wide range of ecosystem services, including clean air and water (Rohilla et al. 2017), support various life forms, improves quality and healthy living, provides recreational opportunities, urban–rural connectivity, strengthens community participation and sustainable transportation (Science for Environmental Policy 2012). It also helps increase life expectancy, levels of physical activity, and physical, psychological, and mental well-being, aiding in temperature regulations, flood management, and aesthetic beautification (Sturiale and Scuderi 2019). It can also generate local employment and enhance land and property values, local economic options, social interactions, people's participation, source of education and moral values while experiencing nature (James et al. 2009). The different forms of GIs are green walls, green roofs, green streets, green belts, green mufflers, green water storms, vertical forests, urban forests, zoological parks and botanical gardens, multifunctional farming, naturalisation of urban rivers, and public premises (Fig. 23.1) (Sturiale and Scuderi 2019). Green walls, also known as living walls, vertical gardens, or plant walls, are a combination of plants and vertical architectural components that are employed in a way to grow particular plants in a growth media consisting of soil, stone, or water that provide the benefits of indoor gardening to reduce noise pollution, aggression, anxiety, mental stress, and fatigue (Manso and Gomes 2015). Green roofs are made of plants that grow on the terrace of a building with an appropriate growth medium and drainage, which provide cooling and insulation, habitats to certain species of pollinators, many insects, and small birds that have lost their natural habitat due to urbanisation and developmental activities, adds aesthetic importance, and a healthy environment to big cities (Rosenzweig et al. 2006). It also aids in reducing stormwater runoff and can hold up to 80% more rainfall, and has a 24% greater capacity for water retention compared to a regular roof (Table 23.1).

Green streets are pathways with suitable vegetation cover, such as aesthetic plants and well-managed stormwater channels (Fig. 23.2) that protect against soil erosion, increase plant cover in urban areas, and provide a progressive approach towards sustainable development (Im 2019). On the other hand, green belts are protected areas around cities for forestry and agricultural practices that prevent urban sprawl, help to conserve wildlife habitats, and effectively promote environmental protection (Ramesh and Nijagunappa 2014). It contributes to the development of satellite cities and provides recreation opportunities to metropolitan residents (Rohilla et al. 2017). An urban forest includes all areas filled with green covers such as trees, shrubs, and palms along streets, roads, yards, watersheds, and protected areas that bring sustainable benefits to the environment, society, and economy of the country (Miller et al. 2015). It helps to withstand natural disasters, mitigate the effects of climate change, support the faster recovery of damages from

Fig. 23.1 Sustainable adaptation strategies for the development of UGSs

Table 23.1 Water retention for traditional standard roof vs. green roof

Rainfall retained	Standard roof (%)	Green roof (%)
Average Retention	24	80
Retention at Peak Runoff	26	74

Source Rosenzweig et al. (2006)

climate change-related risks, reduce pollution, save energy, promote recreation, sequester carbon, improve local food sovereignty, and offer educational opportunities (Govindarajulu 2014). Besides, it improves public health by providing suitable temperature conditions, promotes the healthy practice of walking, inspires others to do so and reduces exposure to harmful vehicular emissions through pollution control (Ramaiah and Avtar 2019). Zoological parks and botanic gardens are scientific institutions and crucial resources for the preservation of nature that provide aesthetic and cultural values, serve as sources of knowledge and fascination for the natural world in our society, and aid in increasing the amount of greenery and habitat for various species in urban areas (Smith 2019). It is important in addressing global climate change, biodiversity conservation, environmental education, sustainability, and human well-being (Krishnan and Novy 2016). Multifunctional farming is another approach to urban agriculture that offers an alternative land use for integrating multiple functions, including improved education and health, food security, socio-economy, and environmental benefits in densely populated areas with additional benefits of microclimate regulations (Skar et al. 2019). Rivers bring fertile soil to the flood plains during flood seasons and may be utilised for food production for urban consumption (Wlodarczyk and Mascarenhas 2016). Naturalisation of the urban river forms a natural defence mechanism against any invasion that supplies food, improves the natural landscape of urban areas, brings more biodiversity into urban areas, purifies the air, helps to fulfil the groundwater loads of the urban areas, and creates opportunities for leisure, sports, and education (Xu et al. 2022). Banks of various water sources such as rivers, lakes, ponds, and channels, along with public premises such as educational institutions, government offices, hospitals, sports complexes, religious places, and cemetery grounds, can be utilised for UGSs development in the city (Cilliers 2015).

Fig. 23.2 Avenue plantations in Ranchi city, Eastern India, with pedestrians and stormwater management

23.5 Case Study: UGSs in Bangalore City, India

Open green spaces are an integral component of a city as they define its character and play an important physical and ecological role (BDA 2017). Cities worldwide emphasise environmental challenges and develop integrated, more sustainable strategies to enrich urban biodiversity (Barrico and Castro 2016). An analysis has been performed emphasising the different types of UGSs in Bangalore, India. The city is situated in the southeast parts of Karnataka state in southern India with a geographical extent of 12°84' to 13°06' N latitude and 77°70' to 77°45' E longitude at an average elevation of 920 m above mean sea level.

The satellite-based Enhanced Vegetation Index (EVI) has been computed to demarcate and classify different types of UGSs in Bangalore city in February 2021 using high-resolution Sentinel 2A datasets having a spatial resolution of 10 m. The majority of the open green spaces are present in the core or central parts of the city in conjunction with the highly dense and heterogeneous built-up landscapes (Fig. 23.3). EVI ranges from -1 to 1, where higher positive values represent the dense vegetation pixels (Liao et al. 2015), which have been used for binary classification to quantify and classify UGSs. Different types of UGSs have been classified, such as avenue plantation, botanical garden (Lalbagh Botanical Garden), green strips developed along power supply lines separating residential areas from other use, parks around residential areas, roadside greenery and plantations, forests viz., Turahalli forests, Turahalli state forests, B.M. Kaval Aagara forests, etc.) (Fig. 23.3). The satellite-based assessment of UGS can provide accurate and better inferences to support the planning process more effectively to improve many social and ecological aspects in the city region (Bhat 2019). Bangalore faces substantial socioeconomic and environmental risks to future growth and prosperity, leading to the development of urban diagnostics for the city (Wankhade and IIHS 2014). GIs offer ecological benefits against a range of psychological, social, environmental, and health impacts and acquiring quantifiable information regarding those GIs contributes to sustainable planning (Gupta et al. 2012).

23.6 UGS Planning and Practices

Planting trees helps improve air quality, abandoned industrial sites can be converted into parks and recreational sites, and degraded wetlands can be restored to prevent floods. Green roofs help to mitigate energy demand in urban areas (ICLEI 2017). The developmental sites should be identified with all the potential to develop a UGS before any development of infrastructure based on a suitable scientific analysis of physiographic and hydrological features of the present and past information about the area (McHarg 1969). City Master Plans, as adopted in Gandhinagar and Chandigarh for better integration of urban greenery, should be implemented by the state and central governments, including housing, services, workplaces, infrastructure,

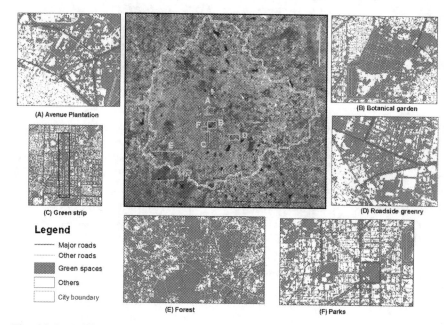

Fig. 23.3 Different types of green spaces in Bangalore city based on EVI **A** Avenue plantation, **B** Botanical Garden, **C** Forests, **D** Roadside greenery, **E** Green strips, and **F** Parks

and institutions, and construction should be avoided in some strategic areas to prevent any effective damage to green space without any compromise (Govindarajulu 2014). Pedestrian corridors are solutions to improve green cover and connectivity in urban streets, thereby improving physical health, social interaction, and mental well-being (Delso et al. 2017). Cities must be planned and prepared by the implementation of complex NbSs comprising regulatory policies and incentive schemes for adaptations to climate change, natural calamities, urban expansion, and pollution control through better urban plans for land-use practices, natural resource utilisation, groundwater consumption, well drainage networks to fulfil ecosystem services (Tozer et al. 2020). Some of the standard practices mentioned below may help in increasing UGSs in the cities—

- Promoting public awareness of the advantages of UGSs through the media, education, communication, and government.
- Increasing options for greening our surroundings by adding more vegetation to paved and compacted areas, filling vacant spaces with appropriate plants, and wherever feasible, trying to make our neighbourhood greener.
- Urban design must be based on ecological principles, with a strong focus on biodiversity enrichment and conservation, as well as the future advantages and disadvantages of urban green areas.

- Urban and peri-urban regions must be protected and have more forest cover through timely tree care, transplantation, and preservation of trees at public premises and building sites.
- City dwellers must practice urban agriculture and horticulture on suitable land for fulfilling sustenance needs and sales.
- UGSs in the roof, terrace, and landscape garden might ensure food security, reduce environmental deterioration, create employment, and generate income.

23.7 Policy Implications for the Development of UGSs

The UN-SDGs comprise a specific goal of Sustainable Cities and Communities (SDG 11) due to the increase in urban areas in terms of size, density, and population. UGSs have become essential in cities to attain urban sustainability and livability (Bush 2020). The Indian urban network consists of over 7000 towns and cities, and the current rate of urban expansion is far exceeding the infrastructure capacity and services to support the present population (ICLEI-South Asia 2015). Despite such a large urban system, none of the major cities has a state of the environment report or action plan for biodiversity and environmental conservation (Rathi et al. 2020). India is facing a high rate of biodiversity loss due to uncontrolled and unplanned urban expansion, land encroachment, lack of urban policy frameworks, high levels of pollutants in the environment, and rapid industrial growth and transportation (ICLEI-South Asia 2015). India is behind in urban development policies, which lack climate adaptation strategies (Rohilla et al. 2017), and the smart city mission of Govt. of India also lacks provisions for UGSs (Sethi 2015). The energy and ecology sector has a total budget of INR 22,535.80 crore, of which there is a provision for ecological restoration with a budget of INR 5697 crore, the ecological infrastructure of 21 crores, and ecological miscellaneous of INR 2080 crore (Ashwathy et al. 2018). UGSs are the elements of the economic transformation; therefore, an integrated approach to social, economic, and environmental concerns in sustainable development policies adds value to green space planning (Cilliers 2015). Present approaches to urban biodiversity, open space, air quality, and health fail to consider or encourage the beneficial role of nature in well-being and achieving government health policy objectives (Brown and Grant 2005). Integrate environmental management policies to regulate services such as biodiversity support, soil conservation, carbon sequestration, climate regulations, and watershed management through the mutual engagement of local communities to create long-term benefits for UGSs (Fig. 23.4) (WHO 2017). Urban planners should maintain the continuous flow of economic development, infrastructure provisions, competing demands of land, and affordable housing for the rising population (Colding 2011). Besides, integrated ecosystem-related approaches to policy implementation are necessary to develop UGSs to address various societal challenges with the support of policymakers, elected officials, bureaucrats, and funding agencies (Cohen-Shacham et al. 2016). Municipalities may request 2–5% of land from residential and commercial areas that must be dedicated for the development of UGSs which can be used for recreational activities and parkland (Maryanti et al. 2017). Green infrastructures and practices such as urban agriculture, horticulture, forestry, watershed conservation, green roofing, plantation, etc. should be promoted by providing incentives to city dwellers by the municipality and private partners (CSE 2021). UGSs are a long-term investment, and the central and local governments have to emphasise developing UGSs and climate change strategies which should include land-use planning and various community services, including recreation, sports, health, family services, and economic development (Bush 2020). Green roofs must be mandatory and rainwater harvesting facilities on all city-owned buildings for new development in all urban planning processes (Mees and Driessen 2011). Promoting the understanding of sustainable accessibility of natural resources, potential impacts of biodiversity loss, and incentive benefits of community participation in the development of UGSs, and conservation of biodiversity may help minimise the negative consequences of urbanisation (ICLEI-South Asia 2015). Rainwater harvesting is the future of water conservation, and it must be promoted to meet the water demands in urban areas with strict enforcement of laws by municipalities to meet the challenging demands of water supply in big cities (Holland-Stergar 2018; CSE 2021).

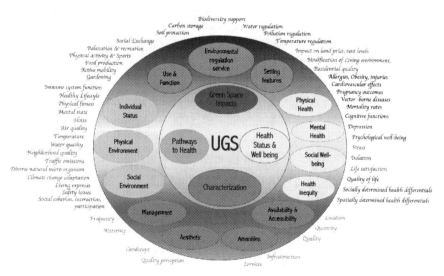

Fig. 23.4 Dimensions of urban green space (*Source* Milvoy and Roué-Le Gall 2015)

23.8 Future Research Prospects and Recommendations

- NGOs and local organisations should come forward for awareness generation among the communities and also strengthen the knowledge of the multiple benefits of UGSs to the policy planners and the communities with more focused research, innovative projects, and extension activities on different forms of UGSs such as pavements, parks, green roofs, canal-side plantations, roadside plantations, and green walls.
- There is an urgent need to determine the quality and quantity of GI required by a particular city to adapt to climate change. Urban planners should incorporate urban GI effectively in the cities.
- All developments must aim to increase the vegetation cover and overall biodiversity for wildlife in the urban areas.
- Integrated approaches to policy implementation with an umbrella concept should cover a wide range of ecosystem-related strategies to address societal challenges and opportunities.
- Social forestry with community participation in urban areas with native trees to be planted to ensure ecological balance.
- Specific policies and an adequate economic budget are needed for smart city missions of Govt. of India with a provision of UGSs.

23.9 Conclusions

Urban regions are becoming more exposed to several environmental concerns, such as air and noise pollution, due to growing populations, limited resources, and the mounting effects of climate change. UGSs and other NbSs provide innovative ways to enhance the health and well-being of urban people by enriching the urban environment, strengthening community resilience, and encouraging sustainable lifestyles. Additionally, UGSs contribute significantly to the livability of urban areas by enhancing the natural beauty and recreational options for urban residents and promoting their general well-being. It also contributes to the carbon sequestration process, aids in reducing the impacts of UHI, improves hydrology by limiting surface runoff, and concurrently recharges the groundwater system. Furthermore, it can serve as a buffer during extreme events like floods since it is a natural stormwater drainage system that reduces the risks of climate-related disasters by providing adaptation to climate change. UGSs are the most sustainable solution to several environmental issues brought on by growing urbanisation, including pollution, climate change, natural disasters, health, and biodiversity loss. Therefore, further study is needed to examine the variety of environmental difficulties that a significant metropolis faces and the role that UGSs play in addressing these challenges. It is becoming increasingly urgent for urban decision-makers and residents to implement policies and practices incorporating nature into their everyday lives as cities play a vital role in a future built on NbSs and ecosystem-based adaptations. Maintaining a sizable section of residential vegetation (1% of the state's area) composed of native plant and animal species may considerably aid in managing and conserving urban ecosystems.

References

Abbasi SA, Chari KB, Gajalakshmi S, Ramesh N, Ramasamy EB (2004) Approaches to greenbelt design. Journal of the Institution of Public Health Engineers, India 2004(3):42–49.

Acton L (2011) Allotment gardens: a reflection of history, heritage, community and self. Papers from the Institute of Archaeology (PIA) 21:46–58.

Ahluwalia IJ, Kanbur R, Mohanty PK (2014) Urbanization in India: Challenges, Opportunities and the Way Forward. SAGE Publishing India, p 358.

Armson D, Stringer P, Ennos AR (2012) The effect of tree shade and grass on surface and globe temperatures in an urban area. Urban for Urban Green 11(3):245–255.

Baggethun EG, Gren Å, Barton DN et al (2013) Urban ecosystem services. In: Elmqvist T, Fragkias M, Goodness J et al (eds) Urbanisation, Biodiversity and Ecosystem Services: Challenges and Opportunities, Springer, Dordrecht, Heidelberg, New York, London, pp 175–251.

Bailey SA, Horner-Devine MC, Luck G, et al (2004) Primary productivity and species richness: relationships among functional guilds, residency groups and vagility classes at multiple spatial scales. Ecography 27:207–217.

BDA (Bangalore Development Authority) (2017) Utilising unused urban space in Bengaluru a field study of Banashankari with policy recommendations. Smart Cities India Foundation. https://www.academia.edu/45673098/Utilizing_Unused_Urban_Space_in_Bengaluru_A_Field_Study_of_Banashankari_with_Policy_Recommendations. Accessed 8 Sept 2021.

Barrico L, Castro P (2016) Urban biodiversity and cities' sustainable development. In: Castro P, Azeiteiro UM, Bacelar-Nicolau P et al (eds) Biodiversity and Education for Sustainable Development, Springer, Cham, pp 29–42.

Beatley T (2000) Green Urbanism: Learning from European Cities, Island Press, Washington, DC, p 51.

Bhat S (2019) The need to evaluate urban green spaces in the city core in Bengaluru. The International Journal of Recent Technology and Engineering (IJRTE) 8(3):908–913.

Blair RB, Launer AE (1997) Butterfly diversity and human land use: species assemblages along an urban gradient. Biological Conservation 80(1):113–125.

Bolund P, Hunhammar S (1999) Ecosystem services in urban areas. Ecological Economics 29(2):293–301.

Braubach M, Egorov A, Mudu P et al (2017) Effects of urban green space on environmental health, equity and resilience. In: Kabisch N, Korn H, Stadler J et al (eds) Nature-Based Solutions to Climate Change Adaptation in Urban Areas, Springer, Cham, pp 187–205.

Brown C, Grant M (2005) Biodiversity and human health: what role for nature in healthy urban planning? Built Environment 31(4):326–338.

Bush J (2020) The role of local government greening policies in the transition towards nature-based cities. Environmental Innovation and Societal Transitions 35:35–44.

Chaudhry P, Tewari VP (2011) Urban forestry in India: development and research scenario. Interdisciplinary Environmental Review 12(1):80–93.

Choudary S, Kumar A (2020) Evaluating the contribution of urban ecosystem services in regulating thermal comfort. Spatial Information Research 29:71–82.

Cilliers EJ (2015) The importance of planning for green spaces. AFF 4(4–1):1–5.

Cobbinah P, Erdiaw-Kwasie MO, Amoateng P (2015) Africa's urbanisation: implications for sustainable development. Cities 47:62–72.

Cohen-Shacham E, Walters G, Janzen C et al (eds) (2016) Nature-Based Solutions to Address Global Societal Challenges, IUCN, Gland, Switzerland, pp 1–61.

Cohen B (2006) Urbanisation in developing countries: current trends, future projections, and key challenges for sustainability. Technology in Society 28:63–80.

Colding J (2011) The role of ecosystem services in contemporary urban planning. In: Niemelä J, Breuste HJ, Elmqvist T et al (eds) Urban Ecology: Patterns, Processes, and Applications, Oxford University Press, Oxford, UK, pp 228–237.

Collins JP, Kinzig A, Grimm NB et al (2000) A new urban ecology. American Scientist 88(5):416–425.

CSE (Centre for Science and Environment) (2021) Legislation on rainwater harvesting. https://www.cseindia.org/legislation-on-rainwater-harvesting. Accessed 18 May 2021.

Delso J, Martín B, Ortega E et al (2017) A model for assessing pedestrian corridors: application to Vitoria-Gasteiz City (Spain). Sustainability 9:1–15.

DGRI (Directorate-General for Research and Innovation) (2015) Towards an EU Research and Innovation Policy Agenda for Nature-Based Solutions & Re-Naturing Cities: Final Report of the Horizon 2020 Expert Group on Nature-Based Solutions and Re-Naturing Cities, European Commission, Brussel, pp 1–70.

Dhaliwal S (2017) Types of vegetable gardens. In: Handbook of Vegetable Crop, Kalyani Publishers, pp 18–33.

Dhyani S, Karki M, Gupta AK (2020) Opportunities and advances to mainstream nature-based solutions in disaster risk management and climate strategy. In: Dhyani S, Gupta AK, Karki M (eds) Nature-Based Solutions for Resilient Ecosystems and Societies, Disaster Resilience and Green Growth, Springer, Singapore, pp 1–26.

Elmqvist T, Zipperer WC, Güneralp B (2016) Urbanisation, habitat loss and biodiversity decline: solution pathways to break the cycle. In: Seto KC, Solecki WD, Griffith CA (eds) The Routledge Handbook of Urbanization and Global Environmental Change, Routledge, London, New York, pp 139–151.

Gill SE, Handley JF, Ennos AR et al (2007) Adapting cities for climate change: the role of the green infrastructure. Built Environment 33(1):115–133.

Goi CL (2017) The impact of technological innovation on building a sustainable city. International Journal of Quality Innovation 3(6):1–13.

Government of India (1927) The Forest Act 1927, Government of India, pp 1–56.

Govindarajulu D (2014) Urban green space planning for climate adaptation in Indian cities. Urban Climate 10:35–41.

Gupta AK, Gotmare S, Nair U et al (2019) Green growth benefits for climate and disaster resilience: concerns for urban and infrastructure systems. Position Paper, NIDM and GGGI, p 17.

Gupta K, Kumar P, Pathan SK et al (2012) Urban neighborhood green index—a measure of green spaces in urban areas. Landscape and Urban Planning 105:325–335.

Haq A, Shah M (2011) Urban green spaces and an integrative approach to sustainable environment. Journal of Environmental Protection 2(5):601–608.

Hernández JGV, Pallagst K, Wielgołaska JZ (2018) Urban green spaces as a component of an ecosystem. In: Dhiman S, Marques J (eds) Handbook of Engaged Sustainability, Springer, Cham, 2:885–916.

Hernández JGV, Wielgołaska JZ (2021) Urban green infrastructure as a tool for controlling the resilience of urban sprawl. Environment, Development and Sustainability 23:1335–1354.

Herzele AV, Wiedemann T (2003) A monitoring tool for the provision of accessible and attractive urban green spaces. Landscape and Urban Planning 63(2):109–126.

Holland-Stergar B (2018) The law and policy of rainwater harvesting: a comparative analysis of Australia, India, and the United States. Journal of Environmental Law and Policy 36(1):127–165.

ICLEI (2017) Nature-Based Solutions for Sustainable Urban Development, Local Governments for Sustainability, Germany, pp 1–13.

ICLEI-South Asia (2015) Urban Green Growth Strategies for Indian Cities, Delhi, India 1:1–137.

Im J (2019) Green streets to serve urban sustainability: benefits and typology. Sustainability 11:1–22.

Ingold CT (1971) Fungal spores: their liberation and dispersal. Clarendon Press, Oxford, p 302.

James P, Tzoulas K, Adams MD et al (2009) Towards an integrated understanding of green space in the European built environment. Urban for Urban Green 8(2):65–75.

Jansson M (2014) Green space in compact cities: the benefits and values of urban ecosystem services in planning. Nordic Journal of Architectural Research 26(2):139–160.

Jim CY (2013) Sustainable urban greening strategies for compact cities in developing and developed economies. Urban Ecosystems 16:741–761.

Kabisch N, Korn H, Stadler J et al (eds) (2017) Nature-based solutions to climate change adaptation in urban areas-linkages between science, policy and practice, Springer Nature, Switzerland, pp 1–337.

Konijnendijk van den Bosch C, van den Bosch M, Nielsen A et al (2013) Benefits of urban parks a systematic review—a report for the international federation of parks and recreation administration, p 68.

Krishnan S, Novy A (2016) The role of botanic gardens in the twenty-first century. CAB Reviews 11(23):1–10.

Kumar A, Pandey AC, Hoda N, Jeyaseelan AT (2011) Evaluation of urban sprawl pattern in tribal dominated cities of Jharkhand State, India. International Journal of Remote Sensing 32(22):7651–7675.

Kumar A, Diksha, Pandey AC et al (2020) Urban risk and resilience to climate change and natural hazards: a perspective from a million-plus cities from the Indian subcontinent. In: Srivastava PK, Singh SK, Mohanty UC et al (eds) Advanced Techniques for Disaster Risk Management and Mitigation. Geophysical Monograph. John Wiley and Sons Ltd, Hoboken, United States, pp 33–46.

Kumar A, Kumar A (2019) Assessing human and carbon footprint of Ranchi urban environment using remote sensing technology. Journal of Urban and Environmental Engineering 13(2):257–265.

Kumar A, Pandey AC (2013) Spatio-temporal assessment of urban environmental conditions in Ranchi township using remote sensing and GIS techniques. International Journal of Urban Sciences 17(1):117–141.

Kumar A, Kumari R, Saikia P (2022) Avenue plantation as a viable carbon confiscation tool: A blueprint for eco-smart cities. In: Dervash M, Wani A (eds) Climate Change Alleviation for Sustainable Progression: Floristic Prospective and Arboreal Avenues as a Viable Confiscation Tool, Chapter 17, CRC Press (Taylor & Francis Group), Boca Raton, USA, pp 333–346.

Lal P, Prakash A, Kumar A (2020) Google earth engine for concurrent flood monitoring in the lower basin of Indo-Gangetic-Brahmaputra plains. Natural Hazards 104:1947–1952.

Lee YY, Fadhil M, Ponraj M et al (2014) Overview of urban heat island (UHI) phenomenon towards human thermal comfort. Environmental Engineering & Management Journal (EEMJ) 16(9):2097–2111.

Levent TB, Nijkamp P (2004) Urban green space policies: a comparative study on performance and success conditions in European cities. Regions and Fiscal Federalism, Porto, Portugal, 25–29 August 2004, pp 1–18.

Liao Z, He B, Quan, X (2015) Modified enhanced vegetation index for reducing topographic effects. Journal of Applied Remote Sensing 9(1):1–21.

Longato D, Geneletti D (2019) Nature-based solutions: new challenges for urban planning. Planning for Transition, Venice, 19–13 July 2019, pp 3785–3792.

Mandal B, Ganguly A, Mukherjee A, Shome D (2019) Assessment and analysis of avenue trees in urban Kolkata: a case study. Journal of Environment and Sociobiology 14(1):7–15.

Manso M, Gomes JC (2015) Green wall systems: a review of their characteristics. Renewable and Sustainable Energy Reviews 41:863–871.

Marois ED, Mitsch JW (2015) Coastal protection from tsunamis and cyclones provided by Mangrove wetlands-a review. International Journal of Biodiversity Science, Ecosystem Services & Management 11(1):71–83.

Maryanti MR, Khadijah H, Uzair AM, et al (2017) The urban green space provision using the standards approach: issues and challenges of its implementation in Malaysia. In: Brebbia CA, Zubir SS, Hassan AS (eds), Sustainable Development and Planning VIII, WIT Transactions on Ecology and the Environment, 210:369–379.

Mayer H (1999) Air pollution in cities. Atmospheric Environment 33(24–25):4029–4037.

McHarg IL (ed) (1969) Design with Nature. John Wiley & Sons Inc., New York.

Mckinney ML (2002) Urbanisation, biodiversity, and conservation. BioScience 52(10):883–890.

McMichael AJ (2000) The urban environment and health in a world of increasing globalisation: issues for developing countries. Bulletin of the World Health Organization 78(9):1117–1126.

Mees HLP, Driessen PPJ (2011) Adaptation to climate change in urban areas: climate-greening London, Rotterdam, and Toronto. Climate Law 2:251–280.

Miller RW, Hauer RJ, Werner LP (2015) Urban forestry: planning and managing urban greenspaces. Waveland Press, Illinois, pp 560.

Milvoy A, Roué-Le Gall A (2015) Develop healthy play areas. Health in Action 434:38–39.

Molnar A, Scherr SJ, Khare A (2004) Who conserves the world's forests? A new assessment of conservation and investment trends*. Forest Trends & Eco-Agriculture Partners, Washington, DC.

Mukherjee M, Takara K (2018) Urban green space as a countermeasure to increasing urban risk and the UGS-3CC resilience framework. International Journal of Disaster Risk Reduction 28:854–861.

Mukhopadhyay P, Revi A (2009) Keeping India's economic engine going: climate change and the urbanisation question. EPW 44(31):59–70.

Ozyavuz M (ed) (2012) Landscape Planning. IntechOpen, UK.

Pancel L (2016) Mixed tree plantations in the tropics. In: Pancel L, Köhl M (eds) Tropical Forestry Handbook, Springer, Berlin, Heidelberg, pp 1549–1560.

Qiu GY, Zou Z, Li X et al (2017) Experimental studies on the effects of green space and evapotranspiration on urban heat island in a subtropical megacity in China. Habitat International 68:30–42.

Ramachandran R (ed) (1999) Urbanization and Urban System in India. Eighth edition, Oxford University Press, New York.

Ramaiah M, Avtar R (2019) Urban green spaces and their need in cities of rapidly urbanising India: a review. Urban Science 3(94):1–16.

Ramesh RM, Nijagunappa R (2014) Development of urban green belts—a super future for ecological balance, Gulbarga city, Karnataka. International Letters of Natural Sciences 27:47–53.

Randhawa GS, Mukhopadhyay A (2001) Landscaping Places of Public Importance. Floriculture in India, Allied Publishers, pp 506–507.

Rathi V, Bangera Y, Shetty N et al (2020) India urban infrastructure report 2020: special focus on Mumbai transport infrastructure with key impact markets. Knight Frank, India, pp 1–85.

Revi A (2008) Climate change risk: an adaptation and mitigation agenda for Indian cities. Environment and Urbanization 20(1):207–229.

Revi A, Satterthwaite DE, Aragón-Durand F et al (2014) Urban areas. In: Barros CB, Dokken VR, Mach DJ et al (eds) Climate change 2014: impacts, adaptation, and vulnerability. Part A: global and sectoral aspects. Contribution of working group II to the fifth assessment report of the intergovernmental panel on climate change, field, Cambridge University Press, United Kingdom, New York, USA, pp 535–612.

Ritchie H, Roser M (2018) Urbanisation. Published online at OurWorldInData.org. https://ourworldindata.org/urbanization Accessed 6 May 2021.

Rohilla SK, Jainer S, Matto M (2017) Green Infrastructure: A Practitioner's Guide, Centre for Science and Environment, New Delhi, pp 1–111.

Rosenzweig C, Gaffin S, Parshall L (eds) (2006) Green roofs in the New York metropolitan region: research report. Columbia University Center for Climate Systems Research and NASA Goddard Institute for Space Studies, New York, pp 1–59.

Roy AH, Wenger SJ, Fletcher et al (2008) Impediments and solutions to sustainable, watershed-scale urban stormwater management: lessons from Australia and the United States. Environmental Management 42:344–359.

Saha T (2017) Urban forestry: importance, strategy, and planning in Indian context. IJHSS 4(1):302–310.

Science for Environmental Policy (2012) The multifunctionality of green infrastructure. In-depth report, DG environmental news alerts, European Commission, pp 1–37.

Sethi M (2015) Smart cities in India: challenges and possibilities to attain sustainable urbanisation. Nagarlok 47(3):1–37.

Sharma D, Tomar S (2010) Mainstreaming climate change adaptation in Indian cities. Environment and Urbanization 22(2):451–465.

Skar SLG, Pineda-Martos R, Timpe A et al (2019) Urban agriculture as a keystone contribution towards securing sustainable and healthy development for cities in the future. Blue-Green Systems 2(1):1–27.

Small C, Nicholls RJ (2003) A global analysis of human settlement in coastal zones. Journal of Coastal Research 19(3):584–599.

Smith P (2019) The challenge for botanic garden science. Plants, People, Planet 1(1):38–43.

Smith WS, da Silva FL, de Amorim, SR, Stefani MS (2018) Urban biodiversity: how can the city do its management? Biodiversity International Journal 2(3):246–251.

Stokes A, Douglas GB, Fourcaud T et al (2014) Ecological mitigation of hillslope instability: ten key issues facing researchers and practitioners. Plant and Soil 377:1–23.

Sturiale L, Scuderi A (2019) The role of green infrastructures in urban planning for climate change adaptation. Climate 7(119):1–24.

Száraz LR (2014) The impact of urban green spaces on climate and air quality in cities. Geographical Locality Studies 2(1):326–354.

Taha H (1997) Urban climates and heat islands: albedo, evapotranspiration, and anthropogenic hat. Energy and Buildings 25(2):99–103.

Ashwathy A, Sreevatsan A, Taraporevala P (2018) An Overview of the Smart Cities Mission in India. Centre for Policy Research, India, pp 1–17.

Thompson CW (2002) Urban open space in the 21st century. Landscape and Urban Planning 60:59–72.

Tozer L, Hörschelmann K, Anguelovski I et al (2020) Whose city? Whose nature? Towards inclusive nature-based solution governance. Cities 107:102892.

Urban Greening Guidelines (2014) Town and Country Planning Organisation, Government of India, Ministry of Urban Development, pp 1–40.

Watkin LJ, Ruangpan L, Vojinovic, Z et al (2019) A framework for assessing benefits of implemented nature-based solutions. Sustainability 11(23):6788.

Wankhade K. IIHS (Indian Institute for Human Settlements) (2014) Future proofing Indian cities: Bangalore diagnostic report. https://www.academia.edu/32683247/Future_Proofing_Indian_Cities_Bangalore_Diagnostic_Report. Accessed 8 September 2021.

WHO (2017) Urban green spaces: a brief for action. World Health Organisation Regional Office for Europe, Copenhagen, pp. 1–22.

Wlodarczyk AM, Mascarenhas JMRD (2016) Nature in cities. Renaturalization of riverbanks in urban areas. Open Engineering 6(1):681–690.

Xu F, Wang Y, Wang X et al (2022) Establishment and application of the assessment system on ecosystem health for restored urban rivers in North China. International Journal of Environmental Research and Public Health 19(9):1–19.

Zhang B, Gao JX, Yang Y (2014) The cooling effect of urban green spaces as a contribution to energy-saving and emission reduction: A case study in Beijing, China. Building and Environment 76:37–43.

Zope PE, Eldho TI, Jothiprakash V (2015) Impacts of urbanisation on flooding of a coastal urban catchment: A case study of Mumbai city, India. Natural Hazards 75(1):887–908.

CHAPTER 24

Urban Civic Services Delivery and Climate Change Challenges: A Study of Two Indian Cities

Ramakrishna Nallathiga and Kala Seetharam Sridhar

24.1 INTRODUCTION

Given the economic and demographic importance of the cities in developing nations, it becomes imperative that we ensure that they have all the ingredients required for such a transformative role. Providing basic civic infrastructure services in cities is vital for them to play a transformative role. The provision of urban infrastructure services is also considered essential in 'sustainable urban development' (UN Habitat 2012). The 'Sustainable Development Goals (SDGs)' have laid down 17 goals to be achieved by all member nations by 2030 in the mission towards sustainable development (UN 2015). Sustainable Development Goal 3 (good health and well-being), SDG 6 (clean water and sanitation), SDG 9 (industry, innovation and infrastructure) and SDG 11 (sustainable cities and communities) are linked to the provision of basic civic infrastructure services as an essential means of achieving larger development goals of countries (ICLEI 2017).

R Nallathiga is currently a Faculty member at NICMAR University, Pune.

R. Nallathiga (✉) · K. S. Sridhar
Centre for Research in Urban Affairs, Institute for Social and Economic Change, Bengaluru, India
e-mail: nramakrishna@isec.ac.in

K. S. Sridhar
e-mail: kala@isec.ac.in

© The Author(s), under exclusive license to Springer Nature Singapore Pte Ltd. 2023
S. Nautiyal et al. (eds.), *Palgrave Handbook of Socio-ecological Resilience in the Face of Climate Change*,
https://doi.org/10.1007/978-981-99-2206-2_24

Global climate change and consequential global changes have acquired the attention of the entire global community. The rising levels of greenhouse gas (GHG) emissions from various development activities—particularly industrial production and service sector expansion—led to the formation of a warm GHG blanket around the earth, leading to the heat trap. As a result, there have been multiple effects on the global environment, such as the surface temperature rise, the sea level rise and an increase in the occurrence of extreme events like floods, droughts and other natural disasters (IPCC 2018). This, in turn, affects human beings' physical and socio-economic environment. As cities are the concentrations of human population and socio-economic activities, they bear the consequences of climate change, apart from contributing to long-term climate change (IPCC 2007); Sridhar (2010) contains evidence on the carbon emissions of India's cities.

An important association between climate change and urbanisation is that the cities with a larger proportion of developed land may turn into 'heat islands' due to the concentration of heat-absorbing materials and reduced evaporative cooling (due to the lack of vegetation), which gets exacerbated by the warming due to climate change. Another important association is in the form of extreme events exacerbated by climate change (IPCC 2018). There is an increasing and intensifying occurrence of extreme events like heat waves, droughts, heavy downpours and coastal (as well as inland) flooding due to climate change. With the rise in the frequency and intensity of extreme weather events, the cities are at a greater risk as they have a greater share of the human population. The vulnerability of cities to climate-related disasters is shaped by the cultural, demographic and economic characteristics of residents, local governments' institutional capacity, the robustness of city infrastructure, built environment, the provision of ecosystem services and human-induced stresses such as the removal of storm buffers, pollution, overuse of water and urban heat island effect.

Current Study

In the above background, it is imperative to understand how the cities are positioned to meet the challenges of climate change impact. Good civic infrastructure services are important not only for achieving larger economic and development goals but also for improving the welfare of citizens—both rich as well as poor; the lack of which results in the search for (costly) alternatives by the rich and the coping with shortfall by the poor (Ahluwalia 2012). The provision of civic infrastructure services in line with appropriate civic service norms/standards makes the cities robust to meet climate change impacts. Any deficiencies in the current infrastructure status may exacerbate the vulnerability of cities to such effects and reduce their ability to support the human population and economic growth. However, there is inadequate attention paid to civic infrastructure service delivery in cities in the context of climate change and its induced impacts. The preparedness of cities for climate change impacts is also not understood well, and few cities have plans to meet the new challenges.

This chapter attempts to understand and assess the current status of civic infrastructure services in two major Indian cities—Pune and Hyderabad, which have recently turned into important centres of economic activities. We make a critical assessment of the current status of urban civic services—water supply, sewerage, roads, solid waste management, storm water drains and street lighting—in the study cities with reference to the challenges of climate change impacts. In fact, sound civic infrastructure service delivery is important for the cities to be resilient to climate-induced challenges such as floods, droughts, heat waves and excess rainfall. The status and performance of civic infrastructure services in the cities are depicted based on the literature and secondary data from pertinent sources while drawing a comparison with civic service delivery norms (wherever possible); we summarise and conclude in the final section.

24.2 Current Status of Civic Infrastructure Services in Study Cities

Water Supply

Water supply is an important civic infrastructure service rendered to citizens and firms, as it is essential for living, working and performing economic activities in a city. The provision of water supply in Indian cities has been improving over time. However, it is still inadequate to the citizen requirements and requires delivery systems, institutions and governance reforms. In this section, we discuss the status of water supply in study cities in terms of institutional arrangement, water sources, water adequacy, water supply infrastructure and major issues.

Institutional Arrangements
Hyderabad

Hyderabad has an organisation of water supply systems existing for a long time due to the historical past. The city's water supply sources—reservoirs impounded on two rivers—were built in the sixteenth century by the erstwhile Nizam rulers. Subsequently, the lakes built around Hyderabad for irrigation and flood protection were also brought into the city water supply sources network, which continued to play that role until some time ago. The Municipal Corporation of Hyderabad (MCH) took over the water supply function in the city after India's independence, which was further changed to the Greater Hyderabad Municipal Corporation (GHMC) in 2008 after the merger of neighbouring municipal bodies into it. However, with the rapid population growth witnessed in the neighbouring areas of the city in the 1980s, the State Government established Hyderabad Metro Water Supply and Sewerage Board (HMWSSB), a parastatal entity, to provide water supply and sewerage in those areas. Subsequently, the HMWSSB services were extended to the MCH area as well. The HMWSSB jurisdiction currently extends over 688 sq km, covering the GHMC and the other regions (villages) outside it.

Pune

The institutional arrangement for the provision of water supply is different in Pune, where it is provided by the Pune Municipal Corporation (PMC) to the citizens of Pune. Pune also has a long history of water supply organisation from the times of Peshwa kings, who ruled with Pune as the Maratha capital. PMC supplies water to the areas under its jurisdiction and cantonment areas (Khadki, Camp and other defence establishments) within the city and some adjoining rural areas, in contrast to Hyderabad's institutional arrangement. Such differences in institutional arrangements may give rise to differences in response to water supply issues that arise due to climate change. Local government, being closer to people and with elected representatives driving its operation, is expected to respond better than a parastatal agency. Pune has, therefore, an edge over Hyderabad in this regard.

Water Sources
Hyderabad

Hyderabad has historical water sources in the form of lakes/reservoir impoundments that were designed as low-cost gravity-based supply schemes, lending credence to the fact that a city usually uses its least expensive water sources first, as Williamson (1988) argued. Increasing population makes it necessary to tap more distant water sources farther away from the city, and requiring water pumping at several locations, thereby raising the marginal costs of water supply, as Williamson (1988) proposed, and as Sridhar and Mathur (2009, 2011) argued with empirical evidence. While the cost of water supply to Hyderabad from existing sources was Rs. 22.28 per kL, the marginal cost of water supply from new sources had risen to Rs. 33 per kL for Krishna Phase I and Rs. 38 per kL for Godavari Phase I; water pricing, however, makes an average cost recovery of Rs. 26 per kL just enough to meet the O&M costs (Rao 2013).

Pune

The water supply sources of Pune city are four major reservoirs—Panshet, Varasgaon, Temgar and Khadakwasla—impounded on rivers by the irrigation department of the State Government. These water supply reservoirs are relatively newly built; marginal costs may not be high and may not rise due to their relative abundance of water. The combined storage capacity of these reservoirs is 29.12 tmc (or 2260 MLD); however, the amount of annual water supply to be released to Pune city is allocated by the Maharashtra Water Resource Regulation Authority (MWRRA) while balancing water allocation to other regional uses. The main water supply sources are the reservoirs of Khadakwasla dam (890 MLD) and Pavanariver dam (90 MLD). Pune city has a different issue regarding the sourcing of water supply—it needs to get a higher allocation of water for city water supply in the event of climate change impacts like warming. Still, the realisation depends upon the water demand of other sectors in the region and the decision of the water regulator, i.e., MWRRA.

Adequacy of Water

Hyderabad

In 2013, the city was receiving a water supply of 340 Million Gallons per Day (MGD) (or, 1287 MLD), which was less than the estimated demand of 480 MGD, thereby giving rise to a demand–supply gap in water (Rao 2013). Although water supply improved in terms of supply (or storage) capacity to 1545 Million Litres per Day (MLD) in 2017, the demand would also have risen due to population growth. The HMWSSB estimates (while considering a higher population of 90 lakhs in urban agglomeration) the water supply per capita of the city to be 143 litres per capita per day (LPCD), which meets the service delivery norm of GoI (2011) and HPEC (2011) at 135 LPCD.

However, water supply is higher in the central city (covering the erstwhile MCH area) at 160 LPCD, compared with that in suburban areas (which include the 12 municipalities that were merged into the GHMC) at 100–130 LPCD (Rao 2013). The citizens in suburban areas also receive water supply in the form of water tankers from both HMWSSB and private suppliers. In addition, individual households, societies or residential complexes in peripheral areas also make groundwater withdraw. As a result, an alarming decline of groundwater levels in all the areas across the city has been a major concern during the 1990s, 2000s and even now.

Pune

Pune has developed a combined water treatment plant capacity of 1289 MLD and 200 MLD treatment plant capacity under construction (PMC 2017). On the whole, it is estimated that the PMC supplies water of 1250 MLD, which includes the supply of 1123 MLD supply within PMC. The piped water supply through water connections covers 94% of the population; only less than 1% is served through tankers. The average per capita water supply in Pune city is reported as 194 LPCD [which is well above the 135 LPCD norm of HPEC (2011) and GoI (2011)] and the average duration of water supply is six hours (PMC 2017). It is interesting to note that Nallathiga and Sridhar (2021) find that more water was supplied in Pune than required, based on a survey of households.

However, there are large variations in water supply (in terms of per capita water supply) between the central city and suburbs in Pune, similar to other cities like Bengaluru (Sridhar and Smitha 2018). While the central city receives a water supply in the range of 200–350 lpcd, the suburban areas receive a water supply in the field of 138–192 lpcd, similar to what is found in Hyderabad, confirming what Williamson (1988) argued. In Pune also, the residents deploy water services of tankers in summer.

Water Supply Infrastructure
Hyderabad

The city water supply network is 4500 km over a coverage area of 688 sq km, which translates into a pipeline network density of 6.54 km per sq km area. In addition, water supply transmission lines extend over 500 km. Therefore, a large amount of older water supply network gives rise to a good amount of water losses or Unaccounted for Water (UFW).

Pune

The city has 85 service reservoirs with a storage capacity of 328 million litres, and 67 water supply zones distribute water to six parts of the city. The leading water supply transmission network is 210 km in length. It is being re-designed to a narrow range of transmission pipeline sizes to overcome the problem of uneven water flows. The distribution network of the water supply system in Pune is 2688 km, which translates to 8.12 km per sq km area (PMC 2017), which is more than Hyderabad's.

Major Issues
Hyderabad

There has been a lot of criticism about the water supply system in Hyderabad. Some point to the excessive attention paid to the service provider in the central city, the neglect of peripheral areas, and the lack of attention paid to resource sustainability (Prakash 2014). Some others also point to the general bureaucratic approach of the HMWSSB and the neglect of social dimensions and approach to the urban poor (Sahu 2019). Moreover, the water supply demand–supply gap has been rising over time due to rapid population growth and scarcer water supply sources. Further, the declining water tables are also pushing the cost of alternates. This might make the city water supply vulnerable to climate change-induced water scarcity.

The older water supply infrastructure in the central city area is giving rise to a large amount of water losses or UFW. Water metering and billing are not giving rise to revenue improvements to GHMC, despite the introduction of IT and automation of the systems. Water metering appears to be not functioning well, and bill collection is inefficient. The UFW rate may be high due to water theft, illegal connections and poor billing and collection. There is scope for improving water supply management to meet future challenges.

Pune

The water supply system in Pune has several issues that require attention, highlighted in PMC (2017). Water supply demand is rising continuously in Pune; water supply sources are not costly but subject to regional water allocation. Despite the finding of Nallathiga and Sridhar (2021) regarding the water supplied in Pune being more significant than the need, the quantity of water supply varies spatially between the central city and suburbs as well as between the northern and southern parts of the city. The topographical differences give rise to such disparity in the water supply.

The old water supply distribution network system in many areas of the city gives rise to a large amount of physical water losses (or, UFW). The fast depletion of reservoirs results in high peak factors and a smaller number of water supply hours to consumers. Distribution infrastructure improvements are important for meeting the water supply challenges due to climate change.

Sewerage

Sewerage (or the conveyance of domestic sewage) is an important service required for the citizens in any city. Unfortunately, sewerage service connectivity is poor in Indian cities. Only a small proportion of cities have sewerage systems in place; whose population coverage in these cities is far less than complete. We discuss sewerage service status in the study cities in terms of sewer pipeline network and treatment capacity. Better sewerage systems—network, coverage and treatment—render the cities better prepared to cope with climate change-induced risks such as frequent and heavy rains.

Pipeline Network
Hyderabad
HMWSSB, a State parastatal, is responsible for sewerage service provision in Hyderabad. Much of the city's main trunk sewerage pipeline (or core network) was laid down before the independence, primarily in the jurisdiction of erstwhile MCH. Some improvements were made in the 1980s and 1990s to the sewerage trunk line, and Sewage Treatment Plants (STPs) were established. Even now, the sewerage network coverage is high in the core/central city, i.e., MCH area (at about 80%), and low in suburban areas (about 30%) (Rao 2013), and therefore does not meet the service delivery norm of 100% sewerage service coverage of the city (GoI 2011; HPEC 2011). The sewerage network coverage is mainly confined to the central city (MCH area). It has not been expanded to the suburban areas (peripheral municipalities) in the GHMC despite their rapid growth. The lack of expansion of the sewerage network is a significant shortcoming in the current sewerage service in Hyderabad.

Pune
PMC, the local government, is responsible for the sewerage system, i.e., collection, transport, treatment and disposal of sewage, unlike in Hyderabad, where a parastatal entity is accountable for the same. The PMC is responsible for the areas under its own jurisdictions and those under Cantonment. The sewerage system coverage of households at 97.6% is satisfactory compared to the GoI (2011) norm/benchmark of 100% (PMC 2012). However, sewage collection efficiency is 70% against the GoI (2011) norm/benchmark of 100%, implying

that more than one-fourth of the sewage is not collected by the system, which enters the surface water bodies of rivers and lakes, polluting them.

Treatment Capacity
Hyderabad

Hyderabad has an installed Sewage Treatment Plant (STP) capacity of 700 MLD, which is much lower than the required capacity of 1200 MLD; it falls short of the GoI (2011) and HPEC (2011) norms of 100% treatment of sewage. There are also several issues associated with sewerage service in Hyderabad city (Rao 2013): the core sewerage network is older than 40 years and not in a position to withstand the current sewage loads, which leads to sewage overflows and pipeline breakdowns; sewerage treatment capacity has not been augmented; the frequent choking of grit chambers is another major issue with the sewerage system.

Pune

Pune has a sewerage network of 1261 km; the city has an installed sewage treatment capacity of 575 MLD and treats 527 MLD of sewage. The city has nine (9) STPs where the sewage is treated. The estimated sewage generation in Pune city is about 744 MLD, which includes 629 MLD from the water supply connected households; it is estimated that the city would generate about 1566 MLD sewage by the year 2041 based on the projection of the city's water supply (PMC 2012). However, as the treatment capacity and sewage treatment are less than the sewage generated, Pune does not meet the GoI (2011) norm of 100% sewage treatment. Also, the sewage treatment plant capacity is 67% of the requirement (or adequacy), which again leads to the pollution of water bodies due to untreated sewage (PMC 2012). Furthermore, wastewater recycling and reuse is done to the extent of 5.4% but not 20% as laid down by the GoI (2011) norm/benchmark.

The lack of adequate networks and sewage treatment in both the study cities are serious environmental issues, which climate change tends to aggravate.

Solid Waste Management

Solid Waste Management (SWM) is an important civic service rendered by municipal governments; it has become more important in the light of increasing population and economic activities, giving rise to more and more waste generation. Efficient SWM is required for the functioning of cities; otherwise, they are prone to public health risks such as the epidemic spread. An important turning point for SWM in India was framing Municipal Solid Waste (Management and Handling) Rules, 2000. We discuss the status of SWM in study cities regarding waste generation, collection, transport and treatment. Better solid waste management—generation, collection, transportation and treatment—render cities less vulnerable to climate change-induced risks like warm surface temperature hazards.

Waste Generation
Hyderabad
Hyderabad city generates solid wastes in an amount of 4200 tonnes per day (TPD), which translates to about 620 gm per capita per day (PCPD), which is well above the national average of 580 gm PCPD (Sajith and Kumar 2018).

Pune
Pune city generates solid waste at an amount of 1347 metric tonnes per day (TPD), which translates to about 430 gm per capita per day (PCPD), which is below the national average of 580 gm PCPD (PMC 2017).

Waste Collection
Hyderabad
Solid waste collection in Hyderabad is organised and carried out predominantly by the GHMC. Still, the Resident Welfare Associations (RWAs)/Non-Government Organisations (NGOs) and rag pickers contribute to some extent. RWAs manage waste collection in 7 circles, and the GHMC operates it in the rest of the 11 circles. The door-to-door collection rate is also estimated to be good at 73% overall but below the GoI (2011) norm of 100% coverage.

Pune
In Pune, waste collection efficiency, i.e., the amount of waste collected to waste generated, is almost 80% (PMC 2017). However, the CDP of Pune mentions that the door-to-door waste collection coverage of SWM service is only 52.7% overall, which reflects the deficiency of coverage when compared to the GoI (2011) norm of 100%; it also mentions that the waste segregation is only 27.96% against the GoI (2011) benchmark of 100%.

Waste Transport
Hyderabad
The GHMC collects the waste from these bins placed in various localities using its hydraulic/mechanical operating trucks. The solid waste transported from the bins is then taken to the waste transfer points by trucks and then to the landfill site through many lorries. Therefore, Hyderabad might be meeting the norm of 100% transportation of waste set by GoI (2011) and HPEC (2011).

Pune
Pune also has a good waste transport system that picks up waste at the household level and then from the depots to take waste to treatment plants in various parts of the city and a landfill area in the south. Therefore, Pune might also be meeting with the norm of 100% waste transportation set by GoI (2011) and HPEC (2011).

Waste Treatment

In Hyderabad, solid wastes are treated at one large dumping site (or landfill site) located about 35 km away in the southeast part of the city. In Pune, the solid wastes collected from households and others are taken to a sanitary landfill site located about 20 km away from Pune. Currently, attempts are being made to decentralise waste treatment by establishing several more treatment plants. Notably, both cities use landfill treatment of wastes, which generates methane gas emissions that present fire hazard risks at the landfill site. Moreover, such risks may get exacerbated by climate change impacts such as rising temperatures, especially during hot summers, as it happens recurrently in Delhi.

Stormwater Drains

Stormwater drains are an essential civic service rendered by any municipal government; they carry rainwater into rivers during monsoons, thereby avoiding flooding. It assumes significance, as continuous water flooding severely damages property and belongings while also causing human mortality and morbidity. Better stormwater drainage systems in terms of drainage networks render the cities lesser vulnerable to climate change-induced risks like frequent and heavy rainfall. Stormwater drain status in the study cities is discussed in terms of flooding incidence and drainage network.

Flooding Incidence
Hyderabad

Flooding in Hyderabad is primarily due to—(i) the lack of stormwater drains in several parts of the city, (ii) their poor design that does not integrate well with overall drainage, (iii) poor maintenance of natural drains, (iv) illegal occupation and choking of drains by structures, silting and other developments and (v) encroachments of both natural drains and their outfall areas like lakes/tanks. The preparation of the GHMC is inadequate as the stormwater drains are not well protected and subject to occupation. Also, drainage channel improvement through de-silting is not made regularly, which becomes critical in the monsoon period. It is the poor drainage network and its maintenance that cost very dear when flooding takes place.

Pune

Flooding in Pune is primarily due to the undulating terrain and high rainfall (average 721 mm). The natural drains discharge stormwater into rivers, which carry a significant amount of water through their river channel. Four rivers flow through Pune with a length of 53.92 km in the city; a much longer river channel through the city, rather than the 10–20 km length observed for other cities, gives rise to the potential for the flooding of areas adjoining the river, in the event of heavy rainfall in a short span of time, as happened in 2017. Stormwater drain coverage has improved over time and has reduced local flooding. However, city-wide flooding risk persists due to the limited conveyance capacity of drainage systems and rivers, which get tested yearly during the monsoon.

Both Hyderabad and Pune exhibit vulnerability of their stormwater drain system to flooding, which might become exacerbated by the climate change that leads to intense rainfalls in these cities. Stormwater drains are poorly designed and maintained in both cities; they also face illegal occupation and choking of gutters by structures, silting and other developments and encroachments of both natural drains and their outfall areas like lakes/tanks. The recent floods and casualties in both Hyderabad (2019) and Pune (2017) are primarily because of the encroachment of natural drains, leaving less scope for the quick discharge of stormwater.

Drainage Network
Hyderabad
Hyderabad has had an excellent underground stormwater drainage system laid down by the Nizam rulers, but it is again confined to the central/core city, as with most other Indian cities. The service coverage does not meet the norm of 100% laid down by GoI (2011). The city has an estimated 800 km length of stormwater drains (GHMC 2010), which is far less than that required for the quick discharge of stormwater. Successive governments have neglected stormwater drains; several localities do not have well-designed and operational drain systems. Stormwater drains are not well integrated with streets and local roads. As a result, even occasional heavy rains during monsoons play havoc with everyday life due to the submergence and flooding of houses, streets and shopping areas. This again falls short of the GoI (2011) norm of no waterlogging incidents to be existent.

Pune
Pune has a good stormwater drainage system consisting of roadside, natural drains (or nallas) and drainage channels. Natural drains and their tributaries form the central system of the primary drainage channel for Pune. There are 228 nallas (natural drains) with a total length of 382.633 km (PMC 2014). The roadside drains exist on major roads and streets, primarily surface drains that carry stormwater into a nearer drainage channel. The roadside drain network covers only 52% of the area (PMC 2014), which is below the norm of 100% laid down by the HPEC (2011). In a drive to move towards an underground drainage system, many open drains have been converted into underground stormwater drains. It is reported that more than 98% of the households have been covered with underground stormwater drains; more than 95% in slum areas and more than 94% in chawls (PMC 2017), which comes close to the GoI (2011) norm.

Roads

Roads are an important civic service that provides connectivity to a city's various areas while enabling people's mobility. Every city aspires to provide motorable road connectivity to households; it also aspires to provide a good road network that provides linkages between various parts of a city. Better road network density and durable pavement render the cities manage their traffic better, thereby reducing traffic congestion and pollution. However, a higher share of durable pavement renders them vulnerable to climate change-induced risks like heat island effects. Hyderabad and Pune have good road network coverage that reaches most households and provides connectivity to various parts of cities. Notably, the public transport services (buses) are run on city-level arterial/sub-arterial and major roads. We discuss the status of roads in study cities in terms of road network status.

Road Network
Hyderabad

Hyderabad had a road network of 6246 km in 2010, which increased to 7158 km in 2017 (estimated to be about 9000 km now). The road density in 2017 stood at 11.45 km per sq km area falls mildly short of the HPEC (2011) norm of 12.25 km per sq km. The GHMC maintains the entire road network, and the maintenance activity includes routine, periodic and curative repairs. In addition, there are several junctions works like flyovers and underpasses, bridges and footpaths and other traffic management works, including signals, which are improved and maintained by the GHMC. Of the total length of the road network 6246 km in 2010, black top/bitumen roads take the bulk share (2280 km), followed by cement concrete roads (2030 km) and metalled roads (480 km) respectively; earthen roads (1180 km) have a good share of the total road network (GHMC 2010). The GHMC, therefore, meets the CRRI (1989) norm of having at least 75–80% motorable road (or all-weather road).

Pune

Pune has a road network of 1922 km, of which 1872 km is within municipal jurisdiction (in terms of development and maintenance); the remaining 50 km other road is a national/state highway outside the municipal jurisdiction. The road density of the city stands at 5.802 km per sq km of area, which is well below the HPEC (2011) road density norm of 12.25 km per sq km area. Of the 1872 km municipal road network, only 57 km length (3%) is built with cement concrete; much of the road network—1330 km (71%)— is built of asphalt/bitumen, and a good amount of roads are also made with water bound macadam (295 km, or 16%) and earthen gravel (190 km, or 10%). However, the ward-wise data shows that more than 98% of the road is paved, which is above the CRRI (1989) norm of at least 75–80%, and about 37.5% of the roads have footpaths (or pedestrian shoulder ways).

Street Lights

Streetlights' provision is an important civic service provided by municipal governments. However, unlike other civic infrastructure services, street lights are considered pure public goods whose benefits are not easily measurable and whose beneficiaries cannot be identified. Street lights give illumination and associated benefits related to safety and crime prevention/reduction to the citizens in a locality; it also aids transportation during night hours. Furthermore, better streetlight systems in terms of energy-efficient street lamps and lesser street lamp post spacing render the cities with better image and preparedness to meet climate change-induced risks like frequent rainfall events. We discuss the status of streetlights in terms of lamp post spacing and energy efficiency.

Lamp-Post Spacing
Hyderabad

The GHMC assumed the role in 2004 electric utility. It manages the O&M of streetlights as well as installs new streetlights. Using the road network length of the year 2017 and the total number of lamp posts, we arrive at the lamp pole density of 47.21 lamp posts per km length, which equals 21.18 m distance between lamp posts; it is much better than the HPEC (2011) norm of 40–45 m distance between lamp posts.

Pune

The PMC assumed the role of operating and maintaining streetlights, and it has an energy wing that takes care of it. The city has 80,623 lamp posts fitted with street lights along the road length of 1450 km, thereby assuming 55.6 lamp posts per km length of the road, which translates to 18.18 m spacing of lamp posts. The PMC reports having achieved 62 lamp posts per km road length across several wards, which translates into a lamp post spacing of 16.129 m. Both the above measures of street lamp post spacing are better than the HPEC (2011) norm of 40–45 m spacing of lamp posts.

Energy Efficiency

Carbon emissions and energy inefficiency contribute to climate change.

Hyderabad

The GHMC undertook a drive to introduce energy efficiency measures in 2017. It has launched a project to replace street lamps/bulbs with energy-efficient LED lamps at a capital cost of Rs. 271.4 crores. In this project, the GHMC incurs no upfront capital costs, but it is assured of up to 55% of energy savings. In addition, the GHMC refurbishes existing street lamps to make them suitable for LED lamps. In 2017 itself, more than 3.1 lakhs of bulbs were replaced by LED lamps, which would have shown a savings of Rs. 5.9 crores in the form of reduced electricity charges in three months.

Pune

PMC is moving towards an energy-efficient light system by installing LED lamps on lamp posts. It entered into a public–private partnership with a private firm to switch towards LED lamps. Under this project, the private firm will incur expenditure towards replacing bulbs or fixtures with proper lux (or light intensity) at no cost to PMC. It will then take care of the operation and maintenance of the light system for the next twelve years. Several wards reported achieving 100% or near 100% conversion to LED lamps.

24.3 Summary and Conclusions

While cities are considered significant contributors to GHG emissions that lead to climate change, they are also the spaces affected by climate change impacts. Here, the cities with robust civic infrastructure fare better than those without it. The robustness of civic infrastructure services can be gauged from their status reflected by various service dimensions that can be measured on different metrics and compared with established norms/standards. The comparative analysis of the status of urban civic services in Hyderabad and Pune points out a differential performance state for each service. A summary of comparative analysis of the current quality of civic infrastructure services and their vulnerability to climate change-induced risks is provided in Table 24.1.

Pune fares better in terms of robust civic infrastructure when it comes to water supply and sewerage services, which are provided by the municipal body, unlike the parastatal providing them in Hyderabad. Poor stormwater drainage in both cities renders them highly vulnerable to flooding, likely more frequently due to intense rains caused by climate change. Although Hyderabad generates more solid waste, waste management is better organised in both cities. Still, the waste treatment systems using landfill sites make them vulnerable to fire hazards from warm summers caused by climate change. The Road network is better in Hyderabad than in Pune on the whole. Better road infrastructure may lead to lower traffic congestion and GHG emissions, but it (together with settlements) may also make the city vulnerable to heat waves formed during summer. Both cities fare better with streetlights by meeting service norms of density and moving towards energy-efficient streetlight systems.

A comparative analysis of the status of civic infrastructure services of the study cities—Hyderabad and Pune—indicates that even such big cities struggle to provide civic infrastructure services at a level that matches civic service norms/benchmarks. Such deficiency in the provision of civic infrastructure service points to some diagnostic factors related to the ULGs of cities—functions, finances, institutions and governance. Despite the 74th CAA, functional devolution is incomplete and financial strength is not achieved. State governments must devolve service delivery functions to ULGs and provide technical and financial support. They must improve ULG finances by giving them more

Table 24.1 Summary of comparative civic service status and vulnerability of study cities

Civic infrastructure service	Current status and vulnerability	
	Hyderabad (GHMC)	Pune (PMC)
Water supply	Older water supply systems, scarce/costly water sources, poor supply infrastructure and spatial inequities of water supply render the city more vulnerable to climate change-induced risks. In addition, inefficient water supply management (though improving) exacerbates the risks further	Older water supply and poor supply infrastructure, but the good source and water supply render the city less vulnerable to climate change-induced risks than Hyderabad. Better water supply management also aids in its preparedness for such threats
Sewerage	Older sewerage network, incomplete coverage and inadequate treatment capacity render the city with a poor service level that makes it vulnerable to climate change-induced risks such as frequent and heavy rains	Older sewerage networks with good coverage, collection and inadequate treatment capacity render better service levels than Hyderabad, but vulnerability to climate change-induced risks remains high
Solid waste management	Higher amount of waste generation, good waste collection and transport, but poor treatment render it more vulnerable to climate change-induced risks like fire hazards during warm periods	The lower amount of waste generation, good waste collection and transport, but poor treatment render it less vulnerable to climate change-induced risks like fire hazards during warm periods
Storm water drains	The older stormwater drainage system, poor management and inadequate coverage render the city more vulnerable to climate-induced risks like frequent and heavy rainfall	The older stormwater drainage system, better management and good coverage render the city less vulnerable to climate-induced risks like frequent and heavy rainfall when compared to Hyderabad
Roads	A good road network with durable pavements renders the city lesser prone to traffic congestion and pollution; however, the higher share of roads with durable pavement causes vulnerability to heat island effects during summer	An inadequate road network with durable pavements renders the city more prone to traffic congestion and pollution; also, the higher share of roads with durable pavement renders it vulnerable to heat island effects during summer

(continued)

Table 24.1 (continued)

Civic infrastructure service	Current status and vulnerability	
	Hyderabad (GHMC)	Pune (PMC)
Streetlights	Better street lamp post spacing and drive towards efficient street lights render the city with a cleaner image, better preparedness and lesser vulnerability	Better street lamp post spacing and drive towards efficient street lights also render the city with a cleaner image, better preparedness and lesser vulnerability

Source Authors based on the above analysis

autonomy to make decisions and providing grant support wherever required (especially for capital development). The institutional capacity and governance of ULGs are low, which can be improved by undertaking necessary reforms. The recent Central government initiatives of dedicated urban missions is a good approach, which needs further strengthening. Such improvements will render the ULGs to provide better civic infrastructure services and prepare them to meet the challenges of climate change-induced risks.

References

Ahluwalia, IJ (2012), Set Standards and Reform to Deliver Urban Public Services, *Financial Express*, June 27.

CRRI (1989), *Capacity of Roads in Urban Areas, Project Report*, Central Road Research Institute, New Delhi.

GHMC (2010), *Annual Budget 2010–11*, Greater Hyderabad Municipal Corporation (GHMC), Hyderabad.

GoI (2011), *Improving Urban Services Through Service Level Benchmarking*, Ministry of Urban Development, Government of India (GoI), New Delhi.

HPEC (2011), *Report of High Powered Expert Committee (HPEC) on Urban Infrastructure and Services*, Government of India, New Delhi

ICLEI (2017), *Cities and Sustainable Development Goals*, International Council of Local Environmental Initiatives (ICLEI), Geneva.

IPCC (2007), *Climate Change 2007: Impacts, Adaptation and Vulnerability. Contribution of Working Group II to the Fourth Assessment Report of the Intergovernmental Panel on Climate Change*, Cambridge University Press, Cambridge, UK.

IPCC (2018), *Summary for Urban Policy Makers: What the IPCC Special Report on Global Warming of 1.5 °C Means for Cities*, UN Inter-Governmental Panel on Climate Change (IPCC), New York.

Nallathiga, R and KS Sridhar (2021), Urban Civic Service Delivery and Norms: A Pilot Study of Two Indian Cities. In: Kala S Sridhar and George Mavrotas (eds) *Urbanisation in the Global South: Perspectives and Challenges*, Routledge Press, New Delhi.

PMC (2012), *Revising/Updating the City Development Plan (CDP) of Pune City-2041*, Report Prepared by Voyants Solutions Pvt Ltd for the Pune Municipal Corporation (PMC), Pune.

PMC (2014), Water Supply System for Pune City, *Detailed Project Report prepared by Studio Galli Ingeneria (SGI) for the Pune Municipal Corporation (PMC)*, Pune.

PMC (2016), *Report of Road Development and Road Maintenance Committee*, Pune Municipal Corporation (PMC), Pune (August 2016).

PMC (2017), *Water Supply for Pune's Future: 24×7 Water Supply System*, Presentation Prepared by the Pune Municipal Corporation (PMC), Pune.

Prakash, A (2014), The Peri-Urban Water Security Problem: A Case Study of Hyderabad in Southern India, *Water Policy* 16(2014): 454–469.

Rao, JS (2013), The Status of Water & Sewerage in Hyderabad Metro City, *Presentation Made at Workshop on Water & Sanitation*, ICRIER, New Delhi (May 6).

Rao, GR, A Naresh, and MG Naik (2019), Management of Water Supply and Sewerage System in Hyderabad, *Journal of Indian Water Works Association* 2: 120–125.

Reddy, BJ (2016), *Administrative and Financial Structure of Greater Hyderabad Municipal Corporation*, Presentation Made at the Dr. MCR HRD Institute, Hyderabad on April 6, 2016.

Sahu, S (2019), Challenges of Water Provisioning in Hyderabad: Evidences from the Field, *Journal of Governance & Public Policy* 9(2): 71–80.

Sajith, S, and AY Kumar (2018), Chapter 8: Evaluating Municipal Solid Waste Management in Indian Cities: A Comparative Assessment of Three Metros in South India, In: J Mukherjee (ed), *Sustainable Urbanisation in India*, Springer Nature Singapore Pte Ltd., Singapore.

Satyanarayana, M (2017), *IT for Management of Water Supply: SCADA in HMWSSB*, Presentation Made on Behalf of HMWSSB, Hyderabad.

Sridhar, Kala S (2010), Carbon Emissions, Climate Change and Impacts in India's Cities, In: 3i Network (ed), *India Infrastructure Report 2010: Infrastructure Development in a Sustainable Low Carbon Economy: Road Ahead for India*, Oxford University Press, New Delhi, pp. 345–354.

Sridhar, Kala S and OP Mathur (2009), *Costs and Challenges of Local Urban Services: Evidence from India's Cities*, New Delhi: Oxford University Press, pp. 1–277.

Sridhar, Kala S and OP Mathur (2011), Pricing Urban Water: A Marginal Cost Approach, In: 3iNetwork (eds), *India Infrastructure Report 2011: Water: Policy and Performance for Sustainable Development*, Oxford University Press, New Delhi, pp. 351–359.

Sridhar, Kala S and KC Smitha (2018), The Geography of Economic Migrants: Characteristics and Location in Bengaluru, In: Irudaya Rajan (ed), *India Migration Report 2017: Forced Migration*, Routledge, London, pp. 176–187.

UN (2015), *Sustainable Development Goals*, The United Nations (UN), New York (accessed at https://sustainabledevelopment.un.org/sdgs).

UN Habitat (2012), *State of the World's Cities 2012/13: Prosperity of Cities*, United Nations Human Settlements Programme (UN Habitat), Nairobi.

Williamson, JG (1988), Migration and Urbanisation, In: Hollis Chenery and TN Srinivasan (eds), *Handbook of Development Economics: Volume I*, North Holland, Amsterdam.

CHAPTER 25

Climate Change and Water Insecurity: Who Bears the Brunt? (A Case of Yelenahalli Village, Bengaluru)

Aakanksha Srisha and S. Yogeshwari

25.1 Introduction

As the story of most Indian cities goes, Bengaluru, the capital of the southern Indian state of Karnataka, has grown exponentially in the last couple of decades. After post-independence, from 1949 to 2007, the city is said to have increased tenfold (Sudhira et al. 2007). The city expanded by another 333% in the year 2007, with the inclusion of 110 villages, seven municipal councils and one town municipal council to the existing Bangalore Mahanagara Palike, creating the Bruhat Bengaluru Mahanagara Palike (BBMP) (Bangalore Water Supply and Sewerage Board 2017). The BBMP is the urban local body responsible for administering a massive population of 11.5 million people and an area of 741 sq. km. The expanded city includes the parliamentary constituencies Bangalore Urban and parts of Bangalore Rural, with the latter housing most of the 110 villages added. The city once known for its pleasant climate, gardens and lakes, paints a different picture today.

Several concerns have been raised about spikes in the average temperature and extreme precipitation conditions, both of which can be attributed to climate change. The KSNDMC used daily rainfall data from the taluks in the state for the period 1960–2017 to compute changes in rainfall patterns

A. Srisha
Bocconi University, Milan, Italy

S. Yogeshwari (✉)
Christ University, Yeswanthpur Campus, Bengaluru, India
e-mail: yogeshwari.s@christuniversity.in

over the years. Dividing the period into two periods—1960–1990 and 1991–2017, it is found that there has been a fall in the mean rainfall in the second period in the Bangalore Rural district and a rise in the mean rainfall in the Bangalore Urban district. There has been a rise in the variability of rainfall in both the districts in the second period along with a statistically significant rise in the average annual temperature in the period 2002–2018 (Karnataka State Natural Disaster Monitoring Centre 2020). Despite the rise in rainfall in Bengaluru Urban district, it has not translated into greater water security due to reasons that will be discussed in the following sections.

Furthermore, a study finds that the paved area in the city has grown by 1028% in 45 years (1973–2017) (Ramachandra et al. 2017). A striking feature of Bengaluru's urbanisation has been the encroachment of lakes. A study by Mundoli et al. (2018) on the peri-urban public commons found that the city has a long history of man-made interconnected lakes that were built by the successive rulers of the region over the centuries. These lakes were a source of occupation, water and served as a means of groundwater recharge and flood control. The city that once was home to 1452 water bodies in 1800 could claim to the existence of only 194 lakes in the year 2006. These built lakes were vital to the water management system of the city that lay on the ridge of four water-sheds-Challaghatta, Hebbal, Kormangala and Vrishabhavathi (Sudhira et al. 2007).

This changing and more variable rainfall pattern, rising temperature, unchecked urbanisation and encroachment of water bodies are of great consequence as a significant section of the city's population relies on groundwater to meet its needs. The availability of groundwater is dependent on factors such as rainfall, rate of extraction, recharge and elevation, among others. Groundwater is said to cater to 85% of the water demands of Bangalore Rural district and 50% of the demands of the Bangalore Urban district (Sekhar et al. 2017). This dependence on groundwater is the result of a rapidly expanding city and the lack of state infrastructure to cater to the population through the laying of pipes. A study conducted in Bengaluru by BUMP[1] finds that the 20% cream of the population of the city consumes 40% of the piped surface water and the next 60% of the population consumes the remaining 60% (Mehta et al. 2013). Therefore, the bottom 20% has no access to this service. Consequently, the central areas of the city have seen a rise in the groundwater levels owing to pipe leakages and minimal extraction of groundwater while the peripheral areas such as the northeast and southeast have seen a steep decline as they are not connected to piped water supply (Sekhar et al. 2017). The lack of access to piped water often manifests itself through various other associated problems such as inadequate consumption of water, contamination of water in addition to the depleting groundwater levels referred to above, which raises concerns about the water security of the population. UN-Water defines *Water Security* as *the capacity of a population to safeguard sustainable quantities of acceptable*

---

[1] Bengaluru Urban Metabolism Project.

quality water for sustaining livelihoods, human well-being, and socio-economic development, for ensuring protection against water-borne pollution and water-related disasters, and for preserving ecosystems in a climate of peace and political stability (UN Water 2013). Closely related to this peri-urban water insecurity is the increased vulnerability to natural disasters as the recent cases of urban flooding have shown. It would not be an overreach to attribute these changes to uncontrolled urbanisation.

The Bangalore Water Supply and Sewerage Board supplies water to the city from the Kaveri River which flows at an approximate distance of 100 kms from the city. The 110 villages were outside the coverage of the BWSSB's services until the Cauvery Water Supply Scheme Stage V (CWSS Stage V) was launched in the year 2018 with funding from the Japan International Cooperation Agency. The process of laying water pipes in these villages is underway and some villages have started receiving water supply two days a week. With the expansion of piped water supply to the 110 villages under CWSS Stage V there has been a change in the water security condition of those residing in these newly included peripheries. The laying of pipes is not an assurance of regular, inclusive and sufficient supply of water which perpetuates the dependence on groundwater.

Due to the cross-subsidisation followed in the water tariff policy of the BWSSB, piped water becomes an unviable option for several large apartment complexes and commercial establishments. In addition to this, several areas of the city that are yet to be covered with piped water supply lie in relative geographical proximity to those areas that have started receiving piped water. Furthermore, the BWSSB has declared that it will be incapable of meeting the needs of the growing population. It estimates that by the year 2051, the utility's supply of water will fall short by 2650 MLD in catering to a projected 20 million large population. This is considering the additional 620 MLD to be added with the execution of CWSS Stage V. Thus, a large section of the city's population is bound to be left behind to meet its water needs through private sources even in the foreseeable future.

The study aims to analyse the socio-economic dimensions of water insecurity in regions that are in a transitional phase between different sources of water. The research aims to understand the efficacy of piped water supply initiatives in improving the water security of different sections of the population. The study has been confined to a case study of the village Yelenahalli that falls inside the Begur ward of the Bommanahalli zone. It is one of the 110 villages that are yet to be completely covered by the BWSSB water supply. The study has been limited to the village and the blocks of the Begur ward falling within a radius of 2 kms of the village. The rationale behind choosing the ward involves the fact that the area is in a transitional phase from being completely dependent on the private sources of water and groundwater to being partially dependent on piped water. Therefore, the analysis uses dependence on private tankers in the area as an indicator of dependence on groundwater. This makes it the ideal location to answer the research question through an in-depth analysis. The study also aims to explore the link between groundwater and pollution of water bodies through secondary sources.

25.2   Review of Literature

The extent of dependence on groundwater and informal sources of water, preservation of water bodies and urban natural disasters are all important pieces of the puzzle that is peri-urban water security, within the context of climate change.

Groundwater, Water Bodies and Urban Natural Disasters

It is documented that the number of water bodies in the city of Bengaluru has fallen dramatically from 1452 in 1800 to 285 lakes in the early 1970s to 194 in 2006 as mentioned in the introduction. A 2009 (Thippaiah) study notes that the existing data on the number of lakes and tanks in the city underplays the extent of the decline in the number of lakes. The study also finds that the encroachment of these lakes for non-agricultural purposes has not only reduced the amount of irrigated and catchment area in the city but has also led to the loss of the livelihoods of many small and marginal farmers. The study also notes the complete disappearance of lakes as a source of drinking water in most cities in the state. The lakes that remain too have witnessed a fall in water storage capacity due to garbage dumping and siltation. Polluted lakes not only affect the water in the lakes themselves but also the groundwater of the surrounding areas and the produce cultivated in the vicinity. A fall in the groundwater levels across the city have also been attributed in some part to the disappearance of lakes in addition to the private bore-wells that make-up for the lack of state-supplied water.

The lakes in the city were interconnected and an important source of groundwater recharge and rainwater harvesting (Ramachandra et al. 2017). A study of 31 lakes in the city finds that 14 lakes have either been converted for other uses or are in a degraded state with considerable pollution. Of the remainder, 6 lakes are in a relatively good condition which implies presence of minimal water and some garbage dumping. The 12 remaining lakes were found to be in a good condition which indicates limited pollution and a considerable level of dependence by locals for washing, grazing, fishing and fuelwood collection (Mundoli et al. 2018). The paper also highlights the city's dependence on lakes such as the Hesarghatta and Y Chetty lakes up until 1930 for meeting its water needs after which the TG Halli and Cauvery schemes diverted river water to the city.

A 2013 study notes that efforts have been made to revive lakes around the city (Sudhira and Nagendra 2013). However, the study finds that the multiplicity of public bodies that are responsible for the preservation of lakes such as the BBMP, BDA,[2] BWSSB, LDA[3] and KSPCB[4] among others, delay the process of lake rejuvenation. The KSPCB's wide jurisdiction has been found to act as a deterrent in the body's ability to focus solely on lake rejuvenation in the city. The LDA's attempts at PPP[5] have been subject to criticism from the citizenry while its collaborations with local communities have proved to be successful.

The study by Sekhar et al. (2017) highlights that Bengaluru previously found that during the months of August–September 2017, with extreme rainfall, there was a rise in the groundwater level through indirect recharge pathways.[6] This shows that despite the increased occurrence of extreme precipitation due to climate change, natural water bodies and artificial water storage mechanisms can aid in replenishing groundwater by reducing the amount of water loss.

A paper on urban flooding in Bengaluru (Avinash 2014) finds a combination of high population density and short spells of high-intensity rains in highly urbanised areas that make these floods dangerous. The lack of space for water to infiltrate the soil results in flooding. The study finds that the 147 low-lying regions of the city are susceptible to inundation due to the insufficient carrying capacity of stormwater drains. These instances of flooding result in massive damage to public and private property, halt transport and result in huge costs to the state. Low-lying areas are low on land value due to their vulnerability to floods and thus are occupied by those with fewer means who end up bearing the brunt of the rains with water flooding their homes and meagre resources. Another report attributes the frequent flooding in the city to the rapid growth of the built-up area, the concretisation of stormwater drains, encroachment of the lake area and the lack of interconnection between the lakes in the city (Ramachandra et al. 2017). The encroachment of valley zones that acted as a path for water to flow between lakes has also contributed to the problem. The water meant to enter the aquifers fails due to the obstacle placed by the built-up area.

A study on the flooding in Chennai in 2015 also notes that the encroachment of riverbanks and low-lying areas, along with an explosion in the population leading to the urban occupation of land, has contributed to worsening the damage caused by the floods (Seenirajan et al. 2015). Another study on the chaotic Mumbai floods in 2005 finds that urban decongestion, removal of solid waste from stormwater drains and robust weather forecasting and alerting mechanisms were necessary to avert similar ordeals in the future in the aftermath of heavy rains (Gupta 2007).

---

[2] Bangalore Development Authority.
[3] Lake Development Authority.
[4] Karnataka State Pollution Control Board.
[5] Public–Private Partnership.
[6] Ponds, rainwater harvesting system, storm water drain, construction pits.

Socio-Economic Implications of Water Scarcity

Another result of the rapid urbanisation these mega cities witness, in addition to the susceptibility to flooding, ironically, is the scarcity of water. This scarcity often targets certain sections of the population that lie outside the limits of state-supplied water.

A study conducted in 2015 on informal water supply in Chennai found a positive correlation between the household's income level and the amount spent on water (Venkatachalam 2015). However, the per capita consumption of public water sources was the highest among the lowest income group as the income rose; the more public water became an inferior good. It was also found that rent-seeking was widespread with free water from the public tap and hand pump.

A study conducted in the Bommanahalli and Byatarayanapura localities of Bengaluru found that between 2007 and 2009, the price charged by private water tankers was three to five times that provided by the BWSSB (Ranganathan 2014a). Water is sold by the pot to the urban poor.

The prices charged by private suppliers have not been found to rise with inflation as they claim that their low-income customer base cannot afford high price hikes. Low-income households are said to consider convenience, reliability, quantity and quality in meeting their water needs. More importantly, they are prepared to allocate 3–5% of their income on the resource, preferably in instalments (Conan et al. 2004). Time is also said to be a constraint in deciding the water source among poor households (Kjellen and McGranahan 2006).

Another study in Bengaluru found that the price charged by private water tankers depended on the quantity of water, the distance travelled and the season (Rajashekar) and not particularly political motives. Nevertheless, middle and high-income houses often resort to constructing sump and overhead tanks due to the irregular supply of utility-supplied water. However, lower-income households that cannot afford these contraptions need to manage with buckets and pots.

In areas with high levels of piped water coverage but discontinuous, periodic (weekly) water supply, the distinguishing feature among low and high-income households is the infrastructure available to store water. Households prefer a piped water connection the most, followed by a private well and a community source or shared connection (Conan et al. 2004). The average willingness of households to pay for piped water connection is said to be higher than the service's operation and maintenance (O&M).

A study conducted between 2007 and 2009 in the Bommanahalli ward found that those who opted to avail of this utility were the middle class who owned unauthorised property in the city's revenue layouts. They sought the pipe connection to secure their land tenure (Ranganathan 2014b). However, several households could not afford the contribution and thus were left out of the facility.

25.3 Methodology

The analysis has been divided into two parts to understand the different dimensions of water insecurity in the city. The first part of the analysis aims to understand the capacity of the remaining lakes in the city to meet the water security needs of the population through secondary data. The second part of the study is designed as a case of a village on the peri-urban border of the city that has survived primarily on groundwater to meet its needs and the extent of the influence that piped water could have in alleviating the water security concerns of different sections of the population.

25.4 Analysis and Results

The State of Lakes in the City

The Karnataka State Pollution Control Board conducts monthly tests on the quality of 106 lakes around the city, the KSPCB classifies lakes into 5 classes (see Table 25.1):

A: Drinking water source without conventional treatment but after disinfection.
B: Outdoor bathing (organised).
C: Drinking water source with conventional treatment followed by disinfection.
D: Propagation of wildlife, fisheries.
E: Irrigation, industrial cooling, controlled waste disposal.

'*' in the table indicates either no record or no water in the lake and NI indicates Newly Identified.

The KSPCB also rates the water in the lakes based on the Water Quality Index (WQI) into:

S = Satisfactory.
US = Unsatisfactory.

An analysis of the monthly data for a period of one year (April 2019–May 2020) shows that 'S' grade or Satisfactory water quality has only been found in three instances. Furthermore, we find that none of the lakes fall under class A or B indicating the complete absence of potable quality water. Even post-treatment, only thrice has the water in lakes has been found to be suitable for drinking. However, it is to be noted that a large number of lakes have been tested very irregularly. After these lakes were brought into regular testing in February 2020, there has been a substantial rise in the number of lakes that are identified as class 'D', accounting for nearly 60% of the lakes in the city that are deemed suitable for fisheries and unfit for consumption purposes. The

Table 25.1 KSPCB classification of lakes into five classes

Class	Apr-19	May-19	Jun-19	Jul-19	Aug-19	Sep-19	Oct-19	Nov-19	Dec-19	Jan-20	Feb-20	Mar-20	Apr-20	May-20
*	16	16	20	17	27	13	25	21	11	9	30	55	12	11
C						1	1		1					
D	38	44	28	38	39	55	40	43	49	50	47	31	66	73
E	16	17	31	26	14	12	15	17	21	24	29	20	28	22
NI	36	29	27	25	25	25	25	25	24	23				
Blank					1									
Grand Total	106	106	106	106	106	106	106	106	106	106	106	106	106	106

Source Karnataka State Pollution Control Board 2023, https://kspcb.karnataka.gov.in/environmental-monitoring/water

next major category is sadly 'E', water that can only be used for industrial cooling and waste disposal. This indicates that the lakes in the city have been nearly eliminated as sources of drinking water and water for household use. This is not just what the data from the past year shows. A look at the data from even as far back as 2015 shows that most if not all the 56 lakes that were studied up until 2018 were classified either as 'D' or 'E'.

This is in sharp contrast with the situation that prevailed less than a century ago when lakes were the primary source of water. When the fact that the lakes are so heavily polluted is considered along with the reality that rainfall is more variable these days, the groundwater levels in the city have fallen steeply and Kaveri water can only meet the need of a section of the population, thus the water security situation in the city looks grimmer. This reveals that the city may face a greater challenge in meeting the water security needs of its burgeoning population in the coming years.

In the following section, an analysis has been performed on the differential impact of piped water supply on varied sections of the population. We find that laying of pipes is not a promise of greater water security through a case study of Yelenahalli.

Case Study of Yelenahalli

The case study of the Yelenahalli village has been structured to take the perspectives of the multiple residents of the locality. A quantitative analysis has been on the responses of the personal interviews addressed to the independent houses. Personal interviews of apartment complexes and commercial establishments have also been analysed. The case has also been supported by the personal interviews of private water tanker operators in the locality. The dependence on private water tankers has been taken as an indicator of dependence on groundwater as a majority of tanker operators in the locality extract water through their bore-wells. For the section of the study with quantitative analysis, a judgement sampling technique has been adopted. A sample size of 56 independent households has been chosen for quantitative analysis. The sample structure of this study is given in Table 25.2.

Table 25.2 Sample structure	Sl. No	Respondent type	Sample size
	1	Independent households	56
	2	Apartment complexes	10
	3	Commercial establishments	5
	4	Private water tanker operators	5
		Total	76

The non-parametric Wilcoxon Signed Rank test has been used to examine a change in the price and quantity of tanker water demanded which is used to indicate an indirect dependence on groundwater. A Generalised Estimating Equation (GEE) has been used to estimate the mean quantity of tanker water ordered across income categories, with and without a BWSSB connection. All quantities and prices of tanker water have been considered in kiloliters (Kl). A GEE accounts for the repeated measures in the data by correcting for covariance present within the subject. Since the correlation between the observations within a subject is of little interest under these objectives, a Generalised Estimating Equation is found to be a better fit than a mixed effects model (Diggle et al. 2002). Repeated measures pre and post the commencement of BWSSB water supply have been taken for the subjects retrospectively.

Equation 25.1 has been estimated using a gamma distribution and an identity link. It is of the form:

$$E(QuantityInKl) = \alpha 2 + \theta 1, 2, 3, 4, 5, 6, 7,$$
$$(HHIncome 1, 2, 3, 4 * BWWSB\ 0, 1) + \varepsilon \quad (25.1)$$

An independent variance structure has been chosen for the GEE.

Furthermore, Kruskal–Wallis tests have been conducted to explore the differences across sub-localities pre and post the laying of BWSSB water pipes in the area. The test has also been used to identify a significant difference in the water storage capacities of households falling under different income categories. These findings have been further corroborated with evidence from personal interviews of the apartments and commercial establishments.

Independent Households, Apartments and Commercial Establishments

It was found that 66.1% of the households interviewed in the study relied on the BWSSB and private tankers for their water needs; 16.1% of the respondents relied solely on private water tankers, 12.5% also relied on personal bore-wells in addition to the two sources mentioned above and 5.4% of the respondents relied on private water tankers and personal bore-wells. The study also finds that private water tends to be several times more expensive than piped water for independent houses as indicated in the literature.

A test of normality has also been conducted on the tanker price per Kl and quantity of tanker water ordered per month in Kl variables. It is found that both sets of prices and quantities are not normally distributed based on the Kolmogorov–Smirnov test of normality. We reject the null hypothesis of the presence of normality with significance values lower than 0.05 in all four distributions (see Table 25.3).

Price and Quantity

As it is found that the price and quantity variables pre and post the laying of Kaveri pipes are not normally distributed, we use a non-parametric related-samples Wilcoxon Signed Rank test to examine the presence of a statistically

Table 25.3 Tests of normality

	Kolmogorov-Smirnov			Shapiro-Wilk		
	Statistic	df	Sig	Statistic	df	Sig
Price Post Kaveri Kl	0.107	56	0.016	0.957	56	0.043
Price Pre Kaveri Kl	0.214	56	0.000	0.934	56	0.004
Quantin Kl Post Kaveri	0.240	56	0.000	0.757	56	0.000
Quantin Kl Pre Kaveri	0.235	56	0.000	0.776	56	0.000

Table 25.4 Wilcoxon signed ranks test on price per Kl and quantity in Kl

	QuantinKlPreKaveri–QuantinKlPostKaveri	PricePreKaveriKl–PricePostKaveriKl
Z	−5.599	−5.818
Asymp. Sig. (2-tailed)	0.000	0.000
	Based on negative ranks	Based on positive ranks

significant difference in the quantity and price of tanker water ordered in the two periods. With a z test statistic of −5.599 based on the negative ranks and with an asymptotic significance value of 0.000, the null hypothesis for the quantity of water ordered is rejected. We find that 41 of the 56 households reduced their monthly tanker water consumption after the laying of Kaveri pipes in the locality. The remaining households that have either a constant or an increased consumption have not availed of the BWSSB water supply. Thus, we find a significant difference in the monthly quantity of private tanker water ordered in those houses that have availed the BWSSB piped Kaveri connection before and after availing the connection.

Furthermore, it is found that there is a significant difference in the prices charged for tankers pre and post commencement of supply of BWSSB water in the area. Of the 56 respondents, 50 cited an increased price, 4 cited decreased prices and 2 cited no change in price. The null hypothesis of no significant median difference is rejected with a z test statistic of −5.818 based on the positive ranks with an asymptotic 2-tailed significance level of 0.000 (see Table 25.4).

Socio-Economic Factors
Since it is established that a majority of the independent households primarily switched to piped water to meet their needs, we examine the socio-economic nuances. In order to analyse the impact of the socio-economic variables on the quantity of tanker water consumed, a GEE has been used.

Equation 25.1 examines the difference in the average quantity of tanker water ordered pre and post availing BWSSB water. The quantity ordered in

Kl is regressed on the interaction effect between household income and the presence and absence of a BWSSB connection. It is found that in the absence of a BWSSB connection, the extent of difference in the quantities of tanker water consumed between the income categories is large. For instance, in the absence of BWSSB, there is a difference of ~ 13Kl between the income category Rs. 100,001 and above and the income category Rs. 10,001–50,000. However, in the presence of BWSSB, there is a difference of ~ 5Kl between the two income categories. This reduced gap in the water consumption of the categories can be attributed to the significant difference in the storage capacity of the two groups that is established through the Kruskal–Wallis test (Table 25.5). This corroborates the evidence put forth by Venkatachalam (2015) that the amount of money spent on water in absolute terms rises with income as is indicated by a larger dependence on tankers prior to the laying of pipes. We also find that the price paid by those in the poorest sections of the locality for private water is higher per Kl.

Through an independent samples Kruskal–Wallis test, the water storage capacity across different income categories is compared. It is proven that the distribution of the storage capacity is different across monthly income categories with a sig. value of 0.000 and a test statistic of 21.234. From the pair-wise comparisons, it is found that the highest income category has the highest storage capacity. There is found to be a significant difference in the storage capacities of the income category of Rs. 100,000 and the two lowest income categories. However, there is no statistically significant difference between the storage capacities of income categories of Rs. 0–10,000 and Rs. 10,001–50,000. Even the second highest income category of Rs. 50,001–100,000 has a statistically significant difference in the storage capacity from the two lowest income categories. The impact of income differences on the water storage capacity supports the argument put forth by Potter et al. (2010) and Rajashekar (2015).

Two independent samples Kruskal–Wallis tests have been conducted to test the differences in the quantities of private tanker water ordered in the presence and absence of Kaveri in the area (Table 25.6). In the period prior to the laying of Kaveri pipes, while it is seen that there is a significant difference in the distribution of the quantity of tanker water consumed across sub-localities, the pair-wise comparisons show insignificant differences. However, when the quantity ordered is looked at post the commencement of Kaveri water supply in these areas, it is found that there is a significant difference between certain areas. This is seen from the adjusted significance values. Sub-localities 1 and 3 have significantly higher tanker water consumption post the laying of pipes in the area. This can be attributed to the fact that Sub-locality 1 is composed of households falling under the lowest category of income who have not availed of the piped water supply service. The fact that low-income households cannot afford to invest in the Beneficiary Capital Contribution required to avail a pipe connection sheds light on the exclusivity of Kaveri water. The houses in Sub-locality 1 were also already availing the municipal water intermittently prior

Table 25.5 Generalised estimating equation results

Parameter		B	Std. Error	95% Wald CI Lower	95% Wald CI Upper	Hypothesis test Wald Chi-Square	Sig
Equation 25.1	Intercept	4.77	1.55	1.74	7.80	9.54	0.00
	[BWSSB = 0] * [HHY = Rs. 0–10,000]	9.85	2.42	5.10	14.60	16.54	0.00
	[BWSSB = 0] * [HHY = Rs. 100,001 & above]	24.34	3.20	18.08	30.60	58.03	0.00
	[BWSSB = 0] * [HHY = Rs. 10,001–50,000]	13.99	2.53	9.04	18.94	30.66	0.00
	[BWSSB = 0] * [HHY = Rs. 50,001–1,00,000]	11.16	1.96	7.32	15.00	32.46	0.00
	[BWSSB = 1] * [HHY = Rs. 100,001 & above]	1.76	2.56	−3.26	6.79	0.47	0.49
	[BWSSB = 1] * [HHY = Rs. 10,001–50,000]	−2.86	1.73	−6.26	0.53	2.73	0.10
	[BWSSB = 1] * [HHY = Rs. 50,001–100,000]	Reference category					
	Scale	0.76					

Dependent variable: Quantin KI

to the laying of pipes. On the other hand, Sub-locality 3 has households that have availed Kaveri water supply but have very poor supply. The intermittent Kaveri water supply is insufficient to even fill the sumps of these houses. As the sub-locality lies on a higher gradient, the flow of water received is below optimal. This is unlike in the case of houses in Sub-localities 2 and 4. Sub-locality 5 has been excluded from the second test as it is yet to receive Kaveri water supply. However, sub-locality 5 has witnessed a rise in the price of private water despite not being provided with piped water.

What is evident from the socio-economic analysis of water security is that income and water security are positively correlated. Low-income households that cannot afford the initial investment in piped water and nor can they afford the storage infrastructure have to face the brunt of a hiked price of private water per Kl. Sadly, neither can they depend on lakes or other water bodies to meet their needs. Those households that lie in areas that are uphill or in areas that have unreliable supply are also worse off as they lose a substantial amount of money as BCC while continuing to pay for private water at a hiked price.

Table 25.6 Kruskal–Wallis test results

Income category pairs	Test statistic	Adj. Sig
Rs. 0–10,000 & Rs. 10,001–50,000	−10.529	1.000
Rs. 0–10,000 & Rs. 50,001–100,000	−28	0.015
Rs. 0–10,000 & Rs. 100,001 and above	−28.773	0.007
Rs. 10,001–50,000 & Rs. 50,001–1,00,000	−17.471	0.021
Rs. 10,001–50,000 & Rs. 100,001 and above	18.243	0.003
Rs. 50,001–1,00,000 & Rs. 100,001 and above	0.773	1.000
Pre-BWSSB		
Sub-locality Pairs	Test statistic	Adj. Sig
1–2	10.037	1.000
1–3	11.310	1.000
1–4	19.617	0.344
1–5	25.083	0.052
2–3	1.272	1.000
2–4	9.58	1.000
2–5	−15.046	0.158
3–4	−8.307	1.000
3–5	−13.774	1.000
4–5	−5.467	1.000
Post-BWSSB		
2–4	1.650	1.000
2–3	19.929	0.006
1–2	−24.000	0.001
3–4	18.279	0.560
1–4	−22.350	0.015
1–3	−4.071	1.000

From the personal interviews of the apartment complexes and the commercial establishments, it is found that the private water prices received by the large and mid-sized apartments are still lower than the average prices received by independent households in the post-Kaveri period. The management of the large and mid-sized apartments tend to be price setters while the independent households are price takers in the private water market. The large and mid-sized apartments tend to have constant prices as long as they have a nearly daily demand for tanker water. Water-intensive commercial establishments tend to receive fairly stable prices.

Private Water Tanker Operators
It is seen that the tanker operators resort to second-degree price discrimination by offering lower prices to customers with a regular demand. However, only apartment complexes fall under their classification of regular customers as they demand multiple loads in a week, often with daily demand. Most independent houses order up to 2–3 loads a week. Thus, apartments have the potential to receive lower prices. While the tanker operators cite reasons such as the threat of the entry of new players that prevents them from hiking prices, they also cite several instances of collusion among the existing players in the market. Thus, there is a chance for collusive decision-making to hike the price due to the commencement of Kaveri water supply in the locality. There is also an explicit understanding among the existing operators that the customers of other operators cannot be catered to by other suppliers.

Following the commencement of BWSSB water supply in the locality, all the tanker operators have witnessed a shrinking business and customer base. Most operators witnessed a halving of their business. They find apartments to be more dependable customers as these buildings often do not avail a connection to BWSSB. Thus, some operators have resorted to investing in tankers with larger capacities such as 12Kl. This makes the independent houses unattractive customers.

These private tanker operators have faced the brunt of depleting groundwater levels, with many having to dig multiple wells before finding a reliable source. They also need to dig deeper to get water as compared to a decade back. All the tanker operators interviewed primarily sourced their water from bore-wells. The tanker operators also invested in water-purifying equipment as they doubted the quality of groundwater that explicitly stated was of low quality. They also spend regularly in conducting tests on the quality of water they extracted. Some of the tanker operators who have been in the business for long have noticed that their electricity expenses have risen steeply over the years due to the lack of water available and extracting it from deep bore-wells.

25.5 Conclusion

These findings imply increased pressure on lower-income households to invest in water storage mechanisms to cope with the intermittent Kaveri water supply and the hiked tanker prices. During this study, the Kaveri water supply in the area has fallen to once a week from bi-weekly. This is bound to pressure households to increase their reliance on tankers. However, the increased reliance is unlikely to cause a downward revision in tanker prices as that is uncharacteristic of their market behaviour. The Beneficiary Capital Contribution along with the other unofficial manual labour expenses tend to be highly extractive considering the poor efficiency in frequency and force of flow of piped water supply and the minimal costs involved in offering a piped connection. Furthermore, the summer months are known to have minimal to no BWSSB water supply. Considering the rise in the base price post the introduction of piped water supply in the locality, there are chances of chagrining exorbitant tanker prices in the summer months, more than those charged during the previous summers. Tanker operators find large apartments to be more viable customers post the introduction of Kaveri water. New entrants into the tanker market are more likely to shift to larger capacities of tankers which can reduce the competition in the market for tankers with smaller capacities that service independent houses. Thus, there is a potential for a further upward revision in tanker prices.

Most of the households in the study area had dried up bore-wells. Even tanker operators are going to depths of greater than 1500 ft to get water. They are aware of the fact that it is an unviable occupation in the long run considering the rising sunk costs. Thus, there can be pressure on them to hike prices in the near future.

In such a situation where the extraction of groundwater and the reliance on Kaveri water is unsustainable, the water bodies and lakes of the city must be turned to for help. However, we find that these too have been misused to our own detriment.

Thus, considering the upward trend in tanker prices, the limited abilities of the BWSSB to cater to the needs of all sections of the population and the poor groundwater levels, there is a need for a policy that encourages sustainable water management at the household and community level. There is a need for stricter enforcement of Rainwater Harvesting (RWH) policies. Community-level innovations that make RWH viable for low-income households is also a necessary step. This has to be done along with measures to rejuvenate lakes through community-led initiatives. Catchment areas need to be cleared and made suitable for percolation of water.

The grim water security situation of the future and the alarm bells ringing with the urban floods herald the need to take concerted measures to adopt sustainable water sourcing practices at a community level, more so on the grounds of equity.

REFERENCES

Avinash S (2014) Flood Related Disasters: Concerned to Urban Flooding in Bangalore, India. International Journal of Research in Engineering and Technology 3(16):76–83.
Bangalore Water Supply and Sewerage Board (2017) Bengaluru Water Supply and Sewerage Project (Phase 3). Bangalore Water Supply and Sewerage Board.
Conan H, Paniagua M, Shrestha A (2004) Bringing Water to the Poor: Selected ADB Case Studies. Asian Development Bank, Manila, Philippines.
Diggle PJ, Heagerty PJ, Liang K et al (2002) Analysis of Longitudinal Data (Second ed.). Great Britain: Oxford University Press.
Gupta K (2007) Urban Flood Resilience Planning and Management and Lessons for the Future: A Case Study of Mumbai India. Urban Water Journal 4(3):183–194.
Karnataka State Natural Disaster Monitoring Centre (2020) Climate Change Scenario in Karnataka: A Detailed Parametric Assessment. Karnataka State Natural Disaster Monitoring Centre 07.
Karnataka State Pollution Control Board (2023) Classification of River, Lake Water Quality under GEMS and MINARS programme, various years. Retrieved from https://kspcb.karnataka.gov.in/environmental-monitoring/water. Accessed on 23 June 2023.
Kjellen M, McGranahan G (2006) Informal Water Vendors and the Urban Poor.
Mehta VK, Goswami R, Benedict EK et al (2013) Social Ecology of Domestic Water Use in Bangalore. Economic and Political Weekly 48(15):40–50.
Mundoli S, Manjunatha B, Nagendra H (2018) Lakes of Bengaluru: The Once Living, But Now Endangered Peri-Urban Commons.
Potter RB, Darmame K, Nortcliff S (2010) Issues of Water Supply and Contemporary Urban Society: The Case of Greater Amman. Jordan 368(1931):5299–5313.
Rajashekar AV (2015) Do private Water Tankers in Bangalore Exhibit "Mafia-Like" Behavior? Massachusetts Institute of Technology.
Ramachandra TV, Vinay S, Aithal BH (2017) Frequent Floods in Bangalore: Causes and Remedial Measures. Indian Institute of Sciences, Energy & Wetlands Research Group 123.
Ranganathan M (2014a) 'Mafias' in the Waterscape: Urban Informality and Everyday Public Authority in Bangalore. Water Alternatives 7(1):89–105.
Ranganathan M (2014b) Paying for Pipes, Claiming Citizenship: Political Agency and Water Reforms at the Urban Periphery. International Journal of Urban and Regional Research 38(2):590–608.
Seenirajan M, Natarajan M, Thangaraj R et al (2015) Study and Analysis of Chennai Flood 2015 Using {GIS} and Multicriteria Technique. Journal of Geographic Information System 9(2).
Sekhar M, Tomer SK, Thiyaku S et al (2017) Groundwater Level Dynamics in Bengaluru City, India. Sustainability 10(2).
Sudhira H, Nagendra H (2013) Local Assessment of Bangalore: Graying and Greening in Bangalore—Impacts of Urbanization on Ecosystems, Ecosystem Services and Biodiversity. In: Anonymous Urbanization, Biodiversity and Ecosystem Services: Challenges and Opportunities. Springer Open, p 755.
Sudhira H, Ramachandra TV, Bala Subrahmanya et al (2007) City Profile Bangalore. Cities 24(5):379–390.
Thippaiah P (2009) Vanishing Lakes: A Study of Bangalore City 17.

Venkatachalam L (2015) Informal Water Markets and Willingness to Pay for Water: A Case Study of the Urban Poor in Chennai City, India. International Journal of Water Resources Development 31(1):134–145.

Water UN (2013) Water Security and the Global Water Agenda.

CHAPTER 26

People's Awareness, Perceptions and Attitudes on Green Buildings: *A Study in Bengaluru*

S. Manasi, Channamma Kambara, and N. Latha

26.1 Introduction

Buildings are liable for approximately 40% of energy use in most countries, resulting in greenhouse gas emissions (World Council for Sustainable Development). The Intergovernmental Panel on Climate Change (IPCC) estimated that by 2050, buildings are projected to emit 3800 megatonnes of carbon, reflecting that the real-estate industry is expected to consume 38% of the global energy. Furthermore, a study by the Energy Information Administration (EIA), US DoE, the growth rate in total energy consumption has been more significant than the population growth rate. It is expected to grow at 1.3%, while the energy consumption rate is expected to grow at 4.3%, demonstrating that the building sector is a major energy consumer.

The International Energy Outlook projections for 2030 of the US Department of Energy informed that China and India account for nearly one-half of the total increase in residential energy use in non-OECD countries. However,

---

S. Manasi (✉) · C. Kambara · N. Latha
Centre for Research in Urban Affairs, Institute for Social and Economic Change, Bengaluru, India
e-mail: manasi@isec.ac.in

C. Kambara
e-mail: channamma@isec.ac.in

N. Latha
e-mail: latha@isec.ac.in

© The Author(s), under exclusive license to Springer Nature Singapore Pte Ltd. 2023
S. Nautiyal et al. (eds.), *Palgrave Handbook of Socio-ecological Resilience in the Face of Climate Change*,
https://doi.org/10.1007/978-981-99-2206-2_26

India's energy demands are expected to overtake the energy demands of China by 2050, with a speedy explosion in the real estate and construction sectors and extensive energy use. Currently, commercial and residential buildings in India account for more than 30% of the country's total electricity consumption. Hence, it is crucial to have more green buildings in the coming years.

26.2 Green Buildings—An Overview

Green buildings are designed to reduce the demand for non-renewable resources and make the best use of utilisation efficiency through reuse, recycling and use of renewable energy sources like sun, water and wind and a healthy indoor environment. In other words, green buildings aim to sustain the natural environment and human health by efficiently utilising the available renewable energy, water and other resources, by incorporating recycled, eco-friendly and reusable materials during the construction and functioning of the building. In recent years, the green building concept has gained significance due to its non-toxic, reduced waste generation and resource-efficient configuration during the operation of the building. As a result, green buildings offer one of the most cost-effective solutions to climate change and can lead to significant environmental, economic and social benefits globally (W. G. B. Council 2020).

Although the "Green Building Movement" has been accelerating since the last decade, the origin of this movement dates back to the nineteenth century. Historic structures like Milan's Galleria Vittorio Emanuele II, London's Crystal Palace, New York's Flatiron Building and the *New York Times* Building were the few first-evergreen buildings constructed with efficient water use, energy and other materials with better design, maintenance and operation. During the 1930s, some new building technologies, such as using local and renewable materials, roof ventilators and underground air-cooling chambers to decrease the structure's impact on the environment and to regulate the building's indoor air temperature, revived. By the time of 1970, events such as the first Earth Day (April 22, 1970), and the establishment of the U.S. Environmental Protection Agency had taken place. The invention of air conditioning, reflective glass and structural steel buildings in America and the negative impact due to the hike in fuel prices by the OPEC influenced a group of forward-thinking people, mainly architects (from the American Institute of Architects), environmentalists and ecologists to initiate an environmental movement known as the modern green building movement.

IPCC reports indicate that energy-efficient buildings and utility systems can reduce energy demands by as much as 40%. Since the approach towards green buildings is holistic, it will address concerns other than a reduction in energy usage wherein they focus on the design right from the planning stage by envisioning sustainable features at the beginning and considering the complete supply chain from the material sourcing, energy modelling, reuse of resources, public facilities and disposal of waste. However, the possibility of using the local resources innovatively is left to the creativity of the concerned designer architect.

Need to Promote Green Buildings

India is the fastest-growing construction market, expanding at almost twice the rate of China's. The country's rapidly expanding population is generating significant housing demand. A staggering 31,000 homes need to be built daily for the next 14 years to keep up with the growing demand. That will be 170 million properties by 2030.

The Building and construction sector, being the largest energy users, contributes a significant share of 40–45% of energy consumption in the country. Moreover, commercial buildings consume more than 50% of all energy consumed by the construction industry. Hence, the building/ construction sector has the potential to save over 50% of energy consumption and can emerge as a critical segment to address the challenges of worldwide energy demand and climate change. As the United Nations Environment Program says, nearly one-third of the total greenhouse emissions come from buildings and "green buildings" are essential to any policies addressing climate change and environmental concerns.

India was ranked second among the top ten countries in the world by the US Green Building Council (USGBC) for LEED[1] outside of the US. In India, the Confederation of Indian Industries (CII), along with the World Green Building Council and the USGBC, established IGBC in 2001, which paved the way for the green building movements to start in the country. India has more than 2030 registered green building projects and over 60 LEED platinum-certified constructions, including hospitals, hotels, colleges and IT parks. Besides, India is leading the green building movement worldwide. The number of green buildings is anticipated to grow to about one lakh by 2025 across India. With the central government's announcement of 100 smart cities, the need for smarter buildings to make cities smarter was stressed by green architects. It is also proposed to create affordable housing a vital part of smart cities.

The various research studies have identified several benefits of green buildings (both tangible and non-tangible), such as resource savings (energy and water), increased property value, decreased maintenance cost, improved occupants' well-being, increased productivity, health benefits, reduced carbon emissions and waste generation. However, as Zuo and Zhen (2014) points out, the existing studies focus on the environmental aspect of green buildings.

Green buildings, with their unique construction features like economic use of resources and energy conservation, support Sustainable Development Goals

---

[1] Leadership in Energy and Environmental Design (LEED) certification, an internationally recognised standard for measuring a building's carbon footprint. (U.g.B. Council, 2020)

in more than one way. They can improve people's health and well-being (SDG 3, Good Health and Well Being); they use renewable energy, which becomes cheaper to run (SDG 7, Affordable and Clean Energy); green building infrastructure creates jobs and boosts the economy (SDG 8, Decent Work and Economic Growth); the design of green buildings can spur innovation and contribute to climate-resilient infrastructure (SDG 9, Industry, Innovation and Infrastructure); they are the fabric of sustainable communities and cities (SDG 11, Sustainable Cities and Communities); resources are not wasted in green buildings as they use "circular" principles (SDG 12, Responsible Consumption and Production); they produce fewer emissions and help combat climate change (SDG 13, Climate Action); green buildings can improve biodiversity, save water and help to protect forests (SDG 15, Life on Land) and through green buildings, a solid and global partnerships can be created (SGD 17, Partnerships for the Goals) (World Green Building Council).

Currently, constructing green buildings is not an obligation and is not incentivised. There are no provisions for creating demand through various measures like tax incentives (tax rebates, etc.), providing improved access to finance for purchasing green buildings through rewards systems, etc. Since making green buildings is not obligatory, construction will take a long time to get implemented. Various initiatives by the government have made it mandatory to teach green features like rainwater harvesting, solid waste management, setting up sewage treatment plants in gated communities, etc. However, there is more to be done in promoting green buildings. Another limitation is the lack of appropriate incentive mechanisms to promote people to opt for green buildings.

Lack of awareness about the green buildings concept and the benefits derived from green buildings is prevalent. There are perceptions that green buildings are expensive. Besides, several of them are sceptical about the sustainability of these buildings in the long run, thus making it a challenge for both green architects and green building developers. Several builders and architects also are unaware of the cost aspects. As highlighted by a few studies, the general assumption is that green buildings are costlier compared to regular conventional buildings. Besides, the extent of investment costs and actual returns derived from the O and M costs are unclear. This may be due to a lack of sufficient technical information regarding the concept of green buildings, which is still in a budding stage in India. Additionally, the immaturity of the market, lack of resources, long gestation period and lack of focus on the lifetime return on investment (ROI) of these buildings are other reasons. Thus, there is a lack of evidence-based information to provide information to customers to adopt green buildings. Furthermore, a lack of sufficient support and consultancy services for the certification process may hinder the development of green buildings. At large, people are unaware of the processes and require systems to be implemented to ensure that it is convenient for users.

The other main constraints of the building sector are insufficient knowledge, building materials and energy-saving appliances. Lack of sufficient skills among the engineers is another issue, where only 6% of the employees have adequate training. Furthermore, a shortage of green-promoting contractors also adds to the problem. Therefore, capacity-building programmes to promote the construction of green buildings are not enough to meet the demands of the increased construction industry market.

Energy efficiency codes by the Indian administration are comprehensive and elaborate, but they need to be adopted more effectively across the states. There are capacity constraints at the local level which need to be addressed. Training and motivating them to promote the energy efficiency code is important. It would be helpful to demonstrate the efficacy of the code by increasing awareness through demonstration projects. There are simple and affordable performance-based systems like Excellence in Design for Greater Efficiencies[2] (EDGE) that make it easier for developers to follow resource efficiency standards and for certifying agencies to confirm compliance.

To sum up, the problems can be seen as a lack of data and comparable data at a micro-level. General awareness of green buildings in India is scarce and scattered, and evidence of improved climate resilience and institutions' role—draws attention to the need for knowledge but limited evidence. There is little evidence of improved climate resilience/measurement of ecological and economic efficiency in the context of green buildings. Several institutional barriers exist to implementing green buildings—capacity, resources, awareness and knowledge and politics. Thus, there is potential and need for increasing strategies, public and private actors committed to action, innovation potential and infrastructure provision.

26.3 Methodology

After an extensive review of the literature, we discussed with officials/personnel of institutions viz., Town Planning Department, Government of Karnataka, BBMP and Corporate companies. Following the pilot survey and improvements in the questionnaire, surveys were conducted at buildings with green features and certified green buildings—LEED certified (G.I. Gateway, LEED Council 2020), GRIHA certified (G.B.I. India 2020) and Biome (green architects-based firm) constructed buildings.

The sample respondents were representative of different age groups, women and the elderly population. Besides, we looked into aspects of preferences, impacts on health, quality of life and related aspects. The methodology includes both qualitative and quantitative data collection. Given the pandemic

---

[2] EDGE is a building design software that empowers the discovery of technical solutions at an early design stage to reduce operational expenses and environmental impacts. Based on the user's information inputs and selection of green measures, EDGE reveals projected operational savings and reduced carbon emissions. This overall picture of performance helps to articulate a compelling business case for green buildings (G.B.C. Inc About Edge, https://www.gbci.org/press.kit).

situation, we had to do the survey personally and through Google Forms. Random sampling was followed in selecting the respondents for the survey.

To understand the perceptions of the employees working in green buildings, we surveyed employees working across various green buildings in Bangalore. For the survey, we covered both certified and non-certified buildings (that had green features) and collected information on the employee's opinions on aspects like health benefits, productivity and barriers in promoting green buildings and their suggestions to promote the green buildings. One hundred employees were interviewed using structured questionnaires to capture their perceptions. These respondents were working at various green office spaces such as EMPRI (24%), TITAN new campus, Electronic City (22%), KPCL office (9%), ATREE (6%), ISEC (11%), Soukya International Holistic Health Centre (13%) and Sobha Forest view, Prestige Bagmane and IIPM (4% each).

For green buildings occupied as houses, a structured interview schedule was canvassed to the residents from independent houses and gated communities/apartments with a sample size of 100. Respondents belonged to NCC Nagarjuna, Salapuriya Serenity, Salarpuriya Greenage, Sobha Dew Flower, Sobha Forest View, Provident Sunworth, Provident Tree, Prestige Monte Carlo, Sriram Suhaana, Sai Vaibhav, GRC Brundavan, Rite Grand, Kristal Casper and Sunny Acura.

26.4 Capturing People's Perceptions

Experiences and Responses from Employees Living in Green Office Spaces

Awareness Levels

Awareness regarding green buildings is essential for environmental consciousness and its multiplier effects in several ways besides adherence to certain norms, its usage and achieving environmental sustainability. However, when we asked for employees' awareness regarding the same, around 56% of them were aware of the concept of green buildings, indicating that there is much more scope for awareness activities to improve among employees.

The employees who were aware had attained knowledge on green building mainly through the architects (15%). 13% indicated awareness which they acquired from the research they carried out, and another 11% through newspapers, magazines and media. One of the respondents was a student who had good knowledge about green buildings and their importance in achieving environmental sustainability, as the subject was introduced in her curriculum.

Motivation

When we asked about what factor influenced employers' motivation to choose the green building as their office space, the majority responded that it was an initiative to save resource usage (52%), followed by their aspiration of working

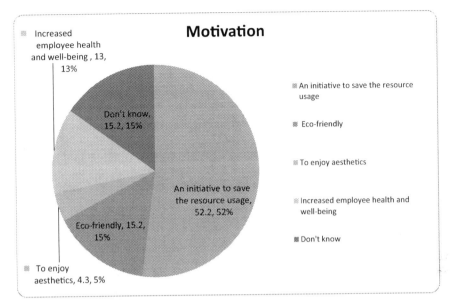

Fig. 26.1 Motivation (*Source* Primary survey 2021)

within the eco-friendly office space. Another 13% opined that working in eco-friendly green buildings increases their health and well-being. Discussion with the Regional Manager, GBCI revealed that the main reason for people to go for green buildings is to take care of their employee's health and well-being (Fig. 26.1).

Materials Used
With respect to materials used in office spaces, the survey results showed that the respondents were largely aware of the materials used in the working spaces—usage of locally available materials (22%), minimum usage of cement (17%) and both used locally available building material and building material specified by rating agency (15%). This shows that most respondents working in office spaces were aware of the green features adopted in the office buildings (Fig. 26.2).

Concerning energy aspects, the respondents opined that more emphasis was given to utilising natural light and highly efficient insulated glass windows (26%). Adopting this green feature, such as natural light, offers immense potential to reduce energy consumption in office spaces. The insulated glass windows and other green aspects include the usage of solar light, which 19% of the employees reported. During discussions with the employees, they expressed happiness with having natural lighting in their workspaces which made them feel good, and good ventilation and were aware that solar lighting would save resources (Fig. 26.3).

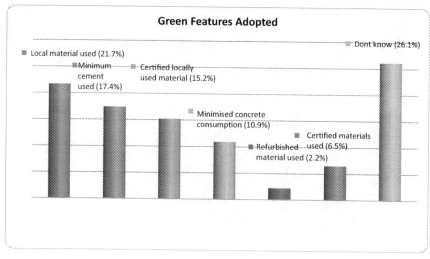

Fig. 26.2 Green features adopted—Office spaces (*Source* Primary survey 2021)

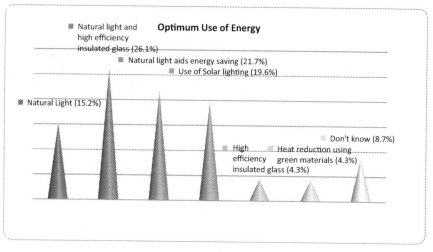

Fig. 26.3 Optimum use of energy

Water Conservation

Water conservation is one of the important aspects of green buildings achieved through harvesting rainwater, conserving stormwater runoff through the landscape and efficient treatment and usage of wastewater. Around 30% of them have reported adopting rainwater harvesting systems and another 41% as rainwater harvesting systems and stormwater conservation through landscape management. It is admirable that in some of the office spaces, native plants have been planted in the landscape. Respondents were aware of the rainwater

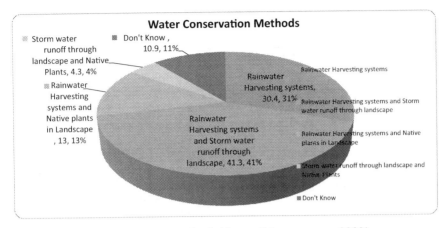

Fig. 26.4 Water conservation methods (*Source* Primary survey 2021)

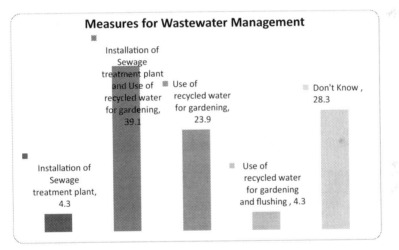

Fig. 26.5 Measures for wastewater management (*Source* Primary survey 2021)

harvesting systems adopted and stormwater runoff through the landscape (41.3%) (Fig. 26.4).

In our survey, most buildings had sewage treatment facilities, and around 39% of the respondents reported using recycled wastewater for gardening purposes in the office buildings. Also, a negligible percentage (4%) use for flushing purposes in the restrooms. However, 28.3% of the respondents were unaware of wastewater management (Fig. 26.5).

Solid Waste Management

To promote effective management of solid waste, almost all office spaces had the facility of collecting dry waste (paper waste, e-waste, etc.) and wet waste (pantry waste) separately. Around 87% of the employees reported separate collection of dry waste and other e-waste. In some of the office spaces, we observed that the organic waste was used to tap or generate gas which is used in the canteen's kitchen. Also, in a few office spaces, organic waste was composted to prepare manure which is later used for plants in their gardens. Largely, the majority of the respondents were aware of water-to-biogas conversion and composting.

Towards Environmental Sustainability
Green buildings use various measures or aspects of sustainability related to indoor air quality such as paints and adhesives that emit a low level of volatile organic compounds. Also, importance has been given to adopting good ventilation, which is a healthy and cost-effective way of saving energy and providing fresh air for the occupants. Hence, good ventilation has become one of the key components of green buildings that is required for certification by the certifying agencies.

Our survey results indicated that respondents were aware of indoor air quality, and good ventilation with increased fresh air as the most common methods adopted in their workplace (35%). The well-ventilated buildings allow the outside ambient weather to provide low humidity and moderate temperature, serving as an alternate cooling source for the buildings throughout. Good ventilation also saves energy costs by minimum use of air conditioners. In addition, few research studies have highlighted that good ventilation helps to keep occupants comfortable and healthy and increases productivity and learning (Fig. 26.6).

Good ventilation with fresh air is also supported by most employees' perception of the availability of enough natural light during the daytime (97.5%). Most respondents opined that the indoor air quality is good inside their office space (78%) and another 11% believe it is acceptable. Regarding temperature inside the office, most of them have experienced a comfortable temperature (80%) (Fig. 26.7).

Awareness About Costs/Savings Among the Employees
The major causes responsible for the growth of green building practices are reducing the negative environmental impacts and improving sustainability. On the other hand, few research studies have shown that in addition to reduced environmental impacts, green buildings also claim to contribute to additional benefits for the organisation, such as reduced operating costs, increased staff productivity and reduction of liabilities associated with poor indoor air quality problems. In our survey results, the respondents at large were not aware of the actual savings in costs, which is understandable as they were part of the organisation and had joined at different points in time. However, we feel that there is scope for making the employees understand the savings by quantifying and displaying it in the office notice boards, which can motivate employees. Quantification and measuring are being done in Titan Integrity Office, however, it is not displayed on notice boards. At large, the employees know that it adds

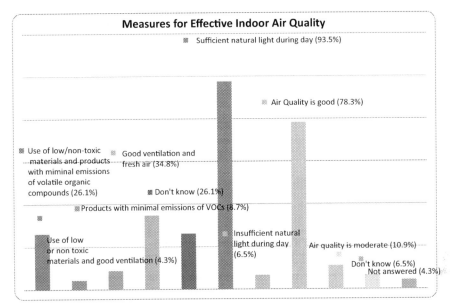

Fig. 26.6 Measures for effective indoor air quality (*Source* Primary survey 2021)

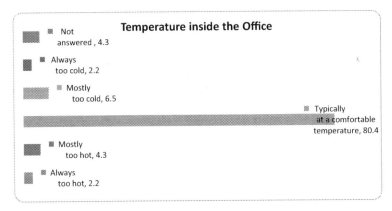

Fig. 26.7 Temperature inside the office

to a reduction in costs, but not actual savings. It was also revealed during a discussion with GBCI, that the myth of green buildings is that increase in the capital cost. Over a period of few years, it is recovered and the payback period has reduced to 1 year from 7 years earlier. When everything is incorporated at the designing stage itself, the cost is less. Even with respect to consultant cost, there is reduction in the costs due to increased number of consultants in the market and increased competition among them.

It is also understandable, except for the employees in the core team of sustainability, 54% were unaware of the investment costs that the company management would have made and the extent of investments made. Therefore, there is scope for making it transparent as well, which makes the purpose of resource savings and efficiency more inclusive and participatory, besides bringing about awareness and consciousness among the employees.

There is a growing demand for certification of operation and maintenance of the buildings also apart from certification for the design of the space.

Awareness of the Benefits of Green Buildings

There are numerous benefits associated with green buildings, such as environmental benefits—less consumption of water, energy, financial and economic benefits and health and productivity benefits. Research studies indicated that the health and productivity gain as the most important benefit of green building and our survey indicated most of the respondents (65%) were aware of the benefits of the green buildings and another 91% believed that green buildings are good at addressing environmental issues (Fig. 26.8).

When we asked which of the green features that were adopted in their building is good at addressing environmental issues, most of them have revealed two features—usage of nature-friendly materials and harvesting of water as the major factors that address environmental issues. Recycling of wastewater and landscape designing were the other factors that addressed environmental issues (Fig. 26.9).

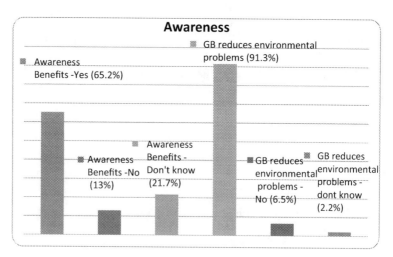

Fig. 26.8 Awareness on benefits (*Source* Primary survey 2021)

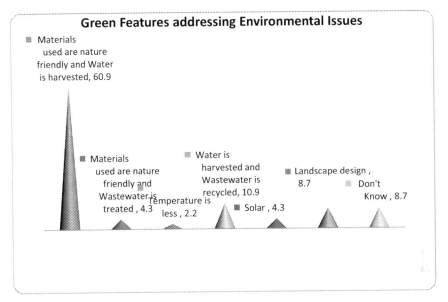

Fig. 26.9 Green features addressing environmental issues (*Source* Primary survey 2021)

Awareness of Health Benefits

The present survey explored the health benefits of the employees working in Green Buildings. Our results indicated that their health was positively influenced by the green features adopted in the office spaces they were working. A majority (74%) of them reported mental relaxation as the greenery is aesthetically appealing. Research has shown that people work more under a comfortable atmosphere, thereby increasing productivity. Around 15% of them have opined that green buildings improve their health and well-being, and another 6.5% believe that it improves employees' productivity due to low levels of indoor air pollution compared to conventional buildings (Fig. 26.10).

Some of the benefits of certification reported during the discussion with GBCI is the improved occupant's health and well-being and also lower operating cost apart from water and energy savings.

26.5 Barriers for the Promotion of Green Buildings

Though green buildings play a significant role in reducing greenhouse gases, carbon footprints and climate change impacts besides their economic and other environmental advantages, promoting green buildings has been difficult due to various barriers. Some of the identified barriers are information and educational, economic, policy and market barriers. Graph 7.18 indicates that among the various barriers that are hindering the development of green

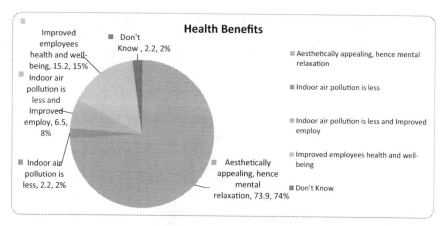

Fig. 26.10 Health benefits (*Source* Primary survey 2021)

buildings, poor awareness and price was the major one with 24%, followed by lack of sufficient information regarding costs aspects, technical information, benefits in terms of costs, etc. Another barrier that the respondents reported was the absence of government initiatives, although the government at the Central and State level have initiated initiatives to promote green buildings.

26.6 Measures to Promote Green Buildings

Several research studies have identified a lack of awareness as one of the major constraints in promoting green buildings. Lack of awareness at various levels and different stakeholders includes—in terms of standards, energy efficiency guidelines, certification and rating programmes among the smaller builders and contractors, insufficient information and understanding on the part of consumers. Therefore, we were interested to know the employees' perception of what could be done or measures that need to be undertaken to promote green buildings. In response to this, the majority of the respondents (46%) emphasised creating awareness among the various stakeholders and also providing sufficient information on cost aspects to the consumers as the most important steps that need to be undertaken to promote green building. However, a few respondents also highlighted the need for some incentive to motivate the public to construct buildings with sustainable features (Fig. 26.11).

26.7 Issues, Suggestions and Expectations

We were interested to know the employees' issues in their office space. Around 29% have not reported any issues. Few respondents have reported having issues in waste management (bad smell, etc.) and frequent repairs in the solar appliances or fittings. Reading suggestions and awareness creation is one

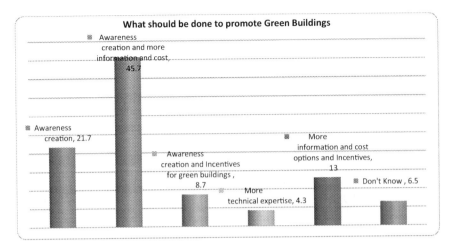

Fig. 26.11 Perceptions of promoting green buildings (*Source* Primary survey 2021)

of the suggestions reported by the employees (11%) and subsidy from the government to promote green buildings (9%) was also suggested.

When we analyse what are the expectations from the government, "provide incentives as a very important initiative to promote green buildings" (13%), followed by "effort of the government to promote in various possible ways" (6.5%) and other negligible percentages of employees (4.3%) expect implementing and regulation of mandatory government policies to promote the green buildings.

26.8 Experiences and Responses from People Living in Green Homes

Awareness About Green Buildings

Of the respondents, only 20% stay in certified buildings. More than 50% are not residing in certified buildings; 28% are unsure if their building is certified or not and this group also consists of respondents residing in independent houses. Of the 20%, 8% reside in LEED-certified residences, whereas GRIHA and IGBC-certified buildings' residents are 1.3%. The rest are aware of certification, but not sure which certified agency has issued it.

The newspapers (24%) followed by television (12%), magazines (11%) and websites (9%). Apart from these sources, builders (9.3%) and architects (9.3%) have also given information about green features in the building/construction to the respondents. Friends and family members, the curriculum in an environmental engineering course, general awareness, real-estate agents are some of the other sources of information for the respondents. Seven per cent of the respondents have gained awareness from family members and friends.

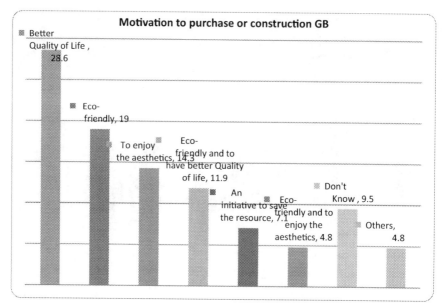

Fig. 26.12 Motivation to purchase or construct a green building (*Source* Primary survey 2021)

The interview of the respondents shows growing awareness about eco-friendly buildings. When we asked about the motivation behind buying a house with green initiatives, the majority of them, around 29%, mentioned the quality of life as an important reason; 19% have given importance to the eco-friendly aspect of green buildings and 14% to aesthetics as motivation to buy/construct green buildings. Seven per cent are conscious of saving energy, whereas close to 5% want an eco-friendly house and aesthetics. Along with them, reasons like the house being close to the office, good construction company, proximity to a hospital and affordability are also considered by around 5% of the respondents (Fig. 26.12).

Types of Green Initiatives Adopted and Awareness Levels

The respondents know the green initiative aspects followed and installed in their residential apartments and houses. It is interesting to see that 16% of the respondents identify rainwater harvesting on their premises. Eight per cent of them know that they have used locally available building materials, refurbished or remanufactured materials and minimised cement/concrete consumption. The respondents are aware that the material used is locally available; hence, carbon footprints are reduced during construction. However, 5% of them are aware of RWH and STP. Very few of them know a rating agency specifies the building materials. Apart from these, there are also respondents (around 4%) who mentioned energy-saving devices, presence of open landscape which naturally recharges groundwater tables, waste segregation, composting, garden waste management, sensor-based lighting in common areas and maximum usage of natural sunlight. However, it is to be noted that around 32% are not aware of the green initiatives adopted in their houses and apartments and 19% have not answered the question. It is to be noted here that when a question was generally asked about the green initiatives adopted, initially, respondents were not very certain about it. In the following tables, they can relate and identify green features in their residences when we ask specific questions.

The respondents are aware of the importance of saving power. Hence, a majority of them are using CFL and LED bulbs. Around 37% are using CFLs, and 29% are using LED bulbs. Nearly 19% use both CFL and LED bulbs, whereas 15% are not aware of the kind of bulbs used in their houses (Fig. 26.13).

Water is the scarcest resource in urban areas. It affects the sustainability of humans in urban settlements. Lack of access to safe water is one of the most important challenges in growing urbanisation. In our study, it is seen that except 9%, all are aware of the measures adopted to conserve water resources at their place. The method of rainwater harvesting is in the lead, with more than 80% of the respondents having it in place at their residence. RWH is the common feature in most houses (51%).

Along with RWH, wastewater reuse is the primary water-conserving method (16%). Other methods include native landscaping, which consists of indigenous trees, plants, grasses and shrubs. As this is adapted to the climate and geography, they are low-maintenance gardens saving water. Similarly, few respondents recognise the existence of rain gardens. These gardens collect water from the roof or driveway and allow soaking into the ground. This becomes more cost-effective and beautiful when planted with grasses and flowering perennials adding to the surrounding aesthetics. In 2.7% of

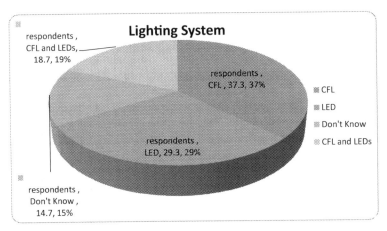

Fig. 26.13 Electricity saving techniques (*Source* Primary survey 2021)

the respondents' houses/apartments, drip irrigation is used for plants to avoid wastage of water. Apart from these, the other methods are using water collected from washing machines to wash the courtyard and using treated water for flushing.

The majority of the respondents are aware of solid waste management and 97% are already segregating the waste for further recycling. As the BBMP has made it mandatory, separate bins for wet, dry and electronic waste are very well managed by the respondents and around 67% reported that the wet waste is being de-composted.

Green buildings ensure indoor air quality (IAQ). The quality of air within and surrounding the building, if polluted by carbon monoxide, radon, volatile organic compounds (VOCs), particulates, microbial contaminants (mould, bacteria) is sure to cause adverse health conditions. Eight per cent of the respondents informed that the materials used in construction are low in toxics or non-toxic. Around 4% stated that the products used resist moisture; hence, no fungus formations can affect breathing. Nonetheless, 33% do not know how indoor quality is maintained in their residences, while 45% did not respond to this question indicating poor awareness in this regard.

Wastewater treatment is a process to remove contaminants from the waste or used water. Sewage Treatment Plant (STP) is installed in apartments with more residential units and is not practical to have in individual houses. Among our respondents, 64% reported having STPs in their apartment premises. 40% are aware that this is reused for flushing and 36% are aware that it is used in watering the garden. Around 20% feel that recycled water is not used, whereas around 1% responded that the recycled water is channelised to rajakaluve. Around 3% of respondents, though they are aware of the wastewater treatment process, do not know the purpose for which it is used.

Landscape design in green buildings is not a mere decorative afterthought. It is multi-functional as it provides various critical green services for a building, including water efficiency, energy efficiency and sustainability. The proper orientation of a building is expected to save energy. The landscape designing will be such that it will help in managing runoff water, filtering stormwater and planting those species of plant that demand less water. In our study, more than 20% have informed their landscaping is done in such a way as to direct the runoff water and thereby manage it. Other methods followed are having plants that consume less water and having plants that filter and clean the stormwater. Around 33% did not have any idea about this aspect.

One of the main advantages of the green building initiative is cost-cutting in the bills. As the buildings have solar and water recycling, this is expected to reduce the costs considerably. Among our respondents, it is seen that only 20% have observed this and at least to this fact. Around 55% deny any cost-cutting happening and the rest 25% do not know about this advantage. There might be many reasons for this. People living in apartments cannot make out the savings in the water bill, as it is included in their maintenance charges. Moreover, in apartments, only those residing on the last floor have access to water heating through solar energy. Hence, not all the units in an apartment are benefitted from solar panels. In the independent residential houses, only a minimum amount of Rs. 60 is deducted from the water bill if they have rainwater harvesting and there is no incentive yet for having solar panels. Hence, not all respondents can understand the savings from solar, rainwater harvesting and STP.

26.9 Information Knowledge Sharing About Green Buildings

We were keen to understand if the information about the benefits of green buildings was provided to beneficiaries. The majority of the respondents opined that they were not informed about the benefits (56%). However, 30.7% were informed about the benefits. During our discussions at the gated communities, we observed that people were not informed in particular, as it has become compulsory to have rainwater harvesting, solid waste management composting plant and sewage treatment plant as an inherent part of the gated communities. So, it is assumed that it is a package. However, the benefits are not discussed in specific. The marketing team of the builders of gated communities, as well as the buyers, mainly focus on the availability of amenities at large, we strongly feel that there is scope for highlighting the green aspects and the benefits derived, which would aid in promotion and awareness of green aspects and resource conservation.

32% opined that they were informed about the extra costs to be paid, given that they are green buildings. Among them, 30.7% were informed about the extra costs and 48% were not informed about it. The percentage of additional costs paid varied from 10 (17.3%), 20 (10.7%) and 50% (1.3%). The rest were unaware and opined that they did not know, while 49% did not answer. The responses indicate that the costs related to green buildings are largely unknown and there is less awareness among the people. However, the respondents who were aware were from certified buildings. Given that all green buildings are not certified, it is understandable that people will not be aware of the specific benefits of green buildings and certification.

Low-Cost Green Buildings

Awareness regarding extra payment for certified buildings among LEED, GRIHA, EDGE or IGBC were examined. 30.7% were aware about it. Although the responsibility of getting the certification is mainly on the construction company, this is not shared widely among the property buyers.

Green buildings can be sustainable only if they are maintained properly. Therefore, responsible operations and maintenance (O & M) are necessary to maintain the lifecycle of the building without compromising the quality which generally costs higher than maintaining regular apartments/houses. In the study, most respondents opined that they paid about 10% extra (22.7%)

for annual maintenance, while 4% felt it was 20%. However, 33% were unaware and a large majority of 38.7% did not answer, indicating that they were not aware of additional costs for maintenance. All this indicates a lack of awareness, and these costs must be made available so that people are aware of facilities' operations and maintenance costs, which would aid in people using resources carefully and becoming more conscious.

Awareness About Benefits

Although people did not know about certifications and standards, they knew the benefits of green buildings. The majority of the respondents living in green buildings have identified positive aspects in terms of health. While aesthetically appealing, mental relaxation was common across all the respondents, they added feel-good factor, physical relaxation (32%), reduced indoor air pollution (34.7%) and good quality of life (20%). Overall, one can witness that living in green buildings was welcome; majorly, more than 80% experienced positive benefits.

Satisfaction

Concerning satisfaction with the features available at their residences, it was observed that a majority (24%) were happy with nature-friendly materials used and well-managed waste and water harvesting systems. The second highest was (13.3%) with satisfaction with water harvesting, wastewater recycling and use of solar power. The rest of it was a combination of various features.

Issues

Majority of the respondents (61%) did not face any issues. However, there was a small percentage who had issues with rainwater harvesting (8%), waste management (5.3%) design (2.7%) and O and M (4%). These are solvable issues; however, it calls for efficiency and support infrastructure.

Expectations and Suggestions

Expectations from the government were to have annual tax rebates for residents of green buildings (17.3%), incentives (12%) and making it mandatory through laws to have self-sustained houses (5.3%).

A majority (73%) had not received any incentives for adopting green features or living in green buildings. However, some respondents had received some benefits (14.7%). These incentives were mainly for solar installations; wherein a few houses have got a 5% discount while purchasing. At the same time, 6 households said they are getting a monthly rebate of Rs. 60 in the electricity bill for the installation of solar panels in their households.

The majority of the respondents (41%) opined that there was a lack of awareness about green buildings combined with other aspects like price, and lack of information on costs as barriers to promoting green buildings. It was a combination of single aspect that people felt was the reasons like lack of incentives, lack of interest by the government and lack of sufficient information among architects.

The respondents feel that the government is not taking enough measures to encourage green buildings in the country. They gave an array of suggestions that they expect the government to implement. Giving annual tax rebates (17%), awareness creation (13%) and providing incentives to buyers and those who construct eco-friendly houses are the most commonly repeated suggestions. In addition, they want the construction of the green building to be mandatory, backed by law and expect the state to set an example to others by constructing more government buildings so that others can emulate them.

26.10 Conclusion

In summation, the awareness among the residents and employees in a green building about green features is identified. It depicts that there is awareness among the majority of the respondents about the concept and features of green buildings. The residents have sourced information from various channels and are aware of the health benefits of green building and hence, they have bought apartments/constructed residences despite the relatively high cost. They are reaping benefits from green buildings, though it is not well quantified. The employees are happy working in green buildings and have directly experienced the benefits of improved energy levels and overall health. However, there are some minor issues with the functioning of green features which has scope for improvement. Realising the benefits and cost of owning an eco-friendly house, both residents and employees suggest tax benefits, other incentives from the government, legal enforcement and efforts to create more awareness among citizens about green buildings.

Although there are challenges along the path to promote green buildings, the concept of green building is catching up gradually as there is a growing awareness among the builders and users. The demand for green buildings is increasing especially in the commercial sector. As 60% of the infrastructure that India needs is yet to be constructed and estimates are that by 2030, 10 billion square metres of construction will happen; it would be valuable to make green buildings. The draft Building Code has incorporated all green features and detailed out all specifications making it easier to adopt. It has also provided specifications regarding certification. It is important to be made effective on the ground and implemented.

It is interesting to note that private developers have been promoting green buildings by giving certain concessions to their buyers. For instance, Good Earth has announced free maintenance costs of residences for three years (*Times of India* 2021).

Awareness creation through Training Programmes is important by making training programmes as a part of the syllabus and tied up with the IGBC with the concerned departments in the Engineering Colleges. IGBC conducts Green Building Training Programmes across the country and has trained several thousand professionals on green building concepts so far. The platform exposes the participants to the latest global trends in green buildings, and concepts and sharing best practices besides providing networking and business opportunities for green equipment. In addition, the advanced training programme on IGBC's Green Building Rating System aims to deal with regional priorities and approaches for "innovative sustainable design strategies".

World Green Building Week (2015) is another event organised by the CII-IGBC in which they are conducting a Green Photography contest to promote positive change. Similarly, an awareness programme on green buildings is organised by the Confederation of Indian Industry (CII) and the Kerala Industrial Infrastructure Development Corporation (Kinfra) where the government would consider granting concessions and incentives for green buildings in Kerala.

Incentives can also make it popular for adaption; residents adopting solar installation are given a subsidy of 5% on their power bills and have made Bangalore one of the cities with the majority of solar installations. When the users see the benefits, the idea tends to spread faster. Although it is important to incorporate all the features of green buildings within a building, even certain features adopted add to the sustainable use of resources. Hence, it must be promoted in all ways possible. In short, both legislation and incentives must be unswerving and effectual.

REFERENCES

G. B. i. India, "What Is GRIHA—Green Buildings in India," [Online]. Available: https://greenbuildingsindia.wordpress.com/tag/what-is-griha/. [Accessed 9 July 2020].

G. I. Gateway, "LEED India: Green Information Gateway," [Online]. Available: http://www.gbig.org/collections/14555. [Accessed 9 July 2020].

Times of India, dated 30.7.2021.

U. G. B. Council, "What Is LEED? | U.S. Green Building Council," [Online]. Available: https://www.usgbc.org/help/what-leed. [Accessed 2 July 2020].

W. G. B. Council, "Advancing Net Zero Status Report 2019," [Online]. Available: https://www.worldGBC.org/sites/default/files/WorldGBC%20ANZ%20Status%20Report%202019_FINAL%20RELEASE_0.pdf. [Accessed 12 April 2020].

W. G. B. Council, "Participating Green Building Councils," [Online]. Available: https://www.worldGBC.org/better-places-people/participating-green-building-councils. [Accessed 12 April 2020].

W. G. B. Council, "The Drive Toward Healthier Buildings 2016," [Online]. Available: https://www.worldgbc.org/sites/default/files/Drive%20Toward%20Healthier%20Buildings%202016_ffff.pdf. [Accessed 16 April 2020].

Zuo, J., and Zhao, Z. Y. (2014). Green Building Research–Current Status and Future Agenda: A review, Renewable and Sustainable Energy Reviews, 30(C): 271–281

CHAPTER 27

Decentralisation in the Urban Sphere: Successes and Failures—(A Note from Experience)

Kathyayini Chamaraj

27.1 Introduction

It was to give back control to local communities over their own planning and development that the Central government passed the 74th Constitutional Amendment Act, better known as the Nagarapalika Act, in 1992. This was the urban equivalent of the parallel legislation for rural areas, the 73rd Constitutional Amendment, well known as the Panchayat Raj Act, which deals with the decentralisation of governance to rural local self-governments. The Nagarapalika Act recognised the urban local bodies as legitimate third tiers of local self-government and gave them Constitutional validity and permanence.

The goal of the 74th Constitutional Amendment (CA) was to make urban local bodies "urban local self-governments" (ULSGs) which gives communities decision-making powers over local assets and resources. But after 30 years, most of the goals have remained largely unfulfilled because the devolution of 3Fs—Funds, Functions and Functionaries—to ensure sustainability and inclusivity, has not happened.

K. Chamaraj (✉)
Executive Trustee, CIVIC-Bangalore, Bengaluru, Karnataka, India
e-mail: kchamaraj@gmail.com

© The Author(s), under exclusive license to Springer Nature Singapore Pte Ltd. 2023
S. Nautiyal et al. (eds.), *Palgrave Handbook of Socio-ecological Resilience in the Face of Climate Change*,
https://doi.org/10.1007/978-981-99-2206-2_27

27.2 Discussion

Weak State Election Commissions

The first mandate of the 74th CA is that urban local bodies should always be governed by an elected body to give them democratic legitimacy. The State Election Commission (SEC) is the constitutional body empowered to conduct the local body elections under the 73rd and 74th CAs, as per Article 243K and Article 243ZA. However, unlike the Central Election Commission and its state branches, and despite what the 74th CA says, the powers of the SECs have been determined by the State governments, which vary in each State.

The control by State governments, such as Karnataka, over the delimitation of wards, the preparation of the reservation of roster, and the issue of notification of elections has resulted in inordinate delays in the conduct of elections. In the case of the delay in the election to the Bruhat Bengaluru Mahanagara Palike (BBMP) [Greater Bengaluru Municipal Corporation] in 2015 and now again since 2020, the State Election Commission expressed its helplessness in conducting the elections before the Courts as only the last task of conducting elections is entrusted to it while the other two preparatory tasks are all in the hands of the State government. It was only after the intervention of the Supreme Court that elections were held for the BBMP. Earlier too, the urban local body elections in Bengaluru were delayed for four years between 2006 and 2010. Administrators were appointed to run it because Bengaluru's municipal boundaries were being extended to create Greater Bengaluru. Lack of an elected body in the third tier, which is closest to the citizen, means that the fora for citizen participation in local decision-making over the local assets and resources do not exist, negating governance of the people, by the people and for the people, which is the essence of democracy.

Non-devolution of 3Fs

The 18 functions of municipalities listed in the 12th Schedule have been devolved on paper only. Among the 18 functions, urban planning itself, for instance, through which one could bring sustainability and inclusivity, is still being handled by parastatals—for example, the Bengaluru Development Authority (BDA) in Bengaluru. These parastatals are accountable to the State government and not the ULSGs. The Metropolitan Planning Committee foreseen in the 74th CA for doing the planning for Bengaluru has been kept dysfunctional. In other districts, District Planning Committees are ineffective. Since there is no effective monitoring mechanism and overseeing by citizens to ensure that what is planned is also implemented, there are tremendous violations of the zoning regulations and building bye-laws leading to flooding and pollution of lakes and groundwater, etc., which is being witnessed year after year. The urban poor who reside mostly in the low-lying areas, lake-beds, buffer zones of lakes and along the stormwater drains suffer the most from the flooding. But as recent floods in Bengaluru demonstrated, even the rich are not spared from floods and had to be rescued in boats.

Parastatals, such as the Bengaluru Water Supply & Sewerage Board (BWSSB) and the Karnataka Urban Water Supply & Sewerage Board, are in charge of water supply and sewerage, which when effectively done will lead to sustainable water solutions. But much of the sewage is not being effectively treated and ends up in the stormwater drains and lakes.

Provision of decent housing for the urban poor through slum improvement, which would lead to inclusivity and less pollution, is handled by the parastatal, Karnataka Slum Development Board and not by the urban local self-governments. Due to these functions not being devolved, cities are unable to work for sustainability.

Some functions that need to be decentralised for local management, such as public transport, do not come under the purview of urban local bodies as they are not listed in the 12th Schedule. Big infrastructure projects, such as the Peripheral Ring Road (PRR) and elevated corridor are planned by parastatals and not by the ULSGs.

Dysfunctional Metropolitan Planning Committees

The 74th CA also included provisions for a District Planning Committee under Article 243ZD for district development planning, including rural and urban concerns. Similarly, under Article 243ZE, a Metropolitan Planning Committee was to be set up for each multi-municipal metropolitan area with a population of 10 lakhs or more. Both these bodies were meant to replace top-down planning by the Central and State Governments for the ULBs and PRIs and bring planning to the district and metropolitan level to be done by the PRIs and ULBs.

In Bengaluru, the MPC was set up after 23 years of the 74th CA through a Court order but has remained dysfunctional. In parallel, the BDA has been conducting public consultations to prepare a Revised Master Plan 2031 for Bengaluru. So, why are there two planning bodies for Bengaluru? In addition, the State government is also making several plans for Bengaluru outside the Master Plan, such as for the steel fly-over, the elevated corridors, etc. The MPC is not taking all these decisions. Also, while BDA prepares a mainly spatial and zoning plan, the MPC must prepare development plans on 15 items, including urban poverty alleviation, etc., along with mechanisms for monitoring and assessing outcomes.

This is the reality, although Article 243ZF of the 74th CAA clearly states that any legal provisions, which are inconsistent with the provisions of the 74th CAA, unless amended or repealed earlier, "shall cease to be valid at the end of one year from the commencement of the 74th CA", i.e., 1994. Hence the power of BDA to prepare the Master Plan for Bengaluru has become automatically void and the planning powers should rest only with the MPC. Implicit in the Constitutional amendment is the objective that the State government should devolve planning powers for metropolitan areas to the third tier of government and desist from itself making plans for them.

No Proper Cost–Benefit Analysis

In the absence of planning by the MPC, the state government through the BDA is doing the planning for widening roads, building ring roads, elevated corridors, etc., which involve large-scale cutting down of trees and increasing concretisation causing heat islands and more air pollution, displacement of the urban poor, etc.

In the case of the Peripheral Ring Road (PRR) that is being planned in Bengaluru, for instance, even the DPR is not in public domain. Along with the Elevated Corridor that is also being planned, it is leading to a concentration of investment in Bengaluru for road infrastructure. In both these projects, there is no transparency about the cost–benefit analysis of these projects in the DPR and on whether these projects are needed at all and whether they will actually produce the outcomes that are sought, that is, to decongest traffic.

While looking at alternatives, for instance, the DPR may look only at whether a proposed road or elevated corridor should go through this or that layout. But whether the outcome could be better achieved by incentivising public transport rather than incentivising private vehicles by building the road or elevated corridor is not analysed in the DPRs. Also, the environmental and health costs of incentivising private vehicles which contribute almost 45% to the air pollution in Bengaluru are not built into the cost–benefit analysis. Whether the projects are in coherence with the Sustainable Development Goals is also not analysed.

Inequitable Development and Sharing of Resources

The larger question that needs to be evaluated at the State level in the cost–benefit analysis would be, whether the money, almost one lakh crore merely to solve the traffic problem of Bengaluru (Rs. 35,000 crore for the elevated corridor alone), is better spent in developing the infrastructure in the Tier 2 and Tier 3 cities and the less-developed Kalyana Karnataka region. The lack of development of Kalyana Karnataka is what causes the huge migration to Bengaluru, necessitating the expansion of road infrastructure in Bengaluru. Investing Rs. 35,000 crore in developing these other cities and areas of Karnataka would stem much of this migration and lead to the overall development of all regions of Karnataka. But in contrast, not even Rs. 3000 crore is being released every year for the development of Kalyana Karnataka. In comparison, it is considered affordable to invest a total of Rs. one lakh crore on road infrastructure in Bengaluru to solve its traffic congestion problem.

A lot of resources that are supposed to be devolved to the ULSGs as per the allocations made by State Finance Commissions, are also not reaching them. Many recommendations of the SFCs remain unimplemented.

Dysfunctional Ward Committees and Area Sabhas

Noting that in very huge urban areas, even the municipality, or third tier, may still be too distant a body for the citizen, the Nagarapalika Act mandated the formation of local area committees, called ward committees, to carry out most of the functions of the municipality. While the Act says that all cities with more than three lakh population shall have ward/s committees, it does not bar smaller cities also from having them. The 74th CA allowed state governments to form ward committees, either for every ward in the urban body or for a group of wards. However, even when the decentralised institutional structure of ward committees is formed, they are dysfunctional.

The Area Sabhas at sub-ward level, parallel to the Grama Sabhas in rural areas, have not been notified in most cities, though built into the Karnataka Municipal Corporations Act as a conditionality under the JnNURM. This is despite High Court orders in Karnataka directing their constitution. Unless Courts chastise the government, there is total neglect and lack of political will to even follow the law and enable citizens to do local planning, monitor implementation of works, conduct social audits of usage of funds and how this is affecting the sustainable development of their ward.

The most compromised arena in these laws is the autonomy of the Ward Committees. Ward Committees are also not equipped with essential functions, rights and powers. Many states have not provisioned Area Sabhas and stay with two-tier system of urban governance. This is against the spirit of the Model Community Participation Law framed by the Centre to strengthen community participation.

None of the states provides financial autonomy to Ward Committees. Most of the functions assigned to Area Sabhas and Ward Committees are either to advise or assist the municipality only. The latest BBMP Act 2020 has reduced ward committees to recommendatory bodies only.

Enlarging Ward Committee Functions

The ward committee should be given the power to decide on the use of public land, properties and resources in the ward, including parks and playgrounds, lakes, use of civic amenity sites, location of public amenities, etc. MLAs, MP's, etc., should not be allowed to interfere in local issues in the ward committees' exclusive domain. Ward committees and Area Sabhas should get all the planned projects for their areas, give consent to projects planned in their ward, and give inputs on CDP, budget, etc. Suppose all such tasks are to be approved and sanctioned only by the Commissioner or higher-level officials, despite having ward committees, the red tape and delays in decision-making will continue.

Unless ward committees have the power to impose financial penalties on ward-level officials, for instance, when they give permissions for building bye-law and zoning violations, which lead to floods, etc., and unless they also have the power to terminate contracts, such as garbage and road contracts, if the contractors fail to perform properly, effective decentralisation cannot happen. Thus, permissions to build roads on lake buffer zones, for instance, in Pattandur Agrahara Lake, create an island and erect a statue in Begur lake are being given.

Area Sabha should have the power to allow/disallow any commercial or industrial activity or any land use conversion or acquisition of any land and the terms and conditions on which it is allowed; to decide the public purpose for which land is being acquired; to allow/disallow removal of a slum/displacement until those being removed are resettled, to get all encroachments removed from public land; have first right and control over all natural resources in its jurisdiction, such as land, CA sites, water, lakes, etc.

Without citizen oversight over these matters, whether solid waste management or building bye-law and zoning regulation enforcement, we know that there is total corruption, and different mafias are in control. The contractor lobby in solid waste management and the real estate lobby are in actual control of Bengaluru's governance, leading to unsustainable development. Also, the regulatory body, the Karnataka State Pollution Control Board, has proven to be helpless in seeing that Bengaluru develops sustainably.

Lack of Database for Measuring Sustainability

There is no index or database to measure sustainability either at the city or ward level. For example, what is the level of air, land and water pollution and the level of housing at the ward level and what needs to be done to bridge the gap to bring about sustainable development? Indicators need to be developed for measuring these at the ward level, and a Ward Sustainability Index developed out of these.

Success on Ward Disaster Management Cells (WDMCs)

When the lockdown was announced due to the pandemic in March 2020, the Disaster Management Authorities at State and District levels were activated. Still, no institutional mechanism was prevalent at the ward level for disaster management. Unless there is a means at the grassroots level to handle disasters and reach out to those affected and the needy with ration kits, medicines, etc., it is very difficult to monitor all that from the district level. But Karnataka is one of the few States which has the mechanism of Ward Disaster Management Cells (WDMCs) built into its Ward Committee Rules of 2016. But these had never been set up. An email by CIVIC to the Chief Justice of the High Court pointing out that the WDMCs had not been set up resulted in them being constituted by BBMP in all 198 wards within two days of the Court's order.

General SOP for WDMCs Absent
Though the WDMCs were set up per the Court order, the sad part was that they had no Standard Operating Procedure (SOP). Noting this, CIVIC and several other CSOs, developed a General SOP for WDMCs, that would ensure "Preparedness, Response, Recovery and Mitigation"—all the four principles of disaster management—and not just a knee-jerk response to COVID. The General SOP for WDMCs would be valid for any disaster, not just COVID. This general SOP was sent to the State government by CIVIC. The effort of CIVIC in preparing the General SOP was appreciated by the Chief Secretary, who forwarded it to the Revenue Secretary (Disaster Management) for the adoption of its relevant points.

Booth-Level Committees Dysfunctional
Though many volunteers were willing to chip in and help at the grassroots during the COVID pandemic, their energies had not been channelised in an organised and systematic manner through training and capacity building of the WDMCs. Further, though a senior official made an effort to set up electoral Booth-Level Committees for every Polling Booth Area (PBA) with the Booth-Level Officer (BLOs) and residents of the Area to handle COVID, it failed to take off as the BLOs did not live in those areas and hence did not come forward to lead the Booth-Level Committees. The then BBMP Commissioner accepted that a resident of the area nominated as Area Sabha Representative (ASR) per the KMC Act could be nominated as the Booth-Level Officer. So it need not always be an official. But since the Urban Development department had not notified every PBA as the "Area" for constituting Area Sabhas, this idea failed to take off. Hence the concept of Booth-Level Committees failed, and there was no way of doing intensive work at the electoral booth level with volunteers to do door-to-door monitoring of those affected by COVID, as was so successfully done in Kerala.

General SOP for WDMCs Acknowledged by NDMA
Meanwhile, during a webinar, Mr. Kamal Kishore, a National Disaster Management Authority member, spoke of the need for a more granular, decentralised management of COVID at levels lower than the state and district levels, at ward and booth levels. Noting this, CIVIC shared its experience of getting WDMCs constituted in Bengaluru and sent him the Draft General SOP for WDMC for adoption at the national level, with suitable modifications, which he promised to look into. It is hoped that the ability to respond to all disasters in a decentralised manner will get institutionalised soon at the national and grassroots level and save the country much chaos, pain, suffering and deaths during disasters.

27.3 Conclusion

Ward 5-Year Vision Plan

The Ward Committees should be asked to prepare a 5-year ward vision plan per a Performance Management System (PMS) **based on human development, social infrastructure and sustainability outcomes at the ward level.** Targets need to be set, and outcome indicators developed for measuring these. Monitoring and review need to be based on performance on those indicators. The Area Sabhas must be involved in setting targets and reviewing municipal performance.

Committee to Be Set up to Review the Functioning of the 74th CA

A committee to review the entire gamut of provisions relating to the implementation of the 74th Constitutional Amendment in its true spirit needs to be set up, headed by a public figure and members devoted to the cause of decentralisation, to review the conformity legislations of States and present a revised set of amendments to the State Acts and the 74th CA.

Organisation of Citizens

Ward committees should foster the organisation of citizens in their areas, as required under the Act, and regularly interact with them. A network of neighbourhood groups/street-wise citizens' committees, which send their representatives to the Area Sabhas, should be actively promoted to create downward accountability mechanisms of the ward committees to the constituents, i.e., the people and community. This will strengthen decentralised governance "of the people, by the people and for the people", which is the essence of democracy.

PART V

Policy Issues and Future Strategies

CHAPTER 28

Integrating Climate Resilience in Sectoral Planning: *Analysis of India's Agriculture Disaster Management Plan*

Sanayanbi Hodam, Richa Srivastava, Anil Kumar Gupta, and Kirtiman Awasthi

28.1 Introduction

Changes in average temperature, seasonal variations and a growing occurrence of extreme weather events, as well as other climate change impacts and slow-onset events, are now occurring around the world and the frequency and severity of it is increasing day by day. (UNFCCC 2021). Therefore, integrating climate change concerns into sectoral planning is necessary and will help build climate resilience. Thus, assessing how climate change and accentuate existing climate-related hazards and introduce new ones and taking measures to address these risk and associated vulnerabilities is part of climate-resilience building. (C2ES 2021). The adoption of the 2030 Agenda for Sustainable Development demonstrates that there is a unique opportunity to build climate resilience for sustainable development by addressing structural inequalities that perpetuate poverty, marginalization and social exclusion, increasing vulnerability to climate hazards (World Economic and Social Survey 2016). Although there is a lot of common sense in the paradigm, there are obvious analytical and practical challenges in applying disaster risk management and sustainable development to disaster reduction (Collins 2018). One of the approach to

S. Hodam · R. Srivastava · A. K. Gupta (✉)
National Institute of Disaster Management, (Ministry of Home Affairs, Government of India), Rohini, Delhi, India
e-mail: envirosafe2007@gmail.com

K. Awasthi
Deutsche Gesellschaft Für Internationale Zusammenarbeit (GIZ), New Delhi, India

© The Author(s), under exclusive license to Springer Nature Singapore Pte Ltd. 2023
S. Nautiyal et al. (eds.), *Palgrave Handbook of Socio-ecological Resilience in the Face of Climate Change*,
https://doi.org/10.1007/978-981-99-2206-2_28

achieve climate resilience is by mainstreaming climate-related risks into sectoral development plans. There are multiple pathways to developmental planning. Among these, sectoral planning has dominated planning landscape in India. Sectoral planning entails the creation and implementation of a set of schemes or programmes directed at the growth of a certain specific sector (s) of the economy such as agriculture, water, health, infrastructure etc. India is a developing country whose majority of the population depends on agriculture and allied activities directly or indirectly. Multiple natural and anthropogenic disasters threaten India's landscape, which is exacerbated by the influence of climate change, which causes massive economic losses (Gupta et al. 2019). Given that 65% of India's landscape is drought prone, 12% is flood prone and river erosion, and the entire coastline is susceptible to cyclones, make them highly vulnerable (NDMP 2019). Climate change affects every facet of our system and environment which is unstoppable. However, we can adopt adaptation measures in order to make it climate resilient. In closely related climate sectors such as agriculture, adaptation is an inevitable path (Mahfoud and Adjizian-Gerard 2021). At the same time, countries around the world are very actively preparing their own adaptation plans (Garschagen et al. 2021). It is found that there are independent policies for adaptation and disaster reduction. Due to the involvement of multiple agencies in similar projects, DRR, including CCA policy, seems extremely challenging (Mall et al. 2019). And therefore, integrating climate change and disaster risk reduction into planning and policies which is also known as climate-resilient pathways needs to be solidified. Climate-resilient pathways include strategies, choices and actions that reduce climate change and its impacts. They also include actions to ensure that effective risk management and adaptation can be implemented and sustained (Denton et al. 2014). This will also provide the opportunity to consider aspects of climate change responses in the planning process, and to look for co-benefits and synergies with sustainable development action (Oliveira et al. 2015). Some of different approaches of integrating Disaster Risk Reduction and Climate Change Adaptation measures are through Legal Policy Framework for Disaster Management, Environmental and Natural Resource Laws in DRR and Integrating CCA and Inclusion of DRR into Development Schemes and Projects (Gupta et al. 2016). As such, a developmental planning focuses on priority sector(s) at the national, state and local level.

India's Prime Minister's Agenda 10 on Disaster Risk Management (Gupta et al. 2016) draws an integrated approach towards implementing the Sendai Framework for Disaster Risk Reduction, Paris Agreement on climate change and the Sustainable Development Goals (SDGs), through its Agenda 1, i.e. all sectors to imbibe the principles of disaster risk management and utilizes the legal mandate under the Disaster Management Act 2005 (NDMA, and the National DM Policy 2009 (NDMA 2009). DM Act under its Section 35 calls for the (i) measures for disaster prevention through risk mitigation, mainstreaming into developmental plans and projects, preparedness and financial resources, at the level of Ministries of Central Government. Sectoral development plans, therefore, take on a critical role in disaster risk management. While sectoral planning is constructed around social, economic and ecological context, any changes in these parameters affect the overall developmental objective (PAGE 2016). At present climate change is seen as a major developmental challenge that is undermining the achievement of sustainable development goals (UN-SDG 2016). The spatial and temporal scales of climate impacts (global phenomenon with long-term) and development pathways (local to regional with short to mid-term developmental objectives) are different (Hoegh-Guldberg et al. 2018). Hence, it may be possible to change development pathways within a shorter time period than to change climate impact pathways (Allen et al. 2018). Since the impact of climate change can be seen in the different sectors of economy, it is important to integrate the risk resilience strategies in the sectoral plans as risks and associated vulnerabilities and potential response are sector specific. Many studies have emphasized the importance of integrating disaster risk reduction into sectoral development plans and policies. The first spatial resilience evaluation in Mexico demonstrates how a mainstreaming approach may be combined with technocratic methodologies to better quantify risk and give decision-makers with a complex tool that is suited to local needs (Moreno et al. 2017). A set of policy briefs aimed at profiling disaster risk management (DRM) policies has also been developed in five Caribbean Development and Cooperation Committee member states: Barbados, Guyana, Saint Lucia, Suriname and Trinidad and Tobago, with the goal of analyzing these policies and their interactions with broader development issues and instruments such as national disaster risk management plans. (Weekes and Bello 2019). Indonesia developed the "Indonesia Climate Change Sectoral Roadmap" in 2009. The goal of this plan was to reinforce the national commitment to climate change described in the "National Action Plan on Climate Change" and "National Development Planning: Indonesia Responses to Climate Change". The sectoral roadmap includes adaptation plans in four sectors: water, marine and fishery, health and agriculture (OECD 2018). The "Action Plan Framework for Adaptation and Mitigation of Climate Change in the Agriculture and Rural Development Sector Period 2008–2020" was adopted by the Vietnamese Ministry of Agriculture and Rural Development. This sectoral framework represents the "National Target Program (NTP) to Respond to Climate Change", a significant set of policies at the national level. In addition, it employs the climate lens by explicitly addressing the need to review the existing legislations in the view of enhancing adaptation (OECD 2009). A study on principles and considerations for mainstreaming climate change risk into national social protection frameworks in developing countries talked about future efforts that should be geared to develop climate-responsive social protection (Aleksandrova 2019). It is now well acknowledged that mitigation must be considered as part of a multi-objective development problem. However, there is a paucity of literature on how to utilize this in policymaking in a practical and effective manner, particularly in emerging economies (Cohen et al. 2018).

Given this, coordinated risk mitigation at the national, sectoral and programme levels is increasingly being emphasized for building climate/disaster risk resilience. Mitigation and adaptation actions that limit the impact of climate change must be incorporated into current development and investment strategies (Warren-Myres et al. 2021). Accordingly, this paper has been prepared with the aim to address the importance of integrating climate change into agriculture sectoral planning to strengthen the Indian Agriculture. The purpose of this paper is to highlight the steps for integrating CCA-DRR into sectoral planning with a special focus on National Agriculture Disaster Management Plan, 2020 (Gupta et al. 2020) for the Ministry of Agriculture Farmers' Welfare (MoAFW), Government of India. In this paper, specifically, the following approaches have been included: (i) Policy support and institutional mechanisms (horizontal/vertical) for guiding and implementing risk resilience strategy (ii) multi-stakeholder engagement (iii) planning that is the formulation of national sectoral strategies (iv) translating national strategy into state/local level action plan(s) (v) mobilizing budgetary support for implementation and (vi) monitoring and evaluation.

28.2 Climate Change Adaptation—Disaster Risk Reduction (CCA-DRR)

This chapter talks about the Climate Change Adaptation–Disaster Risk Reduction (CCA-DRR) integration into sectoral plans for better climate and disaster management systems for agriculture sector. The Ministry of Agriculture and Farmers' Welfare had recently prepared its' own comprehensive National Agriculture Disaster Management Plan (NADMP) as per the mandate of Disaster Management 2005. This NADMP has been prepared as a practical guide, a work agenda and a roadmap for incorporating key aspects of disaster risk reduction (DRR) into the sustainable development agenda of agriculture, especially for crop production, sustainable land management and post-harvest management within scopes and mandate of Ministry of Agriculture and Famer's Welfare. In this plan, there is a chapter which talks about incorporating various parts of DRR into the Ministry's policies, programmes and schemes (Gupta et al. 2020). The article discusses the opportunity mapping of NADMP and mainstreaming climate resilience into the Government's policies and programmes. The framework adopted is shown in Fig. 28.1. Coordinated risk mitigation at the national, sectoral and programme levels is increasingly being emphasized for building climate/disaster risk resilience. Specifically, this approach would include: (i) Policy support and institutional mechanisms (horizontal/vertical) for guiding and implementing risk resilience strategy (ii) multi-stakeholder engagement (iii) planning that is the formulation of national sectoral strategies (iv) translating national strategy into state/local level action plan(s) (v) mobilizing budgetary support for implementation and (vi) monitoring and evaluation.

28 INTEGRATING CLIMATE RESILIENCE IN SECTORAL ...   485

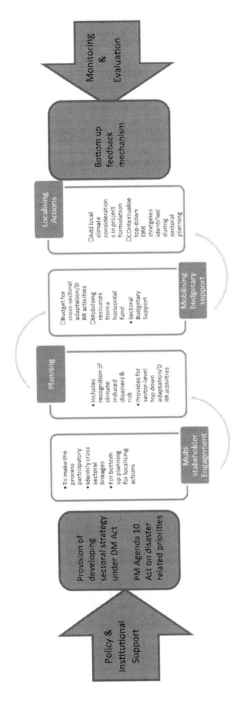

Fig. 28.1 Framework for CCA-DRR integration into sectoral plan based on NADMP experience

28.3 CCA-DRR Integration

The core of integrating adaptation/DRR at the sectoral level is to conduct climate/disaster risk at every stage of strategy/plan development. However, undertaking such process requires robust policy and institutional support and entry points in terms of political priorities. The following explains the different steps involved in CCA-DRR integration into sectoral plans.

Policy and Institutional Support for Integration of Risk Resilience in Agriculture Sector

The institutional and policy support provides entry point and sets out the broad objectives for a given sector. This is the stage at which national legislative and policy goals are transformed into sector-specific policy alternatives, which are then used to build operational strategies and mobilize resources to put them into action. Creating an "action plan" or a "roadmap" to integrate adaptation at the sectoral level is a popular method. These serve as strategies to address adaptation in respective sectors. Prime Minister's Agenda 10 on Disaster Risk Management draws an integrated approach towards implementing the Sendai Framework for Disaster Risk Reduction, Paris Climate Agreement and the SDGs, through its Agenda 1, i.e. All sectors to imbibe the principles of disaster risk management and utilizes the legal mandate under the Disaster Management Act 2005 and the National DM Policy 2009. The Disaster Management Act (Section 35) calls for (i) measures for disaster prevention through risk mitigation, mainstreaming into developmental plans and projects, preparedness and financial resources, at the level of Ministries of Central Government. It also provides for the role of Ministries on disaster prevention, mitigation and preparedness, it's integration and frame policies, rules and regulations in this context (Section 36). Further, it envisages that each Ministry shall draw its plan for prevention and mitigation (accordance with National Plan), integration of risk mitigation into developmental plans and programmes, preparedness and capacity building, financial provisions and roles and responsibilities (Section 37(i)). National Institute of Disaster Management (NIDM) is mandated under DM Act 2005 to support Government/Ministries and related agencies in developing their policies, plans, capacity building and research, etc. Reduction, Paris Climate Agreement and the SDGs, through its Agenda 1, i.e. All sectors to imbibe the principles of disaster risk management and utilizes the legal mandate under the Disaster Management Act 2005 and the National DM Policy 2009. The Disaster Management Act (Section 35) calls for (i) measures for disaster prevention through risk mitigation, mainstreaming into developmental plans and projects, preparedness and financial resources, at the level of Ministries of Central Government. It also provides for the role of Ministries on disaster prevention, mitigation and preparedness, it's integration and frame policies, rules and regulations in this context (Section 36). Further, it envisages that each

Ministry shall draw its plan for prevention and mitigation (accordance with National Plan), integration of risk mitigation into developmental plans and programmes, preparedness and capacity building, financial provisions and roles and responsibilities (Section 37(i)). National Institute of Disaster Management (NIDM) is mandated under DM Act 2005 to support Government/Ministries and related agencies in developing their policies, plans, capacity building and research, etc.

In this context, NADMP of MoAFW was drawn to address multi-hazard risk and vulnerabilities that the agriculture sector faces. This also includes recognition of climate change impact and climate-induced disasters and risk. The objectives were to identify key risks and vulnerabilities, build in sector-level top-down adaptation/DRR activities, and propose a multi-layer approach to resilience building covering all aspects of agriculture system through multi-stakeholder engagement. It includes infrastructure/establishments, resources, people, services and activities associated with the mandates of the MoAFW.

Planning

It is important that such sectoral strategies for climate/disaster resilience identifies key gaps in the knowledge and capacities of the departments, integration of DRR and CCA into developmental planning and assessing technical and functional capacity gaps and training needs of Government officials on climate change in the context of disaster management and climate change impacts. Managing disaster risk and building resilience requires the involvement of multiple agencies at different levels of governance, right from the local government, Panchayati Raj Institutions (PRIs) and communities to the state and central ministries, departments and agencies. These horizontal and vertical linkages along with the role and responsibilities of the concerned departments to deal with the various disasters are explained in the plan.

Mobilization of Funds

Plan highlights the various sources of funds within the ongoing programmes/schemes/plans for disaster management, implementation of DRR in planned schemes and proposed institutional framework for strengthening the budgetary provisions. The DM Act 2005 has mandated upon the Government to ensure that the funds are provided by the Ministries and Departments within their budgetary allocations for disaster management. In India, disaster risk-related schemes and projects are primarily funded through capital and revenue spending by the National and State Governments. Various Ministries play a key role in disaster management as far as specific disasters are concerned. For example, the Ministry of Agriculture and Farmers' Welfare is the nodal ministry for dealing with 4 types of disaster namely drought, hailstorm, pest attack and cold wave/frost and this Ministry is mandated to coordinate relief measures for the mentioned disaster. Similarly, other Ministries like Ministry of Health and Family Welfare, Ministry of Environment, Forest and Climate Change, etc., and other Departments have dedicated schemes, aimed at disaster prevention, mitigation, capacity building, etc. within their particular domain. Budgetary provisions for post-disaster reconstruction activities is normally embedded in sectoral expenditure planning of the national government. This ensures alignment of the objective of "Building Back Better" with the sectoral programmes/schemes. Beside capital and revenue spending, efforts to have flexi fund have also been made by the Government of India to mobilize the resources from external funding agencies for vulnerability assessment, capacity development, institutional strengthening of response mechanism and mitigation measures.

Multi-Stakeholder Engagement

Strengthening climate resilience through mainstreaming is a participatory process and thus multi-stakeholder engagement is an integral part. NADMP being a community document, provided input from relevant stakeholders. Involving the community promotes community ownership of the planning process. Based on the experiences from NADMP development, relevant stakeholders can be involved at various levels as shown in Table 28.1.

Table 28.1 Stakeholder involvement at various levels

Level	Examples of responsible agencies
National level	Ministry of Agriculture and Farmers' Welfare (ICAR Institutes, Central Agricultural Universities, Mahalanobis National Crop Forecasting Centre) Ministry of Jal Shakti (CWC), Ministry of Earth Sciences (India Meteorological Department), Ministry of Science and Technology (Department of Science and Technology and Council of Scientific and Industrial Research), NITI Aayog, Ministry of Home Affairs (National Disaster Management Authority, NIDM and National Disaster Response Force), etc.
State level	State Disaster Management Authority, State Agriculture Department, Department of Rural Development and Panchayati Raj, State Department of Irrigation and Flood Control, State Water Resource Department, Disaster Management Department, State Disaster Response Force, State-Level Research and Technical Institutions, State Level Skill Development Agencies, State Pollution Control Board, State Agricultural Universities, State Drought Monitoring Cell, etc.
District level	District Disaster Management Authority, District Emergency Operation Centre
Village/Block level	Panchayati Raj Institutions, Self Help Groups, Rural Local Bodies

28.4 Integrating Climate and Disaster Resilience at Sectoral Level

India launched National Action Plan on Climate Change (NAPCC) in 2008, with a focus on low-carbon pathways and climate-resilient development (Ghosh 2009). National missions under NAPCC focus on priority sectors such as agriculture, water, energy efficiency, urban habitat, forestry, health and vulnerable regions—Himalayas and coastal regions. Being implemented by the Ministry of Agriculture, the National Mission for Sustainable Agriculture is a key sectoral focused mission. The Mission provides for devising strategies to make Indian agriculture climate resilient. The mission gives special emphasis to dryland agriculture along with strategic initiatives on risk management, access to information and use of biotechnology (MoAFW 2017). National Water Mission is another sector focused Mission under NAPCC. The Mission provides for ensuring integrated water resource management in order to conserve water, improve water use efficiency and ensure more equitable distribution both across and within state (Ghosh 2009). These National strategies are further translated into actionable plans such as State Specific Water Action Plan as part of National Water Mission (NWM 2020).

As climate change is increasingly leading to increased frequency and intensity of climate-induced disasters, climate change impacts are increasingly being framed in the context of risk (NIDM-GIZ 2019). The IPCC has built on key concepts from the disaster risk management discourse and introduced the concept of climate risk in its Fifth Assessment Report (WGII AR5) Emphasizing the link of climate change mitigation, adaptation and sustainable development the IPCC AR5 risk concept serves as a valuable complement to the previously used concept of vulnerability to climate change (IPCC 2014). It broadens the perspective to climate-related impacts triggered by extreme events and slow-onset changes.

Opportunities in the Agriculture Sector

"Strategy for New India at 75', a policy document developed by NITI Aayog (public policy think tank of the Government of India) has delineated its objectives for 2022–2023 for agriculture sector (NITI Aayog 2018), with a focus on doubling farmers' income. This includes the following aspects:

(i) The National Mission on Sustainable Agriculture will be revised to boost agricultural output and help farmers achieve their goal of doubling their income by 2022–2023.
(ii) Leveraging the National Adaptation Fund for Climate Change and other global funds to improve climate change resilience in sectors such as agriculture, forestry, infrastructure and others.

(iii) Low-cost financing, particularly through the Green Climate Fund.
(iv) Establishing e-National Agriculture Markets (e-NAMs) and a unified national market.
(v) Support to "Zero Budget Natural Farming" (ZBNF) approaches, improve land quality, and boost farmer income.
(vi) Modernizing agriculture, enhancing policy and governance and boosting value chain and rural infrastructure.

28.5 Process and Significance of NADMP

NADMP aims at providing a practical guidance. It offers a roadmap for integrating key aspects of DRR into agriculture sector development including crop production, sustainable land management and post-harvest management which is within the scopes and mandate of Ministry of Agriculture and Famer's Welfare. There are 12 chapters (Table 28.2) in NADMP in accordance with National Disaster Management Plan of India (NADMP 2019).

28.6 Analysis of NADMP

According to a growing literature disaster risk reduction, climate change adaptation and sustainable development are all linked. Climate-induced disaster risks have been continuously increasing over recent decades, and extreme weather conditions are projected to intensify further in future with implications on intensity and frequency of disaster risks. This linkage has found place in agendas of major international agreements such as Sendai Framework for Disaster Risk Reduction 2015–2030, Sustainable Development Goals 2030, the Paris Climate Agreement 2015 an UNCCD (Gupta et al. 2020). It is critical to examine present and future issues caused by disasters and climate change in order to accomplish the SDGs. Accordingly, NADMP has been prepared to achieve 8 goals of SDGs—Goal 1, 2, 3, 4, 6, 9, 11 and 13 as shown in Fig. 28.2. For achieving these goals, different climate resilience parameter have been identified and found in different literatures. The same parameters have also been included in the plan as a part of CCA-DRR strategy as given in Table 28.3.

Table 28.2 Contents of NADMP, 2020

Chapter	Title	Discussion
1	Preliminaries	Provides an overview of the agriculture scenario in India, comprising of key challenges of Indian agriculture, including climate change and disaster-related specific challenges, institutional mechanisms to deal with the agriculture sector in India
2	Hazard, risks, vulnerabilities and capacity analysis (HRVCA)	Provides an overview of key elements of vulnerability which the agriculture sector faces such as exposure, sensitivities and adaptive capacity
3	Hazard specific prevention and mitigation measures	Presents the disaster prevention and mitigation measures for each hazard such as drought, flood, cyclone etc., highlighting the roles and responsibilities of agencies at the national and state level
4	Mainstreaming disaster risk reduction	Talks about mainstreaming different aspects of DRR into the plans, policies and schemes of the Ministry
5	Inclusive disaster risk reduction	Emphasizes the importance of Disaster Risk Management in context to the vulnerable groups
6	Coherence of disaster risk management across resilient development and climate change action	Gives an overview of policy framework for disaster management, climate change and Sustainable development
7	Capacity development and communication	Gives an overview of capacity development activities for Disaster Risk Reduction (DRR). It entails the capacity-building themes for disaster management, Role of NIDM in capacity building on disaster risk reduction
8	Coordination—Horizontal and vertical linkages	Presents the institutional framework along with the role and responsibilities of the concerned departments to deal with the various disasters
9	Preparedness and response	Mentions about the preparedness and response measures for various disasters. It details out the standard operating procedures (SoPs), including specific tasks, responsibilities and timeframe for preparedness and response for dealing with various disasters
10	Recovery and reconstruction	Talks about post-disaster need assessment, sustainable recovery framework and build back better approach for agriculture sector
11	Budgetary provisions	It is about financial provisions

(continued)

Table 28.2 (continued)

Chapter	Title	Discussion
12	Plan management	Mentions about plan review and updating mechanisms

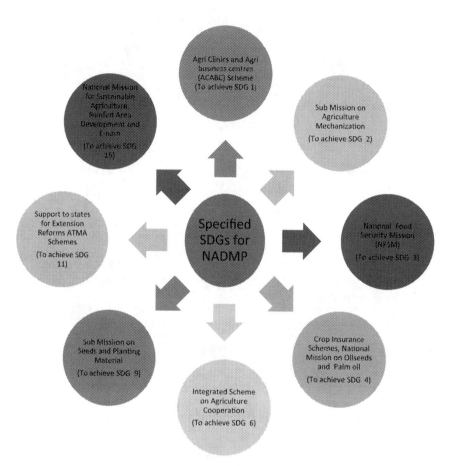

Fig. 28.2 Benefits of NADMP for achieving specific SDGs

28.7 Conclusion

Mainstreaming climate and disaster risks into planning process have multiple pathways and requires coordinated efforts. Integration through sectoral planning is once such approach. However, owing to limited capacities and understanding of climate and disaster risks and DRR and CCA options, development programmes and schemes do not currently take disaster and climate

Table 28.3 Strengthening climate resilience through Gupta et al. (2020)

Climate-resilience parameter	NADMP reference
Risk Assessment (OECD 2018; IDB 2014)	• In the capacity development and the communication chapter, risk assessment is included as a part of capacity building framework • In preparedness and response chapter also, it has been included as a part of emergency function
Adaptation measures (Douxchamps et al. 2017; Takebayashi et al. 2020)	• Providing support in risk assessment has been identified as a major responsibility for central and state agencies • Adaptation measures has been included in the mainstreaming chapter
Climate Vulnerability and capacity analysis (Douxchamps et al. 2017; CARE 2019)	• The plan talks about Hazard Risk Vulnerability and Capacity Analysis (HRVCA) • HRVCA has been identified as sub thematic area for DRR
Early warning (UNDP 2018; OECD 2018)	• Early warning is key for preparedness for any kind of disaster, it has been identified as one of the most important sub thematic area for disaster risk reduction • The same has also been included in the Mainstreaming DRR chapter so that it can be integrated into existing development policies, plans and projects
Prioritizing vulnerabilities (Gerarda 2016; Climate Bonds 2019)	• In the plan, vulnerability mapping has been included as sub thematic areas for Hazard Specific Prevention and Mitigation Measures
Structural and non-structural climate resilience measures (Arsenault et al. 2013; Climate Bonds 2019)	• In the plan, structural and non-structural measures have been identified as a thematic area according to Sendai Framework in the Hazard Specific Prevention and Mitigation Measures chapter • .Under this thematic areas, several sub thematic areas like capacity buildings, programmes and policies of government, conservation structures, etc. have been identified
Stakeholder participation (Climate Bonds 2019; Bajracharya and Hastings 2020)	• In the plan, stakeholders have been assigned roles in various disaster risk reduction and climate adaptation measures

(continued)

Table 28.3 (continued)

Climate-resilience parameter	NADMP reference
System-level climate resilience investments (Climate Bonds Initiative, Climate Resilience Consulting and World Resources Institute 2019; Douxchamps et al. 2017) Awareness generation (IFRC 2011; OECD 2018) Monitoring and evaluation (Katich 2010; OECD 2018)	• Climate resilience investment for agricultural infrastructures esp. like agricultural infrastructure such as irrigation facilities, warehousing and cold storage has been highlighted in the plan • Awareness generation has been identified as a sub thematic area for DRR • In the Recovery and Reconstruction chapter, it is included as a component for recovery framework and also as a part of recovery process

risks into account. The process of NADMP development demonstrates the different steps involved in integration of CCA-DRR into sectoral plans starting from hazard, risk and vulnerability assessment, multi-stakeholder engagement, policy and institutional support, implementation mechanism and monitoring and evaluation framework. For successful, mainstreaming DRR and CCA into development planning can be advanced through appropriate tools and methodologies along with key enablers in place such as institutional and policy support, scientific knowledge and expertise, community participation, capacity building and knowledge networking and monitoring and evaluation. This creates significant opportunities to improve sectoral planning by strengthening climate and disaster resilience such in agriculture sector through NADMP. At the national level, India's commitment to Hyogo Framework for Action, 2005–2015; Sendai Framework 2015–2030, and policy documents like National Action Plan on Climate Change (NAPCC 2008) and State Action Plans of climate change (SAPCC) and state-specific disaster management plans at the state level provide such institutional and policy support for CCA and DRR integration. Furthermore, an array of sectoral initiatives related to water, health, forestry, rural development, etc. provides potential for climate and disaster resilience. However, a lack of effective coordination, both horizontally and vertically limits overall outcome especially on integrating DRR and CCA concerns. Where national governments see adaptation mainstreaming as an appropriate strategy it can be practically achieved through setting the process proposed in this paper in motion. Considering this, NADMP, 2020 also enlist all the roles and responsibilities at central and state level so that the CCA and DRR integration can be practised at ground level. The same has also been highlighted in this article.

REFERENCES

Aleksandrova, M. (2019). Principles and Considerations for Mainstreaming Climate Change Risk into National Social Protection Frameworks in Developing Countries. Climate and Development. https://doi.org/10.1080/17565529.2019.1642180.

Allen, M.R., Dube, O.P., Solecki, W., Aragón-Durand, F., Cramer, W., Humphreys, S., Kainuma, M., Kala, J., Mahowald, N., Mulugetta, Y., Perez, R., Wairiu, M., and Zickfeld, K. (2018). Framing and Context in: Global Warming of 1.5°C. An IPCC Special Report on the Impacts of Global Warming of 1.5°C Above Pre-Industrial Levels and Related Global Greenhouse Gas Emission Pathways, in the Context of Strengthening the Global Response to the Threat of Climate Change, Sustainable Development, and Efforts to Eradicate Poverty. [Masson-Delmotte, V., P. Zhai, H.-O. Pörtner, D. Roberts, J. Skea, P.R. Shukla, A. Pirani, W. Moufouma-Okia, C. Péan, R. Pidcock, S. Connors, J.B.R. Matthews, Y. Chen, X. Zhou, M.I. Gomis, E. Lonnoy, T. Maycock, M. Tignor, and T. Waterfield (eds.)].

Arsenault, R., Brissette, F., and Malo, J.S. (2013). Structural and Non-Structural Climate Change Adaptation Strategies for the Péribonka Water Resource System. Water Resource Management, 27: 2075–2087. https://doi.org/10.1007/s11269-013-0275-6.

Bajracharya, B., and Hastings, P. (2020). Stakeholder Engagement for Disaster Management in Master-Planned Communities. Australian Journal of Emergency Management, 35(3): 41–47.

C2ES. (2021). Climate Resilience Portal. Centre for Climate and Energy Solutions. https://www.c2es.org/content/climate-resilienceoverview/#:~:text=Improving%20climate%20resilience%20involves%20assessing,change%20will%20continue%20to%20accelerate. Last assessed on 9 May 2021.

CARE. (2019). Informing Community-Based Adaptation, Resilience and Gender Equality. A Handbook on Informing Community-Based Adaptation, Resilience and Gender Equality, Version 2. https://careclimatechange.org/wp-content/uploads/2016/06/CARE-CVCA-Handbook-EN-v0.8-web.pdf. Last assessed on 5 May 2021.

Climate Bonds Initiative, Climate Resilience Consulting (CRC) and World Resources Institute (WRI). (2019). Climate Resilience Principles: A Framework for Assessing Climate Resilience Investments. https://www.climatebonds.net/files/page/files/climate-resilience-principles-climate-bonds-initiative-20190917-.pdf. Last Assessed on 7 May 2021.

Climate Bonds. (2019). Climate Resilience Principles Climate Bonds Initiative. https://www.climatebonds.net/files/files/climate-resilience-principles-climate-bonds-initiative-20190917.pdf. Last assessed on 27 September 2020.

Cohen, B., Blanco, H., Dubash, N.K. Dukkipati, S., Khosla, R., Scrieciu, S., Stewart, T., and Gunfaus, M.T. (2018). Multi-Criteria Decision Analysis in Policy-Making for Climate Mitigation and Development. Climate and Development. https://doi.org/10.1080/17565529.2018.1445612.

Collins, A.E. (2018). Advancing the Disaster and Development Paradigm. International Journal of Disaster Risk Science, 9, 486–495. https://doi.org/10.1007/s13753-018-0206-5.

Denton, F., Wilbanks, T.J., Abeysinghe, A.C., Burton, I., Gao, Q., Lemos, M.C., Masui, T., O'Brien, K.L., and Warner, K. (2014). Climate-Resilient Pathways: Adaptation, Mitigation, and Sustainable Development. In: Climate Change 2014:

Impacts, Adaptation, and Vulnerability. Part A: Global and Sectoral Aspects. Contribution of Working Group II to the Fifth Assessment Report of the Intergovernmental Panel on Climate Change [Field, C.B., V.R. Barros, D.J. Dokken, K.J. Mach, M.D. Mastrandrea, T.E. Bilir, M. Chatterjee, K.L. Ebi, Y.O. Estrada, R.C. Genova, B. Girma, E.S. Kissel, A.N. Levy, S. MacCracken, P.R. Mastrandrea, and L.L.White (eds.)]. Cambridge University Press, Cambridge, United Kingdom and New York, NY, USA, pp. 1101–1131.

Douxchamps, S., Debevec, D., Giordano, M., and Barron, J. (2017). Monitoring and Evaluation of Climate Resilience for Agricultural Development—A Review of Currently Available Tools. World Development Perspectives, 5: 10–23. https://doi.org/10.1016/j.wdp.2017.02.001.

Garschagen, M., Doshi, D., Moure, M., James, H., and Shekhar, H. (2021). The Consideration of Future Risk Trends in National Adaptation Planning: Conceptual Gaps and Empirical Lessons. Climate Risk Management, 100357, ISSN 2212-0963. https://doi.org/10.1016/j.crm.2021.100357.

Gerarda, M.S. (2016). Resiliency Planning: Prioritizing the Vulnerability of Coastal Bridges to Flooding and Scour. Procedia Engineering, 5, 340–347. ISSN: 1877-7058. https://doi.org/10.1016/j.proeng.2016.04.086.

Ghosh, P. (2009). National Action Plan on Climate Change. Prime Minister's Council on Climate Change. http://moef.gov.in/wp-content/uploads/2018/07/CC_ghosh.pdf. Last assessed on 16 July 2021.

Gupta, A.K., Chopde, S., Singh, S., Wajih, S.A., and Katyal, S. (2016). Prime Minister's Agenda 10: India's Disaster Risk Management Roadmap for Sustainable Development. Prevention Web, UNDRR. https://www.preventionweb.net/files/51313_51304pmagenda10paper.pdf. Last assessed on 15 June 2021.

Gupta, A.K., Hodam, S., Chary, G.R., Prabhakar, M., Sehgal, V.K., Srivastava, R., Swati, S., and Bhardwaj, S. (2019). Roadmap of Resilient Agriculture in India. Thematic Paper Released on International Symposium on Disaster Resilience and Green Growth for Sustainable Development Organized by Centre for Excellence on Climate Change, NIDM, New Delhi (India), 26–27 September 2019.

Gupta, A.K., Srivastava, R., Hodam, S., Chary, G.R., Sehgal, V.K., Pathak, H., Krishnan, P., Ray, S.S., Singh, K.K., Attri, S.D., and Srivastava, A.K. (2020). National Agriculture Disaster Management Plan. Department of Agriculture Cooperation and Farmers' Welfare, Ministry of Agriculture and Farmers' Welfare, Government of India, New Delhi.

Hoegh-Guldberg, O., Jacob, D., Taylor, M., Bindi, M., Brown, S., Camilloni, I., Diedhiou, A., Djalante, R., Ebi, K.L., Engelbrecht, F., Guiot, J., Hijioka, Y., Mehrotra, S., Payne, A., Seneviratne, S.I., Thomas, A., Warren, R., and Zhou G. (2018). Impacts of 1.5°C Global Warming on Natural and Human Systems. In: Global Warming of 1.5°C. An IPCC Special Report on the Impacts of Global Warming of 1.5°C Above Pre-Industrial Levels and Related Global Greenhouse Gas Emission Pathways, in the Context of Strengthening the Global Response to the Threat of Climate Change, Sustainable Development, and Efforts to Eradicate Poverty [Masson-Delmotte, V., P. Zhai, H.-O. Pörtner, D. Roberts, J. Skea, P.R. Shukla, A. Pirani, W. Moufouma-Okia, C. Péan, R. Pidcock, S. Connors, J.B.R. Matthews, Y. Chen, X. Zhou, M.I. Gomis, E. Lonnoy, T. Maycock, M. Tignor, and T. Waterfield (eds.)]. https://archivepmo.nic.in/drmanmohansingh/climate_change_english.pdf. Last assessed on 29 September 2020.

IDB. (2014). Climate Change Data and Risk Assessment Methodologies for the Caribbean. Technical Note of Inter-American Development Bank Environmental Safeguards Unit, No. IDB-TN-633.

IFRC. (2011). Public Awareness and Public Education for Disaster Risk Reduction: A Guide. International Federation of Red Cross and Red Crescent Societies, Geneva.

IPCC. (2014). Climate Change 2014: Impacts, Adaptation, and Vulnerability. Part B: Regional Aspects. Contribution of Working Group II to the Fifth Assessment Report of the Intergovernmental Panel on Climate Change [Barros, V.R., C.B. Field, D.J. Dokken, M.D. Mastrandrea, K.J. Mach, T.E. Bilir, M. Chatterjee, K.L. Ebi, Y.O. Estrada, R.C. Genova, B. Girma, E.S. Kissel, A.N. Levy, S. MacCracken, P.R. Mastrandrea, and L.L. White (eds.)]. Cambridge University Press, Cambridge, United Kingdom and New York, NY, USA, 688 pp.

Katich, K. (2010). Monitoring and Evaluation in Disaster Risk Management. EAP DRM Knowledge Notes; No. 21. World Bank, Washington, DC. © World Bank. https://openknowledge.worldbank.org/handle/10986/10119. License: CC BY 3.0 IGO.

Mall, R.K., Srivastava, R.K., and Banerjee, T. (2019). Disaster Risk Reduction Including Climate Change Adaptation Over South Asia: Challenges and Ways Forward. International Journal of Disaster Risk Science, 10, 14–27. https://doi.org/10.1007/s13753-018-0210-9.

Mahfoud, C., and Adjizian-Gerard, J. (2021). Local Adaptive Capacity to Climate Change in Mountainous Agricultural Areas in the Eastern Mediterranean (Lebanon), Climate Risk Management, 33, 100345; ISSN 2212-0963. https://doi.org/10.1016/j.crm.2021.100345.

MoAFW. (2017). Annual Report 2016–17 of the Ministry of Agriculture and Farmers' Wefare, Government of India.

Moreno, J.C.V, Ponte, E., Emperador, S., and Noriega, M.O. (2017). An Effective Approach to Mainstreaming DRR and Resilience in La Paz, Mexico. In M. Tiepolo et al. (eds.), Renewing Local Planning to Face Climate Change in the Tropics, Green Energy and Technology. https://doi.org/10.1007/978-3-319-59096-7_14.

NAPCC. (2008). National Action Plan on Climate Change. Prime Minister's Council on Climate Change, Government of India.

National Disaster Management Plan. (2019). A Publication of the National Disaster Management Authority, Government of India. November 2019, New Delhi.

National Water Mission. State Specific Action Plan. http://nwm.gov.in/?q=state-specific-action-plan. Last assessed on 29 September 2020.

NDMA. (2009). National Policy on Disaster Management. National Disaster Management Authority, Ministry of Home Affairs, Government of India. https://ndma.gov.in/sites/default/files/PDF/national-dm-policy2009.pdf. Last assessed on 5 July 2021.

NIDM-GIZ. (2019). Climate Risk Management Framework for India Addressing Loss and Damage; ISBN No. 978-93-82571-25-4

NITI Aayog. (2018). SDG India Index—Baseline Report 2018. Published by NITI Aayog, Government of India.

OECD. (2009). Policy Guidance on Integrating Climate Change Adaptation into Development Co-operation; ISBN No. 978-92- 64-05476-9

OECD. (2018). Climate-Resilient Infrastructure. http://www.oecd.org/environment/cc/policy-perspectives-climate-resilient-infrastructure.pdf. Last assessed on 27 September 2020.

Oliveira, C.R., Oi, C.A., and do Nascimento, M.M.C. (2015). The Origin and Evolution of Queen and Fertility Signals in Corbiculate Bees. BMC Evolutionary Biology, 15, 254. https://doi.org/10.1186/s12862-015-0509-8.

PAGE. (2016). Integrated Planning & Sustainable Development: Challenges and Opportunities State of Indian Agriculture 2015–16: Ministry of Agriculture & Farmers Welfare Department of Agriculture, Cooperation & Farmers Welfare.

Takebayashi, H., Misaka, I., and Akagawa, H. (2020). Chapter—Adaptation Measures and Their Performance. In Hideki Takebayashi and Masakazu Moriyama (eds.), Adaptation Measures for Urban Heat Islands, Academic Press, pp. 9–37; ISBN 9780128176245. https://doi.org/10.1016/B978-0-12-817624-5.00002-6.

UNDP. (2018). Five Approaches to Build Functional Early Warning Systems. United Nations Development Programme (UNDP).

UNFCCC. (2021). What Do Adaptation to Climate Change and Climate Resilience Mean? UNFCCC. https://unfccc.int/topics/adaptation-and-resilience/the-big-picture/what-do-adaptation-to-climate-change-and-climate-resilience-mean. Accessed on 5 June 2021.

UN-SDG. (2016). The Sustainable Development Goals Report of 2016. Published by United Nations. New York.

Warren-Myers, G., Hurlimann, A., and Bush, J. (2021). Climate Change Frontrunners in the Australian Property Sector, Climate Risk Management, Volume 33, 100340, ISSN 2212-0963. https://doi.org/10.1016/j.crm.2021.100340.

Weekes, C., and Bello, O.D. (2019). Mainstreaming Disaster Risk Management Strategies in Development Instruments (II). Policy Briefs for Barbados, Guyana, Saint Lucia, Suriname, and Trinidad and Tobago, Eclacsub Regional Headquarters for the Caribbean. ISSN: 1728-5445.

World Economic and Social Survey. (2016). Climate Change Resilience: An Opportunity for Reducing Inequalities. Published by United Nations, New York; ISBN 978-92-1-109174-8

CHAPTER 29

Scalable Adaptation Model for Sustainable Agriculture Livelihoods Under Changing Climate: A Case Study from Bihar and Madhya Pradesh

Ravindra S. Gavali, V. Suresh Babu, Krishna Reddy Kakumanu, Shrikant V. Mukate, Y. D. Imran Khan, Basavaraj Patil, Utkarsh Ghate, and V. Srinivasa Rao

29.1 Introduction

The Intergovernmental Panel on Climate Change (IPCC) is the United Nations body for assessing the science related to climate change defines climate change as "the state of the climate that can be identified (e.g. using statistical tests) by changes in the mean and/or the variability of its properties, and that persists for an extended period, typically decades or longer (IPCC 2011_factsheet). However, UNFCC states that 'Climate change' means a change of climate which is attributed directly or indirectly to human activity that alters the composition of the global atmosphere and which is in addition to natural climate variability observed over comparable periods" (UNFCC 1992). In the last few decades, various observed phenomena such as rainfall, drought

---

R. S. Gavali · K. R. Kakumanu · S. V. Mukate · Y. D. Imran Khan (✉) · B. Patil · U. Ghate · V. S. Rao
Centre for Natural Resource Management, Climate Change and Disaster Management, National Institute of Rural Development and Panchayati Raj, Hyderabad, India
e-mail: ikeducationkhan08@gmail.com

V. S. Babu
North Eastern Regional Centre of Natural Resource Management, National Institute of Rural Development and Panchayati Raj, Guwahati, India

© The Author(s), under exclusive license to Springer Nature Singapore Pte Ltd. 2023
S. Nautiyal et al. (eds.), *Palgrave Handbook of Socio-ecological Resilience in the Face of Climate Change*,
https://doi.org/10.1007/978-981-99-2206-2_29

and cyclones have frequently depicted climate variability throughout India. In 2019 few states of India like Bihar, Maharashtra, and Kerala, were facing drought-like conditions in one region and flood-prone situations in another. This climate variability must be handled with integrative efforts affecting the community's livelihoods.

United Nations Framework on Convention on Climate change suggests that changes in the physical environment from climate change have significant deleterious effects on natural and manmade communities. It significantly affects the composition, resilience, or productivity of natural and managed ecosystems and the communities' socio-economic status, health, and welfare (UNFCC 1992). It is said that climate change impacts are mostly affecting the resource-poor people, which includes landless communities living in rural areas, small and marginal farmers, tribal communities, and women. Because the available resources are not sufficient to reduce or mitigate or adapt to the altered climatic situations. The farmer communities, especially small and marginal communities especially depend on rainfall to cater to the demand for water for irrigation as they don't have the capital to invest in pipelines or bore well. The livestock rearing by poor farmers doesn't have proper shelters and awareness about feed management and vaccination. Climate change is the result of many actions, erratic rainfall, increased temperature, climatic variability. These factors are hitting hard to the multiple fields including agriculture, water, fishery, livestock, and results into the weakened socio-economic structure of the community. Mitigation efforts of climate change may take a longer time to reverse it, so adaptation is the only choice to save ourselves from the negative impacts of climate change.

Adaptation is the adjustment process to actual or expected climate and its effects (IPCC 2014), which majorly focuses on reducing or eliminating the negative impacts on human beings, livestock, and agriculture, or to take advantage (Noble et al. 2014) through planned or anticipated actions (Mimura et al. 2014; Preston et al. 2013). Many researchers have studied the adaptation capacities of vulnerable communities from different sectors and regions (Mertz et al. 2009; Younus 2010; Pandey et al. 2011; Spires et al. 2014; Archer et al. 2014; Dany et al. 2015; Hammouri et al. 2015; Sud et al. 2015; Upgupta et al. 2015). Climate change is very dynamic and anticipated adaptation plans lag behind the actual consequences in many developed nations (Preston et al. 2010) which are full of resources and developed adaptation capacities (Azhoni et al. 2018). So, in developing nations like India this is a bigger challenge as it has multiple climatic zones, huge population, and limited resources.

Figure 29.1 illustrates that climate change is the result of human interaction with the environment through the overconsumption of resources and a changed or altered climate is responsible for the loss of life, income, and resources. To avoid such losses the only promising way for the present is to adapt the situation through multiple interventions at multilevel in multiscale. So, planning of interventions at a community level will define the scale of adaptation, and adaptation results will define the reduced or mitigated losses

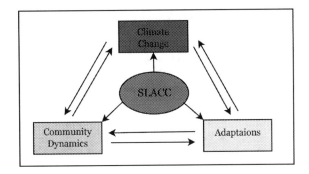

Fig. 29.1 The inter-connectedness of community, climate change, and adaptations and their linkage with the SLACC project

of the community. But one thing must have kept in mind is that adaptations depend on resources available at a community level and the kinds of impacts of climate change. Impact-specific adaptive interventions only help to combat the negative effects of climate change. That is why two arrows are indicating the interdependence among each component of Fig. 29.1. The different components of climate change, community dynamics, and adaptations with their activity and impacts are represented in Table 29.1. SLACC project was initiated with due consideration of all these factors to reduce the impacts of climate change or rural are resource-poor communities of Indian state Bihar and Madhya Pradesh. SLACC project understood this multi-scenario and multi-impact challenge to maintain the livelihoods of rural population.

29.2 About SLACC Project

The SLACC project was funded by the Global Environment Facility (onwards GEF)-administered Special Climate Change Fund being initially implemented in 200 villages of Madhya Pradesh & Bihar, specifically addressing the large-scale proof-of-concept on integrating community-based climate adaptation planning and implementation into livelihood support activities of the Deen Dayal Antodaya Yojna—National Rural Livelihood Mission (onwards DAY-NRLM)/National Rural Livelihood Program (onwards NRLP), Mahila Kisan Sashaktiran Pariyojana (onwards MKSP) and Mahatma Gandhi National Rural Employment Guarantee Scheme (onwards MGNREGS). SLACC addresses the aspects of farm-based livelihoods that may be affected by climate change by helping the community chosen and tested interventions.

Implementation of the SLACC program through DAY-NRLM is because of the strong overlap with the World Bank-supported National Rural Livelihood Project, GOI-supported DAY-NRLM projects, and the State Livelihood Projects. The overlap is primarily in terms of geography of the area, targeted beneficiary groups, strong support from SRLM and well-developed community institutional groups for the anchoring of SLACC interventions.

Table 29.1 Area-wise components and activities to identify the inter-relationship for the effective implementation of program

Area	Impacts of climate change	Components	Activities (SLACC)
Climate change	• Erratic Rainfall • Reduced Rainfall • Increased Temperature • Cyclones and Typhoons	• Climate Change • Climate change Vulnerability	• Weather-Based Agro Advisories • Climate Change Information
Community dynamics	• Loss of Income • Loss of Life • Unaware about climate Change • Poor Knowledge of adaptations	• Socio-economic Stratification • Attitudes and Behavior • Community participation	• Stakeholder Identification • Resource Assessment • Climate Change impact area assessment
Adaptations	• Increased Sustainable Income • Reduced losses • Increased adaptation capability • Increased capacity building • Increased awareness	• Institutional Arrangements • Stakeholder Management • Indigenous Interventions • Non-Indigenous Interventions	• Community developed interventions • Technology-Based Interventions • Financial and Ecological resource management • Convergence programs

Bihar and Madhya Pradesh state units have been implementing the World Bank-supported livelihood projects for more than 5 years. SLACC is complementary to the larger National Rural Livelihoods Project as well as the Bank-supported rural livelihood projects in Bihar and MP, and relies on the same State Livelihood Mission for implementation, they not only have state project-specific social and tribal inclusion plans but also come under the SMF of the NRLP. SMF reinforces the focus on poorest households, and recommends specific interventions for the inclusion of the tribal and non-tribal beneficiaries in community institutions (SHGs, Federations), as well as livelihood financing and promotion interventions. The overall emphasis is on integrating the key SMF interventions in the planning and implementation of the Community Climate Adaptation Plans which will be implemented by field implementation teams/resource agencies. The SMF will rely more on the institutional arrangement created under SLACC for its implementation. Hence, both the state units have good operational familiarity with the core principles and requirements of these frameworks. SMF and the Bihar and Madhya Pradesh social inclusion and tribal development strategies, which are most relevant to the scale and focus of SLACC interventions.

Figure 29.2 depicts that Special Climate Change Funds (SCCF) has been utilized by Global Environment Facility (GEF) through World Bank (WB) and Ministry of Rural Development (MORD) and monitored by National Rural Livelihood Mission (NRLM) in Indian flood-prone areas indicated in blue color (Madhubani and Mandla) and drought-prone areas in red color (Gaya and Sheopur) districts of Bihar and Madhya Pradesh state.

The SLACC project was implemented in 200 villages of about 16 blocks of Bihar and Madhya Pradesh (Fig. 29.3). The states of Bihar and Madhya Pradesh have been identified for implementation of the project based on the readiness of the SMMU, existing capacity and experience in sustainable agriculture, the agro-ecological profile of the state, and anticipated climate change risks. The 200 villages are spread across 8 blocks (sub-district administrative units).

Bihar is 12th largest state of India in terms of size (94,163 sq. km) and 3rd largest by population, 10.38 crores as per Census 2011. It is also one of the poorest and most populous states in India. There are 13.05 and 0.75 million of SC and ST population respectively comprising 15.72 and 0.91 percent of the total population in Bihar. The State has 29 Scheduled Tribes and 94.6 percent of them reside in villages. Bihar is a part of the Gangetic plains and hence

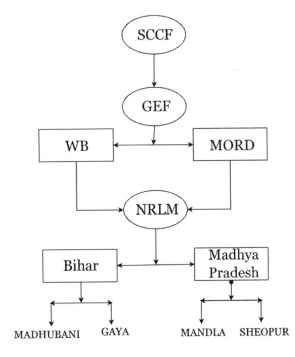

Fig. 29.2 Stakeholder of the SLACC project

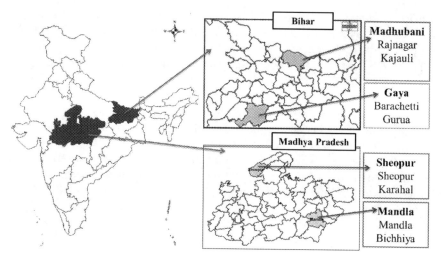

Fig. 29.3 Location map of the SLACC implementation areas

has rich soil and abundant water resources. Agriculture plays a key role in the socio-economic development of the state. Around 90 percent of the population is rural with high dependence on agriculture livelihood. Rural poverty in Bihar is characterized by high levels of landlessness and land fragmentation, high dependency on agriculture, and high levels of seasonal migration. Bihar is also India's most flood-prone state and 76 percent of the population in North Bihar lives under the recurring threat of flood devastation. In 2013 alone floods affected more than 5.9 million people in 37,678 villages in 20 districts in the state. Floods are an important reason for low crop productivity in these areas. Madhya Pradesh comprises 50 districts with a population of 72.6 million (2011 census). Seventy five percent of the total population of the state is rural and is mainly engaged in climate-sensitive activities, namely agriculture and forest-dependent livelihoods. The incidence of poverty in the state is among the highest in the country, with more than 40 percent of people living below the poverty line.

Madhya Pradesh has a large population of rural and tribal people, with high dependence on agriculture, forests, and fishery. The major land use in Madhya Pradesh is agriculture (49.5 percent of geographic area) followed by forest (28.25 percent). Agriculture plays an important role in the livelihoods of the rural poor with nearly 71 percent of the state's population is dependent on it. With two-thirds of the gross cropped area being rainfed, agriculture is vulnerable to climate variability and change. The likely impacts of climate change on the state are: Gradual increase in temperatures and erratic/uneven rainfall, increased intensity of droughts, shifts in cropping and decline in productivity, increased soil erosion, and depletion in the ground water table. Of Madhya Pradesh's 45 districts (2001 census), 14 are categorized as having 'very high' vulnerability of agriculture to climate change, while 16 are categorized as having 'high' vulnerability based on—sensitivity, exposure, and adaptive capacity.

SLACC project was divided into three components along with activities supposed to do and expected outputs and outcomes as depicted in Table 29.2. The component 1 is dedicated to formulate the Community-based Climate Change Adaptations practices which are locale-specific, focus on climate risk management and involve interventions both at the household level and/or community level. Component 2 focuses to build core operational capacity and relevant knowledge base/networks for broader scaling and mainstreaming of climate adaptation interventions. Component 3 is about the Project Management and Impact Evaluation: Funds for implementation of climate adaptation interventions are availed by the SLACC project, NRLP's Community Investment Support, as well as through convergence with other Government programs (such as MKSP, MGNREGS).

29.3 Institutional and Stakeholders Management/Arrangements

Stakeholder Identification

The SLACC project aims to improve adaptive capacity of the rural poor, to climate variability and change and secure and sustain the livelihoods of the poor through community-based interventions on agriculture, land and water, fodder, livestock, fisheries, and other financial and institutional measures. The stakeholders include the institutions and people representing the rural poor, in which majority of the population is directly dependent on climate-sensitive sectors such as agriculture, livestock, and fisheries with limited adaptive capacities. The project specifically targeted to reach out the tribal and other small and marginal farmers women farmers and community leaders which are disadvantaged from social groups. The inclusion of people from Self-Help Groups, Common interest/producer groups such as farmers' groups, livestock rearing groups, and electives such as producer companies is done to ensure its outreach at the largest.

Importance of gender equality is well recognized while the implementation of SLACC project. The inclusion of women farmers especially from poor tribal and scheduled caste is taken on priority due to the higher vulnerability to climate change and related livelihoods impact as they have limited adaptive capacity as compared to the other communities. Women farmers selected as the primary beneficiaries, as well as primary leaders and drivers for assessing, planning, selecting, and implementing SLACC interventions through the women-led village organizations and federations to address the distinct vulnerabilities and capacity needs of women.

Table 29.2 Components and activities of SLACC project

Components	Theme	Activity	Output	Outcomes
1	Community-based climate change Adaptations	• Mobilization and capacity building of community institutions on climate change activities • Community-led adaptation assessment, participatory planning, and implementation of climate adaptation interventions; • Financing community adaptation grants to poor rural households (SHGs/Federations) upon approval of a community adaptation plan; and • Implementation and hand-holding support to community institutions through local resource agencies	• Community utilization of climate financing mechanism for adaptation interventions in 200 community institutions • Community-based climate adaptation measures are implemented by at least 200 community institutions supported by Community Resource Persons; • Enhanced community capacity for planning and implementing climate adaptation plans in 200 community institutions	• Strengthened and diversified livelihoods and sources of income in poor community institutions; • Strengthened community capacities for systematically assessing climate risks and • Planning adaptation interventions for livelihood sustainability

Components	Theme	Activity	Output	Outcomes
2	Scaling and Mainstreaming Community Based Climate Adaptation	• Capacity building of NRLP staff and creation of a cadre of CRPs; • Facilitation of knowledge dissemination on climate adaptation, including policy inputs for scaling-up of the community-based climate adaptation approach within the NRLM • Additional support to National Livelihood Resource Organization (NLRO)	• 200 district and sub-district staff of NRLM trained on climate adaptation • A cadre of 400 trained CRPs • Differentiated IEC and knowledge products on climate adaptation (community adaptation planning tool and manual, CRP training curriculum, web-based inventory of climate adaptation actions, audio visuals) • A website of a consortium of resource organizations on climate adaptation • Seminars for sharing insights/lessons for policy making with Government, donor and NGO • Guidelines on climate change adaptation developed for national livelihoods implementation framework • Policy briefs on themes relevant to climate adaptation and rural livelihoods	• Strengthened operational and adaptive capacity of national and state officials and representatives for integrating climate adaptation into livelihood support activities • Enhanced access to technical information and expertise on climate adaptation and livelihoods • Evidence of climate change mainstreaming into national and state livelihood program frameworks

(continued)

Table 29.2 (continued)

Components	Theme	Activity	Output	Outcomes
3	Project Management and Impact Evaluation	• Establishment of climate adaptation units staffed with full-time professionals within the NMMU and the SMMUs of the participating states • Appointment of state-level implementation teams for providing field implementation support to CRPs and community institutions • Establishment of a monitoring system and evaluation arrangements (baseline, mid-term, and end-of-term)	• Climate adaptation units in NMMU and SMMU • Delivery of services by state-level implementation teams as per agreed Terms of Reference • Evaluation reports (baseline, mid-term, and end-of-term)	• Efficient and effective working of the project as per the proposal

Central Nodal Agencies (CNA)

The National Rural Livelihoods Promotion Society (NRLPS) under MoRD was appointed as the responsible agency for management, coordination, supervision, guidance, and technical support to the state Rural Livelihood Missions of Bihar and Madhya Pradesh for the smooth implementation of SLACC project.

Lead Technical Support Agency (LTSA)

Initially, Lead Technical Support Agencies (LTSA), was Watershed Organization Trust (WOTR) but in later stage National Institute for Rural Development & Panchayati Raj (NIRDPR). National Institute of Rural Development and Panchayati Raj was engaged as a lead technical support agency (LTSA), and fulfilled the following deliverables (i) development of planning and knowledge tools (ii) technical support to states; (iii) training and capacity building and (iv) policy inputs, documentation and sharing lessons. As per the project guidelines climate adaptation experts, having experience in the field of gender, social inclusion, and tribal development was recruited to support the implementation of the SMF.

National Mission Management Unit (NMMU)

The National Social Inclusion expert of the NMMU provided leadership, guidance, and technical support to the SRLMs and state SLACC teams on the implementation of the SMF. They also provided guidance and supported the National and State Climate Adaptation Experts (from the LTSA) to develop knowledge, capacity building, and operational guidance materials and training modules on the SMF and ensure their roll out in the states. They were also responsible for the supervision and mid-term review of SMF implementation.

State Nodal Agency (SNA)

The Madhya Pradesh Rajya Ajeevika Forum (MPRAF) and Bihar Rural Livelihoods Society (BRLPS) appointed as the state nodal agencies to implement SLACC project in Madhya Pradesh and Bihar respectively. The State Climate Adaptation Coordinator assigned for each state to ensure the SMF implementation in project villages, coordination among different thematic state mission management unit (SMMU) teams (social inclusion, gender, tribal development, community institution building, IEC, and knowledge management).

Community Resource Person (CRP)

Project implemented at community level through institutions supported by the NRLM, such as primary federations of women's self-help groups, common interest/producer groups, and producer companies. In Resource Villages the active members of SHGs/Producer Groups developed as Community Resource Persons (Climate-Smart CRPs). These Climate-Smart CRPs encouraged for climate adaptation planning and to provide hand-holding support for the implementation of project actions.

Stakeholder Consultations and Disclosure

A consultation workshop was also held on 17 September 2013 with more than 25 leading government organizations and NGOs that are working on community-based climate adaptation. Extensive consultations and in-depth discussions with the officials from National Rural Livelihoods Promotion Society (NRLPS), MoRD and the State Rural Livelihoods Missions (SRLM) of Madhya Pradesh and Bihar, including the State Mission Directors, the National Mission Director, and the core team (social inclusion, community mobilization, institution building, agriculture and livelihoods). Consultations were also held with officials and NGO partners of the MoRD-supported Women Farmers Empowerment Program (MKSP). Project preparation has been informed by consultations, and field visits to several areas where climate adaptation and farm-based interventions are ongoing, including Bihar, Madhya Pradesh, and Rajasthan, including several tribal areas.

29.4 SLACC INTERVENTIONS IMPLEMENTATION

SLACC project implemented in 793 villages and 30,720 households of Bihar and Madhya Pradesh to adapt the climate smart localized interventions to minimize the impacts of climate change on livelihoods (Table 29.1) (Kakumanu et al. 2020). The number of villages and farmers were selected into three different phases as per the sanctioning of budget from the year of 2016 to 2019. In the year of 2016, 100 villages from Bihar and 50 villages from Madhya Pradesh were selected as resource villages to initiate the interventions implementation. In phase 2, only 50 villages from Madhya Pradesh were selected in the year of 2017. In the last phase of the project both the states have scaled up villages, Bihar scaled up 283 villages and MP scaled up it in 285 villages in the year on 2018 and 2019 respectively. About 15,320 households in Bihar and 15,400 households in Madhya Pradesh benefitted from the SLACC interventions. The interventions developed for implementation are mainly focusing the sectoral activities like agriculture and livelihoods of rural poor farmer. The SLACC interventions are described as below.

Interventions were divided into four components such as Production, Technology, Finance, and Ecology (see Fig. 29.4). As these components are the backbones of the agricultural community to strengthen them to develop resilience against the changing climate. SLACC project stakeholders identified 40 sustainable interventions that can be implemented in Bihar and Madhya Pradesh. These components are interlinked to support sustainable livelihoods through multiple interventions. The mutual dependency of these components has been utilized synergistically to multiply the returns. The identified interventions were implemented at the community and household level through the flagship programs like SLACC and other (NRLM NRLP, MKSP, MGNREGS) funding agencies through the convergence schemes at the state level.

Agriculture is the main source of food, income, and employment for rural populations and also the source of water pollution. Therefore, the majority of the interventions were focused on the improvement in agriculture and livelihoods incomes of the rural poor to alleviate the poverty. The objectives of interventions implementation were to strengthen and stimulate the linkages and information sharing among farmers on climate-resilient high yielding, drought and disease tolerant varieties, crop diversification techniques, and livestock rearing and other alternative livelihoods activities for additional/alternate sources of livelihoods. While the preparation of intervention packages community participation was assured in actions. The CRPs were trained with a set of skills for its outreach in communities through the exposure visits and working knowledge regarding the interventions to be implemented.

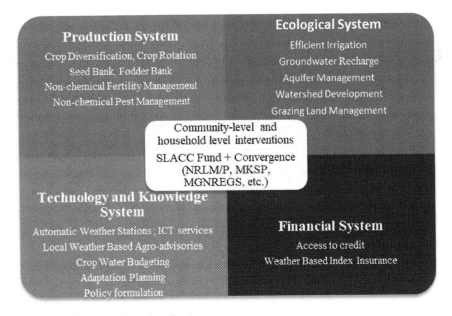

Fig. 29.4 Interventions classification system

Production System

Under the interventions related to production integrated farming system, livestock, fodder crop development, poultry, small ruminants, and alternative livelihoods are being studied.

In the production category, the high-yielding crop varieties have been in place as a means to increase the yields. However, these varieties were susceptible to the changing temperature and water logging levels to cope-up the drought and flood resilience. The strategy on the production front therefore was the introduction of climate-resilient seed varieties replacing the traditional ones having short duration reducing crop period to better fit into the changing climate pattern in the kharif. Simultaneously for the Rabi season heat tolerant wheat and pulse varieties were introduced as a part of climate adaptation strategy.

Production of conventional puddled transplanted rice is facing severe constraints because of water and labor scarcity and climatic changes. Direct-Seeded Rice (DSR) and System of Rice Intensification (SRI) are feasible alternative to conventional puddled transplanted rice with good potential to save water, reduce labor requirement, mitigate greenhouse gas (GHG) emission, and adapt to climatic risks (Kakumanu et al. 2018; Palanisami et al. 2014).

Beside these major interventions, other interventions like soil testing, introduction of climate-resilient verities, zero tillage, intercropping, seed treatment, preparation of organic formulations, micro irrigation, and alternate livelihood practices (FAO 2013; CCAFS 2018), which can enhance the resilience to climate change are demonstrated in the CRPs fields of the adopted villages.

Ecological System

The ecology component of interventions mainly focuses on the ecosystem approach to balance trade-offs among human and nature. For this purpose, several interventions established such as bio fertilizer and pesticides, Integrated Nutrient Management (INM), Integrated Pest Management (IPM), Vermi compost, Azolla, and Green Manuring practices among the SLACC communities.

Since, healthy soil is the foundation for profitable, productive, and environmentally sound agriculture system, soil test is the fundamental step to measure the nutrient composition of the soil. So that the farmers will take necessary measures for getting better yield from the crops through the management of soil health based on the soil health card. Under ecological interventions, the implementation of the Soil health Management practices like Soil Test, Zero Tillage, Compost, and Amrithpani are studied.

Technology and Knowledge System

The technological interventions like Weather-Based Agro Advisory Services, Climate Change Adaptation Plan, Integrated Pesticide Management, Custom Hiring Center, etc., are provided to the farmers.

Technology is the important component to advance in any field. In agriculture sector technology and knowledge used through the weather-based crop advisory and the forecast of climate change among the community members with support from technical agencies like SKYMET, CropIn, and PRAN, for the proper management of the crop and livestock. The community resource persons (CRPs) were assigned to carry the weather alerts to the communities. In addition to this project has armored the rural poor through custom hiring centers (CHCs) and village tool banks (VTBs) for cost and drudgery reduction, time saving, efficiency and effectiveness, and most importantly easy to access.

Custom hiring Centres' (CHCs) managed by CRPs for farm implements (more than 27 different types of farm equipment and machinery) were established in 75 SLACC villages of Bihar and Madhya Pradesh to tide over the shortage of labor and improve the efficiency of agricultural operations. The generated revenue was further utilized for the repair and maintenance of the implements and the remaining amount goes into the revolving fund. The most utilized equipment and machineries are paddy transplanter, reaper, rotovator, zero till drill, power sprayer, power tiller, and chaff cutter mostly used by the women farmers to save the cost of cultivation and time.

Financial System

Finance is a very poorly handled sector among the rural poor. The farmers were made aware of the crop insurance, kisan credit card, and several other important schemes run by state and central Govt. in the state. Farmers were also educated about the importance of crop insurance to save themselves from the worst hits of climate change.

29.5 Important Interventions Adapted Under SLACC

Climate-Resilient Seeds and Breeds

Farmers traditionally grow local varieties of different crops resulting in poor crop productivity due to heat, droughts, or floods. Hence, improved heat and flood-tolerant varieties were introduced to achieve optimum yields despite climatic stresses. This varietal shift was carefully promoted for major crops (Wheat: 1.HD 3118, 2.HD 3059, 3.PBW 373, 4.UP 262, 5.HD 2733, 6.HI 1563, 7.DBW 17Rice: Swarna Sub 1), as well as fodder crops (Napier—CO 4&5) and breeds (poultry—kadaknath, Goatery—Jamnapari), by encouraging village level seed production and linking farmers decision-making to weather-based agro advisories.

Water Saving Technologies

Since climate variability manifests in terms of deficit or excess water, major emphasis was given on implementing water-saving technologies like DSR, SRI, zero tillage, micro irrigation, and other resource conservation practices which also reduce GHG emissions besides saving of water. Impact of climate change on the availability of water for agriculture is the main concern. Due to declining water availability, judicious use of water through micro irrigation can improve the water use efficiency and helps to bring more area under irrigation.

Building Resilience in Soil

Soil health is the key property that determines the resilience of crop production under changing climate. Current soil management strategies are mainly dependent on inorganic chemical-based fertilizers, which caused a serious threat to human health and environment. Improper application of organic & inorganic nutrients contributing to greenhouse gases emissions notably methane and nitrous oxide. If agricultural soils are properly managed with improved nutrient management practices, they have the potential to mitigate and adapt climate change impacts at the same time enhance agricultural productivity. A number of interventions are made to build soil organic carbon, control soil loss due to erosion, and enhance water holding capacity of soils, all of which build resilience in soil. Soil testing has been performed in all villages by establishing mini soil testing laboratories at each block and building capacities of staffs to collect and analyze soil samples and interpret the soil test results to farmers, to ensure balanced use of chemical fertilizers.

Organic agriculture has a greater potential for mitigating climate change, largely due to its greater ability in reducing emissions of greenhouse gases (GHGs) including carbon dioxide, nitrous oxide (N_2O), and methane (CH_4). It also increases carbon sequestration in soils compared with that of conventional agriculture. The project being within National Rural Livelihood Mission, emphasis is laid on promoting organic agriculture.

Farm Machinery (Custom Hiring) Centers

Average operational land holding size in the country is estimated at 1.16 ha. Such farmers cannot invest in costly farm machinery and depend on hiring of implements to carryout agricultural operations in their fields. In rainfed areas, the time for taking up of timely land preparation, sowing, and interculture operations is narrow especially in the low rainfall zones. In high rainfall areas dominated by heavy soils, drainage is more crucial to prevent damage to crop from excess soil moisture in the root zone especially in pulses, oilseeds, and cotton. Adoption of climate-resilient practices such as soil incorporation of legume catch crops and crop residues to improve soil health and resource conservation technologies are linked to timely access to appropriate farm machinery at a reasonable cost. Community managed Custom Hiring Centers (CHC) were established in SLACC villages to access farm machinery/equipment to small and marginal farmers at affordable price to ease timely sowing/planting. This is an important intervention to deal with variable climate like delay in monsoon, inadequate rains needing replanting of crops. Participatory selection of farm machinery led to procuring small farm equipments like Solar Sprayer and Power Sprayer besides heavy machineries like Deep Plough, Integrated Seed Drill, Rain gun, Raised bed Planter, Paddy Transplanter, Reaper, Power Tiller, Weeders, etc.

Improved Water Management Practices

One of the primary effects of climate change is the disruption of the water cycle. Much of the impact of climate change will be felt through changing patterns of water availability, with shrinking water availability in rivers and changing patterns of precipitation increasing the likelihood of drought and flood. Impact of climate change on the availability of water for agriculture is the main concern. Due to decline in the water resources, judicious use of water through micro irrigation can improve the water use efficiency and helps to bring more area under irrigation. Adaption of micro irrigation practices has been one of the major intervention in SLACC villages. Irrigation tanks formed many centuries ago as water harvesting structures to offset the vagaries of monsoons, serve the purpose of collecting and storing not only the rain water, but also the nutrient-rich top soil eroded from their catchment areas. Rain water harvesting through repairing of ponds, restoration of old rain water harvesting structures in dry land/rainfed areas, bore-wells (both construction and recharging), and micro irrigation planning for recharging ground water are taken up for enhancing farm level water storage.

Seasonal Crop Planning and Contingency Plans

To cope with climate variability, seasonal plans for crops along with block level contingency plans for all the 8 blocks are important. Operationalization of these plans during aberrant monsoon years through the CRPs and SRLM staffs help farmers cope with climate variability. In scarce rainfall situation practice of sole cropping is predominant but risky and often results in low yields or sometimes even in crop failure due to erratic monsoon rainfall distribution. In such areas, intercropping is a feasible option to minimize risk in crop production, ensure reasonable returns at least from the intercrop and also improve soil fertility with a legume intercrop.

Climate Change Adaptation Plan

Climate Change Adaptation Planning (CCAP) Tool, a participatory methodology to understand the interrelations between climate impacts and livelihood strategies at the local level. The tool tries to answer the following questions: What are the main livelihoods of the community? What are the community perceptions of their main climate risks to their livelihoods? What are the typical responses of the community to the climate risks identified? What could be No/Low regret responses regarding production, technological, financial, and ecological aspects. The tool helps to articulate climate change concerns and identify measures to address them. Climate Change Adaptation Planning tool is community-engaging, easy-to-use, sensitive enough to capture the different types and degrees of vulnerabilities across communities, and is oriented toward localized adaptive action. It provides an initial understanding of the local context and vulnerability profiles and leads into identifying adaptation measures. The CCAP tool is designed for organizations, district-level authorities, facilitators working closely with government line departments, and Panchayati Raj Institutions (PRIs), who want to implement climate change adaptation projects in rural localities.

Under the SLACC project Climate Change Adaptation Planning has been conducted in all the SLACC villages. Block–wise analysis gives detailed information on major crops and livelihoods in each block. All the climate risks are faced by different crops/livelihoods. Present coping mechanisms and proposed adaptation measures that might help the villagers to minimize the adverse effects of these climate risks on crops and livelihoods.

Weather-Based Agro Advisories

Under climate-resilient agriculture, decision-making and timely advice can help to plan better and manage the risks appropriately. Based on the local weather situation, it will send them timely weather-based agro advisory to manage the crops so that farmers can adopt precautionary measures at the field level. Automatic weather stations (AWS) and automatic rain gauges (ARG) in project villages were established through private partner SKYMET to record real-time weather parameters such as rainfall, temperature, and wind speed, to disseminate weather information among farmers and assist the farmers in taking timely decisions through customized agro advisories services, delivered to registered farmers, in their mobiles, through SMS and an application developed by CROPIN Technology Solutions Pvt. Ltd. The mobile application facilities of two way communication by which a farmer can raise an alert (usually of pest/diseases in his field) and receive customized remedial measure in his mobile, from a subject matter specialist of CROPON, who works at the back end.

Livestock Interventions

Among the agricultural sectors, the livestock sector is also major in contributing to the emission of GHG. Nutrition plays a critical role in making livestock production systems more efficient. There is an urgent need to improve the resource use and production efficiency of livestock production systems, both to improve food security and reduce the intensity of GHG emissions. Use of community lands for fodder production in drought-prone areas, improved fodder varieties (CO 4 & 5) and feed storage methods, feed supplements, micronutrient use to enhance adaptation to heat stress, preventive vaccination, improved shelters for reducing heat/cold stress in livestock, and introduction of stress-tolerant hardy breeds.

29.6 SLACC INTERVENTIONS IMPACT ANALYSIS

NIRDPR as LTSA conducted an impact analysis study to identify the changes that occurred from the implementation of SLACC interventions in the study area. For this study, data was collected from different concerned District Project Managers, Block Project Managers through focus group discussions, and farmers by conducting one-to-one interviews. The selection of farmers to evaluate impact was complex, so 40 farmers were selected from three (i.e. high impact, medium impact, and scale-up villages) project villages and 10 farmers from one non-project (control) village from each block. Thus a total of 400 farmers were sampled from the two states covering 160 farmers from project implementation villages and 40 farmers from non-project (control) village in each State. For the data collection double, difference method was approached in which data collected for the pre and post-project period and compared with the control. Hence, it can be assessed with and without and before and after SLACC project implementation. The data collected from farmers include the crop yield, income, cost, net income & crop area, besides (a) socio-economic profile, (b) climate change awareness levels, (c) SLACC training received, and (d) constraints in adopting/continuing SLACC methods.

Interventions Related to Production

The integrated Farming System is adopted by 53 percent of SLACC and 32 percent of Control farmers. IFS aims to increase income and employment from small holdings by integrating various farm enterprises and recycling crop residues and by-products within the farm itself (Behera and Mahapatra 1999; Singh et al. 2006). Around 67 and 70 percent of SLACC farmers practice, Integrated Pest Management (IPM) and Integrated Nutrient Management (INM) are nil in control villages. Livestock is available with 73 percent of SLACC farmers and 60 percent of control farmers. Still, it is very controversial that fodder crops like Napier, super Napier developed by 40 percent of SLACC farmers and only 10 percent by non-SLACC farmers to meet fodder needs for their cattle. The unawareness among the control farmers increases their vulnerability to changing climate while awareness increases adaptability. Among the SLACC farmers, 20 percent had small ruminants, whereas only 8 percent for control farmers is owed to the easy loan facility to the SLACC farmers to purchase goats and sheep. The rearing of poultry birds is favored by 6 percent of SLACC farmers and 2 percent of control farmers; this is a very low-adopted intervention due to the mythological perceptions of the communities about the vegetarian diet. However, the farmers can be motivated for high-level adoption of this intervention in the future as this activity has a lot of economic and ecological benefits along with a protein-rich diet. Alternative livelihood interventions are encouraged as substitutes or lower-impact livelihood activities under the project to reduce reliance on the local natural resources, provide livelihood enhancement and economic development, and increase local support for conservation (Juliet et al. 2015). Alternative livelihood interventions like Kitchen Gardens, Mushroom, Azolla, and Beekeeping are not adopted by control farmers, whereas 38 percent of SLACC farmers preferred another source of income.

Ecological Interventions

Soil test by 93 percent of SLACC farmers and 12 percent by control farmers. Compost is being practiced by 40 percent of SLACC farmers, whereas only 5 percent are in the control category. The adaptation of organic formulation is important to maintain soil productivity over the period; 71 percent of SLACC farmers are practicing preparation of different organic formulations, whereas the control category found completely away from it. The zero tillage practices are being done by 71 percent of SLACC farmers and 12 percent in control farmers.

Interventions Related to Technology and Knowledge

Weather-Based Agro Advisory Services (WBASS) and Climate Change Adaptation Plan (CCAP) were adopted by 60 and 70 percent of SLACC farmers, respectively. However, control farmers were unaware of it. The establishment of Custom Hiring Centers encouraged mechanization in the agriculture sector. 74 percent of SLACC farmers benefited from the CHCs; however, only 6 percent of control farmers could get this facility. Under the project, 18 percent of farmers practice micro irrigation, whereas only 4 percent control farmers. Seed treatment is done by 85 percent of SLACC farmers, whereas only 7 percent by control farmers. Farm Pond is available to 12 percent of farmers and 2 percent for control farmers.

Financial Interventions

Under financial intervention, farmers are encouraged and provided with loans for better agriculture and livelihood options. 54 percent of SLACC farmers have benefitted from loans for different purposes, including better agriculture, agriculture implements, meeting input costs, and entrepreneurship development. In contrast, only 5 percent of control farmers could get the loans. The literacy levels of the farmers are also affecting their accessibility to loans in all categories of farmers. However, SLACC farmers were helped by the project staff to access the loans. In the same way, 73 percent of SLACC farmers opted for crop insurance with the help of the SLACC project team, whereas 10 percent was in the case of the control category. The convergence of different government schemes is also encouraged under the SLACC project for providing more benefits to the farming community; a total of 19 percent of SLACC farmers are benefiting from the convergence of various schemes like solar irrigation, micro irrigation (drip & sprinkler), farm ponds, poultry, and vaccination of livestock. However, it is only 3 percent under the control category.

The net impacts of SLACC interventions on major crops in Bihar and Madhya Pradesh are represented in Tables 29.3 and 29.4. In Bihar Paddy, Wheat, Mung (green gram), Maize, Paddy, Black Gram, Soybean, Bajra (pearl millet), and Mustard crops in Madhya Pradesh were studied to evaluate the net impact among SLACC and non-SLACC farmers based on the collected data. The results demonstrated that SLACC farmers have less cost of cultivation, more yield, and more income than non-SLACC farmers. Such results are the cumulative effects of different interventions (Production, Ecology, Technology, and Knowledge and Finance).

29.7 Integration of Sustainable Development Goals

It is observed that this project has demonstrated the potential to achieve sustainable development goals. Around 12 sustainable goals covering 38 targets are achieved through this project in a few years in the rural farmer community. The SDGs with their targets achieved are shown in the Fig. 29.5. Localization of SDGs is a very effective way to meet them by 2030 in a sustainable way. This project assured that few SDG targets, such as 1, 2, 3, 4, 5, 6, 7, 8, 10, 11, 12, 13, 15, and 17 seem to be achieved through the implementation of SLACC interventions.

29.8 Training and Capacity Building

As an LTSA, NIRDPR prepared information flip charts (https://bit.ly/3V4kyYB) and a training manual for CRPs in Hindi (https://bit.ly/3XoPwwd) and for mission staff in English (https://bit.ly/3GNpj4B) to train and disseminate knowledge on *production system* such as participatory selection of

Table 29.3 Crop-wise area, cost, yield, income profit details of SLACC and Non-SLACC farmers in Bihar State

District	Block	Season	Category	Crop	Area (Ha)	Cost (Rs./Ha)	Yield (Qt/Ha)	Income (Rs./Ha)	Profit (Rs./Ha)
MADHUBANI	Khajauli	Kharif	SLACC	Rice	0.36	28,000	71	85,200	57,200
			Control	Rice	0.34	30,000	62	74,400	44,400
		Rabi	SLACC	Wheat	0.26	25,000	49	88,200	63,200
			Control	Wheat	0.2	27,500	30	54,000	26,500
		Zaid	SLACC	Mungbean	0.22	20,000	15	45,000	25,000
			Control	Mungbean	0.2	23,500	10	30,000	6500
	Rajnagar	Kharif	SLACC	Rice	0.32	29,500	70	84,000	2525
			Control	Rice	0.26	32,000	64	76,800	44,800
		Rabi	SLACC	Wheat	0.23	26,550	50	90,000	3540
			Control	Wheat	0.2	28,450	28	50,400	21,950
		Zaid	SLACC	Mungbean	0.12	18,750	13	39,000	1500
			Control	Mungbean	0.11	20,650	10	30,000	9350
GAYA	Barachetti	Kharif	SLACC	Rice	0.45	30,820	75	112,225	81,405
			Control	Rice	0.45	29,480	56	84,420	54,940
		Rabi	SLACC	Wheat	0.31	42,411	54	91,857	49,446
			Control	Wheat	0.22	26,800	40	68,340	41,540
		Zaid	SLACC	Mungbean	0.15	33,600	14	84,000	50,400
			Control	Mungbean	0.1	24,300	9	54,000	29,700
	Guruwa	Kharif	SLACC	Rice	0.67	31,021	68	102,510	71,489
			Control	Rice	0.33	26,800	51	76,380	49,580
		Rabi	SLACC	Wheat	0.47	20,100	34	56,548	36,448
			Control	Wheat	0.16	16,951	25	43,282	26,331
		Zaid	SLACC	Mungbean	0.22	36,000	15	90,000	54,000
			Control	Mungbean	0.08	27,000	10	60,000	33,000

Table 29.4 Crop-wise area, cost, yield, and income profit details of SLACC and Non-SLACC farmers in Madhya Pradesh State

District	Block	Season	Category	Crop	Area (Ha)	Cost (Rs./Ha)	Yield (Qt/Ha)	Income (Rs./Ha)	Profit (Rs./Ha)
SHEOPUR	Karahal	Kharif	SLACC	Maize	0.95	12,855	45	49,500	36,645
			Control	Maize	0.60	12,500	30	33,000	20,500
			SLACC	Rice	0.67	13,928	40	72,000	58,073
			Control	Rice	0.48	13,750	28	49,500	35,750
		Rabi	SLACC	Wheat	1.04	16,250	64	95,625	79,375
			Control	Wheat	0.80	15,000	43	63,750	48,750
			SLACC	Black Gram	0.57	10,000	18	38,500	28,500
			Control	Black Gram	0.40	10,000	13	27,500	17,500
			SLACC	Mustard	0.30	11,250	13	37,500	26,250
			Control	Mustard	0.20	10,000	8	22,500	12,500
	Sheopur	Kharif	SLACC	Rice	0.65	47,500	55	115,500	68,000
			Control	Rice	0.60	37,500	45	94,500	57,000
			SLACC	Soybean	0.70	18,750	28	82,500	63,750
			Control	Soybean	0.60	17,500	20	60,000	42,500
		Rabi	SLACC	Wheat	0.56	37,500	65	84,500	47,000
			Control	Wheat	0.40	30,000	26	33,150	3150
			SLACC	Black Gram	0.50	37,500	25	100,000	62,500
			Control	Black Gram	0.40	30,000	18	70,000	40,000
MANDLA	Bichia	Kharif	SLACC	Rice	0.72	17,800	32	56,700	38,900

(continued)

Table 29.4 (continued)

District	Block	Season	Category	Crop	Area (Ha)	Cost (Rs./Ha)	Yield (Qt/Ha)	Income (Rs./Ha)	Profit (Rs./Ha)
		Rabi	Control	Rice	0.64	15,375	22	39,150	23,775
			SLACC	Wheat	0.48	10,750	15	22,500	11,750
			Control	Wheat	0.08	7500	13	18,750	11,250
			SLACC	Bajra	0.20	5000	5	12,500	7500
			Control	Bajra	0.00	0	0	0	0
			SLACC	Green pea	0.50	7375	10	26,093	18,718
			Control	Green pea	0.00	0	0	0	0
Mandla		Rabi	SLACC	Wheat	0.50	11,718	19	32,725	21,008
			Control	Wheat	0.80	7500	8	12,750	5250
			SLACC	Bajra	0.48	20,978	33	81,250	60,273
			Control	Bajra	0.40	6250	8	18,750	12,500
			SLACC	Green pea	0.40	5000	8	17,625	12,625
			Control	Green pea	0.00	0	0	0	0

29 SCALABLE ADAPTATION MODEL FOR SUSTAINABLE … 523

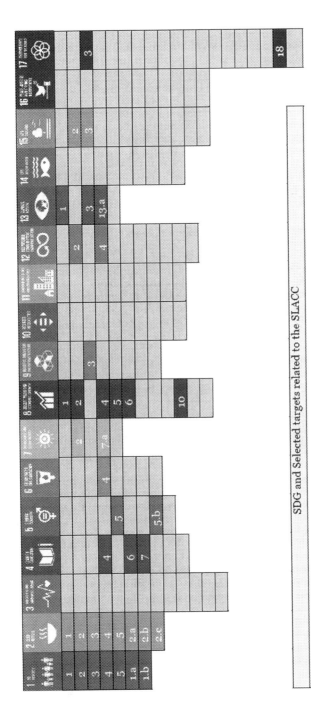

Fig. 29.5 Illustrates all selected SDG goals and their targets that call for action in the SLACC project. As the interventions are classified into four (Production, Ecology, Technology, and Knowledge and Finance) types, SDGs can also be distributed per the respective categories

climate-resilient varieties/breeds; *ecological system* such as tree-based farming, soil moisture conservation; *knowledge system* such as local weather-based agro advisories, and *financial system* such as weather-based insurance, etc., In order to achieve the desired objectives of the CRPs, they are expected to build rapport with the line department officials, understand the village dynamics and climatic conditions of the workplace and its relation with the existing agricultural practices by collecting necessary data to adopt the climate-resilient interventions for sustainable agriculture and livelihoods. NIRDPR trained 200 CRPS (5 batches) and 100 mission staff (3 batches) from Bihar and Madhya Pradesh through classroom sessions and exposure visits to several recognized institutes. The participants were evaluated twice, i.e., before and after the training to know their understanding and unawareness. MANAGE conducted the evaluation and certification of participants. This evaluation found that most participants had enhanced their learnings on climate change and adaptation measures. In addition to that, CRPs were trained to train other people from their locality, so these learnings will not be limited to themselves.

29.9 Conclusion

India has seen a 1.2 °C increase in mean surface air temperature since 1901, and climate change projections up to 2100 indicate an overall increase of 2–4 °C. Climate change is a significant challenge to the disadvantaged population of India as it will affect the yield and income of the small farmer, especially in Rainfed farms, which constitute nearly 58 percent of the net cultivated area in the country and are expected to be significantly impacted by climate change.

To expand demonstrations of appropriate practices and technological interventions have been implemented through the SLACC project in a farmer-participatory mode to enhance their adaptive capacity and coping ability. SLACC covered drought and flood-prone areas of Madhya Pradesh and Bihar, with the agriculture sector at the core of the SLACC project. Capacity building was one of the key components in the SLACC project to enrich the knowledge of different CRPs and Mission staff of both states. NIRDPR developed certificate training courses for CRPs and Mission staff separately for understanding the climate change, variability, interventions, and alternate livelihood activities for climate resilience that can be implemented throughout the country. Each village has placed a trained CRP to support the farmers in climate adaptation planning, implementation, and monitoring. SLACC addressed all aspects of farm-based livelihoods that may be affected by climate change by helping the community to choose interventions for the: production system, knowledge system, ecological system, technology, and financial system.

Climate-smart practices implemented through SLACC, being a cost-effective and ecologically compatible alternative, would be an enabler for the nation in achieving the Sustainable Development Goals by reducing the input costs; this can ensure better income and financial stability, which would, in turn, help alleviate poverty, bring in gender equality and ensure sustainable production. Overall, SLACC has demonstrated the potential to achieve 12 SDGs with 38 targets.

Acknowledgements We thank the Global Environment Facility (GEF) and World Bank for supporting the SLACC Project. We also extend our sincere thanks to DAY-NRLM, Ministry of Rural Development, Government of India, New Delhi and NIRDPR, Hyderabad, for facilitating the effective implementation of the project. The authors would also like to show respect and acknowledge the support of the study area's people who contributed to the study findings and successful execution and completion of the project.

Disclaimer The statements and opinions published in this book are solely those of the individual authors, not the affiliated organization or the funding agencies. The affiliated organization or funding agencies don't give any warranty, express or implied with respect to the material contained herein or for any errors or omissions that may have been made.

REFERENCES

Amariles S, Haman M, Tobón H, van Epp M, Abreu D (2018) CCAFS Web and Statistics Report for 2017. Wageningen, Netherlands: CGIAR Research Program on Climate Change, Agriculture and Food Security.

Archer D, Almansi F, DiGregorio M, Roberts D, Sharma D, Syam D (2014) Moving towards inclusive urban adaptation: approaches to integrating community-based adaptation to climate change at city and national scale. Climate Development 6(4): 345–356.

Azhoni A, Jude S, Holman I (2018) Adapting to climate change by water management organizations: Enablers and barriers. J Hydrology 559: 736–748.

Behera UK, Mahapatra IC (1999) Income and employment generation for small and marginal farmers through integrated farming systems. Indian J Agronomy 44(3): 431–439.

Dany V, Bajracharya B, Lebel L, Regan M, Taplin R (2015) Narrowing gaps between research and policy development in climate change adaptation work in the water resources and agriculture sectors of Cambodia. Climate Policy 16(2): 237–252.

FAO (2013) Sustainability pathways: sustainability assessment of food and agriculture systems (SAFA) http://www.fao.org/nr/sustainability/sustainability-assessmentssafa/en/. Accessed 1 December 2022.

Hammouri N, Al-Qinna M, Salahat M, Adamowski J, Prasher SO (2015) Community based adaptation options for climate change impacts on water resources: The case of Jordan. J Water Land Development 26: 3–17.

IPCC (2014) Climate change 2014: Synthesis report contribution of working groups I, II and III to the fifth assessment report of the Intergovernmental Panel on Climate Change. IPCC, Geneva.

Juliet HW, Nicholas AOH, Dilys R, Rowcliffe JM, Kümpel NF, Day M, Booker F, Milner-Gulland EJ (2015) Reframing the concept of alternative livelihoods. Conservation Biology 30(1): 7–13.

Kakumanu KR, Kaluvai YR, Nagothu US, Lati NR, Kotapati GR, Karanam S (2018) Building farm-level capacities in irrigation water management to adapt to climate change. Irrigation Drainage 67(1): 43–54.

Kakumanu KR, Gavali RS, Babu VS, Rao VS, Ghate U (2020) Capacity building impacts on adaptation of climate-resilient agriculture interventions in India. In: *Climate Change Adaptation and Sustainable Livelihoods*, 1.

Mertz O, Mbow C, Reenberg A, Diouf A (2009) Farmers' perceptions of climate change and agricultural adaptation strategies in rural Sahel. Environmental Management 43(5): 804–816.

Mimura N, Pulwarty RS, Elshinnawy I, Redsteer MH, Huang HQ, Nkem JN, Kato S (2014) Adaptation planning and implementation. In: *Climate change 2014 impacts, adaptation and vulnerability: Part A: Global and sectoral aspects.* Cambridge University Press. pp. 869–898.

Noble IR, Huq S, Anokhin YA, Carmin JA, Goudou D, Lansigan FP, Chu E (2014) Adaptation needs and options. In: *Climate Change 2014 Impacts, Adaptation and Vulnerability: Part A: Global and Sectoral Aspects.* Cambridge University Press. pp. 833–868.

Palanisami K, Ranganathan CR, Nagothu US, Kakumanu KR (2014) Climate change and agriculture in India: studies from selected river basins. Routledge India.

Pandey VP, Babel MS, Shrestha S, Kazama F (2011) A framework to assess adaptive capacity of the water resources system in Nepalese river basins. Ecological Indicators 11(2): 480–488.

Preston BL, Westaway RM, Yuen EJ (2010) Climate adaptation planning in practice: an evaluation of adaptation plans from three developed nations. Mitigation and adaptation strategies for global change 16(4): 407–438.

Preston BL, Dow K, Berkhout F (2013) The climate adaptation frontier. Sustainability 5(3): 1011–1035.

Singh K, Bohra JS, Singh Y, Singh JP (2006) Development of farming system models for the northeastern plain zone of Uttar Pradesh. Indian Farming 56(7): 5–11.

Spires M, Shackleton S, Cundill G (2014) Barriers to implementing planned community-based adaptation in developing countries: A systematic literature review. Climate Development 6(3): 277–287.

Sud R, Mishra A, Varma N, Bhadwal S (2015) Adaptation policy and practice in densely populated glacier-fed river basins of South Asia: a systematic review. Regional Environmental Change 15(5): 825–836.

UNFCCC (1992) United Nations Framework Convention on Climate Change (UNFCCC), United Nations. https://unfccc.int/resource/docs/convkp/conveng.pdf. Accessed 1 December 2022.

Upgupta S, Sharma J, Jayaraman M, Kumar V, Ravindranath NH (2015) Climate change impact and vulnerability assessment of forests in the Indian Western Himalayan region: a case study of Himachal Pradesh, India. Climate Risk Management 10: 63–76.

Younus MAF (2010) Community-based autonomous adaptation and vulnerability to extreme floods in Bangladesh: processes, assessment and failure effects. Dissertation, University of Adelaide.

CHAPTER 30

Delineating Health Sector Resilience in Post COVID-19 Pandemic in the Backdrop of Changing Climate and Disasters

Atisha Sood, Anjali Barwal, and Anil Kumar Gupta

30.1 Introduction

Millions of fatalities due to infectious disease outbreaks have been documented throughout history over the past several centuries. The pandemic caused by the plague in Asia and multiple influenza pandemics that claimed millions of lives are the most well-known in history (Howard, 2020). The pandemics persisted in the present millennium; COVID-19 is the most recent but surely not the last. The lacklustre approach to constructing the capacity to respond to infectious illnesses is one of the causes of the development of pandemics and the delayed response to them (CDCP, 2020).

It has long been accepted that unfavourable socioeconomic conditions increase the risk of contracting infectious illnesses. Although partially accurate, the ongoing COVID-19 in developed nations also reminds us that wealthy nations and populations are not immune to spreading contagious diseases (Tucho & Kumsa, 2021). The premise mentioned above is sufficiently supported by the usage of both naturally occurring agents of mass destruction and weapons of mass destruction created by humans.

The growth of people and infectious illnesses are inextricably linked. This is also acknowledged in Goal No. 3 of the Sustainable Development Goals of the United Nations (UNESCAP, 2016). However, there is an additional

---

A. Sood (✉) · A. Barwal · A. K. Gupta
Government of India, National Institute of Disaster Management, Ministry of Home Affairs, New Delhi, India
e-mail: soodatisha@gmail.com

aspect to development. Technology advancements, the building of new irrigation systems and dams, deforestation, population migration, high population densities, the emergence of urban ghettos, the globalisation of food, and an increase in international travel, are just a few examples of the ecological changes brought on by development activities. All factors aid in the infection's quick spread across the nations. Some of these elements are to blame for the COVID-19 virus' rapid geographic spread across international borders (Radil et al., 2021). Another element that may have contributed to the genesis and spread of various epidemic-prone illnesses is global warming or climate change.

30.2 Sustainable Development and Climate Change

A region's managerial, technological, and economic components must be improved to address socio-environmental issues in a way that promotes the nation's sustainable development. In sustainable development design, three categories of interactions are chosen: (1) Nature and society, (2) Man and society, and (3) Human nature. Here, the ecological study is focused on the first and third categories of interactions. The second category is a research topic in humanities and economics (Dovgal et al., 2020). As a result, the process of designing sustainable development to support economics identifies the steps required to ensure the interdependence of social and natural systems.

The dynamics and laws of nature must be balanced with several considerations while designing sustainable development. Irrational judgments would not affect how socio-natural systems are managed except in this specific situation. The principles of the ongoing operation of this system must be a "machine" system that acts as a supportive "tool" for balancing judgements with natural laws. A single body of scientific knowledge is required to construct such systems, which are unprecedented in human history and allow you to forecast the future evolution of the whole world, a country, a region, an industry, or a corporation (Anker, 2021).

The Global Agenda for Sustainable Development 2030, which world leaders endorsed at the historic UN summit in September 2015, contains the Sustainable Development Goals (SDG), which came into effect on 1st January 2016. Over the next fifteen years, governments will marshal all of their efforts to eradicate all kinds of poverty, confront inequality, and battle climate change in light of these new global goals, led by the tenet of "no one left out." The SDGs are remarkable in that they urge all nations—poor, rich, and middle-income—to take action to advance progress as long as they also safeguard the environment. The SGDs state that efforts to eradicate poverty must be coordinated with plans for economic expansion. They concentrate on a variety of social requirements, such as the need for economic opportunities, social security, and education, as well as the battle against global warming and environmental preservation (Agbedahin, 2019).

The adoption of the 17 SDGs and the development of national frameworks to accomplish them are anticipated by governments even if their implementation is not legally required. Therefore, the primary duty of the nations is to track and evaluate the SDG's implementation's advancement. This necessitates gathering reliable and available data (Leal Filho et al., 2022). Based on the analysis performed at the national level, the monitoring and analysis of the SDG implementation at the regional level will be used for the evaluation and monitoring at the global level.

30.3 Climate Change & Health

Figure 30.1 depicts the main routes relating climate change and health outcomes, which are classified as direct and indirect mechanisms interacting with social dynamics to produce health effects (Rocque et al., 2021). These dangers have social and geographical components, are unevenly distributed worldwide, and are influenced by social and economic growth, technology, and access to health care. Extreme weather changes and the resulting storm, floods, droughts, or heat waves are direct threats. Indirect hazards are mediated by changes in the biosphere (e.g., illness burden and distribution of disease vectors or food availability). In contrast, others are mediated by societal processes (leading, for instance, to migration and conflict). These three pillars, seen in Fig. 30.1, interact with one another and with changes in land use, crop productivity, and ecosystems caused by global development and population processes. Climate change will constrain development ambitions through its effects on national economies and infrastructure, including providing health and other services. In addition, it will have a material and non-material impact on happiness. As seen in Fig. 30.1, societal factors can amplify and modify climatic threats. Policies, laws, and subsidies to ensure enough food supply and affordable pricing, for example, greatly influence the relationships between food production and food security in any country. As a result of the combination of climatic and social processes, vulnerabilities emerge.

30.4 Effect of Covid on the Global Economy and Public Health Systems

The economies of all the nations impacted by pandemics suffer. The poor are the worst affected. The United Nations has acknowledged that the pandemics pose a threat to national security, as has been previously established. According to a thorough study conducted between 1950 and 1991 in 20 countries, including developed, developing, and underdeveloped nations, the prevalence of infectious diseases is on the rise and will eventually lead to an increase in poverty and a gradual deterioration of State capacity in addition to an increase in human mortality and morbidity (Smith et al., 2012). It was discovered that this pathogen-induced economic collapse has a detrimental impact

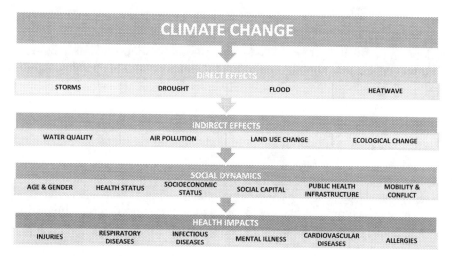

Fig. 30.1 Direct and indirect effects of climate change on health and wellbeing

on state capacity indicators, including fiscal resource, resilience, reach and responsiveness, autonomy, and legitimacy.

Evidence supports the allegations that infectious illnesses pose a real danger to state authority and national security. The incidence of infectious diseases was shown to negatively correlate with the state's capacity to sustain its military forces, which had a detrimental impact on state security (Wilson, 2020).

Equitable distribution of resources among populations and the development of healthy communities require resilience-building strategies designed in partnership with and for these groups. There could be no healthcare system resilience without cross-domain community participation. Cooperation with non-health sectors is also vital in providing the necessary assistance to address the socioeconomic determinants of health. These factors are supported by health equity and results (Fig. 30.2). Positive mental and physical outcomes for all, especially vulnerable and disadvantaged populations, should be the goal of resilient health systems (Barker et al., 2020). COVID-19 deaths have been proportionally higher among older people, minority ethnic communities, socioeconomically disadvantaged people, and low-wage and migratory workers, highlighting the link between equity and healthcare outcomes (Paramoer et al., 2021; Shadmi et al., 2020).

30.5 Adaptive Public Health Sector Reforms

The COVID-19 pandemic has prompted public policy to be reassessed in a paradigm shift regarding how radical action must be dynamically adapted to persistent concerns and new, unprecedented risks to project development. While the crisis has highlighted massive flaw areas inside the healthcare systems and its interconnections with certain other aspects of the economy in nations

Fig. 30.2 Determinants of health systems resilience framework

such as India, it also allows assessing more even people-centric and adaptive changes across the health sector. This advocates for "building back better", drawing similarities with post-disaster rebuilding incorporating many acts and individuals.

Initially, the ability of medical systems in terms of critical infrastructure, human, and financial resources—that allow them to respond to crises without jeopardising other concurrent objectives must be significantly increased (Susskind & Vines, 2020). Unhindered activities in providing vital services for initiatives on infectious disease control such as chikungunya, dengue, tuberculosis, malaria, and HIV/AIDS, safe delivery and antenatal care for expectant mothers, children immunisation and nutrition, and treatment and diagnosis for major non—communicable disease are vital to ensure sustained progress and avoid costly unmet need (Dutta & Fischer, 2021). Effective and feasible methods of integrating incentives and regulatory measures are crucial for correcting supply-side distortions and unbalanced concentrations in the allocation of most medical system inputs.

Moreover, eliminating segmentation inside and across essential healthcare systems' building blocks, such as service delivery, workforce, funding, hospital instruments, and medicine availability, must be given a top focus. An important consideration here is adequate consideration of the "missing-middle" on both the supply and demand sides of the system; enhancing infrastructural faculties in regions so that most healthcare needs can be dealt without having to travel outside of the district; and budgeting for financial risk management for the millions who are not covered by any formal insurance mechanisms, avoiding regressive out-of-pocket funding of medical care (Gauttam et al., 2021). Integrating several plans that lead to a unified system for all Indians, regardless of job or residency status, requires a deliberate, time-bound approach that goes beyond rhetoric.

Ultimately, the epidemic has raised various new problems and considerations for the healthcare system to address. This includes issues such as enforcing physical separation, personal hygiene, and other essential protocols in health facilities with high patient loads; ensuring equitable, transparent systems of access to COVID-19 treatments and vaccine(s); specific needs for other pandemic-induced conditions, particularly mental care needs, even among kids and teens; and accounting for increased vulnerability due to the pandemic's negative economic consequences. This necessitates the government playing effective stewardship responsibilities in coordinating operations between the public and commercial sectors, notably in care delivery and medical supply, as well as among population health and many other non-health government agencies (Blumenthal et al., 2020).

30.6   Lessons Learnt from the Pandemic and the Health system's Response

The most efficient ways to decrease vulnerability are (i) short-term initiatives that integrate and improve fundamental public health measures, such as providing clean water and better sanitation, securing essential health care, and increasing capacity for disaster preparedness and response; and (ii) long-term programs that improve universal healthcare systems and organisations with climate change-related risks integrated into health governance, strengthen health information systems, and ensure effective and accessible delivery of health services.

COVID-19 has rendered governments and the general public more aware of the interdependence between health systems, the home economy, and governance. Government actions shape healthcare infrastructures, rules, and standards, determining access to medicine and treatment, health coverage, and finance. In the short term, government reactions to COVID-19 have resulted in the distinction between lockdown and business as usual and have either destroyed or improved public trust. In the long run, they have affected national decisions on private healthcare vs universal health coverage (UHC) and reinforced or weakened social safety nets that support health and well-being (Greyling et al., 2021).

Overall, COVID-19 reactions saw health policy expanding outside the purview of the Ministries of Health, requiring experience from many other ministries, especially during the initial response (Desson et al., 2020). During the COVID-19 response, countries used whole-of-government initiatives to

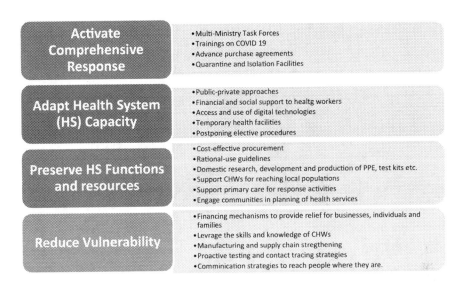

Fig. 30.3 Four resilience elements of highly effective country responses

enhance health systems, particularly those that had already experienced other health-related crises such as Ebola (Fig. 30.3).

In most nations, these decisions were taken by translating evidence-based findings to a policy that retains health system capacity while safeguarding public health and livelihoods. As a result, most nations formed interim COVID-19 advisory panels to help inform government decisions. However, most countries' opinions on these panels were mainly biomedical (Brusselaers et al., 2022).

30.7 Conclusion

The COVID-19 epidemic has impacted societies to a degree that has not been witnessed since the 1918 influenza pandemic. This has had detrimental effects on economic and social difficulties, but on a societal level, it is a "gift in disguise" for environmental issues. The epidemic has profoundly tested global healthcare systems and the individuals they serve. The consequence of a severe shock like the pandemic is to highlight the weakest links in the system and the interdependence of various health, social, and economic components. Because of their creativity and fundamental survival instinct, microorganisms will continue to generate pandemics. Despite the spectacular advancements in epidemiology, disease biology, molecular biology, genomics, and proteomics, it is clear from the COVID-19 outbreak that mankind is still unable to foresee and stop the unexpected development of epidemics and pandemics of infectious illnesses. It is also evident that they have equally devastating socioeconomic repercussions for the affected nations and the entire world, in addition to their terrible impact on human sickness and death.

For the many organisations tasked with reducing the broader adverse effects of pandemics, it is crucial to boost biomedical research, enhance the healthcare delivery system, establish a permanent "watch-dog" body, and develop an improved communication and coordination mechanism.

REFERENCES

Agbedahin, A. V. (2019). Sustainable development, education for sustainable development, and the 2030 agenda for sustainable development: Emergence, efficacy, eminence, and future. *Sustainable Development*, 27(4), 669-680.

Anker, T. B. (2021). At the boundary: Post-COVID agenda for business and management research in Europe and beyond. *European Management Journal*, 39(2), 171–178.

Barker, K. M., Ling, E. J., Fallah, M., VanDeBogert, B., Kodl, Y., Macauley, R. J., ... & Kruk, M. E. (2020). Community engagement for health system resilience: Evidence from Liberia's Ebola epidemic. *Health Policy and Planning*, 35(4), 416–423.

Blumenthal, D., Fowler, E. J., Abrams, M., & Collins, S. R. (2020). Covid-19—Implications for the health care system. *New England Journal of Medicine*, 383(15), 1483–1488.

Brusselaers, N., Steadson, D., Bjorklund, K., Breland, S., Stilhoff Sörensen, J., Ewing, A., ... & Steineck, G. (2022). Evaluation of science advice during the COVID-19 pandemic in Sweden. *Humanities and Social Sciences Communications*, 9(1), 1–17.

Centers for Disease Control and Prevention. (2020). Pandemics influenza. PST pandemics. Atlanta: CDC; 2018.

Desson, Z., Weller, E., McMeekin, P., & Ammi, M. (2020). An analysis of the policy responses to the COVID-19 pandemic in France, Belgium, and Canada. *Health Policy and Technology*, 9(4), 430–446.

Dovgal, O., Goncharenko, N., Reshetnyak, O., Dovgal, G., Danko, N., & Shuba, T. (2020). Sustainable ecological development of the global economic system. The institutional aspect. *Journal of Environmental Management and Tourism*, 11(3), 728–740.

Dutta, A., & Fischer, H. W. (2021). The local governance of COVID-19: Disease prevention and social security in rural India. *World Development*, 138, 105234.

Gauttam, P., Patel, N., Singh, B., Kaur, J., Chattu, V. K., & Jakovljevic, M. (2021). Public health policy of India and COVID-19: Diagnosis and prognosis of the combating response. *Sustainability*, 13(6), 3415.

Greyling, T., Rossouw, S., & Adhikari, T. (2021). The good, the bad and the ugly of lockdowns during Covid-19. *PLoS One*, 16(1), e0245546.

Howard, J. (2020). Plague, explained. Science; 20 August 2019.

Leal Filho, W., Vidal, D. G., Chen, C., Petrova, M., Dinis, M. A. P., Yang, P., ... & Neiva, S. (2022). An assessment of requirements in investments, new technologies, and infrastructures to achieve the SDGs. *Environmental Sciences Europe*, 34(1), 1–17.

Paremoer, L., Nandi, S., Serag, H., & Baum, F. (2021). Covid-19 pandemic and the social determinants of health. *BMJ*, 372.

Radil, S. M., Castan Pinos, J., & Ptak, T. (2021). Borders resurgent: towards a post-Covid-19 global border regime? *Space and Polity, 25*(1), 132–140.

Rocque, R. J., Beaudoin, C., Ndjaboue, R., Cameron, L., Poirier-Bergeron, L., Poulin-Rheault, R. A., ... & Witteman, H. O. (2021). Health effects of climate change: An overview of systematic reviews. *BMJ Open, 11*(6), e046333.

Shadmi, E., Chen, Y., Dourado, I., Faran-Perach, I., Furler, J., Hangoma, P., ... & Willems, S. (2020). Health equity and COVID-19: Global perspectives. *International Journal for Equity in Health, 19*(1), 1–16.

Smith, J. W., le Gall, F., Stephenson, S., & de Haan, C. (2012). People, pathogens and our planet: the economics of One Health. World Bank. Washington, DC.

Susskind, D., & Vines, D. (2020). The economics of the COVID-19 pandemic: an assessment. *Oxford Review of Economic Policy, 36*(Supplement_1), S1–S13.

Tucho, G. T., & Kumsa, D. M. (2021). Universal use of face masks and related challenges during COVID-19 in developing countries. *Risk Management and Healthcare Policy, 14*, 511.

UNESCAP. (2016). Sustainable Development Goal 3: Ensure healthy lives and promote well-being for all ages.

Wilson, J. (2020). Bio threats-infectious diseases and national security.

CHAPTER 31

Climate Change and Environment: Holistic Approaches Towards Climate Resilience

Usha Swaminathan, G. B. Reddy, Inocencio E. Buot, and Siva Ramamoorthy

31.1 Introduction

The World Meteorological Organization (WMO) reports that the hottest decade across the globe has been from 2010 to 2019. The non-tropical countries, including the USA, Australia and tropical countries, including India, are the most affected due to climate change. India is in the top 10 countries to have experienced the hottest climatic conditions since record-keeping began officially. Human activities have contributed to the maximum carbon dioxide and other greenhouse gas emissions besides natural processes. They have been the common factors for continued global warming. Germany, Japan and Phillippines were afflicted by climate change, placing India in the fifth position

---

U. Swaminathan (✉)
School of Social Sciences and Languages,
Vellore Institute of Technology, Vellore, India
e-mail: susha@vit.ac.in

G. B. Reddy
University College of Law, Osmania University, Hyderabad, India

I. E. Buot
Institute of Biological Sciences, University of the Philippines Los Banos, Laguna, Philippines

S. Ramamoorthy
School of Bio-Sciences and Technology,
Vellore Institute of Technology, Vellore, India

© The Author(s), under exclusive license to Springer Nature
Singapore Pte Ltd. 2023
S. Nautiyal et al. (eds.), *Palgrave Handbook of Socio-ecological Resilience in the Face of Climate Change*,
https://doi.org/10.1007/978-981-99-2206-2_31

with a 37,807 million $ deficit and 0.36% less GDP per capita (Global Climate Risk Index 2020). Other contributing factors are land use changes (expansion of agricultural lands, the establishment of industries, illegal mining, using low-lying areas and river banks for various purposes and excessive tourism etc.) in the process of climate resilience. In addition, the Normandy Index (Normandy Index 2020) projects India's risk to climate change as its major shortcoming compared to other countries. The country has experienced a significant heat wave and deteriorating ambient air in the last two years. The Ministry of Earth Sciences (MoES 2020) has recently stated that the average temperature could be elevated by nearly 4.4 °C along with increased duration (approximately double), and the intensity of temperature during summer is predicted to be four times higher by the end of the century.

Given that climate change threatens the healthy survival of living organisms, this chapter looks at the policy initiatives in India through a holistic framework from climate, environmental and development actions that India is taking. The first section looks at policy initiatives at the domestic and international levels; the second section looks at the development actions through the noteworthy schemes and technology plans and how the judiciary has responded. This study has aimed at exploring the optimistic developments in India while limiting to exploring the many issues the country faces, especially in implementing these policies and laws today. For instance, while the judiciary has intervened through progressive judgments, the impact and enforcement of these are questionable, which is not the aim of this study.

31.2 Climate Resilience: Policy Initiatives in India

IPCC defines climate resilience as—*"The ability of a system and its parts to anticipate, absorb, accommodate, or recover from the effects of a hazardous event in a timely and efficient manner, including through ensuring the preservation, restoration, or improvement of its essential basic structures and functions."*

To strengthen climate resilience, India has supported multilateral efforts on climate actions. It ratified the Kyoto Protocol in 2002 and the United Nations Framework Convention on Climate Change (UNFCCC) in 1993. While India is a non-Annex country it does not have legally binding emissions reduction targets. However, it has participated in hosting Annex-1 countries to invest in emission-reducing projects following CDM investments primarily in the energy sector but did not participate in the flexible system for developed countries (emission trading and joint implementation) (Ranganathan and Goyal 2015). The Marrakesh Accords paved the way for creating an Expert Group on Technology Transfer (EGTT), which is a constituent of the technology transfer framework and will be instrumental in its application by keeping in view the technology needs, examining them, and suggesting alternatives to enable and advance technology transfer activities.

According to the Global Climate Risk Index 2020 analysis, extreme climatic conditions and financial deficits increase the vulnerability of underdeveloped and developing countries. Moreover, it may become more frequent and severe due to climate change, and India will suffer from extreme climatic conditions in the coming years. These contributing factors have led to heat waves, floods, a shift in season cycles and human rights violations. Accordingly, India has taken domestic policy initiatives on climate change. India laid down strict domestic policies in 2008, with the National Plan on Climate Change (NAPCC) with eight national missions and three more recently added in 2017. Climate change and energy security have been the focus of NAPCC and have been extended to the state level (SAPCC). In addition, India has made tremendous advancements in promoting clean cooking under the Pradhan Mantri Ujwala Yojana (PMUY), providing free cooking gas (LPG) to 80 million households. India has also set up potential targets to combat climate change in transport, intending to achieve 30% of vehicles to be run by electricity by 2030, along with 100% of electric three-wheelers and two-wheelers by the years 2023 and 2025, respectively. However, India has yet to adopt the legislation for the net-zero emission target. One of the dominant fuels in India to generate power is coal, meeting 50% of its commercial energy requirements. However, in the current scenario, India is a frontrunner in using renewable energy resources and set goal-oriented markers to generate electricity from renewables. In 2018, Delhi's National Electricity Plan (NEP) raised the target of 175 GW (Giga-watts) to be met by 2022 to 275 GW to be achieved by 2027 from renewables, thereby reducing the dependence on fossil fuels. To achieve an ambitious target of 40% of India's electricity to be obtained from non-fossil fuels by the year 2030, the government of India in 2019, raised the initial target of 275 GW to 500 GW by 2028; of which 350 GW and 140GW would be generated from solar energy and wind energy, respectively (Climate Action Tracker 2020).

Unfortunately, India is struggling to meet the 175 GW target by 2022. Many factors have contributed to the slow pace of the self-sufficient energy development strategy. COVID-19 was a sudden blow to India's economy and renewable energy goals. Another factor is the hurdles in the financial sector lagging the growth, such as an interest rate of 12–15% for loans compared to US and European interest rates. An absence of well-systematised and time-oriented plans further aggravated the situation. However, a considerable rise in renewable energy capacity is expected by 2020, regardless of these targets (Jhalani 2021).

On the international front, in 2015, Paris Agreement on climate change, the nationally determined contribution (NDC), was proposed by Delhi to the UNFCCC. A target of 33–35% decline in greenhouse gas (GHG) emissions compared to 2005 levels was set to be achieved by 2030. Furthermore, it committed to augmenting tree cover and forests by 2030, resulting in an increased decline in carbon (around 2.5–3 billion tonnes). Moreover, India is a leader and promoter of renewable energy; in 2018, the Prime Minister of India, Mr. Narendra Modi, along with the President of France, Emanuel Macron, conducted the International Solar Alliance (ISA) conference. At the UN Climate Action Summit in 2019, India proposed the Coalition for Disaster Resilient Infrastructure (CDRI), an association of the world's nations, to foster flexibility of novel and already prevailing set-ups towards weather and calamity risks (EU-India 2020).

Europe (EU) and India in the 2016 conference were noted as the major CO_2 emitters but also as global leaders on climate action resulting in EU-India cooperation and a joint announcement on a partnership on cleaner energy and weather implementation EU and India's intended nationally determined contributions (INDCs). Under the India-EU energy panel, both sides continued the energy dialogue to create an India-EU dialogue on weather alterations and reinforcement in weather and energy study.

During the 2017 conference, the EU-India strengthened their collaboration with a joint declaration on cleaner energy and weather change to nurture EU-India corporate collaboration along with new initiatives and cooperation in green cooling, advanced biofuels, energy storage and solar pumping. In addition, a joint action plan for keen and ecological urban advancement (2019–2020) was also initiated during the summit.

EU supports India's attempts to utilise renewable energy, like wind, solar power and smart grids. The Grid Connected Rooftop and Small Solar Power Plants programme (GCRT) was launched as a partnership between EU-India to aid India's target of establishing 40,000 MWs of photoelectric rooftops by 2022, deriving energy from solar systems. Also, the EU funds energy efficiency projects in buildings that run according to the Energy Conservation Building Code (ECBC). The other projects resulting from the EU-India partnership include India's Solar Park programme and Smart Grid Cooperation, a solar (3 MW) power plant in Bihar (SCOPE-BIG). In addition, under the EU-India urban partnership, there are 70 projects with a funding of €3.7 billion, along with initiatives like the world cities programme, the India-EU Urban Partnership (IEUP), the International Urban Cooperation India (IUC) and the EU-Mumbai Partnership. Also, under its programme on Asia Investment Facility, the EU allotted 31 million € for schemes in India (2016–2020) on green affordable housing, urban mobility, and digital cities, as well as €4.6 million for seven projects (2012–2023). The metro projects in India worth €2 billion have been funded by the European Investment Bank (EIB) and would finance additional projects for smart city transportation up to 1 billion Euros. The Smart Cities Mission (SCM) in 100 cities in India is supported by EU Smart Cities Knowledge & Innovation Programme (SCKIP), developing public transport, water supply and sanitation. (EU-India 2020).

At the 15th EU-India summit held in 2020, alterations in climate were one of the vital themes that confirmed the Paris agreement's implementation. EU-India assured their cooperation in promoting solar energy usage and mobilising private capital for environmentally sustainable investments in the International Platform on Sustainable Finance (IPSF). Furthermore, for effective climate crisis management, EU-India highlighted the importance of the "Leadership Group for Industry Transition" proposed in UN Climate in 2019.

They also approved work at close quarters to protect biodiversity by implementing a post-2020 universal infrastructure to be embraced at the 2021-UN Biodiversity Conference.

31.3 Technological Development Actions Through the Noteworthy Plans, Schemes and Missions

Energy Supply

One of India's main missions has been accessing electricity and clean cooking, wherein electricity was made available to 750 million people between 2000 and 2019 (IEA 2020). The government declared the achievement of electrification of nearly all households by 2020 (IEA 2020). In 2018, the usage of renewable energy was reported to increase (19%), while the use of coal for power generation reduced (73%) along with a 708 gCO_2/kWh fall in emission when compared to 771 g emission in 2015 (Climate Action Tracker 2020). The coal power contribution by India from the global pretext has significantly reduced in two years from 17% in 2018 to 12% in 2020. In 2017, under the central goods and services tax, the taxing of coal was considered for environmental purposes (*Brown to Green: The G20 Transition Towards a Net-Zero Emissions Economy: India* 2019). Moreover, India's subsidies for fossil fuels were reported as 10.8 billion USD in 2017 compared to 11 billion USD in 2008. Also, an additional 10.6 billion USD was allocated for projects related to coal in 2016–2017. The National Electricity Plan (NEP) has proposed a power emission level of 1.0 Gt CO_2e in 2021–2022 and 1.2 Gt CO_2e in 2026–2027 (CEA 2018). The NEP plans to achieve its targets to increase the renewable energy capacity to 175 GW by 2022, making a share of 29–40% renewables, which is projected to become 60–65% by 2030. However, to be compatible with the Paris Agreement, India needs to aim for a sustainable energy stake of 65–80% and 90–100% by 2030 and 2040, respectively (Climate Action Tracker 2020). In 2010, the paramount renewable energy-related policy, "National Solar Mission", was escalated with priority towards regulation in 2015, with a target of 175 GW by 2022, with the administration aiming to raise it to 227.6 GW (Darby 2018). In 2018, India announced a National Wind-Solar Hybrid Policy to facilitate the development of large-scale hybrid net-linked wind-solar photoelectric systems and novel technologies (Economic Times India 2018).

Industry

Accordingly, as per the "National Mission on Enhanced Energy Efficiency", the energy efficiency in industry has been enhanced by the PAT mechanism (Perform, Achieve and Trade), which has set energy targets based on intensity and saved 5.6 GW and 31 $MtCO_2e$ during 2012–2015 (BEE 2018). India is forwarding the "Leadership Group for Industry Transition" to ensure that large-scale industries meet the Paris Agreement (PIB Delhi 2019).

Transport

India aims for 30% of electric vehicles (EVs) by 2030, which is lower than the already announced target of 100% in consensus with intergovernmental targets EV30@30 campaign (Clean Energy Ministerial 2019). Also, the government plans for all-electric two-wheelers by 2026 (Carpenter 2019). In 2019 The Faster Adoption and Manufacturing of Electric Vehicles in India (FAME—II) was initiated to provide incentives to increase the purchase of electric vehicles in domestic transport as an encouragement towards zero emission fuels (Business Today 2019). In 2017, India laid the foundation of its introductory light vehicle fuel efficacy standards with an efficiency of 130 gCO_2/km in 2017 and proposed a low of 113 gCO_2/km in 2022 (Transportpolicy.net 2017). In 2020, Indian railways proposed the completion of its network electrification by 2023 and aimed to achieve its target of zero emissions by 2030 (Cuenca 2020).

Agriculture

In 2012, the National Mission for Sustainable Agriculture (NMSA) was initiated to promote reduced emissions in the agricultural sector. However, it has not achieved its underlined targets (Rattani et al. 2018). Agriculture and electricity are closely linked as electricity is utilised for water pumping in modern irrigation. However, poor power supply to agriculture has resulted in incompetent pumps and an escalated usage of power and water (Sagebiel et al. 2016). To manage inconsistencies in power supply, the Ministry of New and Renewable Energy has initiated the KUSUM scheme to instal grid solar pumps and solarisation of pumps connected to the grid to aid the farmers in selling excess solar power to DISCOMs (Government of India 2019). In addition, National Bank for Agriculture and Development (NABARD) facilitated climate change mitigation and adaptation through several initiatives (NABARD 2019).

Forestry

In 2019, the National Forest policy called for ensuring 1/3rd of India's geographical region be tree-clad and with forests, thereby the target of NDC be backed (by 2030) with a further carbon sink of 2.5–3 Gt CO_2e (Kukreti 2019). Also, the Green India Mission aims to enhance carbon sequestration; however, in 2015–2016, the task was concise by 34% of its goals for plantations. (Rattani et al. 2018).

Other National Strategies

Some of the noteworthy plans, missions and schemes that contributed towards India's climate resilience are the India Cooling Action Plan (2019)—towards energy efficient refrigerants and air conditioners; the Jawaharlal Nehru

National Solar Mission (2015)—more production of solar energy, that is five folds to reach 100,000 MW by 2022; under the National Mission for Enhanced Energy Efficiency—a novel mechanism, the Perform, Achieve and Trade (PAT), which permits trade in energy saving certificates from industrial sectors such as thermal power stations, iron and steel, textiles, pulp and paper among others; the Bureau of Energy Efficiency (BEE)—to support in policy and strategy development for self-regulation and fair market principles with principle aim in being energy efficient keeping in view the complete outline of the Energy Conservation Act, 2001 such as the ECBC in 2017 replacing the 2007 code (can be modified as per local requirements), the National Electricity Plan in 2018. The former is towards reducing emissions of its GDP (35% less than the levels of 2005) by 2030, and the latter is towards improving the electricity sector in India by targeting 275 GW of renewable energy retention by 2027.

EIA

EIA emerged as one of the most effective policies for protecting the environment from anthropological invasions in the form of projects such as mining, dams, thermal power plants, infrastructure projects like ports, airport, highways, and large-scale projects. However, the concern over the consequences of drastic modernisation led to the release of the EIA notification on 27 January 1994 by the Union Ministry of Environment and Forests (MEF), Government of India, under the Environmental (Protection) Act 1986. Twelve more amendments followed it, and the most recent one was released on 13 March 2020, raising severe public concern (CSE 2021).

Even though the policy was implemented to highlight transparency and conservation, the entire draft appears controversial. The most questioned part is post-facto approval given to the projects even if they have started working without practising good environmental protection policies. The gas leak incident in LG Polymers India Pvt Ltd of Visakhapatnam, the oil field incident in Baghjan are examples of disastrous legitimisation. Furthermore, the public hearing period has been reduced to 20 days, contrary to the environmental democracy advocated in previous EIA drafts. The draft even neglects the importance of public consultation in projects like construction, building development, broadening of national highways and inland waterways, and projects regarding national defence and security systems. They also extended the environmental clearance period from 30 to 50 years which will indisputably produce vast ravaging effects on the environment. The compliance report submitted by the cleared project has been extended to one year, which underlines that EIA 2020 is an "Enhancing Industries and Politics", not an 'Environmental Impact Assessment'. In addition, assessing EIA based on single-season data is a clear-cut indication of conspiracy and a lack of basic knowledge on the seasonal variations in biodiversity (Aggarwal 2020; Mishra et al. 2020). The newly implemented policies of EIA will produce far-reaching climatic instabilities and biodiversity effects, and it requires an imperative revision.

In light of the climatic fluctuations and biodiversity loss impacted by the extensive construction works, it is the need of the hour to think, rethink and rethink again before approving massive, devastating environmental policies. A robust well-oriented, multifaceted approach assessing the impact of projects and resilience of the affected environment along with public viewpoints should be a standard to formulate policies like EIA.

31.4 JUDICIAL RESPONSE OVER THE DECADE TOWARDS A HEALTHY ENVIRONMENT

In addition to policies, legislations and schemes, plans and missions, the judiciary in India has played a prominent role in environmental governance. An analysis of the mentioned judgments is depicted in Table 31.1 (Reddy 2012–2018). According to the results of a year-long survey, the Indian judiciary has consistently upheld the right to live in a clean environment. Judiciary has played an active role in the protection of ecology in liaison with the administration of environmental and forest laws. In 2012, some prominent examples were resort establishment cases in Lakshadweep Islands, illegal industrialisation and mining, acquisition of private land for urban development, and practice of *Makara Jyothi* at *Sabarimalai* temple issues which caught the attention of the authorities. The supreme court's proactive role and the central government's constructive role have played an influential role as surveillants and supervisors in safeguarding environmental protection along with speedy decision-making that imposed high precautionary costs. Cases of the immersion of idols in water bodies, brick-kilns, illegal mining, pollution-free environment, anthropocentrism for the preservation of rare animals, sustainable development, the feasibility of nuclear power plants and peoples' sovereignty over natural biological resources were under the scanner in the year 2013.

Given the above cases, the Ministry of Environment, Forests and Climate Change reviewed the existing government policies and strategies along with the Environment (Protection) Act, 1986, Forest (Conservation) Act, 1980, Wildlife (Protection) Act, 1972, the Water (Prevention and Control of Pollution) Act, 1974, and the Air (Prevention and Control of Pollution) Act, 1981. In 2014, the holding up of projects such as dams, power lines, roads and canals etc., for frivolous reasons, declining environment quality, failure in monitoring and enforcement mechanisms under the various environmental laws were observed. Two novel ventures, National Environment Management Authority (NEMA) and State Environment Management Authorities (SEMA), were set up to revamp the environmental clearance proceedings as per the EIA Notification 2006 to process all applications related to ecological authorisation in time. In the forthcoming year, significant judgments, orders, and directions have

Table 31.1 Indian judiciary response over the decade towards climate resilience

Case	Remarks	Reference
T.N. Godavaraman Thirumulpad vs. Union of India	The court directed the state of Chhattisgarh to design a rescue plan to save the red sanders and wild buffalo and take immediate steps to maintain the wild species' genetic purity by ensuring interbreeding between the domestic and wild buffalo does not take place	AIR 2012 SC 1254
Surinder Singh Brar vs. Union of India	The court directed the acquisition of land for urbanisation and environmental protection in which Chandigarh administration allotted the land meant for the development of a technology park to a private agency which in turn was found to be selling the land to other private agencies for profit	[2012] INSC 617 (11 October 2012)
Union Territory of Lakshadweep vs. Seashells Beach Resort	The court intended to conserve and protect Island's marine site and the unique environment by enhancing advancement via scientific principles through a maintainable integrated plan, considering the coast's vulnerability to natural hazards	[2012] INSC 319 (11 May 2012)
Deepak Kumar vs. State of Haryana	The court showed its concern towards serious illegal and unrestricted in-stream, upstream and floodplain sand mining and the degree of degradation of the riverbeds and the other sites of the environment along with being a threat to biodiversity	AIR 2012 SC1389: [2012] INSC 146 (27 February 2012)
Dipak Kumar Mukherjee vs. Kolkata Municipal Corporation	The court targeted the menace of unauthorised and illegal constructions of structures and buildings in different parts of the country, acquiring monstrous proportions	[2012] INSC 607 (8 October 2012)
G. Sundarrajan vs. Union of India	Supreme Court seemed to follow the principle of sustainable development and public trust in permitting the Kudankulam Nuclear Power Plant (KKNPP) to become operational	(2013) 6 SCC 620

(continued)

Table 31.1 (continued)

Case	Remarks	Reference
Pandurang Sitaram Chalke vs. State of Maharashtra	NGT ordered the auction of the illegally mined stone and to use the amount for green belt development in the area and held strictly that mining is carried out only after acquiring environmental clearance	http://www/greentribunal.gov.in/judgement/14-2012(WZ)(App-1ost2013-final-order.pdf
State of Kerala vs. R Sudha	The court dealt with the issues of dumping human excreta, waste and garbage in forests and rivers around Munnar and the failure of the Kerala state government to arrest the ongoing trend	(2013) INSC 715
Association for Environment Protection vs. State of Kerala,	The court exercised the doctrine of public trust. And did not permit the construction of hotels/restaurants over natural ecosystems, river bodies, forests etc., so that they are accessible to the public and not confined to private agencies	(2013) 7 SCC 226
Samaj Parivartana Samudaya vs. State of Karnataka	The Supreme Court pondered the court's contours over the plunder of natural resources, overexploitation, mining, and its environmental impact	(2013) 8 SCC 154
M/s U.A.L. Industries Limited vs. State of Bihar	The court showed concern over the State Pollution Control Board's (SPCB) approval for running an asbestos industry in a residential area	[2014 (4) FLT 237 (Pat., HC)]
Krishan Kant Singh vs. National Ganga River Basin	The court addressed issues related to water pollution in the river Ganga due to the discharge of toxic industrial waste, specifically between Mukteshwar and Narora	2014 ALL (1) NGT Reporter (3) (Delhi) 1
T. N. Godavarman Thirumulpad vs. Union of India	The court addressed the matter regarding tropical forest deforestation in Gudalur and Nilgiri sites by violation of the Forest (Conservation) Act, 1980, as well as the Tamil Nadu Hill Stations (Preservation of Trees) Act, 1955	(2014) 6 SCC 150; [2014 (4) FLT 261(SC)]; AIR 2014 SC 3614

(continued)

Table 31.1 (continued)

Case	Remarks	Reference
Occupational Health and Safety Association Petitioner vs. Union of India	The court showed concern over occupational health services with a lack of an adequate health delivery system and facilities or guidelines regarding occupational safety	(2014) 3 SCC 547
National Highways Authority of India vs. Government of Tamil Nadu,4	The court dealt with the validity of objections claimed by the Public Works Department of Tamil Nadu with respect to the construction of the project "Chennai Port - Maduravoil Elevated Corridor"	(2014) 2 MLJ 280
Shri Krishna vs. Union of India	The court upheld the allocation of forest land for the rehabilitation of Gujjars after obtaining permission from the Central Government	AIR 2015 (NOC) 971 (UTR
Regional Deputy Director vs. Zavaray S. Poonawala	The Supreme Court addressed leopard hunting in Zambia, shot with permission from Zambian authorities. The judiciary clarified the role of Wildlife, CITES and the customs authorities in the arena of wildlife protection towards climate resilience	2015(5) FLT 428(SC); [2015] INSC 27
Arkavathi Kumudavathi Nadi Punschetana Samithi vs. State of Karnataka	A PIL was filed to declare Hesaraghatta Natural Grasslands in Karnataka as a protected area under the Wildlife. However, the court granted "status-quo" to be maintained on the above matter	2015 (5) FLT 328 (Kar. HC)
Chief Conservator of Forests Thiruvananthapuram vs. The Secretary, Paramekkavu Devaswom, Thrissur	In Kerala court passed an order, under the Wildlife Protection Act, for the transportation of an elephant from another state to Kerala; prior permission is mandatory	AIR 2015 (NOC) 916 (Ker)
Indian Hotels Company Ltd vs. Suo Motu	High Court of Gujarat showed concerns about setting up a new tourism site at Chikhal in the south-eastern part of the Gir Wildlife Sanctuary, as well as the hotels which were constructed in violation of the no objection certificate (NOC) issued by the authorities	2015 (5) FLT 489 (Guj. HC)

(continued)

Table 31.1 (continued)

Case	Remarks	Reference
Vardhman Kaushik vs. Union of India	NGT refused to ban ten or more years old vehicles running on diesel though they pollute the clean air and environment	2015 (5) FLT 456 (NGT-PB)
Sarang Yadwadkar vs. District Collector of Pune	NGT initiated the prevention of water pollution at the Ganesh festival every year and to not permit the Plaster of Paris (POP) idols of more than permissible heights along with building artificial tanks based on the minimum water quality	2015 (5) FLT 811 (NGT, WZ Bench, Pune
Antony vs. Corporation of Cochin	High Court of Kerala ordered that any construction within 100 metres from the High Tide Line of the Chilavannor lake, Backwaters would violate notifications (CRZ)	2015 (5) FLT 332 (Ker HC)
A. Paramasivan vs. Tamil Nadu Pollution Control Board	The NGT directed towards stopping unauthorised prawn culture activities in Kaliporamboke by disconnecting electricity and removing encroachment	2016 (6) FLT 1 (N.G.T.-S.Z.-Chennai Bench)
In M.C. Mehta vs. Union of India	The Supreme Court addressed the hardships of residents of Delhi due to high pollution levels (the most polluted city in the world)	(2016) 4 SCC 269
AYUSH Drugs Manufacturers Association vs. State of Maharashtra	The petitioners raised their concern regarding specific rules about the Biological Diversity Rules (2002), the guidelines on access to biological resources and associated knowledge, and regulations on benefits sharing (2014)	AIR 2016 Bom 261
Basheer Mohd vs. District Collector, Thrissur	The court dealt with the quarrying of minerals and blasting operations resulting in property damage, residents' air pollution, and noise pollution by blasting	19 AIR 2017 (NOC) 203 (Ker)
Sunder Singh, President Residents Welfare Association, Issarpur New Delhi vs. State of NCT of Delhi	NGT delivered a noteworthy judgement concerning the revival of water bodies	2017 (7) FLT 28 (NGT-PB-ND)

(continued)

Table 31.1 (continued)

Case	Remarks	Reference
In Pankaj Kumar Mishra vs. Union of India	The NGT addressed severe health hazards and environmental pollution due to the industrial waste of the Singrauli Industrial Area	2017 (7) FLT 86 (NGT Principal Bench New Delhi)
In Zulfiquar Hussain vs. Government of NCT of Delhi	A PIL was filed to the Ministry of Environment, Government of NCT of Delhi, to prohibit nylon kite thread synthetic threads used for flying kites throughout Delhi	18 2017 (7) FLT 57 (Delhi HC)
Secretary, Kerala State Coastal Management Authority vs. DLF Universal Ltd.,	The court dealt with the construction of residential flats of a multi-storeyed complex which fell under the CRZ (coastal regulation zone)	2018 (8) FLT 83 (S.C.); AIR 2018 SCC 389
Goa Foundation vs. SESA Sterlite Ltd	The court made profound observations against illegal and indiscriminate mining resulting in the worst environmental degradation	AIR 2018 SC (Supp) 1269: 2018 (8) FLT 173 (SC)
TPPL vs. State of Chhattisgarh	The high court raised concern regarding the disposal of biomedical waste being generated by labs, nursing homes and hospitals	2018 (8) FLT 347 (Chhattisgarh HC)

been issued by various courts and tribunals in India, and the focus on environmental protection seems to have been shifted to the NGT (National Green Tribunal) Act, 2010. In 2015, the judiciary witnessed cases related to biodiversity, coastal zone regulations, pollution, safeguarding the environment, and forest conservation. The kind of enthusiasm some government offices, NGO's and volunteers have shown regarding the environment during the year 2016 is commendable. Cases on the declaration of sites, lakes and structures which are deserving as heritage sites in India's biodiversity, banning the firecrackers sale in Delhi, and illegal exploitation of groundwater, have been addressed in the year 2016, and compensation has been imposed on the defaulters degrading the environment. Thus, the year witnessed a progressive and active year of environmental protection. In 2017, the union and state governments took no new policy decisions related to environmental protection. The judiciary took up cases concerning the disappearance of lakes, pollution, kite flying hazards and an attempt to release vehicles with old emission standards. Analysis of cases under the scanner shows several proactive directions by the court to safeguard India's feeble ecology and environment. In the last two years, the government has framed new rules and made modifications related to protection from dust, corporate environment, forest policy, rules for bio-waste management and e-pollution. Thereby the tribunals have demonstrated effective strategies to protect our ecology and environment. Cases related to mining, exploitation of natural resources, pollution of groundwater with industrial waste, push-by to sell vehicles with outdated compliance, illegal constructions, violation of coastal zone regulations, banning firecrackers, and safeguarding forest rights have been the focus of the judiciary.

31.5 Conclusion

Looking into all these approaches from the perspective of policies, schemes, plans, technological developments, and the judiciary's role in climate resilience is a question of in-depth analysis that probably has not been considered. Certain researchers analyse that India will suffer due to climate change in the next eighty years. These approaches are more mitigation-centred or adaptation centred. The coronavirus pandemic has halted the economy and sharp short-term emissions reduction. Still, India has not accelerated towards its target of electrical mobility and a decline in coal usage. Ironically, the country's coal capacity would increase from the current use of 200 GW to nearly 300 GW in the years to come. Over the decades, India has wisely progressed through technological developments for mitigation and adaptation with a shortage of supportive domestic policy frameworks. The judiciary's response has been influential in protecting ecology and the environment. It would further pave the way for new policies and strategies, as well as a sense of ownness by the country's people in creating a paradigm shift in this global issue of climate change.

References

Aggarwal M (2020) India's proposed overhaul of environment clearance rules could dilute existing regulations. Mongabay https://india.mongabay.com/2020/03/indias-proposed-overhaul-of-environment-clearance-rules-could-dilute-existing-regulations/. Accessed 18 March 2020.

Bureau of Energy Efficiency (2018) PAT scheme (Perform, Achieve and Trade scheme).

Business Today (2019) Govt notifies FAME-II scheme with Rs 10,000-crore outlay to encourage adoption of electric vehicles. https://www.businesstoday.in/latest/economy-politics/story/govt-notifies-fame-ii-scheme-with-rs-10000-crore-outlay-to-encourage-adoption-of-electric-vehicles-173755-2019-03-09. Accessed 9 March 2019.

Brown to Green: The G20 Transition Towards a Net-zero Emissions Economy: India (2019) https://www.germanwatch.org/en/17200#:~:text=Main%20findings%20of%20the%20Brown%20to%20Green%20Report%202019&text=82%20%25%20of%20the%20G20's%20energy,has%20also%20slowed%20in%202018. Accessed 11 November 2019.

Carpenter S (2019) India's plan to turn 200 million vehicles electric in six years. Forbes Sustainability. https://www.forbes.com/sites/scottcarpenter/2019/12/05/can-india-turn-nearly-200-million-vehicles-electric-in-six-years/?sh=4cabc88715db. Accessed 5 December 2019.

Central Electricity Authority (2018) National Electricity Plan. https://cea.nic.in/wp-content/uploads/2020/04/nep_jan_2018.pdf. Accessed January 2018.

Centre of Science and Environment (2021) https://www.cseindia.org/understanding-eia-383.

Clean Energy Ministerial (2019) EV30@30 campaign | Clean Energy Ministerial |EV30@30 campaign | Advancing Clean Energy Together. https://www.cleanenergyministerial.org/content/uploads/2022/03/ev3030-fact-sheet-june-2019.pdf. Accessed June 2019.

Climate Action Tracker (2020) Paris agreement compatible sectoral benchmarks: elaborating the decarbonisation roadmap. https://climateactiontracker.org/documents/754/CAT_2020-0710_ParisAgreementBenchmarks_SummaryReport.pdf. Accessed August 2020.

Cuenca O (2020) Indian Railways targets net zero emissions by 2030. Int. Railw. J. https://www.railjournal.com/technology/indian-railways-to-achieve-net-zero-emissions-by-2030/. Accessed 16 July 2020.

Darby M (2018) India's power minister says the country can smash its 2022 renewable power goal. Will it happen? Climate Home News. https://www.climatechangenews.com/2018/06/13/india-says-will-smash-2022-renewable-power-goal-will-happen/. Accessed 13 June 2017.

Economic Times India (2018) Government announces national wind-solar hybrid policy. The Economic Times India. Accessed 14 May 2018.

Eckstein D, Künzel V, Schäfer L, Winges M (2019) Global climate risk index 2020. Bonn: Germanwatch, pp 1–44. https://www.germanwatch.org/sites/germanwatch.org/files/20-2-01e%20Global%20Climate%20Risk%20Index%202020_13.pdf.

EU-India: Cooperation on Climate; European Parliamentary Research Service (2020) https://www.europarl.europa.eu/RegData/etudes/BRIE/2020/659348/EPRS_BRI(2020)659348_EN.pdf. Accessed November 2020.

Government of India (2019) KUSUM Scheme.

International Energy Agency (2020) India 2020—energy policy review. India 2020. https://doi.org/10.1787/8550daed-en.

Jhalani N (2021) India looks likely to miss the 2022 renewable energy targets. Mongabay. https://india.mongabay.com/2021/06/india-looks-likely-to-miss-the-2022-renewable-energy-targets/. Accessed 3 June 2021.

Krishnan R, Sanjay J, Gnanaseelan C, Mujumdar M, Kulkarni A, Chakraborty S (2020) Assessment of climate change over the Indian region: a report of the ministry of earth sciences (MOES), government of India. Springer Nature, p 226.

Kukreti I (2019) Draft National Forest Policy cleared; Cabinet to take decision. Down to Earth. https://www.downtoearth.org.in/news/forests/draft-national-forest-policy-cleared-cabinet-to-take-decision-67945. Accessed 26 November 2019.

Lazarou E (2020) Mapping threats to peace and democracy worldwide: Normandy Index 2020 https://doi.org/10.2861/068647.

Mishra A, Mohanbabu N, Anujan K (2020) Draft EIA 2020 undercuts India's biodiversity and climate goals. Down to earth. https://www.downtoearth.org.in/blog/environment/draft-eia-2020-undercuts-india-s-biodiversity-and-climate-goals-73201. Accessed 3 September 2020.

NABARD (2019) Other climate change initiatives of NABARD. National Bank for Agriculture and Rural Development.

Press Information Bureau (PIB) (2019) Leadership group to drive industry transition to low-carbon economy, MoEFCC. https://pib.gov.in/PressReleasePage.aspx?PRID=1585963. Accessed 24 September 2019.

Ranganathan K, Goyal MK (2015) Clean development mechanism–an opportunity to mitigate carbon footprint from the energy sector of India. Curr. Sci. 109(4): 672–678.

Rattani V, Venkatesh S, Pandey K (2018) India's National Action Plan on Climate Change needs desperate repair. Down to Earth. https://www.downtoearth.org.in/news/climate-change/india-s-national-action-plan-on-climate-change-needs-desperate-repair-61884. Accessed 31 October 2018.

Reddy GB (2012–2018) Environmental Law, Indian Law Institue, Annual Report.

Sagebiel J, Kimmich C, Müller M, Hanisch M, Gilani V (2016) Background of the agricultural power supply situation in India and Andhra Pradesh. In Enhancing Energy Efficiency in Irrigation, Springer, pp 7–17. https://doi.org/10.1007/978-3-319-22515-9.

Transportpolicy.net (2017) India: light-duty: fuel consumption.

CHAPTER 32

Green Finance for a Greener Economy

Meenakshi Rajeev and Oisikha Chakraborty

32.1 Introduction

Threats of climate change and the rising level of the planet's temperature call for investment in green products and technologies to reduce environmental damage. For such investments to take place, finance is a prerequisite, and each financing instrument or endeavour, be it public or private, initiated for a green cause falls under what is termed green finance. Thus, green finance refers to the allocation of capital towards protecting the environment, clean energy, green building, and climate change mitigation (Frimpong et al. 2021). According to the UN definition, "Green financing is a method to increase financial flows from banking, micro-credit, insurance and investment, public, private and not for profit sectors to sustainable development priorities" (www.unep.org). Given that green initiatives can take place only with adequate finance, green financing has become an important area of research, and the current paper is an exploratory work in this area.

Several products have been developed to meet the demand for green consumption and investment in green causes. Such causes range from reducing emissions to using renewable resources and leaving lower carbon footprints.

M. Rajeev (✉)
Centre for Economic Studies and Policy, Institute for Social and Economic Change, Nagarabhavi, Bengaluru, India
e-mail: meenakshi@isec.ac.in

O. Chakraborty
University of Hyderabad, Hyderabad, India

© The Author(s), under exclusive license to Springer Nature Singapore Pte Ltd. 2023
S. Nautiyal et al. (eds.), *Palgrave Handbook of Socio-ecological Resilience in the Face of Climate Change*,
https://doi.org/10.1007/978-981-99-2206-2_32

Green bonds are the most common and prominent instruments among various financing options. Green bonds as the name suggests are bond instruments operated in the debt market where the proceeds are utilized for green projects that are energy-saving (The Global Commission on the Economy and Climate 2016). Within the category of bonds, to serve climate finance, instruments like climate bonds are being developed, which are though relatively new, their prevalence is increasing. On the other hand, green products developed by microfinance institutions concentrate on small business owners to help them take up green investments. In addition, there are also green retail banking products such as green car loans, housing loans, etc. In order to mitigate the risks of climate change there are also insurance products.

Even though the number of green financing products has increased over the years, funds generated fall much shorter than required. Various risks, including a lower rate of returns, inadequate technology, and high operational risk, present with green endeavours, which make private players reluctant to invest in green projects (Rawat 2020). At the same time, public institutions alone cannot generate sufficient funds, thereby offsetting the efforts of moving towards a green economy. This supply–demand gap, however, varies across countries.

Various nations have varied experiences with respect to green financing. China, among all the developing countries, is at the forefront of implementing green finance. In the case of India, public sector units, including public banks, play a significant role in generating finances for green projects. Similar to the global trends, India's power generation sector is the primary beneficiary of green finance. However, given the commitment of the Indian Government, the supply of green funds remains far short of demand.

This exploratory paper based on secondary sources looks at the various financing instruments for green causes and their utilization by selected Asian countries including India. We classify the products in three major heads, viz., green deposits and credits; green bonds and green insurance. The paper highlights the major financing challenges faced by India in meeting its green commitments. With these objectives, the paper unfolds as follows. The next section provides a brief review of the literature. It is followed by a discussion of the major green financing products. We look at green investment through various financial products by selected Asian countries in Sect. 32.4. Section 32.5 is devoted to the Indian experience and a concluding section comes at the end.

32.2 Review of Literature

Given the growing importance of Green Finance, various scholars have spoken about it, and hence the literature is vast. A comprehensive review of the literature is beyond the scope of this study. Below are a few contributions on the issue of green finance. The review of the literature is divided into three major sections. The first section deals with the importance of green finance products, the following section discusses the various products of green finance, and the third section delves into the various green finance initiatives undertaken globally.

Importance of Green Finance

As climate change has become a global concern, various studies have tried to establish the importance of green finance and the need to scale it to reduce carbon emissions significantly. For example, some comprehensive studies by, Ticci and Gabbi (2018), Jones (2015), Wang and Zhi (2016), Noailly and Smeets (2016), Li and Jia (2017), Haque and Murtaz (2018), Jayathilake (2019), Zhou et al. (2020), Meo and Karim (2021), and Huang and Zhang (2021) all showed that green finance is the most efficient financial strategy for reducing carbon emissions and it is useful in mitigating conflicts between economy and environment. A recent addition to the literature regarding the increasing health crisis and environmental degradation was discussed by Klioutchnikov and Kliuchnikov (2021). The huge body of literature shows that green finance products are essential in achieving a carbon–neutral state by funding numerous green technology projects.

Therefore, the following section reviews the literature on green finance tools which are developed to achieve sustainable development goals.

Green Finance Products

An increasing body of literature assesses the numerous products of green finance. Green finance includes public funds, venture capital, equity, debt, pension, green infrastructure bonds, etc. (Volz et al. 2015). According to Raberto et al. (2019), different green finance products with specific terms and conditions are emerging in the banking sector to enable growth in energy-efficient production technology.

Various scholars have also looked into the emerging importance of green bonds as a product heavily used to finance green projects. In one such study carried out by Reboredo (2018), the ease of implementation of green bonds for the renewable energy sector and the co-movement between green bonds and financial markets were majorly discussed. Also, the need of expanding green finance products like green bonds was discussed by Hafner et al. (2020). Similarly, a study on green-traded stocks and bonds was conducted by Ziolo et al. (2019). In addition to bonds, Pham (2021) emphasized the role of green equity and talked about how these unconventional green assets can benefit environmental performance.

Along with the importance and consequences of issuing green bonds, the demand and supply side of European green bonds was studied by Gianfrate and Peri (2019). Further, a structural model for green bonds was developed by Agliardi and Agliardi (2019) to explain the dynamics of green bond price mechanics. Additionally, a few works in the literature talk about the implementation of green finance as a public policy tool (Ng 2018; Flammer 2020). The authors emphasized that green bonds are instrumental in achieving sustainable development objectives. In this regard a study conducted a socio-ecological impact assessment of green bonds (Zerbib 2019). The various literature shows that the positive impacts of green bonds is coming to light with growing awareness of climate change.

There are various other instruments like green loans, green leases, green equities, etc., which are studied by scholars. A paper by (Li et al. 2018) attempts to explain the effectiveness of green loans in promoting green innovation. Diverse literature on green finance products was reviewed by Frimpong et al. (2021), which stated that green loans are for those start-ups and small firms involved in environment-friendly projects. Another mechanism to facilitate green loans was the Green Bank Network (GBN) developed by various countries to facilitate knowledge transfer and mobilization of funds for clean energy projects around the globe (Ngan et al. 2019). Similarly, another tool, "enVinance" was developed to promote green financing among commercial banks (Oh and Kim 2018).

Given these instruments and the role of green finance for sustainable development, undoubtedly, green finance tools have been taken up by countries for reducing carbon emissions and adopting green technologies.

Global Experience in Green Finance

There needs to be significant investment in related projects to achieve a green economy. As mentioned above, different products are developed to facilitate green finance and many countries have adopted them to achieve their long-term sustainable goals. There is a body of literature that discusses green finance adoption across the globe. This subsection is further divided into two sections. The first section describes green finance adoption in developed countries, and the second section discusses green finance endeavours in developing countries.

Developed Countries

Developed countries have undertaken major steps to curb pollution and used green finance products at their nascent stage. Here we discuss various developments in the arena of green finance in developed countries by taking a few examples. Firstly, in the European Union, Quantitative Easing programs were undertaken to support environment-friendly projects (McDaniels et al. 2016). This study provides an overview of contributions made by different banks in Germany, France, and the United Kingdom. Secondly, a comprehensive research on the development of green finance in Germany was presented by Schäfer (2018). Thirdly, a thorough investigation of the challenges and opportunities of green finance in Italian biomass production was studied by Falcone and Sica (2019). These are only a few selected research in the area.

Further, a study on sustainable investment, banking, climate risk, insurance, and financial market innovation in the USA, revealed various developments and issues in the sustainability framework (UNEP 2016a, b). Similarly, a report on the Green Bond market in Denmark, Finland, Iceland, Norway, and Sweden discussed the increasing usage of green bonds and continuous innovation in the Nordic market to support climate action (Climate Bonds Initiative 2018). Thus from the existing literature, we get specific insights into green finance status in in various developed nations.

Developing Countries
There is limited literature that discusses the green finance endeavours of developing countries. A few of those studies are briefly reviewed in this section. It is often mentioned that developing countries are exploiting natural resources to achieve higher-digit growth. Therefore, it becomes important to mention the study by Pigato (2018), which states that the growing reliance on fiscal instruments for green financing projects is the need of the hour. While exploring various literature, it was found that China, among all the developing nations, is at the forefront of implementing green finance tools. For instance, Ren et al. (2020) and Yu et al. (2021), through empirical evidences, assessed green innovations by Chinese firms and showed the progress made by China.

Similarly, the rapid expansion of the Chinese green bond market was studied by (Elliot and Zhang 2019). The assessment of institutional arrangements for climate change and finance in Indonesia was reviewed by Maulidia and Halimanjaya (2014). Another report assessing climate finance in Indonesia was made by Tänzler and Maulidia (2013). A study on green energy development and energy mix in Vietnam was conducted by (Nguyen 2018). Further, the background of green finance in Malaysia was investigated by Ngan et al. (2019). This study assessed the risk for green finance in the Malaysian Palm Oil Biomass Industry. The contribution of private commercial banks in Bangladesh to the development of green finance was studied by Zheng et al. (2021). Similarly, Bangladesh's flagship green fund programs, i.e., the Bangladesh Climate Change Trust Fund and Bangladesh Climate Change Resilience Fund, were investigated by Hossain (2018). To study the state of green finance in the Republic of Korea, Oh and Kim (2018) analysed the GHG emissions, Green Finance, National Policies for climate change, and energy mix. For Africa, a comparative analysis of Nigeria's performance using the SDG Index and Dashboard Indicators Framework was done by Onanuga and Onanuga (2019). Also, an analysis of green financing and climate change mitigation of N-11 and BRICS countries from 2005 to 2019 was done by (Nawaz et al. 2021). A few studies on green financing in India have been discussed in section 32.5 where we take up the case of India.

To sum up, the findings on the importance of green finance, its instruments, and the acknowledgement of significant usage of these innovative green financial products across the globe undoubtedly point to bringing the unconventional green financial product to the forefront of the financial sector to achieve more than just a sustainable economy.

32.3 GREEN FINANCIAL PRODUCTS

The emergence of green financial products and services has enabled an awareness of the severity of environmental challenges (UNEP 2007). We would discuss here three major categories of green finance products: retail banking products, investment funding, and green insurance.

Credit-Based Green Products for Consumers (Retail Banking Products)

Loans are today increasing given to consumers for green products. Green home mortgages are loans provided for houses built through energy-saving technology. These loans are given at a reduced rate of interest or without collateral. In addition, there are subsidized loans to install environmentally friendly devices such as solar panels or, loan provision for cars using energy-saving technologies. These loans are available in developed countries like the USA and as well as in developing countries like India.

Another interesting initiative under this category is the fleet loan (in the USA) where the trucking companies are given loans to introduce environmentally friendly energy-saving devices in their transport. Green credit and debit cards are other useful devices for financing green projects because a specific part of the proceeds from the card usage goes for green causes; for example, Robobank of Europe donates the proceeds to WWF (UNEP 2007). Another innovative credit instrument is green loans for entities requiring finances for suitable green assets. Any entity which follows eco-friendly operation techniques or judiciously uses some non-renewable sources of energy and has a decarbonization goal is eligible for green loans (Green Loan Principle 2021) The purpose is to raise capital. But as mentioned green loans are different from traditional loans because the loan amount is specially used for climate-friendly or resource savings investments. This instrument is necessary because not all projects can be financed with just subsidies, and larger funding is required for capital to build various green projects. For example, in Germany, with the help of green loans, an energy-efficient youth hostel was constructed (Landesbank Baden-Wurttemberg n.d.). There are a large number of green loan providers across the globe (for example see Nefco and Swedbank (https://www.nefco.int and https://www.swedbank.com)).

There exist microfinance instruments used by banks to fund small enterprises that usually do not get access to large commercial banks. Green microfinance programs exist to help small business owners become economically self-sufficient. Some examples of green microcredit schemes are Grameen Shakti Renewable Energy Project (Bangladesh) and Alterna Savings (Canada). The Alterna Microloan Fund Program promotes green projects as the borrowers use the capital in environment-friendly projects (Rouf 2012). EcoMicro is financed by Nordic Development Fund and another instrument viz., Multilateral Investment Fund to assist MSMEs and low-income households in Central and South America to utilize cleaner energy and adopt an energy-efficient mechanism (Inter-American Development Bank 2014) are also important initiatives.

In addition to credit-based products, there are also deposit-based products that in turn are loaned out for green causes. In this regard Green Deposit which is a unique green finance product that allows the money depositor to help the environment by providing financial assistance for green projects. An example of green deposits is a program undertaken by Westpac in Australia called the Land Care Term Deposit, which supports sustainable cultivation practices. This scheme enables the clients to deposit their money into a green deposit account, and depending on the account's balance, the bank makes annual donations towards sustainable agriculture (UNEP 2007). Another initiative by ShoreBank in the USA was called EcoDeposit. These special deposits provided financial assistance to local energy-efficient projects or companies aiming to reduce waste and conserve the ecology (UNEP 2007).

Similarly, we see that under its new green deposit program, Deutsche Bank provides required liquidity or loans to clients to help them meet their ESG goals (Deutsche Bank 2021). In this regard Citi Treasury and Trade Solution launched green time deposits for short-term and long-term investments to provide liquidity for environmentally friendly projects (Citigroup 2020). These emerging products attract customers and draw the attention of a wide range of investors, which help banks raise profits and contribute towards sustainable development.

Investment Funding

As mentioned earlier, there are various debt instruments available to finance green projects, prominent among of them is green bonds. Green bonds have the same underlying mechanism as regular bonds, where proceeds are solely applied to finance partially or fully the new or existing clean projects. They can be issued by the government and such bonds are known as sovereign green bonds. The proceeds from the sale of these bonds are utilized for large-scale green projects. There is a rise in the demand for such bonds committed to serving both the social and environmental causes of developing and developed countries. Therefore, the green bond market is expected to grow many folds as many countries indulge in green bonds to fulfil their commitments towards climate change. Many governments backed financial companies, private banks, or other private entities issue these bonds for energy-saving or clean technologies. Figure 32.1 represents the number of green bonds issued by selected emerging markets in 2020. Looking at the figure below, we see that green bonds are prominently used in these emerging countries. China is at the forefront of issuing green bonds among the selected countries, while Poland and

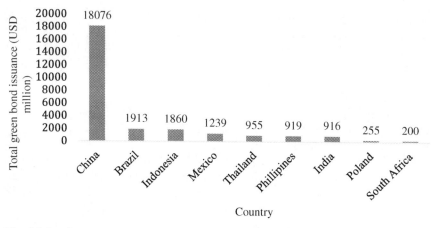

Fig. 32.1 Green Bond Issuance in a selected emerging market, 2020 (USD million) (*Source* Compiled from International Finance Corporation 2020)

South Africa are new issuers and have issued the lowest amount of green bonds. Secondly, among the various developing nations, we see China, Brazil, Indonesia, Mexico, Thailand, and the Philippines have issued green bonds more than India.

Another debt instrument is the Masala bonds, mainly issued by Indian entities. Like any other regular bonds, they are initiated to raise capital. The only difference being that they are denominated in Indian rupees and issued in overseas markets to help various entities raise funds in their local currency and avoid exchange rate risks. In this respect, International Finance Corporation developed a new form of masala bond, namely the green masala bond, primarily for providing substantial funds to green projects in India. Since then, YES Bank, Punjab National Bank, and Axis Bank have used green masala bonds for specific climate financing purposes (IFC 2017).

Additionally, equity-based instruments generate funds for green causes through the equity market. For example, Sukuk is an Islamic financial certificate that complies with the Sharia law (Climate Bond Initiatives 2019). The first green Sukuk in the world was introduced in Malaysia by Tadau Energy in July 2017 (Climate Bond Initiatives 2019). Green assets are also created through public–private partnerships; the private entity usually manages the assets. There have been various joint ventures between private companies for green causes, and venture capital investments have taken place for financing potential green projects that are innovative or have begun as pilot projects (Climate Bond Initiatives 2019).

Green Insurance

The third important category is the green risk mitigating instruments. One such green risk mitigating instrument is Green insurance; it is one of the popular green finance tools. For example, there is green auto insurance in Europe and North America, where one pays for insurance depending on the extent of driving (pay as you drive). In addition, Europe and North America give discounts on insurance for fuel-efficient vehicles. Another important insurance is crop insurance, which is provided with high subsidies in many developing countries, including India. Similarly, another example of green insurance is InsuResilience which promotes the development of climate risk insurance products to protect poor households from climate change and natural disasters (InsuResilience Global Partnership 2018).

Furthermore, the global contribution towards climate finance is increasing through the development of various green products. Figure 32.2 presents the trend in international climate finance from 2013 to 2018 for alternate years. Since 2013 global climate finance has been observed to have an upward trend, but there has been a slight fall in 2016. Moreover, with the help of the disaggregated figure, we see a decline from 2017 to 2018 to the tune of USD 66 billion. Global climate finance comprises private and public finance. The contributions from public and private actors were USD 282 billion and USD 330 billion, respectively, in 2017. But, the assistance from public and private actors fell in the consecutive year to USD 224 billion and USD 323 billion, respectively. Hence the total global climate finance in 2018 was lesser than in 2017. Through this, we found that globally, public finances are less than private finance in the climate finance arena.

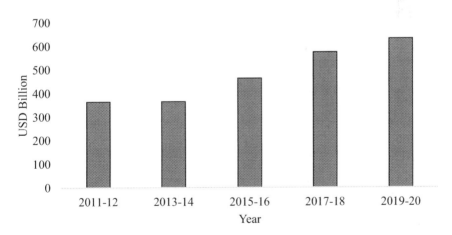

Fig. 32.2 Total global climate finance flows from 2011 to 2020 (USD billion) (*Source* Compiled from Buchner et al. (2021) and Buchner et al. (2019))

32.4 Green Financing in Asia

This section discusses a few green finance initiatives undertaken for long-term ecological preservation by selected Asian economies.

China

China is rapidly evolving its green financial sector and has, from time to time, developed its financial sector to bring green finance to the forefront. For example, China's Banking and Insurance Regulatory Commission has requested the banks to stop lending to companies that are not pro-environment and has repeatedly asked banks to develop innovative green financial products. Since then, the country has seen the development of various green finance products such as green bonds, green insurance, green trust, green leasing products, etc. (Jun et al. 2019). Among all these green products, green bonds and green loans have been used extensively since their introduction in the country. China boosted the use of green loans by reducing the size of capital reserve requirements for banks engaged in providing green loans. Due to this, the total balance of green loans by the end of 2019 was USD 1.50 trillion (Choi et al. 2020). By the end of 2019, China had issued USD 55.8 billion of green bonds, which made it the largest issuer of labelled green bonds (Climate Bonds Initiative & China Central Depository and clearing Research Center 2020).

Further, China took careful measures in developing detailed policies with respect to green credits, green bonds, environmental information disclosures, etc. The government bodies provided interest subsidies, risk compensation, etc., to investors of green loans or bonds. For example, the Local Government of Jiangsu, a province in the PRC, provided an interest subsidy of 30% towards green bonds and other green assets (Jun et al. 2019). Shenzhen, a city in China, offers up to 50% of interest subsidies on green loans specifically to borrowers who commit to reducing carbon emissions.

To sum up, China has taken significant measures to provide investment support in the form of green finance products to cut carbon emissions and build a clean energy system.

Bangladesh

Bangladesh is a developing economy and is one of the most vulnerable nations to climate change. It has great potential for improving the renewable energy sector, but till now, the share of renewable energy in the total energy portfolio is seen to be negligible. It is felt that the country needs financial resources and technical know-how to develop green projects to a desired level.

The idea of green finance in Bangladesh is still at a primitive stage. To finance sustainable practices, the country is trying to bring greater attention to modern green financial products such as green bonds, green equity, and green debentures. To do so, the Bangladesh Bank (BB) announced a Green Transformation Fund to support a transition to green manufacturing practices. The total disbursement of this fund amounted to USD 63.31 million by the end of March 2020 (Bangladesh Bank 2020). In 2015 the Central Bank also announced using foreign exchange in green bonds, making it the first Central Bank to do so. It also became one of the first central banks to promote the usage of its reserve for investing in green bonds (CBI 2015).

Apart from banks, non-bank financial institutions such as the Infrastructure Development Company Limited (IDCOL), provided refinancing facilities to households ready to switch to clean energy sources (UNEP 2015). Banks and other financial institutions utilized 133.18 million takas and 0.22 million takas respectively from the climate risk fund during Jan to March 2020 (Bangladesh Bank 2020).

Indonesia

Here we discuss a few efforts undertaken by Indonesia to combat climate change by promoting green finance. The Government of Indonesia funded various environment-friendly projects like green tourism, renewable energy generation, waste management to name a few. Indonesia issued sovereign green Sukuk and raise USD 2 billion through it in February, 2019. Green bonds are still at the early stage of development in the financial market of Indonesia. It was seen that in 2019 Rakyat Indonesia TBK became the first bank to issue a green sustainable bond. Also, the Indonesian Ministry of Finance Climate Budget Tagging mechanism in 2017 successfully contributed USD 5.7 billion towards climate change (Keuangan 2019). The country also devised an investment instrument, viz., the Carbon Trust Fund, in 2010 to provide financial support for energy-efficient and renewable energy based projects (Liebman et al. 2019). Of late, even the public and private banks of the country are trying to take significant steps to include green aspects in their lending and investing activities because sustainable financing has became mandatory in 2018.

But it is observed that these efforts and incentives are not enough to meet Indonesia's sustainable development goals and require a significant number of investments to meet their future and current demands. In other words, Indonesia lags in providing enough financial support to green causes. Hence, it must take necessary measures to promote essential tools for enhancing green finance.

Philippines

The Philippines has performed well in terms of including sustainable finance with the help of its banks, businesses, and government. It has used green debt and equity instruments to expand financing options. Some of the instruments include green bonds, green loans, credit guarantees, and funds for investing in a green cause (Davidson et al. 2020).

The Philippines mostly uses green bonds to finance its sustainable practices among the various products. In early 2016 it issued close to USD 226 million worth of green bonds (Davidson et al. 2020). By the end of 2019, it was reported that the green bond proceeds were used for financing the renewable energy sector, green buildings, waste management, pollution prevention, etc. In addition, the country saw that major banks like AC Energy, Rizal Commercial Bank, BDO Unibank, and Bank of the Philippines Island issued green bonds (Rimaud et al. 2020). It is to be noted that the private sector primarily drives the green bond market in the Philippines.

The banking sector in the Philippines are known for providing green loans for developing green projects. For example, the Bank of the Philippines Island developed a program called Sustainable Energy Finance which provides capital to projects having energy efficiency and climate resilience (Davidson et al. 2020). Similarly, the Development Bank of the Philippines started to offer loans to green projects under its umbrella program Green Financing Program (GFP) (DBP n.d.).

As it is one of the most vulnerable countries to natural calamities, it has made efforts to develop catastrophe insurance with the combined efforts of multilateral institutions and international markets. It has also come up with microinsurance products to address climate risk. This product is made available to households that are poor and vulnerable to climate-related disasters (UNEP 2016a, b).

In addition, Fig. 32.3 shows the green bond issuance in a few of the selected Asian countries. They use this debt instrument widely to bring in potential investment for environmental or climate-friendly projects. The green debt market in these selected countries, i.e., Malaysia, Philippines, Singapore, and Thailand are still at an early stage of development.

32.5 Green Financing and India

Climate change has profound implications for the Indian economy concering income and employment reduction. The climatic aberrations and uncertainty created in the process can lead thousands below the poverty line. According to the estimation of the World Bank, India's GDP can decline by 2–3% due to the ill effects of climate change (World Bank 2018). As a response to such threats, Indian Government has declared to reduce greenhouse gas emissions to 33–35% (Jain 2020) through various measures such as increasing its forest covers and using more non-fossil fuels by 2030. These decisions were conveyed in the United Nations Convention on Climate Change (UNFCCC) held in Paris in 2015. However, huge amounts of resources are necessary to make a successful attempt to achieve the above-said goals. Some estimates show that India might require approximately USD 206 billion (2014–2015 prices) between 2015 and 2030 (Jain 2020) in sustainable agriculture, forestry, infrastructure, and

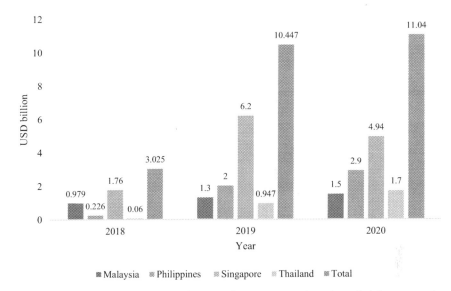

Fig. 32.3 Total green bond and green loan issuance in selected Asian countries (2018–2020) in USD billion (*Source* Compiled from Climate Bonds Initiative (2019, 2020 and 2021) ASEAN Sustainable Finance State of the Market. Available via https://www.climatebonds.net/files/reports/cbi_asean_sotm_2019_final.pdfhttps://www.climatebonds.net/files/reports/asean-sotm-2020.pdfhttps://www.climatebonds.net/files/reports/asean_sotm_18_final_03_web.pdf Accessed on 12 Nov 2022

water resources. In addition, India requires USD 2.5 trillion from 2015 to 2030, or roughly USD 170 billion per year (see Fig. 32.4), to address and build a climate-resilient environment (Acharya et al. 2020).

In the case of India, the National Action Plan on Climate Change was created in 2008, but some scholars opined that the country has yet to chalk out a proper financing roadmap to meet its agreed goals (Rattani 2018). While several green products are developed over the years, supply of funds remains far short of demand. A study conducted by Climate Policy Initiatives (CPI) revealed that India successfully generated only USD 18 billion in climate investments in 2018. Still, the annual fund requirement is about nine times higher, estimated at USD 160 billion (Jena and Purkayastha 2020). It is evident that state finance will never be sufficient to meet the requirement and one needs to have both public and private investments. Domestic financing sources are necessary but may not be enough, and India may need international funding to meet its huge demand (Sinha et al. 2020).

Over the years, several instruments have been utilized in India, including debt, equity, retail loans, and insurance. Among the instruments, debt is the predominant one. Grants in aid and budgetary allocations for addressing

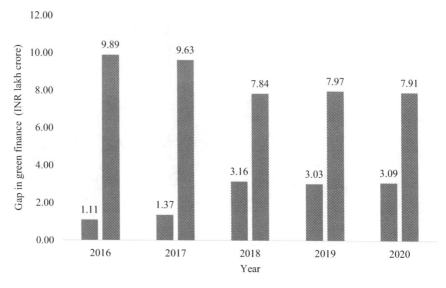

Fig. 32.4 Year wise green finance investment and the gap between actual and estimated green finance investment required to meet the current Nationally Determined Contributions (NDCs) in India *Note* The required green finance from 2016 to 2030 was estimated to be INR 11 lakh crores per year (*Source* Khanna et al. (2022))

the climate change challenges contribute substantially to India. Public Sector Units (PSUs) are other institutions through which funds are channelized in India towards green causes (Acharya et al. 2020). India has adopted several measures to increase investments in green projects, leading to structural and fundamental changes in the financial system.

Until now, India's commitment to green finance relies on using bonds, loans, and deposits. Among these, green bonds are the most prominent ones. Green bonds were issued in India primarily by public sector banks. In 2015 and 2016, the public banks issued green bonds worth USD 350 million and USD 75 million, respectively. On the other hand, private banks issued USD 209.2 million in 2015 and USD 548.5 million in 2016 (Olaf and Vasundhara 2020), which means that the contribution of private banks in issuing green bonds increased with time. Thus, the total green bond issuance has increased significantly since 2015 in India. For example, an Indian commercial bank, YES Bank, in 2015, reported that it had issued green bonds worth USD 49.2 million, and the proceeds from the bonds would be used to finance renewable energy projects. Further, it was reported that in 2020 State Bank of India raised USD 100 million in green bonds through private placement (IBEF 2021).

Along with this, PNB Housing Finance, EXIM Bank, IDBI, and NTPC played a pioneering role in developing the Indian green bond market. Now with the help of disaggregated figure (see Fig. 32.5), we look into the annual issuance of green bonds in India. It must be noted that India, from 2014 to August 2020, has issued USD 28.3 billion (USAID 2021). Institutional investors, project developers, and financiers need continuous participation to encourage the usage of this debt instrument because long-term debt capital is critical for the growth of India's green projects.

India also benefited from another type of bond, the green masala bond. It is very similar to the regular type of bond. A couple of differences between them are that they are issued specifically to fulfil financial requirements of green projects; they are issued in the overseas market and are denominated in Indian currency. For instance, the Indian Renewable Energy Development Agency (IREDA) declared to use proceeds from green masala bonds to finance renewable energy projects (CBI 2017).

Another new and innovative product recently launched in India called the Green Deposit. The deposits are committed towards a green portfolio on behalf of the clients. It was launched in India by HSBC for corporate clients. (HSBC n.d.). Like other green finance products, the proceeds from green deposits are used to invest in environment-friendly projects. Therefore, entities facing hindrances in attaining sufficient funds for their green projects can use this product.

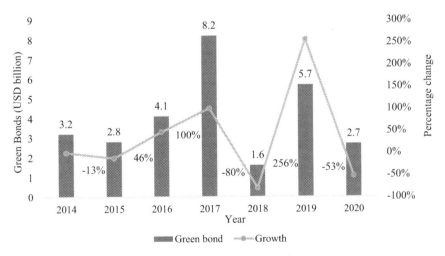

Fig. 32.5 India's annual green bond issuance (USD billion) and its percentage change from 2014 to 2020 (*Note* 2020 figure is calculated up to August; *Source* India's transition to a high performing, low emission, energy-secure economy, USAID 2021)

Additionally, green loans are also prevalent in India. These green loans are helpful for financing green-certified projects. Besides HSBC, Credit Agricole and SBI launched a Euro 50 million green loan facility (Credit Agricole n.d). In the year 2020, another set of green loans was issued in India by the Development Bank of Singapore Limited (DBS) (DBS 2020).

While India has introduced several green finance products, it faces several challenges in raising funds for green causes.

Challenges Faced by India

The major challenges faced are limited participation of investors in sustainable projects, insufficient knowledge, no clear definition of green finance, lack of awareness about climate change, and limited technical proficiency (Sarangi 2018). As the investment required to mitigate climate change is enormous, the risk associated with such investments is also huge. Investors also fear the risk of greenwashing, which is the process of providing false information about how the products or projects are environmentally sound (UNEP 2017).

Further, it faces the problem of high borrowing costs because of asymmetric information, higher risk perception, and false claims on environment clearance. Lack of familiarity or asymmetric information within the financial sector leads to a very slow pace in increasing the volume of green finance. India also fails to have a national measurement platform that provides services such as reporting and verification to track climate finance (Ghosh et al. RBI Bulletin 2021). Through periodic assessments, it has been found that the required amount of funds are not transferred to non-conventional sources of energy even though priority sector lending towards the renewable energy sector was suggested by the RBI (Ghosh et al. 2021).

Given the early stage of the renewable energy sector, the investment is relatively lower as the creditworthiness of such projects is unknown. In other words, such projects are new, and hence no historical data is available to check such industries' performance. As a result, investors tend to invest in low-risk and high-rating bonds or government bonds. Adding to this is the unfamiliarity with green bonds and loans; based on this premise, it becomes challenging to convince investors regarding the purpose of green financial products. Since investors are sceptical, it becomes challenging to channel funds into sustainable projects. Moreover, India has still been unable to improve its long-term financing instruments and lacks fineness in the domestic bond market (Sarangi 2018).

With growing advancements in non-conventional sources of energy, the demand for dynamic technology is also increasing. The evolving technologies need the support of well-equipped banking facilities. The banks and financial institutions need to accept such transition in the economy and cater to the requirement of a changing demand (Umeshwaram and Seth 2015) even though initially there may be some risks.

32.6 Conclusion

Environmental degradation has given rise to the dangerous impacts of climate change. Countries across the globe today have realized this threat and started to combat the risks of climate change. For example, the governments of various countries under the Paris Climate Agreement have agreed to reduce greenhouse gas emissions. Excessive usage of fossil fuels is one important factor in global warming. Thus, countries, including India, are trying to reduce their dependence on fossil fuels. The reason for shifting towards renewable energy is to reduce emissions and maintain energy self-sufficiency and security by having a diverse energy source (Sachs et al. 2019a, b). However, to take up such environment-friendly measures and invest in sustainable technologies, one needs sufficient funding, and this type of purposeful financing is termed green financing. This exploratory article discusses the importance of green financing and various instruments to fund environment-friendly, green activities. Based on various secondary sources, we discussed consumption-based retail activity funding such as green housing or car loans and green deposits and credit cards. The article also talks about various debt and equity-based instruments that can be used to raise funds for investment in green technologies. Further we talked about green finance tools that are risk mitigating such as the green insurance.

Countries across the globe have used these financing options to fund their green activities, and we look at the experiences of some Asian developing countries, including China. China is seen to be at the forefront when taking environment-sustaining measures and funding. The paper finally takes up the case of India to look at the green financing instruments utilized by the country and the demand–supply gap of green funds that persists. Some of the major challenges India faces in mobilizing finance for its green activities have been discussed in detail in the paper. We highlight below a few selected measures that India can adopt to fulfill its commitments towards green causes.

India can take up various policies for generating finances to address the climate challenges it faces. For instance, to improve green finance in India, we need to reduce the information gap and increase awareness related to green finance. Coordination among various stakeholders will help India to improve its investment in green causes, and it will help India to bring green finance into mainstream banking operations. Secondly, the market needs to be informed about the positive impact of green finance on the environment and highlights climate change's negative effect on the financial sector. These additional disclosures and reduction in transaction costs will encourage the participation of potential investors in the Indian green finance debt market. Thirdly, some developments along the lines of the Green Infrastructure Investment Coalition (GIIC) will help the government, investors, asset managers, etc., of all the developing countries like India to come together and promote large-scale green financing endeavours (UNEP 2016a, b). A green framework, especially for the insurance industry like the Sustainable Insurance Policy, will be extremely useful in addressing the ESG framework in India (UNEP 2016a, b). Including certain renewable energy sources in the priority sector lending will help India to achieve its climate change commitments. India should encourage commercial banks to provide green credit, debit cards, and green mortgages so that the proceeds push liquidity in enterprises promoting sustainability through their products. India needs to solve the exchange rate risks arising due to the volatility in the foreign exchange market as it requires support from international finance to achieve its green endeavours (Nelson and Shirmali 2014).

To sum up, we saw that India had launched various successful green finance products and made them available in the market. We also explored the measures taken in seeking investment to advance sustainable development. However, India is still at the nascent stage of transitioning into a green economy. It faces a lot of challenges while promoting and implementing the products of green finance. Hence, India needs to walk a long way to align finances with sustainable development goals the country wishes to achieve.

References

Acharya M, Sinha J, Jain S, Padmanabhi R (2020) Landscape of Green Finance in India. Climate Policy Initiative. Available via CPI. https://www.climatepolicyiniti ative.org/wp-content/uploads/2020/09/Landscape-of-Green-Finance-in-India-1-2.pdf

Agliardi E, Agliardi R (2019) Financing Environmentally-Sustainable Projects with Green Bonds. Environment and Development Economics 24(6):608–623

Bangladesh Bank (2020) Sustainable Finance Department Bangladesh Bank. Quarterly Review Report on Green Banking Activities of Banks and Financial Institutions and Green Finance activities of Bangladesh Bank. Available via. https://www.bb.org.bd/pub/quaterly/greenbanking/greenbanking_janmar2020.pdf

Buchner B, Clark A, Falconer A, Macquarie R, Meattle C, Tolentino R, Wetherbee C (2019) Global Landscape of Climate Finance, 2019. Climate Policy Initiative. Available via CPI. https://www.climatepolicyinitiative.org/wp-content/uploads/2019/11/2019-Global-Landscape-of-Climate-Finance.pdf. Accessed on: 12 November 2022

Buchner B, Naran B, Fernandes P, Padmanabhi R, Rosane P, Solomon M, Stout S, Strinati C, Tolentino R, Wakaba G, Zhu Y, Meattle C, Guzmán, S (2021) Global Landscape of Climate Finance 2021. Climate Policy Initiative. Available via. https://www.climatepolicyinitiative.org/wp-content/uploads/2021/10/Full-report-Global-Landscape-of-Climate-Finance-2021.pdf

Climate Bonds Initiative (2015) Bangladesh Bank to Fund Climate Investment with Forex Reserves. https://www.climatebonds.net/2015/10/bangladesh-bank-fund-climate-investment-forex-reserves

Climate Bonds Initiative (2017) IREDA USD 300 m Green Masala Bond Launch at LSE: Climate Bonds Certified!. https://www.climatebonds.net/2017/09/ireda-usd300m-green-masala-bond-launch-lse-climate-bonds-certified

Climate Bonds Initiative (2018) The Green Bond Market in the Nordics. https://www.greenfinanceplatform.org/research/green-bond-market-nordics

Climate Bonds Initiative (2019) ASEAN Green Finance State of the Market 2019. CBI and Asia Development bank. https://www.climatebonds.net/files/reports/cbi_asean_sotm_2019_final.pdf

Climate Bonds Initiative (CBI) & China Central Depository and Clearing Research Center (CCDC) (2020) China Green Bond Market 2019. https://www.climatebonds.net/files/reports/cbi_china_sotm_2021_06d.pdf

Citigroup (2020) Citi Launches Green Deposits, a New Sustainable Investment Solution. https://www.citigroup.com/citi/news/2020/201123c.htm

Choi J, Escalante D, Larsen ML (2020) Green Banking in China- Emerging Trends. Climate Policy Initiative. https://www.climatepolicyinitiative.org/wp-content/uploads/2020/08/Green-Banking-in-China-Emerging-Trends-1.pdf

Credit Agricole (n.d) First Green Loan Executed for an Indian Bank with State Bank of India. Available online. https://www.ca-cib.com/pressroom/news/first-green-loan-executed-by-indian-bank-with-state-bank-india

Davidson K, Gunawan N, Almeida M, Boulle B (2020) Green Infrastructure Investment Opportunities: Philippines 2020 Report. Climate Bond Initiative and Asian Development Bank. https://doi.org/10.22617/TCS200335-2

Deutsche Bank (2021) Deutsche Bank Launches Green Deposit for Its Corporate Client. Available via. https://www.db.com/news/detail/20210331-deutsche-bank-launches-green-deposits-for-its-corporate-clients

Development Bank of Singapore (2020) Available online. https://www.dbs.com/newsroom/DBS_grows_its_sustainable_financing_footprint_with_maiden_green_loans_in_India_totalling_over_INR_10_and_5_billion

Development Bank of Philippines (n.d) Green Financing Program. Available via. https://www.dbp.ph/wp-content/uploads/2021/01/Green-Financing.pdf

Elliott C, Zhang LY (2019). Diffusion and Innovation for Transition: Transnational Governance in China's Green Bond Market Development. Journal of Environmental Policy and Planning 21(4):391–406. https://doi.org/10.1080/1523908X.2019.1623655

Falcone PM, Sica E (2019) Assessing the Opportunities and Challenges of Green Finance in Italy: An Analysis of the Biomass Production Sector. Sustainability 11(517):1–14. https://doi.org/10.3390/su11020517

Flammer C (2020) Green Bonds: Effectiveness and Implications for Public Policy. Environmental and Energy Policy and the Economy 1(1):95–128. https://doi.org/10.1086/706794

Frimpong I, Adeabah D, Ofosu D, Tenakwah E (2021) A Review of Studies on Green Finance of Banks, Research Gap and Future Directions. Journal of Sustainable Finance & Investments 11(1):1241–1246. https://doi.org/10.1080/20430795.2020.1870202

Ghosh S, Nath S, Ranjan A (2021) Green Finance in India: Progress and Challenges. RBI Bulletin. https://www.rbi.org.in/Scripts/BS_ViewBulletin.aspx?Id=20022

Gianfrate G, Peri M, (2019) The Green Advantage: Exploring the Convenience of Issuing Green Bonds. Journal of Cleaner Production 219:127–135. https://doi.org/10.1016/j.jclepro.2019.02.022

Green Loan Principle (2021) Guidance on Green Loan Principle. In: Advancing the Corporate Loan Market. Loan Syndications & Trading Association. Available via https://www.lsta.org/content/guidance-on-green-loan-principles-glp/

Hafner S, Jones A, Anger-Kraavi A, Pohl J (2020) Closing the Green Finance Gap – A Systems Perspective. Environmental Innovation and Societal Transitions 34:26–60. https://doi.org/10.1016/j.eist.2019.11.00

Haque MS, Murtaz M (2018) Green Financing in Bangladesh. Paper presented in the International Conference on Finance for Sustainable Growth and Development, Chittagong, Bangladesh, 82–89 March 2018

Hossain M (2018) Green finance in Bangladesh: Policies, Institutions, and Challenges, ADBI Working Paper, No. 892 Available via Asian Development Bank Institute (ADBI). https://www.adb.org/publications/green-finance-bangladesh-policies-institutions-challenges

HSBC (n.d) Green Deposits. Available online. https://www.business.hsbc.co.in/en-gb/green-deposit

Huang H, Zhang J (2021) Research on the Environmental Effect of Green Finance Policy Based on the Analysis of Pilot Zones for Green Finance Reform and Innovations. Sustainability 13:3754. https://doi.org/10.3390/su13073754

India Brand Equity Foundation (2021). https://www.ibef.org/industry/banking-india.aspx

InsuResilience Global Partnership (2018) Available Online. https://www.insuresilience.org/wp-content/uploads/2018/11/Flyer_InsuResilienceGlobalPartnership_2018.pdf

Inter-American Development Bank (2014) The IDB's Multilateral Investment Fund Program EcoMicro Wins United Nation Climate Solutions Award. Available Online. https://www.iadb.org/en/news/news-releases/2014-12-10/mif-program-ecomicro-receives-unfccc-award%2C11013.html

International Finance Corporation (2017) Masala Bond Program- Nurturing a Local Currency Bond Market. International Finance Corporation World Bank Group. Available Online. https://www.ifc.org/wps/wcm/connect/be6c1b09-9e08-4fc4-963045ecc078cc88/EMCompass+Note+28+Masala+Bond+Program+FINAL.pdf?MOD=AJPERES&CVID=lDmkW5e

International Finance Corporation (2020) Emerging Market Green Bonds Report 2020. International Finance Corporation World Bank Group. Available via. https://www.ifc.org/wps/wcm/connect/publications_ext_content/ifc_external_publication_site/publications_listing_page/emerging-market-green-bonds-report-2020. Accessed on: 12 November 2022

Jain S (2020) Financing India's Green Transition. Observer Research Foundation. ORF Issue Brief No. 338. Available Online. https://www.orfonline.org/research/financing-indias-green-transition-60753/

Jayathilake S (2019) Impact of Green Financing for the Corporate Governance in the Banking Industry OIDA International Journal of Sustainable Development 12(11):23–30. http://www.ssrn.com/link/OIDA-Intl-Journal-Sustainable-Dev.html

Jena LP, Purkayastha D (2020) Accelerating Green Finance in India: Definition and Beyond. Climate Policy Initiatives. Available Online. https://www.climatepolicyinitiative.org/wp-content/uploads/2020/07/Accelerating-Green-Finance-in-India_Definitions-and-Beyond.pdf

Jones AW (2015) Perceived barriers and policy solutions in clean energy infrastructure investment. Journal of Cleaner Production 104:297–304. https://doi.org/10.1016/j.jclepro.2015.05.072

Jun M. A, Jialong L, Zhouyang C, Wenhong X (2019) Green Bonds. In: Schipke A, Rodlauer M, Longmei Z (eds) The Future of China's Bond Market. International Monetary Fund, Washington, DC, p 155–168

Keuangan OJ (2019) Sustainable Finance in Indonesia: How the Financial Services Industry Contributes to Environment, Social and Governance Issues. Nomura Journal of Asian Capital Market 4(1). Available via. http://www.nomurafoundation.or.jp/en/wordpress/wp-content/uploads/2019/09/NJACM4-1AU19-03.pdf

Khanna N, Purkayastha D, Jain S (2022). Landscape of Green Finance in India. Climate Policy Initiative. Available via CPI. https://www.climatepolicyinitiative.org/wp-content/uploads/2022/08/Landscape-of-Green-Finance-in-India-2022-Full-Report.pdf

Klioutchnikov I, Kliuchnikov O (2021) Green Finance: Pandemic and Climate Change. E3S Web of Conferences. 234. 00042. https://doi.org/10.1051/e3sconf/202123400042

Landesbank Baden-Wurttemberg (n.d.) Available Online. https://www.lbbw.de/articlepage/experience-banking/green-loans_9uz7816wn_e.html

Liebman A, Reynolds A, Robertson D, Nolan S Argyriou M, Sargent B (2019) Green Finance in Indonesia Barriers and Solutions. https://doi.org/10.1007/978-981-10-8710-3_5-1

Li W, Jia Z (2017) Carbon Tax, Emission Trading, or the Mixed Policy: Which is the Most Effective Strategy for Climate Change Mitigation in China? Mitigation and Adaptation Strategies for Global Change 22(6):973–992. https://doi.org/10.1007/s11027-016-9710-3

Li Z, Liao G, Wang Z, Huang Z (2018) Green Loan and Subsidy for Promoting Clean Production Innovation. Journal of Cleaner Production 187:421–431. https://doi.org/10.1016/j.jclepro.2018.03.066

Maulidia M, Halimanjaya A (2014) The Coordination of Climate Finance in Indonesia. ODI Centre for Policy Research, Jakarta. Available via. https://odi.org/en/publications/the-coordination-of-climate-finance-in-indonesia/

McDaniels J, Robins NN Strauss D, Thoma J, Dupre S (2016) Building a Sustainable Financial System in the European Union United Nations Environment Programme. https://doi.org/10.1089/jwh.2006.0192

Meo S, Karim MZA (2021). The Role of Green Finance in Reducing CO_2 Emissions: An Empirical Analysis. Borsa Istanbul Review. https://doi.org/10.1016/j.bir.2021.03.002

Nawaz MA, Seshadri U, Kumar P (2021) Nexus Between Green Finance and Climate Change Mitigation in N-11 and BRICS Countries: Empirical Estimation Through Difference in Differences (DID) Approach. Environmental Science and Pollution Research 28:6504–6519. https://doi.org/10.1007/s11356-020-10920-y

Nefco (n.d) Accelerating the Green Transition. Available via. https://www.nefco.int/

Nelson D, Shrimali G (2014) Currency Exchange Risks in Renewable Energy Financing. Energetica India, 4–6

Ng AW (2018) From Sustainability Accounting to a Green Financing System: Institutional Legitimacy and Market Heterogeneity in a Global Financial Centre. Journal of Cleaner Production 195:585–592. https://doi.org/10.1016/j.jclepro.2018.05.250

Ngan SL, Promentilla MA, Yatim P, Lam H (2019) A Novel Risk Assessment Model for Green Finance: the Case of Malaysian Oil Palm Biomass Industry. Process Integration and Optimization for Sustainability 3. https://doi.org/10.1007/s41660-018-0043-4

Nguyen TC, Chuc AT, Dang LN (2018) Green Finance in Viet Nam: Barriers and Solutions. ADBI Working Paper 886. Tokyo: Asian Development Bank Institute. Available. https://www.adb.org/publications/green-finance-viet-nam-barriers-and-solutions

Noailly J, Smeets R (2016) Financing Energy Innovation: The Role of Financing Constraints for Directed Technical Change from Fossil-Fuel to Renewable Innovation. EIB Working Papers. European Investment Bank. https://ideas.repec.org/p/zbw/eibwps/201606.html

Oh D, Kim SH (2018) Green Finance in the Republic of Korea: Barriers and Solutions. ADBI Working Paper 897. Tokyo: Asian Development Bank Institute. Available via ADBI. https://www.adb.org/sites/default/files/publication/469261/adbi-wp897.pdf

Olaf W, Vasundhara S (2020) Sustainable Energy through Green Bonds in India. Center for International Governance Innovation and Gateway House. https://doi.org/10.13140/RG.2.2.30215.01446

Onanuga O, Onanuga A (2019) Carbon Fiscal Instruments and Green Finance: An Aid to the Success of SDGs in Nigeria?. Journal of Innovation in Business and Economics 3. https://doi.org/10.22219/jibe.v3i02.7895

Pham L (2021) Frequency Connectedness and Cross-Quantile Dependence between Green Bond and Green Equity Markets. Energy Economics 98(105257). https://doi.org/10.1016/j.eneco.2021.105257

Pigato M (ed) (2018) Fiscal Policies for Development and Climate Action: Policy Summary for Finance ministeRs. World Bank. Avialable VIA WB. https://resource-cms.springernature.com/springer-cms/rest/v1/content/40190/data/References+Basic+Style

Raberto M, Ozel B, Ponta L, Teglio A, Cincotti S (2019) From Financial Instability to Green Finance: The Role of Banking and Credit Market Regulation in the Eurace Model. Journal of Evolutionary Economics 29(1):429–465. https://doi.org/10.1007/s00191-018-0568-2

Rattani V (2018) Coping with Climate Change: An Analysis of India's National Action Plan on Climate Change. In Centre for Science and Environment. Available via http://cdn.cseindia.org/attachments/0.55359500_1519109483_coping-climate-change-NAPCC.pdf

Rawat S (2020) Recent Advances in Green Finance. International Journal of Recent Technology and Engineering. 8(6):5528–5533. https://doi.org/10.35940/ijrte.F9980.038620

Reboredo JC (2018) Green Bond and Financial Markets: Co-Movement, Diversification and Price Spillover Effects. Energy Economics 74:38–50. https://doi.org/10.1016/j.eneco.2018.05.030

Ren X, Shao Q, Zhong R (2020) Nexus between Green Finance, Non-Fossil Energy Use, and Carbon Intensity: Empirical Evidence from China Based on a Vector Error Correction Model. Journal of Cleaner Production 277. https://doi.org/10.1016/j.jclepro.2020.122844

Rimaud C, Siva H, Almedia M, Whiley A, Tukiainen K (2020) ASEAN Green Finance State of Market 2019. Climate Bond Initiative. Available via CBI. https://www.climatebonds.net/files/reports/cbi_asean_sotm_2019_final.pdf

Rouf KA (2012) Green Microfinance Promoting Green Enterprise Development. International Journal of Research Studies in Management 1(1): 85–96. https://doi.org/10.5861/ijrsm.2012.v1i1.32

Sachs J, Thye W, Yoshino N, Taghizadeh-Hesary F (eds) (2019a) Handbook of Green Finance. Sustainable Development. Springer, Singapore. https://doi.org/10.1007/978-981-10-8710-3_35-1

Sachs JD, Woo WT, Yoshino N, Taghizadeh-Hesary F (2019b) Importance of Green Finance for Achieving Sustainable Development Goals and Energy Security. Handbook of Green Finance 3–12. https://doi.org/10.1007/978-981-13-0227-5_13

Sarangi GK (2018) Green Energy Finance in India: Challenges and Solutions. ADBI Working Paper 863. Tokyo. Asian Development Bank Institute. Available. https://www.adb.org/publications/green-energy-finance-india-challenges-and-solutions

Schäfer H (2018) Germany: The 'Greenhorn' in the Green Finance Revolution. Environment: Science and Policy for Sustainable Development 60(1):18–27. https://doi.org/10.1080/00139157.2018.1397472

Sinha J, Jain S, Padmanabi R (2020) Landsacpe of Green Finance in India. Climate Policy Initiative. Available via CPI. https://www.climatepolicyinitiative.org/wp-content/uploads/2020/09/Landscape-of-Green-Finance-in-India-1-2.pdf

Swedbank (n.d) Green Loans. Swedbank. Available Online. https://www.swedbank.com/corporate/advisory-services-and-investment-banking/loans-and-syndications/green-loans.html

Tänzler D, Maulidia M (2013) Status of Climate Finance in Indonesia: Country Assessment Report. GIZ and Adelphi. Avialable via. https://www.adelphi.de/en/system/files/mediathek/bilder/indonesia_climate-finance-report_giz-adelphi.pdf

Ticci E, Gabbi G (2018) Sustaining Sustainable Development: Re-thinking the Role of the Financial System. 2014. Available via. http://www.congressi.unisi.it/fessud siena/wp-content/uploads/sites/24/2014/04/Gabbi.pdf

The Global Commission on the Economy and Climate (2016). The 2016 New Climate Economy Report. In: The Sustainable Infrastructure Imperative: Financing for Better Growth and Development. Washington, DC. Available via. http://newclimateeconomy.report/2016/wp-content/uploads/sites/4/2014/08/NCE_2016Report.pdf

UNEP (2015). Designing a Sustainable Financial System in Bangladesh. UNEP. BB. https://wedocs.unep.org/bitstream/handle/20.500.11822/7422/-

UNEP (2016a) The State of Sustainable Finance in the United States. Available via UNEP. https://wedocs.unep.org/bitstream/handle/20.500.11822/9828/The_state_of_sustainable_finance_in_the_United_States2016aThe_State_of_Sustainable_Finance_in_the_US.pdf.pdf?sequence=3&amp%3BisAllowed=

UNEP (2016b) Green Finance for Developing Countries. Needs, Concern and Innovation. UNEP. Available via UNEP. https://www.greengrowthknowledge.org/sites/default/files/downloads/resource/Green_Finance_for_Developing_Countries-1.pdf

UNEP (2017) Guidelines for Providing Product Sustainability Information. UNEP. Available via UNEP. https://wedocs.unep.org/bitstream/handle/20.500.11822/22180/guidelines_product_sust_info.pdf?sequence=1&isAllowed=y

United Nation Climate Change (n.d) Green Finance and the Aggregation of Swedish Local Government Investment Projects. Avialable via UNCC. https://unfccc.int/climate-action/momentum-for-change/financing-for-climate-friendly-investment/green-finance-and-the-aggregation-of-swedish-local-government-investments-projects

United Nations Environment Program (UNEP) (2007) Green Financial Products and Services Current Trends and Future Opportunities in North America. A Report of the North American Task Force (NATF). United Nations Environment Program Finance Initiative. Available via. https://www.unepfi.org/publications/green-financial-products-and-services-current-trends-and-future-opportunities-in-north-america/

Umamaheswaran S, Seth R (2015) Financing Large-Scale Wind and Solar Projects–A Review of Emerging Experiences in The Indian Context. Renewable and Sustainable Energy Reviews 48:166–177. https://doi.org/10.1016/j.rser.2015.02.054

USAID (2021) India's Transition to a High performing, Low Emission, Energy-Secure Economy, USAID. Available via USAID. https://pdf.usaid.gov/pdf_docs/PA00X9CM.pdf

Volz U, Böhnke L, Knierim K, Richert K, Röber GM, Eidt V (2015) Financing the Green Transformation: How to Make Green Finance Work in Indonesia. Palgrave Macmillan: Basingstoke, UK. https://doi.org/10.1057/9781137486127

Wang Y, Zhi Q (2016). The Role of Green Finance in Environmental Protection: Two Aspects of Market Mechanism and Policies. Energy Procedia 104:311–316. https://doi.org/10.1016/j.egypro.2016.12.053

World Bank (2018) Available on. https://www.worldbank.org/en/news/press-release/2018/06/28/climate-change-depress-living-standards-india-says-new-world-bank-report

Yu CH, Wu X, Z Dayong, Chen S, Zhao, J (2021) Demand for Green Finance: Resolving Financing Constraints on Green innovation in China. Energy Policy 153(C). https://doi.org/10.1016/j.enpol.2021.112255

Zerbib OD (2019) The Effect of Pro-environmental Preferences on Bond Prices: Evidence from Green bonds. Journal of Banking & Finance 98:39–60. https://doi.org/10.1016/j.jbankfin.2018.10.012

Zheng GW, Siddik A, Masukujjaman M, Fatema N, Alam S (2021) Green Finance Development in Bangladesh: The Role of Private Commercial Banks (PCBs). Sustainability 13:795. https://doi.org/10.3390/su13020795

Zhou X, Tang X, Zhang R (2020) Impact of green finance on economic development and environmental quality: A study based on provincial panel data from China. Environmental Science and Pollution Research 27:19915–19932. https://doi.org/10.1007/s11356-020-08383-2

Ziolo M, Filipiak BZ, Bak I, Cheba K (2019) How to Design more Sustainable Financial Systems: The Roles of Environmental, Social, and Governance Factors in the Decision-Making Process. Sustainability (Switzerland) 11(20):560. https://doi.org/10.3390/su11205604

CHAPTER 33

Climate Change: A Major Challenge to Biodiversity Conservation, Ecological Services, and Sustainable Development

Shilky, Subhashree Patra, Pawan Ekka, Amit Kumar, Purabi Saikia, and M. L. Khan

33.1 Introduction

The most challenging environmental problem that humanity now confronts is climate change (Gills and Morgan 2020). Rapid and severe changes in temperature, rainfall, evapotranspiration patterns, and a rise in the frequency of extreme events are now prevailing in many ecosystems (IPCC 2018). The primary direct cause of biodiversity loss and worldwide changes in ecosystem services may be climate change and its effects (MEA 2005). Allowing the earth to warm above 1.5 °C will have negative consequences on humans and biodiversity in terms of increased incidences of drought, floods, heat waves, and sea level rise (IPCC 2018). Furthermore, global warming is increasing the likelihood, timing, severity, and geographic distribution of zoonoses, epidemics, and other public health risks through a network of interrelated causative

---

Shilky · S. Patra · P. Ekka · P. Saikia (✉)
Department of Environmental Sciences, Central University of Jharkhand, Ranchi 835205, India
e-mail: purabi.saikia@cuj.ac.in

A. Kumar
Department of Geoinformatics, Central University of Jharkhand, Ranchi 835205, India

M. L. Khan
Department of Botany, Dr Harisingh Gour Vishwavidyalaya (A Central University), Sagar 470003, India

© The Author(s), under exclusive license to Springer Nature Singapore Pte Ltd. 2023
S. Nautiyal et al. (eds.), *Palgrave Handbook of Socio-ecological Resilience in the Face of Climate Change*,
https://doi.org/10.1007/978-981-99-2206-2_33

factors (Patz et al. 1996). Extreme weather conditions may alter the dynamics of ecosystems from high mountains to coastal marine systems by resetting successional clocks and causing regime transitions from grasslands to shrublands which were evident in California during El Nino events (Hobbs and Mooney 1995). Effects of extreme weather effects are expected to occur more frequently and adversely interfere with several ecological processes (Jentsch and Beierkuhnlein 2008). Growing public unhappiness with how national governments are carrying out mitigation and adaptation measures are seen in the wildfires that have spread from California to Australia, Siberia to the Amazon, and the floods that have affected Mozambique and other countries (Ossebaard and Lachman 2021). Additionally, a number of illnesses that were thought to be eradicated are reappearing in areas with shifting climatic conditions that favour their reemergence (Adedeji 2014). Besides, the connection between the climate catastrophe and the coronavirus disease (COVID-19) epidemic is undeniable (Ossebaard and Lachman 2021).

The ability of an ecosystem to adapt to climate change is determined by the diversity of species that coexist in it (Perring 2010). The most serious ecological consequence of climate change is the rapid trend of changing temperatures which influences plant distribution and abundance patterns as physiological constraints of each species (Lavergne et al. 2010). Climate change is pushing a species to shift outside its climatic niche, which may prevent it from remaining suited to the specific environmental factors in a certain region (Bellard et al. 2012). Mountains and hills and their constituent plants may be among the most vulnerable to the effects of climate change worldwide (Thuiller et al. 2005). Some animal species' seasonal cycles change the timing of events, such as earlier nesting and egg-laying by Tree swallow (*Tachycineta bicolor*) (Broecker 1975).

On the other hand, coral bleaching has resulted as a consequence of climate change in the majority of marine ecosystems due to the failure of the coral animal and algal partner relationship (Lovejoy 2008). All levels of biodiversity, from the level of the organism to the level of the biome, are anticipated to be impacted by the numerous repercussions of climate change (Parmesan 2006). Nevertheless, the United Nations Sustainable Development Goals cannot be achieved without ecosystem services, which are the benefits that humans obtain from ecosystems (Costanza et al. 1997). It has become more challenging to stabilise and adapt to climate change (SDG 13) or protect biodiversity (SDGs 14 and 15) because it is quite challenging to address the causes individually despite the fact that they are intimately connected (IPCC 2018; IPBES 2019).

33.2 Major Challenges Caused by Climate Change

Understanding the consequences of climate change depends on indicators such as species extinction, distribution, altitudinal shifts, phenological changes, and changes in the structure and composition of forest communities (Ammer 2019). The majority of extinctions connected to climate change have included changed species interactions, which usually underlie unanticipated reactions to climate change (Cahill et al. 2013). Despite the absence of evidence of current extinctions brought on by climate change, studies indicate that in the next decades, climate change may exceed habitat loss as the largest worldwide danger to biodiversity (Leadley et al. 2010). Poor people will be particularly affected by hazard management and soil and water control at low-elevation coastal zones and dryland margins as a result of the ecosystem-related effects of climate change (Hallegatte et al. 2016). Multi-species studies have demonstrated parallel changes in feeding and predation susceptibility, species diversity, and abundance (Queirós et al. 2015). Warming can impact the growth and recruitment of plants that are subject to herbivory, but it can also mitigate its effects by accelerating plant growth rates (Holmgren et al. 2006).

Additionally, increased temperatures enable herbivores to feed for longer periods of time or in previously inaccessible places (Traill et al. 2010). As a consequence of increasing snowfall due to changing climatic conditions, *Canis lupus* (wolves), for example, tend to hunt in larger groups, which decreases the population of *Alces alces* (moose) and increases the amount of their winter feed, *Abies balsamea* (Post et al. 1999). In general, many species have advanced in phenology to varying degrees, although some have shown no apparent changes and others have had delayed seasonal phenologies (Both et al. 2009). Rare or endemic species with small latitudinal ranges are particularly vulnerable to losing fertility as ambient temperatures rise, and in many cases, they may also be unable to shift their distribution range to keep up with changing climates because they lack the genetic diversity and gene flow needed to adapt to new stressors (Hoffman 2010). Climate parameters like temperature, growing degree days, water availability, and potential evapotranspiration are regulated by plant physiology in terms of their impact on survival, growth, development, motility, and reproduction (Kearney and Porter 2009). Temperature-induced infertility has serious economic consequences in tropical areas (Peña et al. 2019). In particular high temperatures, heat stress is dangerous to the genetic material and related reproductive physiology (Paupière et al. 2014). Because spermatogenesis occurs at temperatures below body temperature, animals have evolved lower genitalia (Moreno et al. 2012). Increased temperatures may change animal colouring patterns and the roles of predators and prey during crypsis (Traill et al. 2010). Higher microhabitat temperatures may affect body colourations, which may affect habitat selection and predator avoidance behaviour, and the positive relationship between temperature and colour brightness in the *Hyla cinerea* (tree frog) (King et al. 1994).

33.3 Climate Change Impacts on Ecosystem and Ecosystem Services

Ecosystem services act as a boon to humankind by supplying important commodities and services necessary for sustaining life on the earth through biodiversity and ecological functions (Cardinale et al. 2012). Changes in biodiversity significantly impact ecosystem services, including the provisioning, regulating, supporting, and cultural advantages marginalised people acquire from ecosystems (Perring 2010). The survival of ecosystem services depends on biodiversity because it maintains climate resilience (see Fig. 33.1) and promotes human well-being (Jetz et al. 2019). Indicators of healthy ecosystems include the diversity, abundance, and distribution of species, which are greatly impacted by climate change, disrupting ecosystem stability and altering most of its components (Isbell 2010). The two main worldwide factors causing a disturbance in terrestrial ecosystems are climate change and human activity (Franklin et al. 2016). Because climate change affects tree recruitment, growth, and death, it directly affects species distributions and the composition of forests (Fadrique et al. 2018). New habitats and communities are formed in response to the effects of a changing climate on flora and fauna (Pedrono et al. 2016). Forests, especially mountainous forest ecosystems, are significantly impacted by the intensity of climate change (Seidl et al. 2017). It also modifies the occurrences of disturbance events and regimes by affecting the timing, magnitude, frequency, and length of disturbance events and the biotic interactions on a particular landscape (Dale et al. 2001). Numerous abiotic pressures, interactions between organisms, and other ecological disturbances are the major factors affecting ecosystems' sensitivity and response to climate change (Malhi et al. 2020). The ability of ecosystems to deliver ecosystem services is demonstrated by the ecological processes that provide the resilience of ecosystems in the form of dynamic equilibrium (Pedrono et al. 2016). Variations in temperature and precipitation are disrupting the organisation of biotic communities and associated ecological processes in a number of ways, which negatively affect the supply of ecosystem services (Pedrono et al. 2016). Some species can adapt to the changing environment due to their climatic range or ability to disperse, while many others face extinction (Malcolm et al. 2006). Globally, the ecosystem is being quickly replaced by changes in CO_2 concentration, precipitation, ocean chemistry, water balance, temperature, severity, and frequency of extreme weather events (Malhi et al. 2020). Other severe occurrences brought on by climate change include cold waves, droughts, heat waves, and floods, which have a significant impact on ecosystem health (Handmer et al. 2012). Climate change also hampers the availability, continuity, and dispersion of ecosystem functioning (Nelson et al. 2013). The influence of climate change directly impacts natural disasters, food, fodders, and water availability, which play a significant role in human well-being (Palomo 2017). Crops and plants are subjected to moisture stress due to climate disruptions, particularly during and after the monsoon season

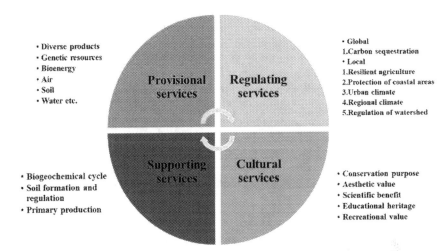

Fig. 33.1 Ecosystem services provided biodiversity for climate change resilience (*Source* Locatelli 2016)

(Chauhan et al. 2022). One of the main consequences of climate change is the expansion of agricultural areas to meet the food requirement of the growing human population which affects carbon sequestration and biodiversity loss (Malhi et al. 2020). The major effect of climate change on ecosystems and ecosystem services are summarised in Table 33.1.

33.4 Climate Change Impacts on Biodiversity

Overexploitation of species, the spread of invasive plant species, climate change, forest degradation, and the elimination of unique ecosystems are the main drivers of biodiversity loss (Butchart et al. 2010). Anthropogenic activities are degrading soil and cause significant changes in environmental conditions, simultaneously can harm flora, fauna, habitat, and landscape (Misra et al. 2009). Increased extinction rates will result from the present rate and severity of climate change, which will likely be greater than certain species' capacity to endure and adapt to new climatic conditions (Bellard et al. 2012). All levels of biodiversity have been impacted by climate change during the past century through altering latitudes, precipitation regimes, species distributions, and assemblages (Fig. 33.2) (Araújo and Rahbek 2006). It has harmed ecosystem resilience and functioning by reducing genetic diversity and ecological networks through fast migration and directed selection (Botkin et al. 2007). Climate change has also had an influence on the food web, resulting in a decline in species fitness and strength, which has an indirect effect on ecological interactions for habitat and niche (Gilman et al. 2010). About 9650 interspecific systems linked to 6300 species are on the verge of extinction as a result of the severe impacts of climate change (Koh et al. 2004). The rate of

Table 33.1 Effect of climate change on ecosystem and ecosystem services

Effects	Changes and impacts	Source
Species extinction	Distribution, morphology, population structure, mortality rate	Pedrono et al. (2016)
Ecosystem changes	New habitat and species communities	
Coastal flooding	Rising sea level, small islands disappear	
Reduction of freshwater sources	Lack of replenishment of groundwater, water deficit in inland areas	
Crop yield reduction	Increase in pesticides and pathogens	Raza et al. (2019)
Disruption in human health	Increase in communicable disease	Forster et al. (2020)
Disruption in plant health	Diseased leaves, fire-prone conditions, pest, pathogens, and changes in plant physiology	FAO (2021)
Disruption in animal health	Negative impact on the livestock health and welfare, increase the risk of death	Lacetera (2019)
Loss of biodiversity	Affect all the levels of biodiversity, its structure, and function	Sintayehu (2018)
Economic crisis	Hindrance of growth with rising operation costs	Ecker et al. (2020)
Increase in cyclonic events	Intense and frequent cyclones result in the loss of human population and environment	Pedrono et al. (2016)

extinction of both plant and pollinator diversity was increased because of the mismatch between plants and pollinators caused by the phenological shifts in angiosperms brought on by climate change, significantly altering the structure and functions of the ecosystem (Rafferty and Ives 2011). The higher levels of biodiversity in plant communities have the ability to significantly impact a biome's integrity in terms of resilience, ecotype features, distribution changes, desertification, and frequency of disasters, which are impacted by climate change (Bellard et al. 2012). Climate change influences organisms' ability to adapt to environmental conditions and eventually pushes them outside of their climatic niche (Salamin et al. 2010). Changed vegetation structure, species turnover, and the sifting of biomes across vast landscapes are some of the responses that biodiversity at various latitudes has begun to exhibit in response to the huge global range shift (Zimmermann et al. 2013). Native species moving to new environments open up vacant niches for invasive species, which respond favourably to these changes, invade, and adapt (Jarnevich et al. 2014). Climate change enhanced the competitive ability of invasive species to grow, reproduce, and spread successfully through adaptive traits, genetic variability, and physiology (Finch et al. 2021). It has severely impacted reproduction and the accumulation of excess energy in species like birds and fishes by causing habitat loss and migratory periods (Robinson et al. 2005). Climate change impacts insect pest phenology, survival, growth rate, distribution, and demography and also reduces host plant resistance, which makes it easier for pests to

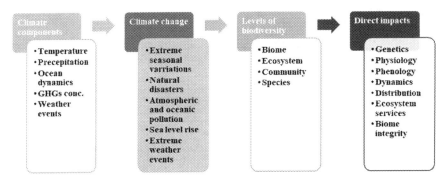

Fig. 33.2 Climate change and its effect on different levels of biodiversity

spread disease and harm delicate plant species, while some pests are negatively impacted by a lack of hosts for food and reproductive success (Shrestha 2019). A shift in climatic patterns, a prolonged change in climate elements, and a loss of resilience caused the population of some species to progressively decline, eventually resulting in the extinction of those species and the collapse of entire ecosystems (Gerlach 2010). Regional working groups can assess and classify the climate change risk for the IUCN-Red Listed species to address current extinction issues (Urban et al. 2016). Climate change poses serious threats to biodiversity, endangering total biomes, ecosystem health and services, food security, natural resources, productivity, and economic stability, making it a serious concern for environmental sustainability (Khan et al. 2021).

33.5 Adaptations and Mitigation Measures to Achieve Sustainable Development

It is essential to understand mitigation and adaptation measures in order to address climate change-related vulnerability in the larger context of fostering long-term sustainability (Sarkar 2012). Sustainable development can take advantage of new opportunities that climate change presents by integrating government policies, the scientific community, research and analysis, social efforts, and sufficient feasible strategies for building resilience, mitigating damage, and adapting to climate change (Denton et al. 2014). It is the primary strategy for reducing climate change through developing pathways, opportunities, and action capabilities for risk management through socioeconomic, cultural, biophysical, and institutional contexts (Bizikova et al. 2007). It is the process of creating climate-resilient changes by fusing innovation and development for problem-solving with efficiency in mitigating and coping with climate change (Abisha et al. 2022). Mitigation measures are widely used at the international, national, and local levels by making investments in the fields of cutting-edge technologies and infrastructures for increasing energy efficiency through renewable sources (Laukkonen et al. 2009). Social justice and environmental integrity are intertwined in sustainable adaptation, which is crucial for achieving sustainable development in relation to combating climate change (Eriksen and Brown 2011). This will help to reduce energy consumption and protect the country and its citizens from economic crises (Laukkonen et al. 2009). The aspect of adaptability that is needed for sustainable development is the combination of local change management strategies with decisions and regulations that may favour certain developments over others (Eriksen and Brown 2011). Scientists and decision-makers reinforce mitigation efforts to reduce greenhouse gas emissions into the atmosphere by maintaining the quality and increasing the number of sinks like forests (Harry and Morad 2013). Reduced dependency on expensive fuels, lower energy costs for the economy, less resource depletion, and decreased pollution are all accelerated by the introduction of efficient innovation and its successful implementation with better skills in many industries (Swart et al. 2003). The US government's Green Deal Policy aims to create a carbon-neutral society by 2050 in an effort to address climate change and pollution by strengthening green technology and sustainable energy efficiency measures including sustainable industries, hydrogen fuel cells, etc. (Streimikiene et al. 2021). Considering the consequences of climate change on the environment and society, numerous strategies are recorded by government bodies for completing climate-compatible development and sustainable development (Fig. 33.3) (Alemaw and Simatele 2020). The solutions that receive the most support on a worldwide level for improving the conceptual framework through sustainable and economical approaches are ecosystem-based adaptations (EbAs) (Muthee et al. 2021). It considers equitable maintenance of water, land, and biological resources through integrated planning and governance (CBD 2021). Mangrove conservation and reforestation for erosion management and storm protection, mixed farming for soil fertility and water conservation, and terrace farming to gain soil moisture and minimise runoff are some of the EbAs implemented to combat climatic hazards (CEP 2015). Additionally, new energy incentives are adopted as part of National Adaptation Plans to advance adaptive measures for climate change mitigation (Munang et al. 2013).

33.6 Opportunities to Improve Climate Change Resilience

Climate change imposes a variety of threats to the biosphere, natural resources, environment, and sustainability initiatives (Moore and Schindler 2022). Climate change resilience is *'the capacity to withstand shocks and disruptions resulting from climatic risk factors in order to sustain the ecosystem's identity while actively using new chances for learning and adaptation'* (Barbés-Blázquez et al. 2017). It is a complicated system of socio-ecological features that restores linear ecosystem dynamics to their pre-disturbance steady state by boosting diversity, stabilising variables, and enhancing connection (Biggs et al. 2015).

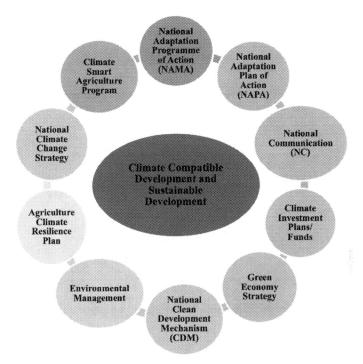

Fig. 33.3 Policies on climate-compatible development and sustainable development (*Source* Alemaw and Simatele 2020)

Strengthening community resilience is needed immediately to prevent the devastating consequences of climate change on biodiversity, environmental services, and sustainable development (Edwards and Wiseman 2011). Beyond adaptation and resistance to climate change, conservation is also essential for the preservation of biodiversity (Sgrò et al. 2011). In situ restoration of the landscape using evolutionary principles as a response to climate change sustains biodiversity processes and creates adaptive tolerance to climatic extremes by enhancing the persistence of species diversity and distribution through time (Reusch and Wood 2007).

The global food system and land transformation associated with agriculture contribute to 35 and 24% of all greenhouse gas (GHG) emissions, respectively (Vermeulen et al. 2012). Food and agricultural systems offer an opportunity to investigate and identify solutions for reducing the effects of climate change through organised procedures such as pre-production, production, processing, transport, consumption, and waste generation (Niles et al. 2018). It assembles all factors, including the environment, people, inputs, processes, infrastructure, institutions, etc., to create resilient, sustainable, and effective processes for food production, processing, distribution, preparation, and consumption (HLPE 2014). Mitigating climate change by reducing carbon emissions requires shifting to renewable energy sources, improving nutrient efficiency, reducing GHG emissions, expanding perennial crops and organic production, and making optimal use of water resources (Sommer and Bossio 2014). In order to reduce the emission of GHG, it will be crucial to adopt sustainable horticulture and livestock production through managed grazing, waste management, maintaining animal health, utilising the ecosystem services for climate resilience, and crop diversity, as well as avoiding energy use beyond the needs (Vieux et al. 2012). It is urgent to develop climate-friendly approaches like expanding market opportunities for local products, reducing food and agricultural waste, carbon pricing, climate funding, technologies to capture GHG, and stronger public policies on land use change to balance other goals beyond biodiversity, resilience, and sustainability (Niles et al. 2018).

33.7 Recommendations and Future Research Prospects

The most crucial and urgent measures to meet the current and future constraints in food production, enhance productivity, and address the challenges of existing climate change are capturing human attention in the present world. Improving socio-economy and community development requires evaluating the social and societal implications of coping with and mitigating the consequences and challenges caused by climate change. Communities should be given incentives to build resilience for addressing the effects of climate change and the catastrophes that result from it. EbAs are among the adaptive strategies with the fastest pace of growth globally due to their affordability, many benefits, and range of applications in different landscapes. It comprised incorporating ecosystem services and biodiversity into the overall adaptation strategy in an effort to build secure, well-balanced systems by maximising benefits and reducing conflicts. A shift away from reliance on natural resources towards fewer fossil fuels and more renewable energy will be essential for maximum resilience at the grassroots. The emphasis should be on studying the implications of individual actions across scales and sectors in order to maximise co-benefits, avoid unintended consequences, and comprehend net impacts across several decision-making domains. There must be support for flexible and adaptable decision-making and risk management in the face of uncertainty by regularly assessing and modifying research priorities. Sustainable development may result from raising resilience to support communities and habitats and empowering local communities to develop enduring solutions. Scientific research and green initiatives development must be given utmost priority for combating the consequences of climate change by.

- Enumerating regenerative techniques for stabilising climate conditions, maximising nature's capacity for carbon sequestration, and promoting the adoption of clean, smart energy policies.

- Facilitating a conversion to environmentally friendly agricultural systems, maintaining healthy edaphic conditions, and increasing mangrove plantations.

33.8 Conclusion

Climate change is primarily the result of human activities leading to the occurrences of extreme weather events, biodiversity loss, and limited ecosystem productivity. Changes in biodiversity have a significant impact on ecosystem services, including the provisioning, regulating, supporting, and cultural benefits that people derive from ecosystems. The survival and well-being of humans depend on ecosystem services which in turn depend on biodiversity. Proactive adaptation strategies are required to diminish the adverse effects of climate change and disasters on ecosystems since they can reduce the risk of climate-related catastrophes and increase ecosystem resilience. Furthermore, a detailed understanding of the ecosystem functions can help to achieve long-term sustainability and human well-being. Strict policy implementation and sustainable uses of natural ecosystems may help to address the current extinction risk of various endemic and RET plant species. EbAs are effective and may be applied globally to improve resilience, reduce damage, and adapt to climate change. Nevertheless, sustainable development can take advantage of new opportunities that climate change presents by integrating government policies, the scientific community, research and analysis, social efforts, and sufficient feasible strategies for building resilience, mitigating damage, and adapting to climate change. Therefore, it is urgent to develop climate-friendly approaches like expanding market opportunities for local products, reducing food and agricultural waste, carbon pricing, climate funding, technologies to capture GHGs, and stronger public policies on land use change to balance other goals beyond biodiversity conservation, ecosystem resilience, and sustainability.

References

Abisha R, Krishnani KK, Sukhdhane K et al (2022) Sustainable development of climate-resilient aquaculture and culture-based fisheries through adaptation of abiotic stresses: a review. J. Water Clim. Change 13(7):2671–2689.

Adedeji O (2014) Global climate change. J. Geosci. Environ. Prot. 2(2):114.

Alemaw BF, Simatele, D (2020) Integrating climate change adaptation and mitigation into sustainable development planning: the policy dimension. In: Matond JI, Alemaw BF, Sandwidi WJP (eds) Climate variability and change in Africa. Springer, Cham, pp 191–208.

Ammer C (2019) Diversity and forest productivity in a changing climate. New Phytol. 221(1):50–66

Araújo MB, Rahbek C (2006) How does climate change affect biodiversity? Science 313:1396–1397.

Bellard C, Bertelsmeier C, Leadley P et al (2012) Impacts of climate change on the future of biodiversity. Ecol. Lett. 15(4):365–377.

Berbés-Blázquez M, Mitchell CL, Burch SL et al (2017) Understanding climate change and resilience: assessing strengths and opportunities for adaptation in the Global South. Clim. Change 141(2):227–241.

Biggs R, Schlüter M, Schoon ML (2015) Principles for building resilience: sustaining ecosystem services in social-ecological systems. Cambridge University Press, Cambridge.

Bizikova L, Robinson J, Cohen S (2007) Linking climate change and sustainable development at the local level. Clim. Policy 7(4):271–277.

Both C, van Asch M, Bijlsma RG et al (2009) Climate change and unequal phenological changes across four trophic levels: constraints or adaptations? J. Anim. Ecol. 78(1):73–83.

Botkin DB, Saxe H, Araujo MB et al (2007) Forecasting the effects of global warming on biodiversity. Bioscience 57(3):227–236.

Broecker WS (1975) Climatic change: are we on the brink of a pronounced global warming? Science 189(4201):460–463.

Butchart SH, Walpole M, Collen B et al (2010) Global biodiversity: indicators of recent declines. Science 328(5982):1164–1168.

Cahill AE, Aiello-Lammens ME, Fisher-Reid MC et al (2013) How does climate change cause extinction? Proc. R. Soc. B. 280(1750):20121890.

Cardinale BJ, Duffy JE, Gonzalez A et al (2012) Biodiversity loss and its impact on humanity. Nature 486:59–67.

CBD (2021) Ecosystem Approaches. Available at URL. https://www.cbd.int/ecosystem/. Accessed on: 1st September 2022.

Chauhan AS, Singh S, Maurya RKS et al (2022) Impact of monsoon teleconnections on regional rainfall and vegetation dynamics in Haryana, India. Environ. Monit. Assess. 194(7):1–29.

Core Environment Program (CEP) (2015) Ecosystem-based Approaches to Address Climate Change Challenges in the Greater Mekong Subregion. Greater Mekong Subregion Environment Operations Center, Bangkok, Thailand. Available at URL. https://www.adb.org/sites/default/files/publication/158165/ecosystem-based-approaches-gms.pdf. Accessed on: 1st September 2022.

Costanza R., d'Arge R, De Groot R et al (1997) The value of the world's ecosystem services and natural capital. Nature 387(6630):253–260.

Dale VH, Joyce LA, McNulty S et al (2001) Climate change and forest disturbances: climate change can affect forests by altering the frequency, intensity, duration, and timing of fire, drought, introduced species, insect and pathogen outbreaks, hurricanes, windstorms, ice storms, or landslides. BioScience 51(9):723–734.

Denton F, Wilbanks TJ, Abeysinghe AC et al (2014) Climate-resilient pathways: adaptation, mitigation, and sustainable development. In: Field CB et al (eds) Climate change 2014: impacts, adaptation, and vulnerability. Part A: Global and sectoral aspects. Contribution of Working Group II to the Fifth Assessment Report of the Intergovernmental Panel on Climate Change, Cambridge University Press, Cambridge, United Kingdom and New York, USA, pp 1101–1131.

Ecker UK, Butler LH, Cook J et al (2020) Using the COVID-19 economic crisis to frame climate change as a secondary issue reduces mitigation support. J. Environ. Psychol. 70:101464

Edwards T, Wiseman J (2011) Climate change, resilience and transformation: challenges and opportunities for local communities. In: Weissbecker I (ed) Climate change and human well-being. International and cultural psychology. Springer, New York, NY.

Eriksen S, Brown K (2011) Sustainable adaptation to climate change. Clim. Dev. 3(1):3–6.

Fadrique B, Báez S, Duque Á et al (2018) Widespread but heterogeneous responses of Andean forests to climate change. Nature 564(7735):207–212.

FAO (2021) Plant health and climate change. IPPC Secretariat, FAO, Rome, Italy. Available at URL. https://www.fao.org/3/cb3764en/cb3764en.pdf. Accessed on: 21 August 2022

Finch DM, Butler JL, Runyon JB et al (2021) Effects of climate change on invasive species. In: Weynand TM, Finch TP, Miniat DM et al (eds) Invasive species in forests and rangelands of the united states: a comprehensive science synthesis for the United States Forest Sector, Poland. Springer, pp 57–83.

Forster PM, Forster HI, Evans MJ et al (2020) Current and future global climate impacts resulting from COVID-19. Nat. Clim. Change 10(10):913–919.

Franklin J, Serra-Diaz JM, Syphard AD et al (2016) Global change and terrestrial plant community dynamics. Proc. Natl. Acad. Sci. U.S.A. 113(14):3725–3734.

Gerlach J (2010) Climate change, species extinctions and ecosystem collapse. Phelsuma 17A:13–31

Gills B, Morgan J (2020) Global climate emergency: after COP24, climate science, urgency, and the threat to humanity. Globalizations 17(6):885–902.

Gilman SE, Urban MC, Tewksbury J et al (2010) A framework for community interactions under climate change. Trends Ecol. Evol. 25(6):1–7.

Hallegatte S, Vogt-Schilb A, Bangalore M et al (2016) Unbreakable: building the resilience of the poor in the face of natural disasters. World Bank Publications.

Handmer J, Honda Y, Kundzewicz ZW et al (2012) Changes in impacts of climate extremes: human systems and ecosystems. In: Managing the risks of extreme events and disasters to advance climate change adaptation: Special report of the Intergovernmental Panel on Climate Change, Cambridge University Press, pp 231–290.

Harry S, Morad M (2013) Sustainable development and climate change: Beyond mitigation and adaptation. Local Econ. 28(4):358–368.

HLPE (2014) Food losses and waste in the context of sustainable food systems. A report by the High-Level Panel of Experts on Food Security and Nutrition of the Committee on World Food Security, Rome 2014, pp 1–115. Available at URL. https://www.fao.org/3/i3901e/i3901e.pdf. Accessed on 12 August 2022.

Hobbs RJ, Mooney HA (1995) Effects of episodic rain events on Mediterranean climate ecosystems. In: Time scales of biological responses to water constraints: the case of Mediterranean biota. SPB Academic Publishing Bv, pp 71–85.

Hoffmann AA (2010) Physiological climatic limits in Drosophila: patterns and implications. J. Exp. Biol. 213(6):870–880.

Holmgren M, López BC, Gutierrez JR et al (2006) Herbivory and plant growth rate determine the success of El Niño Southern Oscillation-driven tree establishment in semiarid South America. Glob. Chang. Biol. 12(12):2263–2271.

IPBES (2019) Summary for policymakers of the global assessment report on biodiversity and ecosystem services of the intergovernmental science-policy platform on biodiversity and ecosystem services. Díaz S et al (eds). IPBES Secretariat, Bonn, Germany.

IPCC (2018) Global warming of 1.5°C. An IPCC special report on the impacts of global warming of 1.5°C above pre-industrial levels and related global greenhouse gas emission pathways, in the context of strengthening the global response to the threat of climate change, sustainable development, and efforts to eradicate poverty. World Meteorological Organization, Geneva, Switzerland.

Isbell F (2010) Causes and Consequences of Biodiversity Declines. Nature Education Knowledge 3(10):54.

Jarnevich CS, Holcombe TR, Bella E et al (2014) Cross-scale assessment of potential habitat shifts in a rapidly changing climate. Invasive Plant Sci. Manag. 7(3):491–502.

Jentsch A, Beierkuhnlein C (2008) Research frontiers in climate change: effects of extreme meteorological events on ecosystems. C. R. Geosci. 340(9–10):621–628.

Jetz W, McGeoch MA, Guralnick R et al (2019) Essential biodiversity variables for mapping and monitoring species populations. Nat. Ecol. Evol. 3(4):539–551.

Kearney M, Porter W (2009) Mechanistic niche modelling: combining physiological and spatial data to predict species' ranges. Ecol. Lett. 12(4):334–350.

Khan N, Jhariya MK, Raj A et al (2021) Soil carbon stock and sequestration: implications for climate change adaptation and mitigation. In: Ecological intensification of natural resources for sustainable agriculture Springer, Singapore, pp 461–489.

King RB, Hauff S, Phillips JB (1994) Physiological color change in the green treefrog: responses to background brightness and temperature. Copeia 2(16):422–432.

Koh LP, Dunn RR, Sodhi NS et al (2004) Species coextinctions and the biodiversity crisis. Science 305:1632–1634.

Lacetera N (2019) Impact of climate change on animal health and welfare. Anim. Front. 9(1):26–31.

Laukkonen J, Blanco PK, Lenhart J at al (2009) Combining climate change adaptation and mitigation measures at the local level. Habitat Int. 33(3):287–292.

Lavergne S, Mouquet N, Thuiller W et al (2010) Biodiversity and climate change: integrating evolutionary and ecological responses of species and communities. Annu. Rev. Ecol. Evol. Syst. 41:321–350.

Leadley P, Pereira H, Alkemade R et al (2010) Biodiversity scenarios: projections of 21st century change in biodiversity and associated ecosystem services. In: Secretariat of the Convention on Biological Diversity (ed. Diversity SotCoB). CBD Secretariat, Montreal, Technical Series no. 50, pp 1–132.

Locatelli B (2016) Ecosystem services and climate change. In: Potschin M, Haines-Young R, Fish R et al (eds) Routledge handbook of ecosystem services. Routledge, London and New York, pp 481–490.

Lovejoy T (2008) Climate change and biodiversity. Rev. Sci. Tech. (International Office of Epizootics) 27(2):331–338.

Malcolm J, Liu C, Neilson R et al (2006) Global warming and extinctions of endemic species from biodiversity hotspots. Biol. Conserv. 20:538–548.

Malhi Y, Franklin J, Seddon N et al (2020) Climate change and ecosystems: threats, opportunities and solutions. Philos. Trans. R. Soc. B 375(1794):20190104.

MEA (2005) Millennium ecosystem assessment: synthesis report. Island, Washington, DC.

Misra S, Maikhuri RK, Dhyani D et al (2009) Assessment of traditional rights, local interference and natural resource management in Kedarnath Wildlife Sanctuary. Int. J. Sust. Dev. World 16(6):404–416.

Moore JW, Schindler DE (2022) Getting ahead of climate change for ecological adaptation and resilience. Science 376(6600):1421–1426.

Moreno RD, Lagos-Cabré R, Buñay J et al (2012) Molecular basis of heat stress damage in mammalian testis. In: Nemoto Y, Inaba N (eds) Testis: Anatomy, Physiology and Pathology, pp 127–155.

Munang R, Thiaw I, Alverson K et al (2013) Climate change and ecosystem-based adaptation: a new pragmatic approach to buffering climate change impacts. Curr. Opin. Environ. Sustain. 5(1):67–71.

Muthee K, Duguma L, Nzyoka J et al (2021) Ecosystem-based adaptation practices as a nature-based solution to promote water-energy-food nexus balance. Sustainability 13(3):1142.

Nelson EJ, Kareiva P, Ruckelshaus M et al (2013) Climate change's impact on key ecosystem services and the human well-being they support in the US. Front. Ecol. Environ. 11(9):483–893.

Niles MT, Ahuja R, Barker T et al (2018) Climate change mitigation beyond agriculture: a review of food system opportunities and implications. Renew. Agric. Food Syst. 33:297–308.

Ossebaard HC, Lachman P (2021) Climate change, environmental sustainability and health care quality. Int J Qual Health Care. 33(1):mzaa036.

Palomo I (2017) Climate change impacts on ecosystem services in high mountain areas: a literature review. Mt Res Dev. 37(2):179–187.

Parmesan C (2006) Ecological and evolutionary responses to recent climate change. Annu. Rev. Ecol. Evol. Syst. 37:637–669.

Patz JA, Epstein PR, Burke TA et al (1996) Global climate change and emerging infectious diseases. JAMA 275(3):217–223.

Paupière MJ, van Heusden AW, Bovy AG (2014) The metabolic basis of pollen thermo-tolerance: perspectives for breeding. Metabolites 4(4):889–920.

Pedrono M, Locatelli B, Ezzine-de-Blas D et al (2016) Impact of climate change on ecosystem services. In: Torquebiau E (ed) Climate change and agriculture worldwide. Springer, Dordrecht, pp 251–261.

Peña ST, Stone F, Gummow B et al (2019) Tropical summer induces DNA fragmentation in boar spermatozoa: implications for evaluating seasonal infertility. Reprod. Fertil. Dev. 31(3):590–601.

Perrings C (2010) Biodiversity, ecosystem services, and climate change. The Economic Problem/Environment Department Papers 120(1):39.

Post E, Peterson RO, Stenseth NC et al (1999) Ecosystem consequences of wolf behavioural response to climate. Nature 401(6756):905–907.

Queirós AM, Fernandes JA, Faulwetter S et al (2015) Scaling up experimental ocean acidification and warming research: from individuals to the ecosystem. Glob. Chang. Biol. 21(1):130–143.

Rafferty NE, Ives AR (2011) Effects of experimental shifts in flowering phenology on plant–pollinator interactions. Ecol. Lett. 14(1):69–74.

Raza A, Razzaq A, Mehmood SS et al (2019) Impact of climate change on crop adaptation and strategies to tackle its outcome: a review. Plants 8(2):34.

Reusch TB, Wood TE (2007) Molecular ecology of global change. Mol. Ecol. 16:3973–3992.

Robinson RA, Learmonth JA, Hutson AM et al (2005) Climate change and migratory species. A Report for Defra Research Contract CR0302, The Nunnery, Thetford, Norfolk: British Trust for Ornithology, pp 1–304.

Salamin N, Wüest RO, Lavergne S et al (2010) Assessing rapid evolution in a changing environment. Trends Ecol. Evol. 25(12):692–698.

Sarkar AN (2012) Sustainable development through pathways of mitigation and adaptation to offset adverse climate change impacts. In: Leal FW (ed) Climate Change and the Sustainable Use of Water Resources. Springer, Berlin, Heidelberg, pp 515–554.

Seidl R, Thom D, Kautz M et al (2017) Forest disturbances under climate change. Nat. Clim. Change 7(6):395–402.

Sgrò CM, Lowe AJ, Hoffmann AA (2011) Building evolutionary resilience for conserving biodiversity under climate change. Evol. Appl. 4:326–337.

Shrestha S (2019) Effects of climate change in agricultural insect pest. Acta Sci. Agric. 3(12):74–80.

Sintayehu DW (2018) Impact of climate change on biodiversity and associated key ecosystem services in Africa: a systematic review. Ecosyst. Health Sustain. 4(9):225–239.

Sommer R, Bossio D (2014) Dynamics and climate change mitigation potential of soil organic carbon sequestration. J. Environ. Manage. 144:83–87.

Streimikiene D, Svagzdiene B, Jasinskas E et al (2021) Sustainable tourism development and competitiveness: the systematic literature review. Sustain. Dev. 29(1):259–271.

Swart R, Robinson J, Cohen S (2003) Climate change and sustainable development: expanding the options. Clim. Policy 3(1):19–40.

Thuiller W, Richardson DM, Pyšek P et al (2005) Niche-based modelling as a tool for predicting the risk of alien plant invasions at a global scale. Glob. Chang. Biol. 11(12):2234–2250.

Traill LW, Lim ML, Sodhi NS et al (2010) Mechanisms driving change: altered species interactions and ecosystem function through global warming. J. Anim. Ecol. 79(5):937–947.

Urban MC, Bocedi G, Hendry AP et al (2016) Improving the forecast for biodiversity under climate change. Science 353(6304):aad8466.

Vermeulen SJ, Campbell BM, Ingram JS (2012) Climate change and food systems. Annu. Rev. Environ. Resour. 37:195–222.

Vieux F, Darmon N, Touazi D et al (2012) Greenhouse gas emissions of self-selected individual diets in France: changing the diet structure or consuming less? Ecol. Econ. 75:91–101.

Zimmermann NE, Normand S, Psomas A (2013) Climate change and range shifts in mountain plants of the European Alps and Carpathians. Acta Biol. Crac. Ser. Bot. 55(1):18.

CHAPTER 34

Green Social Work; A Call for Climate Action

Jose John Mavely, Rosemariya Devassy, and Kiran Thampi

34.1 Introduction and Background

The exhibition of greediness by exploiting the resources of the environment unfavourably and utilising those resources for massive development, human beings have deracinated the equilibrium of the green planet. The extensive use of forest resources and pollutants has shifted the balance of nature, resulting in unfavourable atmospheric conditions and disasters like floods, forest fires, hurricanes, and droughts. The increase in greenhouse gases, more than 50%, has led to global warming and subsequent consequences in the climate system (Riebeek, 2010). In Brazil, the drought decreased potable water supplies to 10% of capacity and negatively impacted crop harvests (Nobre et al., 2016). Where as in China, the city of Henan, a large producer of soybeans, barley, and rice, was hit the worst by the drought, resulting in a considerable decrease in global exports.

J. J. Mavely · K. Thampi
Rajagiri College of Social Sciences, Kalamassery, Kochi, Kerala 683104, India
e-mail: josejohnmavely@gmail.com

K. Thampi
e-mail: kiran@rajagiri.edu

R. Devassy (✉)
Rajagiri Family Counselling Centre, Kalamassery, Kochi, Kerala 683104, India
e-mail: 5rosemariya@gmail.com

© The Author(s), under exclusive license to Springer Nature Singapore Pte Ltd. 2023
S. Nautiyal et al. (eds.), *Palgrave Handbook of Socio-ecological Resilience in the Face of Climate Change*,
https://doi.org/10.1007/978-981-99-2206-2_34

The sustainable development goals see climate action as an important goal (SDG 13) and aim to effectively address the developing countries' adaptation to climate change and promote low-carbon development. The year 2019 was the second warmest year recorded, and the atmosphere's carbon dioxide and the level of greenhouse gases have increased to new records (NASA, 2020). The Paris Agreement looks forward to strengthening countries' capacities to deal effectively with the impact of climate change by initiating suitable technologies, capacity, and financial flows (Obergassel et al., 2016).

Social work aims to work with the "individual in the environment" from a human rights and social justice standpoint (IASSW). Green Social work is an emerging area of professional practice among social workers. Green social work aims to enhance people's and the planet's well-being for a better future (Dominelli, 2013). Green social work focuses on environmental problems and strives for environmental justice. The impact of climate change is unjust and unequal. People with fewer resources will be the most affected. People affected by climate-induced disasters will require social work assistance to cope with the consequences.

From the literature, it was found that social workers were less involved in climate change activities (Cumby, 2016). Therefore, one of the significant questions was *whether Green Social Work could address the concern of climate change*. This paper focuses on the importance of green social work in addressing the issue of climate change and would contribute to the existing studies related to social work practice in climate change.

34.2 METHODOLOGY

This paper is developed using the systematic review design, and the secondary data were collected through repeatable analytical methods. The narrow question for the article was defined, and a clear plan for developing the paper was constructed. Initially, the keywords in the titles were repeatedly searched in different databases, including Scopus-indexed journals, book chapters, articles, newspaper cuttings, and blogs. The screening was done in two phases. First, the titles and abstracts with similar themes and information were screened. Then, in the second phase, the full text of the sorted title and abstracts is reviewed repeatedly for evidence. Finally, two reviewers did the screening process under the tiebreaker's supervision. First, common themes were identified from the evidence gathered. Then, the reviewers chose the themes relevant to the developing paper together from the specified theme.

Inclusion Criterion:

- To be included, the papers need to be focused on ("Green Social Work" AND "Climate Action" OR "Environmental Justice").
- Papers, reviews, or empirical reports of relevant research were included in the study.

- During the scoping process, it became apparent that the number of relevant empirical investigations was minimal. Therefore, grey literature was included.
- The chosen studies were restricted to the English language.

Exclusion Criterion

- Those studies whose scope did not include other professions or areas of social work practice.
- Government papers, policy documents, and theoretical materials were excluded.

34.3 RESULTS

Green Social Work in Reducing Climate Stress

Climate stress is an important issue. Naturally, the reason a person comes up with climate stress would be climate change. Climate stress often varies from person to person. We see the most rural, poor farmers and their families most affected by climate change.

The three most critical climate-related risks to address in a changing climate stress situation are physical risk, transition risk, and liability risk (C.F. Powers, 2018). *Physical risk* refers to the risk related to the physical impacts of climate change, with flood damage induced by severe weather occurrences serving as an example. *Transition risk* is connected with policy and regulatory changes in response to climate change, such as implementing new carbon prices under the Paris Agreement to cut emissions. *Liability risk* is linked with legal actions seeking compensation for insufficient disclosure of climate-related concerns or insufficient preventative measures, such as pollution. Climate tension is an important issue. Naturally, the reason a person comes up with climate stress would be climate change. Climate stress often varies from person to person. It can be seen that rural, poor farmers and their families are most affected by climate change.

Case Example for Green Social Work in the Context of Kenya (C.F. Powers, 2018)

This case shows a collaborative technique for investigating green social work issues in local communities, as well as quick results from an international social worker's examination into the consequences of climate change in two rural Kenyan villages, Mutito and Wamunyu.

Wamunyu and Mutito are semi-arid, rural, and underdeveloped regions. Access to productive land underpins the majority of community members' livelihood strategies. The settlements were semi-arid, with two rainy seasons before the climatic changes witnessed in recent decades, which allowed community members to engage in regular farming. However, surviving in these places has gotten more challenging because of climatic change. First, there were droughts and desertification due to the absence of rain. The flash floods begin to erode the soil's top layer, diminishing the land's production. The rich people of the town adapted by establishing irrigation systems and bore wells, whereas the livelihoods of the impoverished residents were ruined. Famine and food insecurity are becoming more and more common. During times of drought, companies were closed, and kids were taken out of school to collect water, affecting the opportunities of the community. Poor community members in Mutito and Wamunyu found it challenging to identify influential people willing to engage them continuously on these issues.

Green social workers, including those outside Kenya, addressed these gaps. The green social workers put forward an advisory board, selecting marginalised communities that badly affected by climate change. Members and the foreign social worker connected while working on different projects, and over several years, their connection grew to include trust and understanding. Members of the advisory board agreed to lead the project after discussing the requirements for effective partners (e.g., payment, work hours, and location of workplace). The international social worker highlighted concrete benefits over hypothetical or trickle-down benefits, which was crucial to the effective formation of the advisory board. Variations of the Advisory Board Model could be deployed within social work on issues related to the environment in cross-cultural settings to move micro-level socio-political power structures via the self-determination and empowerment of the local community.

Green Social Work for Sustainable Rural Development

Climate change is having a huge impact on our rural areas. This is because climate change is affecting the livelihoods of humans and food security. Climate change is affecting the rural economy as well as the agricultural sector (Sutariya et al., 2020). A clear example is that India is a world-renowned country for a variety of reasons. Therefore, the impact of climate change on agriculture will have a proportionate effect on India's rural development.

The results of one article (Wu et al., 2022) emphasise the need for social work in rural communities experiencing changing climate and environmental concerns. Environmental justice in rural communities should accompany sustainable social work practices exploring social work research, client practices, and social policies. Also, individuals living in rural or agricultural areas are more likely to be exposed to toxins due to farming operations, resulting in significant health impacts. Another paper (Dominelli & Ku, 2017) deals with how environmental social work provides an inclusive approach to reflect and redesign growth and environmental development for transformative change with community, institutional, and government support.

The study (Dominelli & Ku, 2017) showed how social workers transformed a village in China from marketplace-driven advancement to people-centred and environmentally friendly progress through green social work. As a result, cross-disciplinary participatory action research was initiated to encourage local villagers to return to organic farming, produce arts and crafts, and use local capital for urban green consumption. This will help people to generate additional income, preserve and revitalise their cultural pride and identity, protect soil and seed, and foster community participation.

Green Social Work for Community Action

The Council of Social Work Education recently included the concept of environmental justice along with human rights and social and economic justice (*Committee on Environmental Justice | CSWE*). Social workers being change agents could utilise the profession effectively to promote environmental justice. The person-in-environment outline of social work education emphasises the importance of system perspectives and the essential expertise required to engage, assess, and intervene at various levels. Green social work promotes the skills of community organisation of the social worker and orients the community about climate actions and policies.

Case Example for Green Social Work in the Context of Sri Lanka
One paper (Dominelli, 2014) deals with how green social work could be counted in social work training to promote inclusion and thus overcome the crisis. The research was done in the Indian Ocean Tsunami affected South Western Sri Lanka in 2004 and used the ethnographic methodology. The results point to the adaptation from cleaning the remains of environmental disasters to getting along with people, engaging with them and utilising those engagements to build a social relationship, laying the bases for social work practice. This helped the survivors as the social workers could respond to the psychosocial need, health, housing, and livelihood empowerment. Thirty residents who attended the focused group discussion in 2012 said that 11 families haven't received funding for replacing their fishing nets. Since each net created jobs for eight people, the village had lost 88 jobs, forcing residents to either remain jobless in their hometown or travel to Colombo or another location in search of employment, making overcrowding in poor-quality housing in Colombo. Social workers can encourage people to take action on housing-related concerns and create environments supporting health and well-being. The relationship between consumption, production, and reproduction, as well as the extent to which sustainable development might encourage more long-term investments in society, can be better understood by the public with the aid of social workers.

UNICEF has taken effective measures to help the youth to take action to defend the future of our planet. Green social work promotes climate action by levitation the youth voices on the climate crisis and social workers focus to organise youth to address the issue of climate change as Timioci has rightly said "*The Sea is swallowing villages, eating away at shorelines, withering crops.*

Relocation of people ... cries over loved ones, dying of hunger and thirst. It's catastrophic. It's sad ... but it's real".

The policies stand as the algorithm for achieving a goal. An effective climate policy would address the levels of planning and actions taken by the government to address the issue of climate change. Even though policymakers at the global level are designing various action plans to reduce the production of Green House Gas (GHG) emissions, more research is needed in the climate policy (Konidari & Mavrakis, 2007). The Intergovernmental Panel on Climate Change also addressed the need for "rapid and far-reaching changes" to climate actions. It is also important that the policies for initiating climate actions should be based on the factor of equity and should promote human well-being (Klinsky, 2016).

The social worker professional commitment also extends to having concrete knowledge about climate change and its effect on people. Green social work makes social workers think from an inclusive environment perspective. The community affected by climate change end up with fewer resources; environmental justice is essential for such communities and is guided through effective policies and policy responses. Various researchers have also endorsed the profession of social work to be effective in helping people to understand their issues, serving the community and individuals in promoting sustainable energy production, consumptions, and designing effective policies through research.

34.4 DISCUSSION

Climate change is a major problem that the world faces in this era. Increased global warming has created a shift in the normal rhythm of nature and has started to affect the overall plan through disasters and catastrophes. For instance, the sea level has been raised by 20 cm and is expected to rise from 30 to 122 cm by the year 2100. Leaders and policymakers strive to create action plans and policies to help the planet cope with this issue. Considering climate change as a serious concern, the Kyoto protocol was adopted in 1997 to sensitise industrialised countries to reduce and limit greenhouse gas production. Moreover, the sustainable development goal put forward by United Nations sees climate action as an essential indicator. Thus, climate actions to mitigate climate change impacts are vital to protect the planet's flora and fauna.

The Need for Environmental Justice

Without action, the world will rapidly move towards complex catastrophes that will be uncontrollable by human technology. The forest fire that happened in Australia due to extreme weather conditions was a tragedy that burned millions of acres of forest, killing billions of animals and releasing pollution into the air. Environmental justice is only possible through practical actions.

Rural development programs are the key funding instrument for sustainable management of natural resources and climate action. There is a great need to understand and study the problems and challenges in rural areas related to climate change. The highlighted significant challenges include poverty, illiteracy, unemployment, and homelessness. Even though the majority of current explanations of climate change and development centre on the agriculture sector, investments in each area do not address the issues of climate change.

Social work has always been concerned with contemporary social problems and has continually shifted to address and meet the needs of vulnerable people. In the present world, social workers focus more on the person-environment perspective, which contemplates the individual as an active participant in a macro system. One of the major problems evident in the social work profession is the lack of focus on the "environment" perspective, which has reduced social workers' contribution towards environmental justice. Social workers are excellent at engaging various populations, facilitating discourse, and conducting empirical research. These abilities might be applied successfully to pursue inquiry with climate change deniers by asking questions and seeking constructive dialogue as a springboard for debating the scientific consensus on climate change.

Factors of Environmental Justice

From the analysis, we could infer that environmental justice is multidisciplinary and includes different dimensions.

Environmental justice could be achieved by optimum utilisation of resources. Social workers could effectively deal with the increased climate stress through the primary methods of casework and group work. Green social work plays an important role in disaster mitigation and extending support to the affected people's rehabilitation.

Climate action is a significant indicator of sustainable development goals (SDG 13). Organic farming is environmentally friendly and could support SDG 13 adequately. This could replace conventional farming methods, threatening the planet by releasing pollutants into the soil and air. Organic farming also supports the green social work perspective, which emphasises improving the planet's well-being of flora and fauna.

A social worker could effectively resolve an environmental problem observed in a community through community organisation. First, the social worker could study and research these issues. The objectives and needs to tackle this issue will be prioritised by discussing with the community. The community has identified internal and external recourses to manage the environmental issue. The social worker develops the will and confidence of the community to tackle the problem. In the end stage, the community itself plans for an action to reduce or nullify the problem through collaborative and cooperative participation, thus promoting environmental justice (Reisch & Wenocur, 1986). This could be seen in Chennai, where an environment-based

NGO called Exnora achieved a target of zero solid waste through community participation (Lakshmi, 2017). Social action plays a significant role in moulding effective responses and actions to environmental deprivation, climate change, and disasters. The article "*Environmental Justice and Social Work Education: Social Workers' Professional Perspectives*" discusses social justice education to enhance its ability to find solutions to environmental justice and stability (Nesmith & Smyth, 2015). In other words, low-income families and individuals who are more likely to experience or be affected by environmental degradation are the ones who suffer the most from environmental injustice. A climate action strategy is quite diverse. Some plans are motivating brochures, while others are detailed implementation plans with defined aims, specific goals, and well-structured techniques.

Environmental Justice Through Green Social Work

Green social work plays a vital role in preserving the fair dispersal of the Earth's resources to fulfil human needs while simultaneously promoting the present and future well-being of people and the planet. Due to this, international social work organisations have raised the need for social workers to incorporate environmental perspectives into their professional practice. It emphasises the commitment of social workers to enhance the citizens' well-being and help them meet their needs. Furthermore, it highlights a holistic need for knowledge about communities and people. Since climate change is a threat to the community and the citizenry, social workers also need substantial expertise about climate change and its effect on people.

According to Dominelli's conception, it tries to solve these challenges by advocating for a radical shift in how individuals understand their community's social foundation and connections with each other. As a result, social work can take the lead in addressing the human impacts of environmental change through (a) disaster readiness and response, population resettlement, and community organising; (b) development focusing on local and regional capacity to respond to global change in the environment of urban areas. (c) Mitigation and Advocacy; and (d) Participation in practice to address the central causes of ecological change.

The famous authors, (Sugirtha & Little Flower, 2018), have written about the role of green in social work. Conscientious people on conserving natural resources and reducing carbon emissions strive to achieve carbon neutrality. She also notes that climate change perspectives result in social and structural inequality. Another significant result is identifying and preserving the rights of the native people and the underprivileged groups in danger of suffering more from the consequences of climate change and developing and raising awareness of effective mitigation and adaptation strategies, particularly among disadvantaged and vulnerable communities. Advocating for policies that incentivise climate change mitigation, green technologies, and sustainable development, participating in post-disaster restoration and recovery activities to reduce the effects of environmental racism-combating the impacts of climate change, and contributing to sustainable development.

34.5 Conclusion

The environmental issues of the twenty-first century are intensifying. The environmental crisis stems from natural disasters such as earthquakes and human-made disasters rooted in industrialisation processes that take nature for granted. Exploiting its generosity has contaminated the planet's air, land, and water and threatened plant, animal, and human species. The social sciences have been reluctant to react to the difficulties presented by environmental crises and have seen the resulting environmental injustices as the exclusive domain of the physical sciences.

Green social work is an evolving trend in the social work practice, and due to this reason, there is very little literature related to green social work. This paper will contribute to the ongoing studies and research on green social work and its importance in promoting climate action. This paper ensures the role of green social work in the twenty-first century and the distribution of the Earth's assets equitably to meet human needs, organise the people in the community, and help them promote sustainable development in their community. Green social work promotes the skills of community organisation of the social worker and helps the community understand climate actions and policies. Social work methods can be used effectively. It can help to organise people in the community and promote them in their community to raise awareness about the issues related to climate change.

References

Brown, K. (2020, January 15). NASA, NOAA analyses reveal 2019 second warmest year on record [Text]. NASA. http://www.nasa.gov/press-release/nasa-noaa-analyses-reveal-2019-second-warmest-year-on-record.

C.F. Powers, M. (2018). Green social work for environmental justice: Implications for international social workers. In The Routledge handbook of green social work. https://www.routledge.com/The-Routledge-Handbook-of-Green-Social-Work/Dominelli/p/book/9781315183213.

Committee on Environmental Justice | CSWE. (n.d.). Retrieved September 24, 2022, from https://www.cswe.org/about-cswe/governance/commissions-and-councils/commission-on-global-social-work-education/committee-on-environmental-justice/

Cumby, T. (2016). Climate change and social work: Our roles and barriers to action. https://scholars.wlu.ca/etd/1828

Dominelli, L. (2013). Environmental justice at the heart of social work practice: Greening the profession. International Journal of Social Welfare, 22. https://doi.org/10.1111/ijsw.12024.

Dominelli, L. (2014). Promoting environmental justice through green social work practice: A key challenge for practitioners and educators. International Social Work, 57(4), 338–345. https://doi.org/10.1177/0020872814524968.

Dominelli, L., & Ku, H.-bun. (2017). Green social work and its implications for social development in China. China Journal of Social Work, 10(1), 3–22. https://doi.org/10.1080/17525098.2017.1300338.
IASSW. (n.d.). International Association of Schools of Social Work (IASSW). Retrieved September 24, 2022, from https://www.iassw-aiets.org/.
Klinsky, S. (2016). Why equity is fundamental in climate change policy research. Global Environmental Change, 4.
Konidari, P., & Mavrakis, D. (2007). A multi-criteria evaluation method for climate change mitigation policy instruments. Energy Policy, 35(12), 6235–6257. https://doi.org/10.1016/j.enpol.2007.07.007.
Lakshmi, A. P. (2017). EXNORA-A case study on effective community participation in solid waste management in Chennai. Indian Journal of Environmental Protection, 37, 754–757.
Nesmith, A., & Smyth, N. (2015). Environmental justice and social work education: Social workers' professional perspectives. Social Work Education, 34(5), 484–501. https://doi.org/10.1080/02615479.2015.1063600.
Nobre, C., Marengo, J., Seluchi, M., Cuartas, L., & Alves, L. (2016). Some characteristics and impacts of the drought and water crisis in Southeastern Brazil during 2014 and 2015. Journal of Water Resource and Protection, 08. https://doi.org/10.4236/jwarp.2016.82022
Obergassel, W., Arens, C., Hermwille, L., Mersmann, F., Ott, H., Wang-Helmreich, H., & Kreibich, N. (2016). Phoenix from the Ashes—An analysis of the Paris agreement to the United Nations framework convention on climate change (Vol. 28).
Reisch, M., & Wenocur, S. (1986). The Future of community organization in social work: Social activism and the politics of profession building. Social Service Review, 60(1), 70–93. JSTOR.
Riebeek, H. (2010, June 3). Global warming. https://earthobservatory.nasa.gov/features/GlobalWarming.
Sugirtha, T., & Flower, F. X. L. (2018). Global warming, climate change and the need for green social work.
Sutariya, S., Ankur, H., & Meherbanali, M. (2020). Impact of climate change on agriculture.
Wu, H., Greig, M., & Bryan, C. (2022). Promoting environmental justice and sustainability in social work practice in rural community: A systematic review. Social Sciences, 11(8). https://doi.org/10.3390/socsci11080336.

PART VI

Conclusion

CHAPTER 35

Epilogue

Sunil Nautiyal, Anil Kumar Gupta, Mrinalini Goswami, and Y. D. Imran Khan

Climate and climate change are complex subjects with immense contemporary importance. With the growing evidence of the impacts of climate change, more attention to research and development has been gained through adaptation research in the last three decades. While a simultaneous line of scientific advancements is going on scientific underpinning of climate change, climate change scenario analysis, vulnerability assessment, and formulation of plausible solutions. The research going on across the globe with regard to solution-seeking aims at mitigation and adaptation to reduce the adverse impacts of climate change. In the discussion of reducing the impacts of climate change, the most frequently mentioned terms with variations in contexts and definitions are "resilience" and "adaptation". Resilience can be defined as the capacity of ecological, economic, and social systems to cope with disturbances or quickly recover from the consequences of the disturbances. Whereas, adaptation means adjustments in those systems as a response to changing climate, where changes in structure, functions, processes, and practices in any of the systems are brought in through an organic or institutional approach.

S. Nautiyal (✉) · M. Goswami · Y. D. Imran Khan
Centre for Ecological Economics and Natural Resources, Institute for Social and Economic Change, Bengaluru, India
e-mail: sunil@isec.ac.in

A. K. Gupta
National Institute of Disaster Management, Ministry of Home Affairs, Government of India, Delhi, India

© The Author(s), under exclusive license to Springer Nature Singapore Pte Ltd. 2023
S. Nautiyal et al. (eds.), *Palgrave Handbook of Socio-ecological Resilience in the Face of Climate Change*,
https://doi.org/10.1007/978-981-99-2206-2_35

There are numerous instances across geographical locations and social and economic strata where adaptation strategies and resilience-building activities are implemented and provide learning for reducing climate change-related disturbances at local, regional, and national levels. Case studies are extensively used and believed to be particularly useful for identifying effective adaptation options and pathways. The findings and discussions put forward in the chapters, largely based on micro-level case studies provide an understanding of interactions among ecology, society, and economy under different conditions of changing climate. The chapters provide insights to understand both existing and required actions as adjustments to the changes by different actors of human sub-systems at different scales and contexts. Highlights of the learning drawn from the chapters are discussed below.

35.1 Section A

Indian Himalayan Region is an important area where ecological impacts of climate change are quite visible and inevitable and also have a massive significance in terms of supporting life and livelihoods of millions. Adverse effects on the habitat suitability of keystone species have been observed in the region. It is found alarming for the well-being of local communities that are already affected, and further loss of keystone species worsens the scenario. A better understanding of local knowledge practices will be necessary to further refine the approach of future predictions and implement necessary actions at the local level will further help in recognising native species that are relevant and effective for local and regional environments for supporting ecosystem services.

It is necessary to understand the relationship between fluctuations of climatic parameters and crop yields at landscape levels than national or regional levels in order to instrumentalise efficient adaptation techniques. The importance of promoting climate resilience agriculture in climate-sensitive regions in India and other developing countries has been discussed in the chapters. Climate resilience agriculture has gained relevance as studies identified climate stresses on agriculture where a noticeable decline in crop productivity was observed. The consistent negative impact of temperature on cereal crop yields in semi-arid and arid regions demands special attention in this regard. Impacts of climate variability on paddy and millet production in the ecologically fragile Uttarakhand region have also been reported. It can be highlighted that shifts from practices of hilly cultivation of traditional crops to paddy and wheat have aggravated the impacts of crop loss.

It is validated that farmers' perception is important in formulating an adaptation plan. Farmers of Odisha with are hugely affected by drought with consequences such as crop loss, financial loss, loss of savings in banks, and inability to meet basic family needs. Animal husbandry has been assessed as a more climate-resilient livelihood alternative in these regions. One understudied aspect of livestock rearing in India is pastoralism and pastoral communities. Dhangar, a nomadic pastoral community dependent on Savannah type of grasslands, has observed profound changes in terms of their livelihoods. The impacts are observed in terms of degrading grasslands, forests, and water availability which can be attributed to unusual monsoon cycles, increasing temperature, and changing land use. Moreover, socio-political, geographical, and economic affect herders' age-old livelihood.

Forest fires in tropical deciduous forests are identified to have strong linkages with meteorological parameters and thus the changing climate. These fires are attributed to the degradation of natural forests, greenhouse gas emissions, and loss of biodiversity in the Central Indian region. Forest fires in the region have shown an increasing trend in recent years. The results demonstrate the potential of forest fire analysis with geospatial techniques as an effective tool in assessing "where and when" forest fires will most likely occur and further use of these developing preventive measures.

The drinking water crisis has become a severe problem and moving towards a water-scarce status with changing rainfall patterns and unsustainable exploitation of resources resulting in a low per capita water availability in India. Assessment of people's willingness to pay for improved water supply in a city of southern India has shown a positive response which can provide a direction to improve water supply in terms of reducing loss, resilient infrastructure, and conservative use of supplied water. This provides doable policy insights to integrate the concepts of payments for ecosystem services in a robust way.

35.2　Section B

In India, subsistence farmers living at the bottom of the economic ladder bear the highest brunt of climate change as they depend heavily on climate-sensitive resources. Field investigation in Palghar district, Maharashtra, India, suggested that the declining rainfall has adversely impacted crop yield in recent years, where the adoption of agricultural practices such as crop rotation, crop diversification, organic manures, and storage mechanism, exhibits sustainable approaches to deal with the impacts. Research shows that barriers to the adoption of modern technology as adaptation mechanisms can be overcome by enabling easy access to credit and saving facilities.

Climate Smart Agriculture practices and technologies provide traction of their impacts on climate change adaptation. Insights into their nexus in India with a comparison to South Africa as a typical example of developing countries in Asia and Africa unravels the trend in feminisation of agriculture, and the ongoing legislation and program initiatives in India and South Africa to address challenges related to climate change. The literature revealed that both countries hold immense potential for CSA, but the initiatives to implement the CSA concept are still in their execution stages. With both indigenous approaches and research-based interventions possessing CSA qualities, it is suggestive that scaling up CSA will necessitate intersectional and multi-level efforts to allow the design, implementation, and monitoring of context-specific approaches towards integrated prioritisation of CSA.

Extreme meteorological and hydrological phenomena have brought out the severity of losses throughout the state of Kerala, where agriculture and allied sectors are significant as they provide a considerable part of the state's economy as the local people directly or indirectly depend on agriculture for their livelihood. The need for climate-resilient agriculture systems is realised to compete with the calamities and their consequences. Climate-resilient crop varieties such as Pokkali, which is flood and saline resistant and, therefore, can act as a potentially flood-resilient crop across the state, more specifically in the coastal regions. Pokkali variety has the potential to sustain in flood situations which no other rice variety has shown. Therefore, Pokkali may play a significant role in ensuring food security during a disaster (especially flood) and act as a nature-based solution for building flood-resilient agriculture.

Another chapter highlights how the traditional type of kitchen garden (such as *Gharbari* in Odisha) if systematically promoted can support climate-smart agricultural practices. Such models have not only resulted in catering to the nutrition requirement of the household but also helped the tribal women reduce dependency on the market, increase monetary savings, seed conservation, protection of agro-biodiversity, and make them self-reliant. A survey of such kitchen gardens in twenty villages of Kalahandi district of Odisha revealed that the tribal communities have become innovators of Climate Resilient and Adaptive farming. It shows how tribal households have produced multiple crops annually without harming nature, soil, and the ecosystem.

As an adaptive solution to the increasing water crisis, a Group Micro-Irrigation model (GMI) was implemented by Watershed Organisation Trust, an NGO based in Pune in three locations in the semi-arid landscapes of Maharashtra. Farms of these group members were connected to micro-irrigation, and appropriate climate-resilient practices were advised. A mental model of farmers is drawn to highlight their thinking process and identify the enabling and disabling factors for GMI adoption. It has been found that a sense of water security through cooperation, systematic operations, key persons' assistance, etc., were the critical enablers for the adoption of GMI; whereas, climate-related stresses and their impacts were the primary disabling factor. It emphasised on the importance of these factors in sustainability and adaptation planning with water-sharing interventions.

Field-based research in Tamil Nadu to ascertain the demand and supply of water through a scientific and systematic process of "Water Budgeting" attempted to analyse the influence of climate-resilient agriculture on a longitudinal pattern. Water budgeting at the micro-level enables the farmers to be aware of the vagaries of climate change and adapt to climate-resilient agriculture using their traditional wisdom. At the macro level, it could contribute to the policy discourses on disaster management and other cross-cutting domains to mitigate the hardships of climate change. The methodology is replicable and the results could be effectively used as a strategy to diversify farm-based initiatives and adapt to climate change.

Along with climate-resilient village, the carbon–neutral village is also an emerging concept. Attempts are underway in various places in India including, The Union Territory of Ladakh (Districts of Leh & Kargil) and Majuli district in Assam to convert them into carbon–neutral regions. An assessment has been done to identify the challenges faced by a village in Kerala in achieving carbon neutrality in a growth-centred development model. The discussion highlights an operational model and the need to adopt a sustainable and simplistic development model within the Gandhi-Kumarappa model of development and social solidarity framework.

Native people of Indian Himalayan Region have come up with traditional remedies to safeguard this ecological treasure; however, the inclusion and implementation of nature-based solutions (NbS) and policies for sustainable development are necessary. A study focussing on the phytosociological aspects of vegetation of the diverse flora of the higher altitude forests of the IHRs has shown the existing and potential climate change adaptation strategies and nature-based solutions. The findings reiterate the urgent need to restore the balance between economic interest and ecological perspectives of the forests of IHRs to provide livelihood options to a vast forest-dependent population.

35.3 Section C

The impacts of climate change or natural disasters are seen at local levels, which draw the emphasis on local adaptation and mitigation strategies in the global discussion on climate change. Traditional Ecological Knowledge (TEK), called Local Ecological Knowledge (LEK), refers to the knowledge base developed over generations through direct interactions with the natural environment. The people living in eco-sensitive areas like Sundarbans are generally governed by environmental determinism making the local knowledge extremely relevant for adaptation. The results show that TEK is concentrated among the natural resource-dependent (ntfp, forest, and fishing) people. TEK is losing its significance in the case of agriculture. It also reflects on how religious beliefs in the conservation and management of forest resources.

Agrobiodiversity and the associated indigenous knowledge systems are important components of the global agri-food system to remain resilient to climate shocks and to ensure dietary diversity. Across the globe, we have pockets of traditional biodiversity that conserve these systems but are increasingly challenged by Global Environmental Change. A study in Kerala looked into the indigenous agriculture and food practices of three Adivasi communities along with their perceptions of the local climate against the recorded climate. The study reveals that the three communities relate to local flora and fauna differently. Their traditional knowledge of food, farming, and the use of medicinal plants are documented which shows changes in ecology influence their choices.

Inference was drawn that framing effective strategies to recover the biotic and abiotic components of the sand dunes of coastal regions and socio-cultural marine environment is the need of the hour. The local fishermen's Traditional Ecological Knowledge (TEK) on seasonal change and floristic diversity of coastal regions of Tamil Nadu and Pondicherry was documented, which enables understanding of climate-related problems and the relevance of TEK in withstanding impacts and maintaining sustainable livelihoods.

35.4 Section D

Urban regions are becoming more exposed to several environmental concerns, such as air and noise pollution, due to growing populations, limited resources, and the mounting effects of climate change. In this regard, Urban Green Spaces (UGSs) act as a comprehensive tool for environmental sustainability, regulating the urban ecosystem services and microclimatic conditions, protecting biodiversity, and providing various socio-ecological benefits to urban dwellers. Research provided a framework for UGSs in rapidly growing cities for social, economic, and environmental sustainability. Planning strategies like City Master Plans, landscape approaches, government capacity plans, etc., are adapted to manage and develop UGSs. The role of UGSs as nature-based solutions in mitigation and adaptation against natural disasters, climate resilience, and societal challenges as well as restoration of degraded ecosystems has been underlined.

The pressure of global climate change on resources, rising extreme weather events, and demand for energy resources pose critical stress on the functioning of a city. Challenges regarding civic services delivery in Indian cities demand the need for climate-resilient infrastructure and to move towards a low carbon emission path. Analysis of the current status of core urban civic services—water supply, sewerage, waste management, drainage, roads, and streetlights—in two major Indian cities—Hyderabad and Pune—in light of climate change-induced challenges, revealed their vulnerability to climate change-induced risks like high rainfall and warm temperatures. The study points to the need to improve urban civic infrastructure systems in India by reforming and strengthening the city governments to cope with climate change-induced challenges.

The Intergovernmental Panel on Climate Change (IPCC) estimated that by 2050, buildings are projected to emit 3,800 megatonnes of carbon. Green buildings are designed to reduce the demand for non-renewable resources and make the best use of utilisation efficiency through reuse, recycling, and reducing the usage of renewable energy sources like the sun, water, and wind and a healthy indoor environment. Assessment of people's perceptions/experiences living and working in green buildings in Bengaluru city showed people's good awareness levels, satisfactory experiences, and preferences to live/work in green buildings were apparent. Discussions also brought in the significance of the ecology-societal interface in promoting green buildings and building a resilient society through appropriate policy interventions.

The role of governance in implementing climate resilience-building and adaptation strategies is very crucial for the growth of a sustainable city. The 74th Constitutional Amendment or Nagarapalika Act aimed to give urban local bodies constitutional status and make them self-governing institutions. Experience shows most of its goals have not been fulfilled even after 30 years. Weak State Election Commissions, non-devolution of the 3Fs—Funds, Functions, and Functionaries, dysfunctional District/Metropolitan Planning Committees, Ward Committees, and Area Sabhas are in evidence. Enlarging ward committee functions to take control of decision-making over local assets and resources is needed. Developing indicators and a database for measuring sustainability and developing a Ward Sustainability Index out of these is assessed to be a critical need.

35.5 Section E

Mainstreaming actions on climate change adaptation and disaster risk reduction in the planning process have multiple pathways and require coordinated efforts. Adaptation and mitigation measures for extreme climate events need to be solidified and integrated into planning and policies at the sectoral level. Chapters focus on agricultural planning in this regard highlight the steps for integrating Climate Change Adaption-Disaster Risk Reduction (CCA-DRR) into sectoral planning in the context of National Agriculture Disaster Management Plan (NADMP) for the Ministry of Agriculture Farmers' Welfare (MoAFW), Government of India. Specific emphasis includes: (i) policy support and institutional mechanisms (horizontal/vertical) for guiding and implementing risk resilience strategy, (ii) multi-stakeholder engagement planning that is formulation of national sectoral strategies, (iii) translating national strategy into state/local level action plan(s), (iv) mobilising budgetary support for implementation, and (v) monitoring and evaluation.

One significant policy initiative of Indian government, Sustainable Livelihoods and Adaptation to Climate Change (SLACC) was reviewed for its projects in Bihar and Madhya Pradesh States of India by the World Bank and National Rural Livelihoods Mission (NRLM) under the supervision of Ministry of Rural Development, Govt. of India. The main objective of the SLACC project was to improve the adaptive capacity of the rural poor to climate variability and change affecting farm-based livelihoods, through community-based interventions. The conclusion was drawn as a demonstration of outstanding results in some key areas that helped resource poor farmers to cope-up the climate change. At large scale this project has developed the enormous potential to fulfil sustainable development goals. The implementation of the SLACC project in other states with local stakeholders can be fruitful to reduce the climate change impact by improving the adaptive capacity of resource poor communities.

Technologies have been introduced to endure the requirements of our current and forthcoming generations. The rise and fall in planning, development in trade and commerce, town planning, and architecture have contributed to factors affecting the environment paving the way for today's climate change. Since every technological development has its pros and cons towards eco-sustainability, it is imminent that all the stakeholders become more responsible in every approach. To address the socioeconomic challenges, it is necessary to recognise that technology and policy framework must work in tandem. A holistic view of technological developments, legal approaches, judicial responses, and governmental actions for environmental protection in India, reflects that through many schemes and plans and technological interventions, the country has succeeded in lower carbon emissions. Yet it is more towards the energy than the agricultural and transport sectors. Similarly, the judiciary has disposed of cases appropriately and eco-friendly decisions. From a holistic standpoint, all of these approaches share commonalities, such as the desire to improve the natural environment and to be compassionate towards living beings.

Mobilisation of finances through products such as green loans, green bonds, green mortgages, green deposits, etc., to fund green projects is necessary for creating a green economy and is required to ease climate change adaptation and mitigation. Identification of green finance products developed by public and private financial institutions was done and grouped into categories: credit-based green products for consumers, investment funding, and green insurance. The chapter enumerated how Asian countries like China, Bangladesh, Indonesia, Philippines have created green financing channels. India's stature in green finance and the challenges faced are discussed. Lack of awareness and slow rate of returns in investments on green projects are identified as major barriers in bringing finance products into the mainstream. There is a need for better rendezvous with international private and public investors to stimulate global investment in green finance products to achieve a greener economy in India. To improve green finance in India, there is a need to reduce the information gap and increase awareness related to green finance.

Green social work is a division of social work that accord with the environment and its influence on the human population. With an aim of securing the flora and fauna through policy changes and social transformation, green social work enhances the well-being of the people and planet. The scope of green social work in promoting climate action has been identified. It is derived from the discussion that the green social work could play an effective role in endorsing climate action through environmental justice.

35.6 Way Forward

1. To restore the degraded landscapes and ecosystems in IHR, it is important to acknowledge and implement a unified climate-sensitive restoration approach for transformative alterations to sustainable land restoration for developing socio-ecological resilience against climate change and growing disaster risks.
2. Forests of the Western Himalayas are highly disturbed as compared to the Eastern Himalayan forests due to the higher population density in the Western Himalayas. Similarly, forests of the Eastern Himalayas are decreasing at a higher rate due to the traditional Jhum cultivation, higher impacts of ongoing global climate change, and higher livelihood dependency of the forest-dwelling people. Large-scale forest protection, conservation, and restoration efforts are urgently required in the IHRs to avoid the upcoming disturbance-driven biodiversity losses in IHRs. It is also necessary to devise a national strategy considering the multiple vulnerabilities faced by the forests of IHRs to adapt to climate change impacts and also to further amplify ecological integrity, functionality, and sustainability.
3. Proactive adaptation strategies are required to mitigate the adverse effects of climate change and the prevent degradation of ecosystems since they can reduce the risk of climate-related catastrophes and increase ecosystem resilience. Furthermore, a detailed understanding of the ecosystem functions at the micro-level can help to achieve long-term sustainability and human well-being. Strict policy implementation and sustainable uses of natural ecosystems may help to address the current extinction risk of various endemic and RET plant species. Ecosystem-based adaptations are effective and may be applied globally to improve resilience, reduce damage, and adapt to climate change.
4. Integrated climate-sensitive restoration planning in IHR is expected to streamline efforts that pursue to disclose the associations between local socio-ecological and economic interlinkages and their interdependencies. Distinguishing the intricacy and the diversity of factors that govern the dynamics of forest landscape restoration efforts can only be transformed and revolutionised by enabling and instigating more dialogue, and involvement of communities in improving the health of natural forest ecosystems for developing societal resilience to climate change. Practical approaches to protect the existing natural forest ecosystems are only possible by supporting assisted natural regeneration, reducing pressure from soil, developing seed banks, and ensuring livelihood diversification of forest-dependent marginalised communities.
5. Sustainable crops, such as finger millet in dry conditions, should be identified and promoted as a possible alternative adaptation strategy in the long run to reduce the dependency on irrigation in arid and semi-arid regions.

6. It is important to understand the impacts of climate change on crop productivity for food security and sustainable livelihoods. Temperate regions are highly sensitive to climate change; a minor shift in climate conditions of the mountainous region would affect the crop yield by altering soil degradation, nutrient availability, etc. The impact of climate change can be minimised through appropriate crop management and adaptation strategies.
7. Study found that animal husbandry is less affected by rainfall and that animal husbandry farmers are more economically resilient even in drought-hit years. Thus, in a drought-prone area, animal husbandry alone or in combination with other occupations can provide financial assurance to the farmers.
8. The Dhangars of the Deccan plateau mainly depends on grasslands which are essential as they are carbon sequesters, so protecting grasslands will help in going towards climate action. Therefore, the involvement of Dhangars and other rangeland users in the planning and policy processes will allow embracing their perspectives and make more inclusive policies. In this context, the state's response in addressing issues of pastoralists is needed. Indigenous knowledge plays a crucial role in coping with climate change and variability. Therefore, while addressing adaptation needs, considering the traditional knowledge of Dhangar pastoralists becomes a vital strategy. Herders have exhibited Ecosystem-based Adaptation (EbA) by developing resilience by coping with temporal and spatial variability despite inadequate policy support. Therefore, a reasonable and sustainable way forward lies in adopting the EbA approach to build more resilience among herders. The EbA approach will help improve the stakeholders' adaptive capacity through training, constant monitoring, livelihood diversification, and institution building (both customary and formal). Also, national-level programs like National Livestock Mission, and Rashtriya Gokul Mission-2014, need to include the EbA approach for the sustainable development of pastoralists and livestock holders. In the present situation, the Department of Animal Husbandry and Dairying, Govt. of India intends to initiate a particular cell for pastoralists, so such special programs are required to address the present climatic and land use-related issue associated with mobile pastoralism.
9. The analysis of forest fire hotspots and their relationship with meteorological parameters give a better comprehension of future forest fire events that will help in the control, prevention, and mitigation of forests. A special focus should be made on the fire hotspot areas to alert about fire trends to the nearest respective administrative headquarters that will help to take action to control the extent of fire damage. The application of remote sensing and GIS techniques can be scientifically used to study forest fire. In addition, there is an urgent need to

implement a forest fire policy by the government. Thus, an integrated approach would help with its prevention and mitigation.
10. Effective policies and programmes should be implemented to overcome the issues in water resource management. Water budgeting and willingness to pay for improved water supply provided doable policy insights to integrate the concepts of payments for ecosystem services in a robust way. The exercise of water budgeting should be repeated at periodical intervals, involving cultivators (including women farmers and agricultural labourers), elected members and officials of PRIs, officials from agriculture and irrigation departments, and other key stakeholders. The results of the micro-level study and intervention could be effectively used as a strategy to scientifically diversify the farm-based initiatives as well as to mitigate the hardships of climate change to a certain extent, leading to climate-resilient agriculture practices.
11. Study on barriers to adaptation recognises a growing need to go beyond piecemeal solutions by considering the broader perspective of sustainable development with a thrust on sustainable management of natural resources, diversification of farming systems, and building social capital and cohesion in the rural regions of developing countries. In particular, to address the climatic risks, stakeholders need to accept adaptation measures, such as strategising crop diversity, revamping irrigation facilities, and improving the storage mechanisms of perishable foods. By removing the market inefficiencies, such as the provisioning of credit and saving facilities and linking the input and output markets, the farmers can be encouraged to take up modern technologies that can increase yield and profitability and offset many of the uncertainties associated with extreme events. Crop diversification considering the ecology, society, and markets should be prioritised.
12. To scale up the adoption of the best CSA practices to fulfil the growing food demand, agriculture should undergo a transformation that handles simultaneously the dual obstacles of food insecurity and climate change. There are multiple CSA innovations that have been engineered, tried, field tested, and commended under five entry points viz; crop management, land and water management, agroforestry, and integrated food-energy systems that may be replicated and upscaled. Building a cohesive partnership between agricultural worth chain actors such as smallholder farming communities, agro-enterprises, and national research institutions will establish imperative pathways for sustainable adoption of best CSA practices geared towards attaining food security, productivity, resilience, and sustainable livelihoods. Formulating favourable policies and legislation that addresses CSA synchronisation and mainstreaming across different socioeconomic sectors so as to inclusively propel agricultural transformation and advancement at the state (micro), regional (meso), and international (macro) spheres is also warranted. Furthermore, elevating CSA best

practices among assorted stages of the agricultural value chains requires the application of innovative mechanisms to funding as a result of the various structures, operations, and administration within agro chains.

13. It is essential to identify more nature-based solutions such as promotion of climate-resilient crops (agro-ecological coastal farming of Pokalli rice—a flood-resistant variety) and practices to enhance traditional capacity to withstand climate-related disturbances.

14. Adoption of kitchen garden or home garden models (e.g.—Gharbari in Odisha) at the household level can be promoted strengthening resilience through climate-smart village and climate-adapting agriculture practices is essential for rural India. We should focus to build community capacity on Weather Smart, Soil Smart, Water Smart, Carbon Smart, Energy Smart, Knowledge Smart, and Market Smart strategy. All the flagship programmes, schemes, and projects must be driven to focus on above said themes.

15. In the Indian context, it is more of an urgency than just an exploratory phenomenon considering the level of vulnerability from hazards emanating from climate change. India is a country of diverse beliefs, social practices, customs, and knowledge besides having diversity in climatic, topographical, and ecological aspects. A thorough assessment is needed to understand the diverse perceptions, beliefs, and conceptual understanding of the environmental factors that affect people and in particular farmers in India. There should be a gradual shift from a mere descriptive approach to a predictive approach to inform policy. Conducting such studies have the potential to generate salient approaches that when integrated will benefit policy and practices, raising it from sub-optimal levels to more impactful ones.

16. The concept of carbon neutrality and actions towards achieving it in a simplistic development framework should be encouraged for Indian villages. It should collate learnings and experiences from such ongoing efforts in Kerala, Assam, and Ladakh.

17. Work towards co-existence and community-based biodiversity conservation should gain more attention.

18. More research and scientific validation on livestock food and fodder with bio-available nutrients for improved and sustainable livestock rearing should be encouraged.

19. TEK and IKS can be applied for the socio-ecological development at large. Agrobiodiversity emerges as a key theme through which the IKS could be applied for socio-ecological development. Globally, there is a shift in the thinking of key stakeholders of agri-food systems towards agroecology. This should be an avenue for leveraging the IKS. Appreciating the IKS by explicit acknowledgement and awareness building of its ecosystem services could be one way of reversing the erosion of IKS. Some of the existing neglected or underutilised institutional and policy mechanisms could be deployed for this purpose. Likewise, an evaluation

of ecosystem services of the resources could be looked into within the Payments for Ecosystem Services framework.
20. UGSs contribute significantly to the livability of urban areas by enhancing the natural beauty and recreational options for urban residents and promoting their general well-being. It also contributes to the carbon sequestration process, aids in reducing the impacts of UHI, improves hydrology by limiting surface runoff, and concurrently recharges the groundwater system. Furthermore, it can serve as a buffer during extreme events like floods since it is a natural stormwater drainage system that reduces the risks of climate-related disasters by providing adaptation to climate change. UGSs are the most sustainable solution to several environmental issues brought on by growing urbanisation, including pollution, climate change, natural disasters, health, and biodiversity loss. Therefore, further study is needed to examine the variety of environmental difficulties that a significant metropolis faces and the role that UGSs play in addressing these challenges.
21. Urban Local Bodies can play an important role in urban sustainability and climate resilience building. There is a need to improve ULBs' finances by giving them more autonomy to make decisions and providing grant support wherever required (especially for capital development). The institutional capacity and governance of ULGs are low, which can be improved by undertaking necessary reforms. The recent Central government initiatives of dedicated urban missions is a good approach, which needs further strengthening. Such improvements will render the ULGs to provide better civic infrastructure services and prepare them to meet the challenges of climate change-induced risks. It is recommended that an expert committee needs to be set up to review the functioning of the 74th CA. A Ward 5-year Vision Plan for making the ward inclusive and sustainable needs to be drawn up. Organisation of citizens into neighbourhood groups will strengthen decentralised governance of the people, by the people, and for the people. Targets need to be set, and outcome indicators developed for measuring these. Monitoring and review need to be based on performance on those indicators. The Area Sabhas must be involved in setting targets and reviewing municipal performance. A committee to review the entire gamut of provisions relating to the implementation of the 74th Constitutional Amendment in its true spirit needs to be set up, headed by a public figure and members devoted to the cause of decentralisation, to review the conformity legislations of States and present a revised set of amendments to the State Acts and the 74th CA.
22. Since, green buildings are not mandatory, there is no compulsion to adopt them. Further, there is a need for specific initiatives that aid access to finances to purchase green buildings. Nevertheless, at large, green buildings are gaining significance, and implementing better policies

can promote them even further, given the economic and environmental benefits and the threats of climate change. More awareness creation through Training Programmes is important by making training programmes as a part of the syllabus and tied up with the IGBC with the concerned departments in the Engineering Colleges. Incentives can also make it popular for adoption.
23. Integration of disaster risk reduction (DRR) and Climate Change Adaptation (CCA) initiatives through sectoral planning is one important approach. For success, mainstreaming DRR and CCA into development planning can be advanced through appropriate tools and methodologies along with key enablers in place such as institutional and policy support, scientific knowledge and expertise, community participation, capacity building & knowledge networking, and monitoring & evaluation. Furthermore, an array of sectoral initiatives related to water, health, forestry, rural development, etc., provides potential for climate and disaster resilience. However, a lack of effective coordination, both horizontally & vertically, limits the overall outcome especially on integrating DRR and CCA concerns.
24. To expand demonstrations of appropriate practices and technological interventions have been implemented through the SLACC project in a farmer-participatory mode to enhance their adaptive capacity and coping ability. SLACC addressed all aspects of farm-based livelihoods that may be affected by climate change by helping the community to choose interventions for the: production system, knowledge system, ecological system, technology, and financial system. Climate-smart practices implemented through SLACC, being a cost-effective and ecologically compatible alternative, should be widely implemented with more locally validated and demand-driven actions in other climate-sensitive regions of the country.
25. In mobilising green finances, India should take up various policies to address the challenges it faces, which include the reduction of the information gap and increasing awareness related to green finance. Coordination among various stakeholders will help India to improve its investment in green causes, and it will help India to bring green finance into mainstream banking operations. Secondly, the market needs to be informed about the positive impact of green finance on the environment and highlights climate change's negative effect on the financial sector. These additional disclosures and reduction in transaction costs will encourage the participation of potential investors in the Indian green finance debt market. Thirdly, some developments along the lines of the Green Infrastructure Investment Coalition (GIIC) will help the government, investors, asset managers, etc., of all the developing countries like India. India should encourage commercial banks to provide green credit, debit cards, and green mortgages so that the proceeds push liquidity in enterprises promoting sustainability through

their products. To sum up, we saw that India had launched various successful green finance products made available in the market.
26. Green social work promotes the skills of community organisation of the social worker and helps the community understand climate actions and policies. Social work methods can be used effectively. It can help to organise people in the community and promote them in their community to raise awareness about the issues related to climate change.

INDEX

A
Aal, 338
Adaptive capacity, 3, 143
Adivasi communities, 352
Adivasis, 284
Agricultural Labourer, 79
Agricultural production, 230
Agricultural productivity, 162
Agriculture, 504
Agriculture-based economy, 213
Agriculture economy, 82
Agriculture labour, 73
Agriculture production, 40
Agrobiodiversity, 362
Agro-climatic regions, 210
Agroforestry, 2, 176, 322
Agroforestry-based natural resilience, 323
Ahir Dhangars, 86
Amphan, 15
Angiosperm, 372
Animal husbandry, 73, 82, 305
Anthropogenic, 39, 100
Aquifers, 175
Asian open-bill, 116
Automatic rain gauges (ARG), 516
Automatic weather stations (AWS), 516
Avenue plantation, 391

B
Banj oak forests
　Cinnamomum tamala, 27
　Lyonia ovalifolia, 27
　Neolitsea pallens, 27
　Prunus cerasoides, 27
　Pyrus pashia, 27
　Rhododendron arboreum, 27
Behavioural aspects, 241
Bengaluru Water Supply & Sewerage Board (BWSSB), 473
Bihar Rural Livelihoods Society (BRLPS), 509
Biodiversity, 24, 580
Biodiversity Act, 363
Biodiversity conservation, 117, 587
Biodiversity loss, 320
Biodiversity Management Committee (BMC), 288
Biodiversity Register, 274, 285
Biofertilisers, 169
Bioindicators, 352
Biological invasion, 393
Biological resources, 362
Bio-physical vulnerability, 348
Bonbibi, 345
Botanical gardens, 396
Bottom-up approach, 334
Bruhat Bengaluru Mahanagara Palike (BBMP), 472
Building Back Better, 488

© The Editor(s) (if applicable) and The Author(s), under exclusive license to Springer Nature Singapore Pte Ltd. 2023
S. Nautiyal et al. (eds.), *Palgrave Handbook of Socio-ecological Resilience in the Face of Climate Change*,
https://doi.org/10.1007/978-981-99-2206-2

C

Carbon emissions, 99, 279
Carbon footprint, 280, 553
Carbon-neutral, 279
Carbon sequestration, 162, 586
Central Himalayas, 26
Circular economy, 6
Climate action, 557
Climate capitalism, 279
Climate change, 1, 24, 50, 125, 257, 320
Climate change adaptation, 334
Climate Change Adaptation-Disaster Risk Reduction (CCA-DRR), 484
Climate Change Adaptation Planning (CCAP), 516
Climate change policies, 168
Climate change programs, 168
Climate finance, 561
Climate-friendly approaches, 587
Climate-related disasters
  economy, ecology and society, 1
Climate resilience, 481, 550
Climate-resilient agriculture, 5, 162, 238, 274
Climate resilient technologies, 2
Climate Smart Agriculture, 162
Climate Smart Agriculture Practices, 225
Climate-smart alternatives, 3
Climate-Smart CRPs, 510
Climate-smart village, 5
Climate variation, 126
Climatic variability, 500
Coastal areas, 210
Coastal ecosystems, 368
Coastal sand dune, 384
Coexist, 578
Colonial waterbirds, 113, 117
Community-based adaptation, 5, 144
Community Based Adaptation (CBA), 32
Community Based Disaster Risk Management (CBDRM), 13
Community Based Disaster Risk Reduction (CBDRR), 13
Community Climate Adaptation Plans, 502
Community Resource Persons, 510
Conservation biology, 120
Contingent Valuation Method (CVM), 126
Cormorants, 116
COVID/COVID-19, 87, 477, 530
Crop calendar, 338
Crop diversification, 2, 145, 155, 322
Crop diversity, 145
Cropping patterns, 148
Crop rotation, 2, 322
Crop rotation practices, 148
Crop yield, 49
Custom hiring centers (CHCs), 513
Cyclones, 340

D

Darwinism, 334
Deccan plateau, 96
Deciduous broadleaf forests, 104
Deforestation, 320
Dhangar, 85
Disaster Management Act 2005, 482
Disaster Risk Management, 482
Disaster risk reduction (DRR), 4, 323, 394
Disaster risks, 23
Diversification, 333
Diversify crops, 143
Drinking water, 136
Drought, 71
Drought crop, 49
Drought proofing, 291
Dryland landscapes, 85

E

EcoDeposit, 559
Eco-friendly office space, 453
Ecological footprints, 393
Ecological integrity, 325
Ecological model, 116
Ecological systems, 53
Ecology, 120, 510
Ecosystem-based adaptation (EBA), 5, 96, 404, 584
Ecosystem-based disaster risk reduction (EcoDRR), 31
Ecosystem goods, 368
Ecosystem resilience, 587

INDEX   623

Ecosystems
  bays, 368
  estuaries, 368
  mangroves, 368
  marshes, 368
  reefs, 368
  sand dunes, 368
  seagrass, 368
  wetlands, 368
Ecosystem services, 23, 393
Endemic species, 321
Energy efficiency, 451
Energy Information Administration (EIA), 447
Enhanced Vegetation Index (EVI), 399
Environment, 599
Environmental determinism, 348
Environmental Impact Assessment Regulations, 287
Environmental justice, 598, 599
Environmental preservation, 528
Erratic rains, 263
Estuarine, 336
Ethnobotanical, 333
Ethnomedical applications, 368

F
Finance, 510
Fisherman community, 377
Fishing, 342
Flood, 340
Floodplain, 202
Flood resilience, 210
Flora and fauna, 580
Focus Group Discussion (FGD), 215, 260
Fodder-related crops, 92
Food and Agriculture Organization (FAO), 145
Food security, 77, 146, 257
Forest fire hotspots, 100

G
Garhwal, 27
Garlic (*Allium sativum*), 357
Gavli Dhangar, 86
Gene flow, 579

Genetic diversity, 579
Geological sensitivity, 30
Geophysical origins, 11
Gharbari, 215
Gharbari Model, 224
Ghongadi, 95
Global Climate Risk Index, 538
Global warming, 528
Good civic infrastructure, 412
Grassland ecosystems, 91
Grazing, 85
Green bonds, 554
Green buildings, 449
Green economy, 323
Green finance, 553
Green-gas emissions, 162
Greenhouse gas (GHG), 99, 258
Green Infrastructure Investment Coalition (GIIC), 569
Green Infrastructures (GIs), 389
Green initiative, 462
Green insurance, 561
Green investments, 566
Green Loan Principle, 558
Green roofs, 402
Green Social work, 594
Green streets, 396
Ground water (GW), 50, 231, 265
Groundwater management, 238
Group Micro Irrigation (GMI), 232, 238

H
Hatkar Dhangar, 86
Heat resistant seeds, 49
Himalayan highlands, 31
Hindukush Himalayas, 27
Hippophae salicifolia, 26
Historical Transect, 260
Human development, 478
Human Development Report (HDR), 277
Human nature, 528
Human settlements, 116
Human-wildlife coexistence, 112
Human-Wildlife Conflict (HWC), 111
Human-wildlife interaction, 6

Hyderabad Metro Water Supply and Sewerage Board (HMWSSB), 413
Hydrological cycle, 392, 393

I
Immunisation, 531
Indian Himalayan Regions (IHRs), 314
Indian Meteorological Department (IMD), 42
Indian Renewable Energy Development Agency (IREDA), 567
Indian Sundarbans, 332, 335
India's climate-friendly projects, 567
India's rural development, 596
Indicator species (riparian), 26
Indigenous communities, 332
Indigenous knowledge, 96
Indigenous Knowledge Systems (IKS), 351
Indigenous people, 322
Industrialisation, 383
Influenza pandemic, 533
Innovative sustainable design strategies, 468
In situ restoration, 585
Institutionalization, 15
Integrated Nutrient Management (INM), 512
Integrated Pest Management (IPM), 512
Intergovernmental Panel on Climate Change (IPCC), 447
International Platform on Sustainable Finance (IPSF), 540
Intertidal zone, 368
Invasive species, 582
Irrigation practices, 148
IUCN, 583

J
Jal Shevak, 242
Jhum cultivation, 320
Juvenile fish, 344

K
Karl Pearson correlation (KPC), 103
Karnataka, 40

Karnataka Urban Water Supply & Sewerage Board, 473
Kattu Mangappuly, 357
Kattunayakan community, 354
Kerala Institute of Local Administration (KILA), 295
Kharif crops, 43
Khutekar, 95
Kokkare-Bellur, 113
Koonthankulam, 113
Kuruma community, 354
Kyoto protocol, 598

L
Land and water Management, 217
Land Degradation Neutrality, 32
Landholding size, 357
Landscape design, 464
Land transformation, 585
Leadership Group for Industry Transition, 540
Lead Technical Support Agencies(LTSA), 509
Leaf litter, 31
Leaves of Colocasia, 357
Liability risk, 595
Livelihood, 74
Livelihood diversification, 2, 34
Livestock, 85, 299
Livestock products, 77
Local Ecological Knowledge (LEK), 331
Low-lying areas, 433

M
Madhya Pradesh Rajya Ajeevika Forum (MPRAF), 509
Maharashtra, 86
Maladaptation, 94
Man and society, 528
Marginalized communities, 34
Marine seaweeds, 384
Market linkages, 238
Meenangadi, 278
Mental Models, 240
Metropolitan Planning Committee, 472, 473
Microclimate condition, 100

INDEX 625

Micro irrigation, 515
Millennium Development Goal, 126
Millet, 50
Ministry of Environment, Forest and Climate Change, 488
Ministry of Health and Family Welfare, 488
Ministry of Rural Development (MORD), 503
Mitigation, 109
Mixed cropping, 2
Mixed-farming, 322
Mixed forests, 104
Mixed plantations, 391
Mobility Mapping, 260
Multifunctional farming, 396
Munda community, 347

N
Nanominerals, 301
National Action Plan for Climate Change, 175
National Action Plan on Climate Change, 483
National Agriculture Disaster Management Plan (NADMP), 484
National Bank for Agriculture and Development (NABARD), 542
National Development Planning, 483
National Disaster Management Authority, 477
National Green Tribunal Act, 287
National Institute for Rural Development & Panchayati Raj (NIRDPR), 509
Nationally determined contribution (NDC), 539
National Mission for Sustainable Agriculture (NMSA), 542
National Mission on Sustainable Agriculture, 175
National Mission on Sustaining Himalayan Ecosystem (NMSHE), 25
National Rural Livelihood Mission (NRLM), 503
National Rural Livelihoods Promotion Society (NRLPS), 509

National Social Inclusion, 509
National Solar Mission, 543
National Target Program (NTP), 483
Native plants, 454
Natural disasters, 11
Nature and society, 528
Nature based Solutions (NbS), 32, 96, 210
Nelapattu, 113
Net Water Storage (NWS), 266
Net-Zero Emissions, 541
Nilgiri Biosphere Reserve, 354
Nomadic system, 85
Non-communicable disease, 531
Non-Government Organisations (NGOs), 419
Non-timber forest products (NTFPs), 320
Normandy Index, 538
Novel bioactive compounds, 368
Nutritional value, 67
Nutrition security, 257

O
Odisha, 72
Off-seasonal rainfall, 263
Open-landscape, 356
Organic farming, 199
Organic manures, 145
Ornithologists, 112
Outside forces/independent variables, 241

P
Panchayati Raj Act, 274
Pandemics, 527
Paris Climate Agreement, 569
Pastoralism, 85
Pastoralists, 85
Pastoral livelihood system, 86
Pasturelands, 93
Pelicanry, 114
Peripheral Ring Road (PRR), 474
Peri-urban regions, 392
Phenological changes, 579
Phenology, 323
Phenotypic plasticity, 322

626 INDEX

Pokkali, 198
Pollinators, 582
Poverty alleviation, 323
Pradhan Mantri Ujwala Yojana (PMUY), 539
Production, 510
Psammophytes, 368

Q
Quercus leucotrichophora, 26

R
Rainfed cultivation, 40
Rainwater harvesting (RWH), 444, 463
Rangeland, 91
Regenerative techniques, 586
Remote sensing, 100
Resident Welfare Associations (RWAs), 419
Resource Mapping, 260
Resource-poor people, 500
Restoration, 32
Rural livelihood, 502

S
Sacred groves, 293
Salinity, 198
Savannah type, 96
Seasonal Calendar, 260
Semi-arid landscapes, 85
Semi-nomadic system, 85
Semi rid region, 40
Sewerage, 417
Shegar Dhangar, 86
Smallholder farmers, 145
Social-ecological system (SES), 335
Social forestry, 403
Social infrastructure, 478
Social Mapping, 260
Social-religious practice, 333
Social work, 594
Societal resilience, 34
Socio-economical ecosystems, 368
Socio-cognitive-behavioural aspect, 251
Socio-ecological, 30
Socio-economic, 1, 142

Socio-economics/Dependent variables, 241
Socio-political, 96
Soil carbon, 177
Soil conservation, 172
South East Indian coastal region, 369
Special Climate Change Funds (SCCF), 503
Spot-billed pelican, 113
Stakeholders, 275
State Action Plans of climate change (SAPCC), 494
State Election Commission (SEC), 472
Statistical models, 40
Stormwater drains, 433
Suburban areas, 415
Sundari, 344
Surface water, 266
Sustainability, 456, 587
Sustainability outcomes, 478
Sustainable agricultural practices, 144
Sustainable agriculture, 145
Sustainable Cities, 401
Sustainable development, 3
Sustainable Development Goal (SDG), 32, 411, 527
Sustainable future, 7
Sustainable income generation, 162
Sustainable land management (SLM), 97

T
Tamil Nadu, 127
Taungya, 323
Technology, 510
Telineelapuram, 113
The global food system, 585
Theory of Planned Behaviour (TPB), 252
Thumpa (*Leucas cephalotes*), 357
Top-down conservation, 334
Traditional agricultural practice, 149
Traditional Ecological Knowledge (TEK), 331
Traditional Ethnobotanical Knowledge (TEK), 367
Traditional farmers, 143
Traditional knowledge, 5
Traditional mixed-cropping, 322

INDEX 627

Traditional production system, 85
Trans-disciplinary approaches, 120
Transhumant system, 85
Transition risk, 595
Trans-theoretical Model (TTM), 252

U
Urban flooding, 433
Urban Green Spaces (UGSs), 389
Urbanisation, 6, 392, 412

V
Value-Belief-Norm (VBN), 252
Van Panchayats, 33
Village Development Committee, 216
Village Panchayat (VP), 261
Village tool banks, 513
Vulnerabily assessment, 488

W
Ward Disaster Management Cells (WDMCs), 476
Wastelands, 93
Wastewater reuse, 463
Water budgeting, 260, 274
Water conservation, 172
Water-conserving method, 463
Water security, 229
Water sharing, 244
Watershed-based activities, 273
Watershed management, 401
Watershed Organization Trust (WOTR), 509
Water-smart approache, 175
Water Stewardship, 89
Weather Based Agro Advisory Services (WBASS), 518
Western Ghats, 201
Western Ghats Ecological Expert Panel, 354
Western Ghats Ecology Panel, 287
Wetlands, 201, 392
Wildlife habitats, 392
Willingness to Pay, 127
Woody savannas, 104
World bank, 564
World Green Building Week, 468
World Health Organization (WHO), 367
World Meteorological Organization (WMO), 537